T0281436

The revolution in elementary particle physics sparked by the unearthing of the bizarre J/Ψ particle in 1974 and followed by the discovery of the equally mysterious τ and Υ particles, led to a beautiful interweaving of theory and experiment, culminating in the Salam–Weinberg theory of electroweak interactions and the quantum chromodynamic (QCD) theory of strong interactions. The extraordinary prediction of the W and Z^0 bosons was fulfilled in 1983, and it is now possible to produce Z^0 in millions. The emphasis today is on refined testing of the detailed quantitative predictions of the theories, and, to match this, more sophisticated calculations are demanded. This book presents, in two volumes, a comprehensive and unified treatment of modern theoretical and experimental particle physics at a level accessible to beginning research students. The emphasis throughout is on presenting underlying physical principles in a simple and intuitive way, and the more sophisticated methods demanded by present-day research interests are introduced in a very gradual and gentle fashion.

Volume 1 covers electroweak interactions, the discovery and properties of the 'new' particles, the discovery of partons and the construction and predictions of the simple parton model. Volume 2 deals at some length with CP violation, but is mainly devoted to QCD and its application to 'hard' processes. A brief coverage of soft hadronic physics and of non-perturbative QCD is included.

This work will provide a comprehensive reference and textbook for all graduate students and researchers interested in modern particle physics.

CAMBRIDGE MONOGRAPHS ON
PARTICLE PHYSICS,
NUCLEAR PHYSICS AND COSMOLOGY

4

AN INTRODUCTION TO GAUGE THEORIES AND MODERN PARTICLE PHYSICS, VOLUME 2

CAMBRIDGE MONOGRAPHS ON PARTICLE PHYSICS, NUCLEAR PHYSICS AND COSMOLOGY

1. K. Winter (ed.): *Neutrino Physics*
2. J. F. Donoghue, E. Golowich and B. R. Holstein: *Dynamics of the Standard Model*
3. E. Leader and E. Predazzi: *An Introduction to Gauge Theories and Modern Particle Physics, Volume 1: Electroweak interactions, the 'new particles' and the parton model*
4. E. Leader and E. Predazzi: *An Introduction to Gauge Theories and Modern Particle Physics, Volume 2: CP violation, QCD and hard processes*

AN INTRODUCTION TO GAUGE THEORIES AND MODERN PARTICLE PHYSICS, VOLUME 2

CP-violation, QCD and hard processes

ELLIOT LEADER
Birkbeck College, University of London

ENRICO PREDAZZI
University of Torino

Published by the Press Syndicate of the University of Cambridge
The Pitt Building, Trumpington Street, Cambridge CB2 1RP
40 West 20th Street, New York, NY 10011-4211, USA
10 Stamford Road, Oakleigh, Melbourne 3166, Australia

First published 1996

A catalogue record for this book is available from the British Library

Library of Congress cataloguing in publication data

Leader, Elliot, 1935–
An introduction to gauge theories and modern particle physics /
Elliot Leader, Enrico Predazzi.
p. cm. – (Cambridge monographs on particle physics,
nuclear physics, and cosmology; 3–4)
Includes bibliographical references and index.
Contents: v. 1. Electroweak interactions, the "new particles"
and the parton model – v. 2. CP-violation, QCD and hard processes.
ISBN 0 521 46468 4 (v. 1) – ISBN 0 521 46840 X (pbk. : v. 1). –
ISBN 0 521 49617 9 (v. 2). – ISBN 0 521 49951 8 (pbk. : v. 2)
1. Gauge fields (Physics) 2. Particles (Nuclear physics)
I. Predazzi, Enrico. II. Title. III. Series.
QC793.3.G38L43 1996
539.7′54–dc20 95-25233 CIP

Vol. 1 ISBN 0 521 46468 4 hardback

ISBN 0 521 46840 X paperback

Vol. 2 ISBN 0 521 49617 9 hardback

ISBN 0 521 49951 8 paperback

Set of two vols.

ISBN 0 521 57780 2 hardback

ISBN 0 521 57742 X paperback

Transferred to digital printing 2004

TAG

Dedication

To our children, Darian, Francesca, Imre, Irene and Valentina.

Perché si scrive?

> ... Per insegnare qualcosa a qualcuno. Farlo, e farlo bene, può essere prezioso per il lettore, ma ... l'intento didattico corrode la tela narrativa dal di sotto, la degrada e la inquina: il lettore che cerca il racconto deve trovare il racconto, e non una lezione che non desidera. Ma appunto, le eccezioni ci sono, e chi ha sangue di poeta sa trovare ed esprimere poesia anche parlando di stelle, di atomi, dell'allevamento del bestiame e dell'apicultura...

Why does one write?

> ... To teach something to someone. To do this and do it well can be valuable for the reader but ... the didactic intention corrodes the narrative canvas from underneath, degrades it and contaminates it: the reader who looks for a story must find a story and not a lesson he does not want. But, of course, exceptions there are, and whoever has the blood of a poet will find and express poetry also when talking of stars, of atoms, of cattle breeding and of the raising of bees...

Primo Levi

Contents: Volume 2
CP violation, QCD and hard processes

Contents: Volume 1
Electroweak interactions, the 'new particles' and the parton model

Preface

For a book of its genre, our previous book, *An introduction to gauge theories and the "new physics"* (1982) was a great success. It was not, alas, sold in airport lounges, but it did run to two additional printings (1983, 1985), and to extensively revised editions in Russian (1990), and in Polish (1991). More importantly, it seemed to achieve the principal goal which we had set ourselves, namely, to present a *pedagogical* account of modern particle physics with a balance of theory and experiment, which would be intelligible and stimulating for both theoretical and experimental graduate students. We did *not* try to write a profound book on field theory, *nor* a treatise on sophisticated experimental techniques. But we did wish to stress the deep, intimate and fruitful interaction between theoretical ideas and experimental results. Indeed, for us, it is just this aspect of physics which makes it seem so much more exciting than say pure mathematics. Our greatest pleasure came from the favourable reaction of students who were working through the book and from those reviewers who caught what we hoped was its essential flavour—'the writing creates the feeling of an active progression of ideas arising from the repeated interaction of theoretical prejudice with experimental observation', 'unlike most textbooks, it is highly readable, and makes everything appear simple and obvious'. Well, the last comment is surely an exaggeration but that was our aim.

In thinking about a second edition we were faced with a serious conceptual problem. Ten years ago we were in a state of excited expectation. A beautiful theory had been created and led, via the simplest of calculations, to absolutely dramatic experimental predictions; principally the existence and basic properties of the heavy vector bosons $W^{\pm}$ and Z^0. A host of interesting new phenomena could be studied with no more effort than the calculation of a lowest order Born diagram. Much of the new physics could be discussed and understood from rather qualitative arguments. That idyllic situation is much changed now.

After the few years during which the experimentalists were struggling to demonstrate the very existence of these new phenomena, when the world of physics was electrified by the discovery of *one single* W or Z event, we have moved into an era when LEP is mass-producing millions of Z^0s!

Consequently, and unavoidably, the physics emphasis has changed drastically. Now it is the fine quantitative detail, the precise width and line shape of the Z^0, precision measurements of forward-backward asymmetries and branching ratios etc. which are under scrutiny experimentally. And to match that, more sophisticated and vastly more complicated theoretical calculations are demanded. Thus we have passed from a simple heuristic era to one of demanding quantitative rigour.

One consequence is that instead of a second edition we have ended up with a large new two-volume book!

The above does not mean that the subject has become boring and moribund. On the contrary, great theoretical issues are at stake. For the earlier Born approximations did not really test the deeper *field-theoretic* aspects of the theory, whereas the present comparisons between theory and experiment *are* sensitive to these elements. They play a rôle almost like that of the Lamb shift in establishing the validity of quantum electrodynamics (QED).

Faced then with the need to introduce and discuss these more extensive calculations, and given that schoolchildren, so it seems, are taught about quarks and gluons, there was the temptation to abandon the long and leisurely historical introduction to partons, quarks and gluons that we had provided in the first edition.

We have not done so for the following reason. The introduction of the new level of 'elementarity', the quarks and gluons, beyond the level of mesons and baryons, is, we believe, of a fundamentally new kind, both physically and philosophically in the modern history of science. For in all previous cases the *ultimate* proof of the existence of a hypothesized constituent was to produce and identify that constituent in the laboratory, for example, via a track in a bubble-chamber; or, where a neutral particle was involved, like the Λ^0, to have very obvious and incontrovertible evidence of its propagation and decay into charged tracks.

Now, for the first time, we are postulating the existence of constituents which, according to the present interpretation of the theory, can never be truly freed, which can, *in principle*, never be seen as free particles in the laboratory. This means that we really must provide convincing evidence for these constituents and must examine very critically the steps that lead to our postulating their existence.

It is on these grounds we have decided to retain the detailed discussion of the historical process leading to the belief in the parton picture.

The major *new* features of this book are:

1. We give a detailed explanation of higher order electroweak effects (the so-called radiative corrections).

2. We provide a much expanded discussion of quark mixing (the Kobayashi–Maskawa matrix), of K^0–$\bar{K}^0$ and B^0–$\bar{B}^0$ mixing and new sections on both the phenomenology and dynamics of CP violation.

3. The sections on charm and beauty and on jet physics have been totally revised in order to take into account the mass of data that has accumulated over the past few years.

4. We have enlarged the treatment of deep inelastic lepton–hadron scattering in three directions. Catalyzed by the major discoveries of the European Muon Collaboration there is a more detailed treatment of both polarization effects and of nuclear effects. And in anticipation of HERA we discuss Z^0–γ interference which will be important for large Q^2 physics.

5. The treatment of QCD corrections to the simple parton model is presented in much more detail and a new chapter is devoted to the derivation of the parton model from field theory.

6. Also linked to the coming into operation of HERA we discuss in more detail the 'low x' region in deep inelastic scattering. Here the usual evolution equations break down and one approaches the non-perturbative region of QCD, creating a tremendous challenge to theory.

7. A brief discussion of elastic and soft reactions is provided so as to give the reader at least an inkling of the remarkable $\bar{p}p$ physics being carried out at the CERN collider and at the Fermilab Tevatron. In particular, we discuss the impact of the very unexpected result of the UA4 experiment at CERN.

8. An introduction is given to the many and varied attempts to deal with the non-perturbative or confinement region of QCD, especially to the lattice approach and to the sum rule method, and to the exciting new ideas about baryon and lepton number violations. Necessarily the treatment of these is rather brief and not very comprehensive. We seek to convey the basic ideas and methods.

9. The Appendix has been much enlarged. It now contains a more detailed specification of the Feynman rules for electroweak theory and QCD, and a discussion of the relations between S-matrix, transition amplitudes and Feynman amplitudes. There are also sections on CPT invariance and on the operator form of Feynman amplitudes

or effective Hamiltonians, both important for CP violation. The evaluation of matrix elements of conserved currents, much used in deep inelastic scattering and in the study of the Kobayashi-Maskawa matrix, is explained in some detail. Finally, a complete list of cross-section formulae for all $2 \to 2$ partonic reactions is given.

It is fascinating to step back and to view what has been achieved in the past decade and where we now stand. Electroweak theory and quantum chromodynamics have been remarkably successful, even at the level of the detailed questions now being examined. But essential to their present formulation and to their success are two objects, two crucial ingredients, the top quark and the Higgs meson, which have not yet been found. The Higgs meson is central to the symmetry breaking mechanism which gives mass to the W and Z bosons. And the top quark plays a vital rôle in the cancellation of anomalies and in the higher order calculations of small but significant effects beyond the Born approximation. Indeed the success in comparing these small corrections with experiment has led to quite tight limits on what mass the top quark could have. Most recently, at the Dallas Conference on High Energy Physics in August 1992, the following remarkably narrow limits for m_t were reported:

$$\begin{aligned} 107 \le m_t \le 143 \text{ GeV}/c^2 &\quad \text{for} \quad m_H = 60 \text{ GeV}/c^2, \\ 126 \le m_t \le 162 \text{ GeV}/c^2 &\quad \text{for} \quad m_H = 300 \text{ GeV}/c^2, \\ 143 \le m_t \le 179 \text{ GeV}/c^2 &\quad \text{for} \quad m_H = 1 \text{ TeV}/c^2. \end{aligned}$$

In a way the sense of expectation for the discovery of 'top' and the 'Higgs' is almost as strong as was the expectation of the W and Z a decade ago.

Acknowledgements

We are enormously indebted to Nora Leonard for her intelligent and skillful handling of a large part of the LaTeX typesetting of our manuscript. We are grateful to Roy Abrahams for his preparation of many of the diagrams. Thanks are also due to Giuseppe Ferrante for help with the typing and to Nestor Patrikios for aid with computer generated figures.

Finally, there are many colleagues and friends, too many to mention individually, to whom we are indebted for providing data and comments and for drawing our attention to misprints and errors in the first edition.

Notational conventions

Units

Natural units $\hbar = c = 1$ are used throughout.

For the basic unit of charge we use the *magnitude* of the charge of the electron: $e > 0$.

Relativistic conventions

Our notation generally follows that of Bjorken and Drell (1964), in *Relativistic Quantum Mechanics.*

The metric tensor is

$$g_{\mu\nu} = g^{\mu\nu} = \begin{pmatrix} 1 & 0 & 0 & 0 \\ 0 & -1 & 0 & 0 \\ 0 & 0 & -1 & 0 \\ 0 & 0 & 0 & -1 \end{pmatrix}.$$

Space-time points are denoted by the contravariant four-vector x^μ ($\mu = 0, 1, 2, 3$)

$$x^\mu = (t, \boldsymbol{x}) = (t, x, y, z),$$

and the four-momentum vector for a particle of mass m is

$$p^\mu = (E, \boldsymbol{p}) = (E, p_x, p_y, p_z),$$

where

$$E = \sqrt{\boldsymbol{p}^2 + m^2}.$$

Using the equation for the metric tensor, the scalar product of two four-vectors, A, B, is defined as

$$A \cdot B = A_\mu B^\mu = g_{\mu\nu} A^\mu B^\nu = A^0 B^0 - \boldsymbol{A} \cdot \boldsymbol{B}.$$

γ-matrices

The γ matrices for spin-half particles satisfy

$$\gamma^\mu\gamma^\nu + \gamma^\nu\gamma^\mu = 2g^{\mu\nu}$$

and we use a representation in which

$$\gamma^0 = \begin{pmatrix} I & 0 \\ 0 & -I \end{pmatrix}, \quad \gamma^j = \begin{pmatrix} 0 & \sigma_j \\ -\sigma_j & 0 \end{pmatrix}, \quad j = 1, 2, 3,$$

where σ_j are the usual Pauli matrices. We define

$$\gamma^5 = \gamma_5 = \mathrm{i}\gamma^0\gamma^1\gamma^2\gamma^3 = \begin{pmatrix} 0 & I \\ I & 0 \end{pmatrix}.$$

In this representation one has for the transpose T of the γ matrices:

$$\gamma^{j\mathrm{T}} = \gamma^j \qquad \text{for} \quad j = 0, 2, 5,$$

but

$$\gamma^{j\mathrm{T}} = -\gamma^j \qquad \text{for} \quad j = 1, 3.$$

For the Hermitian conjugates † one has

$$\gamma^{0\dagger} = \gamma^0, \qquad \gamma^{5\dagger} = \gamma^5,$$

but

$$\gamma^{j\dagger} = -\gamma^j \qquad \text{for} \quad j = 1, 2, 3.$$

The combination

$$\sigma^{\mu\nu} \equiv \frac{\mathrm{i}}{2}[\gamma^\mu, \gamma^\nu]$$

is often used.

The scalar product of the γ matrices and any four-vector A is defined as

$$\not{A} \equiv \gamma^\mu A_\mu = \gamma^0 A^0 - \gamma^1 A^1 - \gamma^2 A^2 - \gamma^3 A^3.$$

For further details and properties of the γ matrices see Appendix A of Bjorken and Drell (1964).

Spinors and normalization

The particle spinors u and the antiparticle spinors v, which satisfy the Dirac equations

$$\begin{aligned} (\not{p} - m)u(p) &= 0, \\ (\not{p} + m)v(p) &= 0, \end{aligned}$$

respectively, are related by

$$\begin{aligned} v &= \mathrm{i}\gamma^2 u^*, \\ \bar{v} &= -\mathrm{i}u^{\mathrm{T}}\gamma^0\gamma^2, \end{aligned}$$

where $\bar{v} \equiv v^\dagger\gamma^0$ (similarly $\bar{u} \equiv u^\dagger\gamma^0$).

Note that our spinor normalization differs from Bjorken and Drell. We utilize

$$u^\dagger u = 2E, \qquad v^\dagger v = 2E,$$

the point being that the above can be used equally well for massive fermions and for neutrinos. For a massive fermion or antifermion the above implies

$$\bar{u}u = 2m, \qquad \bar{v}v = -2m.$$

Cross-sections

With this normalization the cross-section formula (B.1) of Appendix B in Bjorken and Drell (1964) holds for both mesons and fermions, massive or massless. This is discussed in our Appendix 2—see in particular eqn (A2.1.6).

Fields

Often a field like $\psi_\mu(x)$ for the muon is simply written $\mu(x)$ or just μ if there is no danger of confusion.

In fermion lines in Feynman diagrams the arrow indicates the direction of flow of *fermion number.* Thus incoming electrons or positrons are denoted as follows:

Normal ordering

In quantum field theory the products of operators that appear in Lagrangians and Hamiltonians should really be *normal ordered* with all creation operators to the left of all annihilation operators. With the exception of Sections 5.1.1 and 19.4 this is irrelevant throughout this book, so the normal ordering symbol is never indicated.

Group symbols and matrices

In dealing with the electroweak interactions and QCD the following symbols often occur:

- n_f = number of flavours
- N which specifies the gauge group $SU(N)$—Note that $N = 3$ for the colour gauge group QCD;
- the Pauli matrices are written either as $\boldsymbol{\sigma}_j$ or $\boldsymbol{\tau}_j$ ($j = 1, 2, 3$);
- the Gell-Mann $SU(3)$ matrices are denoted by $\boldsymbol{\lambda}^a$ ($a = 1 \ldots 8$);
- for a group (G) with structure constants f_{abc} one defines $C_2(G)$ via

$$\delta_{ab} C_2(G) \equiv f_{acd} f_{bcd}$$

and one often writes

$$C_A \equiv C_2[SU(3)] = 3.$$

If there are n_f multiplets of particles, each multiplet transforming according to some representation R under the gauge group, wherein the group generators are represented by matrices $\boldsymbol{t}^a$, then $T(R)$ is defined by

$$\delta_{ab} T(R) \equiv n_f Tr(\boldsymbol{t}^a \boldsymbol{t}^b).$$

For $SU(3)$ and the triplet (quark) representation one has $\boldsymbol{t}^a = \boldsymbol{\lambda}^a/2$ and

$$T \equiv T[SU(3); \text{triplet}] = \tfrac{1}{2} n_f.$$

For the above representation R one defines $C_2(R)$ analogously to $C_2(G)$ via

$$\delta_{ij} C_2(R) \equiv t^a_{ik} t^a_{kj}.$$

For $SU(3)$ and the triplet representation one has

$$C_F \equiv C_2[SU(3); \text{triplet}] = 4/3.$$

The coupling in QCD

Finally, to conform with recent review articles, we have changed the sign convention for the coupling in QCD, i.e. our new g is minus the g used in the first edition. Since everything of physical interest in QCD depends upon g^2 this is essentially irrelevant.

Colour sums in weak and electromagnetic current

Since the weak and electromagnetic interactions are 'colour-blind' the colour label on a quark field is almost never shown explicitly when dealing with electroweak interactions. In currents involving quark field operators (e.g. in Sections 1.2 and 9.3) a *colour sum is always implied.* For example, the electromagnetic current of a quark of flavour f and charge Q_f (in units of e) is written

$$J^\mu_{\text{em}}(x) = Q_f \bar{q}_f(x) \gamma^\mu q_f(x)$$

but if the colour of the quark is labelled j ($j = 1, 2, 3$) then what is implied is

$$J^{\mu}_{\text{em}}(x) = Q_f \sum_{\substack{\text{colours} \\ j}} \bar{q}_{f_j}(x)\gamma^{\mu} q_{f_j}(x).$$

Note added in proof: the discovery of the top quark (?)

On Tuesday 26 and Wednesday 27 April 1994 nearly simultaneous press conferences in the USA, Italy and Japan announced that evidence for the top quark had been found by the CDF collaboration at Fermilab. In fact most of the information had by then already appeared in a long article in *The New York Times* and rumours about the discovery had been circulating for some time.

The collaboration was at pains to insist that they were not claming the *discovery* of top, but only some *evidence* for it. There are several reasons for their prudence. The main one is that the $t\bar{t}$ cross-section measured, $13.9^{+6.1}_{-4.8}$ pb, is a factor of three larger than the theoretical expectations. Also the other collaboration at Fermilab studying this question (the DO collaboration) is not yet willing to make any public statement, but it seems that their $t\bar{t}$ cross-section would be considerably smaller. The calculated $t\bar{t}$ cross-section at Fermilab energies is a sensitive function of m_t, varying from 20 pb for $m_t = 120$ GeV/c^2 to 4 pb for $m_t = 180$ GeV/c^2.

Interestingly, the theoretical estimates of m_t, based on the calculation of radiative corrections to various high precision LEP electroweak measurements, have been moving towards higher and higher values. At the 1990 International High Energy Physics Conference at Singapore the best value was given as $m_t = 137 \pm 40$ GeV/c^2, whereas recent studies suggest $164 \pm 16 \pm 19$ GeV/c^2. The value for m_t given by CDF is compatible with these higher values, namely $m_t = 174 \pm 13^{+13}_{-12}$ GeV/c^2. Such a high value for m_t has significant implications for the radiative correction calculations and in particular for the theoretical understanding of CP violation, as discussed in Chapter 19.

The result reported by CDF is a 2.5 sigma effect, which becomes less then a 2 sigma effect when their data is combined with the DO data, Clearly, therefore, much better data are needed before a totally convincing picture can emerge. The evidence for top, if confirmed, would be one more very strong factor in favour of the standard model. Ultimately, however, there will still remain the crucial question as to the existence or not of the Higgs boson. And it may be some time before we have a definitive answer to *that* question.

By May 1995 the evidence for the top quark is firmer. The new combined CDF and DO results have increased the significance of the signal to more than 4 sigma. The masses quoted are:

$$\text{CDF:}\quad 176 \pm 8 \pm 10\ \text{GeV}/c^2$$
$$\text{DO:}\quad 199^{+19}_{-21} \pm 22\ \text{GeV}/c^2.$$

Note added in proof: the demise of the SSC

At various points in this book we have talked rather optimistically about future accelerators, in particular about the gigantic, 54 miles in circumference, Superconducting Super Collider (SSC) which was to be built at Waxahachie in Texas and which would have produced 20 TeV + 20 TeV proton–proton collisions. The energy densities attainable would have matched those found in the universe as close as 10^{-13} seconds to the big bang, providing extraordinary possibilities for testing not just the standard model of elementary particle interactions, but the whole picture of the evolution of the universe.

Alas, on 22 October 1993, after a long and agonizing period of indecision, the US House of Representatives voted to end the funding of the SSC project. At this pont, 2 billion US dollars had already been spent and about one-fifth of the tunnel completed.

In the words of Hazel O'Leary, the US Secretary of State for Energy, 'this decision by Congress ... is a devastating blow to basic research and to the technological and economic benefits that always flow from that research.'

There is still hope that the somewhat more modest European project for a Large Hadron Collider (LHC) with 8 TeV + 8 TeV proton–proton collision, will go ahead. A final decision was due during 1994. After much procrastination the CERN Council finally voted in favour of the project in December 1994. The construction will proceed in stages, with full-scale operation planned for 2008!

18

Determination of the Kobayashi–Maskawa matrix

The Kobayashi–Maskawa matrix, which delineates what combinations of quark fields are operative in the charge changing weak interaction, was introduced in Chapter 9. Its elements V_{ij} have occurred repeatedly in the preceding chapters and are crucial in providing quantitative results from the theory. We here address the issue of measuring these matrix elements and of summarizing the present knowledge of them.

In the SM the weak currents are all expressed as KM matrix elements multiplying currents built up from the quark fields which, for brevity, we denote by $u, d, \ldots, \bar{u}, \bar{d}, \ldots$ In some cases the latter currents can be identified as the conserved Noether currents corresponding to symmetries of the strong interactions, for example, isotopic spin invariance. In that case, as explained in Appendix 3, the matrix elements (at least the forward ones) of the conserved currents are known exactly. Thus the weak transition amplitude appears as a V_{ij} multiplied by a known matrix element, thereby allowing the measurement of that particular V_{ij}. To the extent that the symmetry is not perfect and that one is seeking very precise knowledge of the V_{ij}, one must attempt to correct for the symmetry breaking. This is a highly technical procedure for which the reader should consult Leutwyler and Roos (1984) and Donoghue, Holstein and Klimt (1987), and the major review article by Paschos and Türke (1988). We shall only deal with the case of perfect symmetry and will illustrate the approach in β-decay type reactions.

If the current which V_{ij} multiplies cannot be related to a conserved current one must try to look at situations where the weak interactions with the hadron can be reasonably clearly interpreted in terms of a direct interaction with an almost free quark, which is not masked by strong interaction effects. The deep inelastic reactions discussed in such detail in the previous few chapters are precisely of this type.

All the above yields information about the *magnitude of* the V_{ij}. We stressed, in Chapter 9, that the generalization from the Cabibbo theory to the KM theory introduces a phase which allows the remarkable possibility of providing a natural mechanism for CP violation as seen for example in the $K^0 - \bar{K}^0$ system. We therefore discuss CP violation, and $B^0 - \bar{B}^0$ mixing and the theoretical interpretation of these phenomena in the following chapter.

18.1 KM matrix elements from β-decay reactions

Written out in detail the electroweak current h_+^μ responsible for CC reactions, given in (9.3.10), is

$$\begin{aligned} \mathrm{h}_+^\mu &= \bar{u}\gamma^\mu(1-\gamma_5)[V_{ud}d + V_{us}s + V_{ub}b] \\ &\quad + \bar{c}\gamma^\mu(1-\gamma_5)[V_{cd}d + V_{cs}s + V_{cb}b] \\ &\quad + \bar{t}\gamma^\mu(1-\gamma_5)[V_{td}d + V_{ts}s + V_{tb}b]. \end{aligned} \tag{18.1.1}$$

Amongst the above quark currents only those involving u, d or s can be connected to conserved Noether currents related either to isospin invariance (a very good symmetry) or to $SU(3)_F$ flavour invariance (not such a perfect symmetry). Thus only the KM elements V_{ud} and V_{us} can be found by the conserved current approach. The rest must be found either with the help of a dynamical model or from the analysis of deep inelastic scattering data.

In order to measure any V_{ij} we try to look at the simplest possible reactions in which it plays a rôle; and that implies trying to minimize the effect of the strong interactions. The obvious choice theoretically is the semi-leptonic reactions. From an experimental point of view the most accurate information will come from nuclear β-decay and other β-decay type reactions like $\mathrm{n} \to \mathrm{p} + \mathrm{e}^- + \bar{\nu}_\mathrm{e}$, $\Lambda \to \mathrm{p} + \mathrm{e}^- + \bar{\nu}_\mathrm{e}$, $K^+ \to \pi^0 + \mu^+ + \nu_\mu$, $\pi^+ \to \pi^0 + \mathrm{e}^+ + \nu_\mathrm{e}$, etc.

Consider the generic case

$$\mathrm{A} \to \mathrm{B} + \ell + \bar{\nu}_\ell.$$

In the SM the reaction proceeds, in lowest order, via the Feynman diagram in Fig. 18.1 and thus involves the hadronic matrix element $\langle \mathrm{B}, \boldsymbol{p}' | \mathrm{h}_+^\mu | \mathrm{A}, \boldsymbol{p} \rangle$. In general such a matrix element depends upon various form factors [functions of q^2, the square of the four-momentum transfer; see, for example (A3.25) in Appendix 3], some linked to the vector part V_+^μ, others to the axial vector part A_+^μ of h_+^μ. Moreover the Lorentz structure of the matrix element depends upon the spin of the hadrons. [For details see Bailin (1982).]

In the case that the current we are dealing with corresponds to a conserved Noether current we know its *forward* hadronic matrix elements

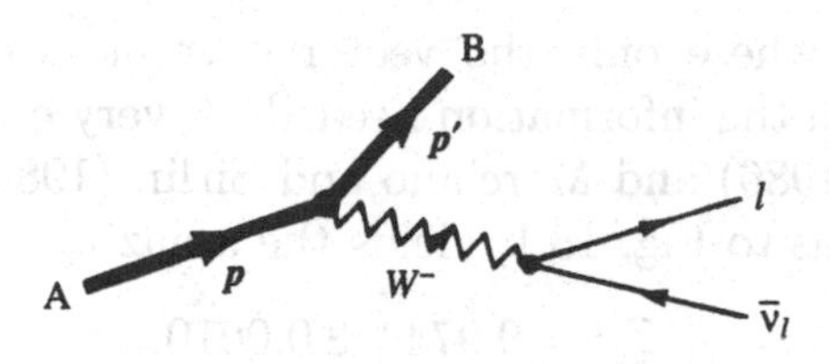

Fig. 18.1. Lowest order Feynman diagram for β-decay.

exactly, as explained in Appendix 3. But this only applies to the *vector* part. The axial currents are not conserved. There are then two possible strategies: utilize nuclear β-decay transitions in which the axial vector contribution is suppressed or measure a sufficient number of parameters, (lepton energy spectrum, correlations with neutron polarization etc.), to be able to evaluate the ratio of axial to vector contributions experimentally. In either case the absolute scale of the hadronic matrix element is set by

(relevant V_{ij}) $\times$ (forward matrix element of conserved vector current)

of which the latter is supposed to be known exactly. As mentioned earlier, to the extent that the symmetry is not perfect one must attempt to correct for symmetry breaking. For the *vector* currents one is aided by the Ademollo–Gatto theorem (1964) which guarantees that the first order perturbative corrections vanish. For details of these corrections the reader should consult the references given at the beginning of this chapter.

Let us illustrate the approach for V_{ud}. The part of h_+^μ responsible for the transition $\mathrm{n} \to \mathrm{p}$ is

$$V_{ud}\bar{u}\gamma^\mu(1-\gamma_5)d = V_{ud}[V_+^\mu + A_+^\mu] \tag{18.1.2}$$

where V_+^μ is the isospin raising current

$$V_+^\mu = V_1^\mu + \mathrm{i}V_2^\mu,$$

the $V_j^\mu (j = 1, 2, 3)$ being the three components of the conserved isospin current. From isospin invariance one then has, from (1.3.17),

$$\begin{aligned}\langle \text{proton}, \boldsymbol{p}'|V_+^\mu|\text{neutron}, \boldsymbol{p}\rangle &= \langle \text{proton}, \boldsymbol{p}'|V_3^\mu|\text{proton}, \boldsymbol{p}\rangle \\ &\quad - \langle \text{neutron}, \boldsymbol{p}'|V_3^\mu|\text{neutron}, \boldsymbol{p}\rangle\end{aligned} \tag{18.1.3}$$

so that for the forward matrix element ($\boldsymbol{p}' = \boldsymbol{p}$)

$$\langle \text{proton}, \boldsymbol{p}|\text{vector part of } \mathrm{h}_+^\mu|\text{neutron}, \boldsymbol{p}\rangle = V_{ud}2\mathrm{p}^\mu \tag{18.1.4}$$

where we have used (A3.20) of Appendix 3.

The most accurate determinations of V_{ud} come from $0^+ \to 0^+$ super-allowed Fermi transitions in nuclear β-decays involving $I = 1$ nuclear isospin multiplets where only the vector current is operative, so that (18.1.4) provides all the information needed. A very careful study by Sirlin and Zucchini (1986) and Marciano and Sirlin (1986), which included radiative corrections to Fig. 18.1, yields the amazingly precise result

$$|V_{ud}| = 0.9744 \pm 0.0010. \tag{18.1.5}$$

The current responsible for strangeness-changing β-decays ($\Delta Q = \Delta S = 1$) is

$$V_{us}\bar{u}\gamma^\mu(1-\gamma_5)s = V_{us}\left\{V_4^\mu + \mathrm{i}V_5^\mu + A_4^\mu + \mathrm{i}A_5^\mu\right\} \tag{18.1.6}$$

where the $SU(3)_F$ octet of vector and axial-vector currents V_j^μ, A_j^μ ($j = 1, \cdots, 8$) is given in (1.2.21). To the extent that the hadrons belong to $SU(3)_F$ multiplets we again have precise results for the forward matrix elements of $V_4^\mu + \mathrm{i}V_5^\mu$. For spin $\frac{1}{2}$ hyperon β-decays the forward matrix elements are given by eqns (A3.25, 27 and 29) of Appendix 3. Thereby one finds, for example,

$$\begin{aligned} \langle \mathrm{p}|\text{vector part of } \mathrm{h}_+^\mu|\Lambda\rangle &= -\sqrt{\tfrac{3}{2}}V_{us}2\mathrm{p}^\mu \\ \langle \mathrm{n}|\text{vector part of } \mathrm{h}_+^\mu|\Sigma^-\rangle &= -V_{us}2\mathrm{p}^\mu \\ \langle \Lambda|\text{vector part of } \mathrm{h}_+^\mu|\Xi^-\rangle &= \sqrt{\tfrac{3}{2}}V_{us}2\mathrm{p}^\mu. \end{aligned} \tag{18.1.7}$$

Similar considerations apply to mesonic strangeness-changing β-decays. For example, for the forward matrix elements:

$$\begin{aligned} \langle \pi^0|\text{vector part of } \mathrm{h}_+^\mu|K^+\rangle &= \frac{1}{\sqrt{2}}V_{us}2\mathrm{p}^\mu \\ \langle \pi^-|\text{vector part of } \mathrm{h}_+^\mu|K_\mathrm{L}^0\rangle &= V_{us}2\mathrm{p}^\mu. \end{aligned} \tag{18.1.8}$$

However in the case of strangeness-changing β-decays the breaking of perfect $SU(3)_F$, as witnessed by the mass differences within an $SU(3)_F$ hadronic multiplet, is far too large to ignore. This is perhaps not so critical for the baryon octet, but it is manifestly nonsense to pretend that $m_K = m_\pi$! So the 'forward' matrix elements in the meson case cannot correspond to $q^2 = 0$ since $\mathrm{p}_\pi^\mu \neq \mathrm{p}_K^\mu$ etc. One obvious improvement is the replacement

$$2\mathrm{p}^\mu \to \mathrm{p}_K^\mu + \mathrm{p}_\pi^\mu \tag{18.1.9}$$

in (18.1.8) and an analogous replacement in (18.1.7). But other effects too, for example the mass differences between the u, d and s quarks, are important, and lead to corrections of about 3%. These corrections are, in fact, considerably larger than the radiative corrections to Fig. 18.1.

Ultimately, after a very detailed study, the result is

$$|V_{us}| = 0.220 \pm 0.002. \tag{18.1.10}$$

Since we now have confidence that there are only 3 generations, and since the KM matrix is unitary, we may take

$$|V_{ud}|^2 + |V_{us}|^2 + |V_{ub}|^2 = 1 \tag{18.1.11}$$

so that we obtain from (18.1.5 and 10)

$$|V_{ub}| = 0.046 \pm 0.030 \text{ [unitarity]} \tag{18.1.12}$$

which is not very precise, but does indicate how small $|V_{ub}|$ is likely to be. We shall soon consider another method of evaluating V_{ub} which suggests a value well below the central value in (18.1.12).

One might be tempted to try to measure the other V_{ij} by looking at a suitable β-decay reaction. For example $|V_{cs}|$ controls the Cabibbo allowed $\Delta S = \Delta C = \Delta Q$ decays $D^+ \to \bar{K}^0 \ell^+ \nu_\ell$ and $D^0 \to K^- \ell^+ \nu_\ell$. However, in the absence of a conserved current one is forced to a dynamical model of the decay and there is then a problem [see Section 13.2.1]; the simplest 'spectator' mechanism leads to equal D^0 and D^+ lifetimes—far from the measured ratio given in Table 13.1.

There is somewhat more hope of studying V_{cb} and V_{ub} from the semi-leptonic decays of bottom mesons. Here the decaying bottom quark is so much more massive than the other quarks involved that the spectator model (with phase space and QCD corrections) might be expected to be reliable. The matrix element V_{cb} is responsible for the decays $B \to D\ell\nu$ or $B \to D^*\ell\nu$ in which a bottom particle turns into a charm particle, and these are well studied. But the search for charmless decays, controlled by V_{ub}, has been difficult (recall that we know, from (18.1.12), that $|V_{ub}|$ is very small) and such events have only recently been positively identified by the ARGUS collaboration (Albrecht *et al.*, 1990) and the CLEO collaboration (Fulton *et al.*, 1990).

We follow the discussion of Kim and Martin (1989) and study the inclusive lepton energy spectrum in $B \to \mathrm{X} + \ell + \nu$ when ℓ is a specific lepton. At the quark level one has $b \to c + \ell + \nu$ and $b \to u + \ell + \nu$. The semi-leptonic width in the corrected spectator model is then, analogous to (4.2.3),

$$\Gamma(B \to \mathrm{X}\ell\nu) = \frac{G^2 m_b^5}{192\pi^3}\left\{|V_{cb}|^2\Phi(\epsilon_c) + |V_{ub}|^2\Phi(\epsilon_u)\right\} \tag{18.1.13}$$

where $\epsilon_c = m_c/m_b, \epsilon_u = m_u/m_b$ and the phase-space and QCD corrections are contained in

$$\Phi(\epsilon) = [1-8\epsilon^2+8\epsilon^6-\epsilon^8-24\epsilon^4 \ln\epsilon]\left\{1 - \frac{2\alpha_s}{3\pi}\left[\left(\pi^2 - \frac{31}{4}\right)(1-\epsilon)^2 + \frac{3}{2}\right]\right\}. \tag{18.1.14}$$

It happens that to a good approximation

$$\Phi(\epsilon_u) \approx 2\Phi(\epsilon_c)$$

so that

$$\Gamma(B \to X\ell\nu) \equiv \frac{BR(B \to X\ell\nu)}{\tau_B} \approx |V_{cb}|^2 \frac{G^2 m_b^5}{192\pi^3} \Phi(\epsilon_c) \left\{ 1 + 2 \left| \frac{V_{ub}}{V_{cb}} \right|^2 \right\}. \tag{18.1.15}$$

The branching ratios and lifetime are quite well determined:

$$\begin{aligned} \mathrm{BR}(B \to X\mu\nu) &= (10.3 \pm 0.5)\% \\ \mathrm{BR}(B \to Xe\nu) &= (10.7 \pm 0.5)\% \\ \tau_B &= (1.29 \pm 0.05) \times 10^{-12}\mathrm{s} \end{aligned} \tag{18.1.16}$$

so that, knowing how small $|V_{ub}|$ is expected to be, the above fixes $|V_{cb}|$ quite accurately *provided* we are confident in our value of m_b. Since m_b appears raised to the fifth power, this is clearly a dangerous source of error!

To overcome this problem one fits also the energy spectrum, in the rest frame of the b, of the lepton ℓ, since this is a sensitive function of ϵ_c and ϵ_u and therefore of m_b. Moreover if

$$x \equiv 2E_\ell/m_b \tag{18.1.17}$$

then the end-point of the spectrum is at

$$x = x_c = 1 - \epsilon_c^2 \qquad \text{and} \qquad x = x_u = 1 - \epsilon_u^2 \tag{18.1.18}$$

for the two types of decay respectively. In principle, therefore, all events with $x > x_c$ must come from $b \to u$ transitions, allowing a separate determination of $|V_{ub}|$. In practice the spectrum of leptons from $b \to c$ transitions dies off very rapidly beyond $E_\ell = 2$ GeV and leptons from $b \to u$ could have energies up to $E_\ell = 2.7$ GeV. But the smallness of the $b \to u$ rate makes the interpretation of the signals very difficult. Allowance must be made for 'continuum' events which could produce leptons without B mesons, and for leptons arising from $b \to J/\Psi X$ followed by $J/\Psi \to e^+e^-$ etc. As an example Fig. 18.2 shows the ARGUS lepton momentum spectrum (a) from the $\Upsilon(4S)$ and (b) estimated from the 'continuum'. The small excess of events in the region $2.3 \le p_\ell \le 2.6$ GeV/c, after subtracting the continuum events etc., is interpreted as due to $b \to u$ transitions. Table 18.1 taken from Albrecht *et al.* (1990) gives some idea of the delicacy of the whole procedure. Fig. 18.3 shows the CLEO lepton spectrum including an enlargement of the region $p_\ell > 2.3$ GeV/c.

Finally various attempts are made to correct for the fact that the b is bound inside the B meson. For example in Altarelli *et al.* (1982) the b is given a Fermi momentum with a gaussian distribution with a mean $\langle p_F \rangle$ which is determined from the fit to the data.

Ultimately, one finds

$$|V_{cb}| = 0.049 \pm 0.006 \tag{18.1.19}$$

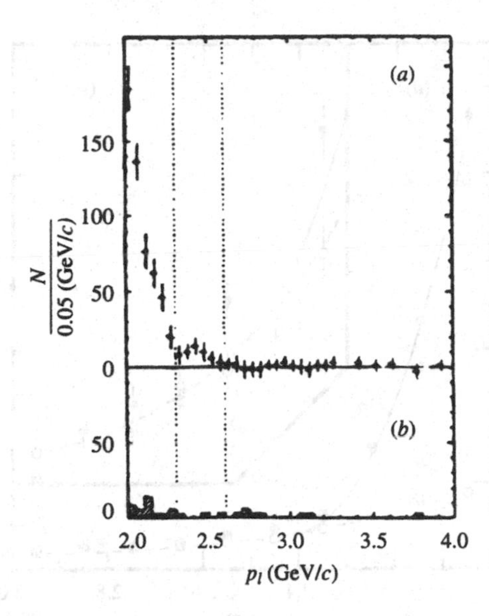

Fig. 18.2. ARGUS lepton momentum spectrum (a) for $\Upsilon(4S)$ and (b) as estimated for 'continuum' events.

	Single leptons		Dileptons	
	e	μ	e	μ
$\Upsilon(4S)$	31	29	11	10
Backgrounds:				
continuum (scaled)	6.8	14.5	1.0	1.0
$b \to c$	4.0	4.4	1.1	1.2
J/ψ	0.6	0.4	0.2	0.1
fakes	0.9	1.7	0.7	1.4
Total background	12.3 ± 2.8	21.0 ± 4.0	3.0 ± 1.0	3.7 ± 1.1
Signal	18.7 ± 6.2	8.0 ± 6.7	8.0 ± 3.5	6.3 ± 3.4

Table 18.1. ARGUS: observed single lepton and dilepton events in the range $2.3 \leq p_\ell \leq 2.6$ GeV/c and estimated backgrounds. (Albrecht *et al.*, 1990)

with $\langle p_F \rangle \approx 0.45$ GeV/c and effective masses 'm_c' $= 1.5$ GeV/c^2 and 'm_b' $= 4.8$ GeV/c.

The value for $|V_{ub}|$ is rather model dependent, but it seems reasonable to conclude from the ARGUS and CLEO data that

$$\frac{|V_{ub}|}{|V_{cb}|} = 0.11 \pm 0.01 \tag{18.1.20}$$

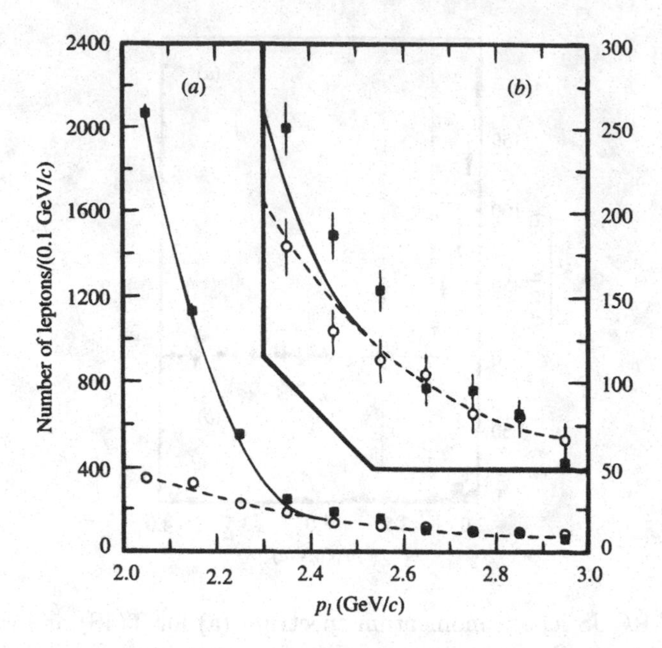

Fig. 18.3. CLEO lepton spectrum showing excess over background for $p_\ell > 2.3$ GeV/c with enlarged scale. Solid points are spectrum at $\Upsilon(4S)$; open points are off resonance.

implying the very small value

$$|V_{ub}| \approx 0.005 \tag{18.1.21}$$

which is barely compatible with (18.1.12).

A comparison of (18.1.5, 10, 19 and 21) provides the justification for the order of magnitude estimates given for the first row and third column of the KM matrix in (9.2.6).

18.2 KM matrix elements from deep inelastic scattering

The cross-section per nucleon for *inclusive* deep inelastic neutrino and antineutrino scattering on an isoscalar target $\mathrm{N_0}$ was given in (17.2.5). Upon examination of the structure of U and D in (17.2.3) one can see which terms are responsible for the production of a charm particle in the final state. Thus, if we study the *semi-inclusive* processes

$$\begin{aligned} \nu + \mathrm{N_0} &\to \mu^- + c + \mathrm{X} \\ \bar{\nu} + \mathrm{N_0} &\to \mu^+ + \bar{c} + \mathrm{X} \end{aligned} \tag{18.2.1}$$

above the charm production threshold, the cross-section for $Q^2 \ll M_W^2$ is given by

$$\frac{\mathrm{d}^2\sigma}{\mathrm{d}x\mathrm{d}y}(\nu \mathrm{N_0} \to \mu^- c\mathrm{X}) = \frac{G^2 s}{2\pi}\xi_c \left\{[u(\xi_c)+d(\xi_c)]|V_{cd}|^2 + 2s(\xi_c)|V_{cs}|^2\right\} \tag{18.2.2}$$

where, as explained in Section 16.4,

$$\xi_c = x(1+m_c^2/Q^2). \tag{18.2.3}$$

For the $\bar{\nu}$ reaction simply replace u, d by $\bar{u}, \bar{d}$.

Thus, provided one can identify semi-inclusive charm production, one can learn about $|V_{cd}|$ and $|V_{cs}|$. Let us concentrate on the neutrino induced reaction. Then the signal for charm production is the appearance of oppositely charged muon pairs arising from the decays

$$c \to s + \mu^+ + \nu_\mu \quad \text{and} \quad c \to d + \mu^+ + \nu_\mu \tag{18.2.4}$$

yielding a μ^+ in addition to the μ^- coming from the leptonic vertex. There are, of course, other possible sources of a μ^+, for example from π and K decays in the hadron shower, but because of the large difference in mass between initial and final quarks in (18.2.4), the μ^+ from c decay will generally be much more energetic than those from π or K decay. The background from the non-charm events is estimated by Monte Carlo methods. As an example in a recent Fermilab experiment (CCFR, 1990b) with a mean neutrino energy of 160 GeV, 1552 ν_μ induced pairs were selected and the background was estimated to be 102 events.

In principle the most direct way of obtaining information on the KM matrix elements is the following. Let us, in (18.2.2), split the u and d contributions into valence and sea contributions and assume $s(x) = \bar{s}(x)$. Then for charm events

$$\frac{\mathrm{d}^2\sigma}{\mathrm{d}x\mathrm{d}y}(\nu \mathrm{N_0} \to \mu^-\mu^+\mathrm{X})_c = \frac{G^2 s\xi_c B_c}{2\pi}\left\{[u_V + d_V + \bar{u} + \bar{d}]|V_{cd}|^2 + 2s|V_{cs}|^2\right\} \tag{18.2.5}$$

and

$$\frac{\mathrm{d}^2\sigma}{\mathrm{d}x\mathrm{d}y}(\bar{\nu} \mathrm{N_0} \to \mu^+\mu^-\mathrm{X})_c = \frac{G^2 s\xi_c B_c}{2\pi}\left\{[\bar{u} + \bar{d}]|V_{cd}|^2 + 2s|V_{cs}|^2\right\} \tag{18.2.6}$$

where B_c is the branching ratio for $c \to \mu^+\nu_\mu\mathrm{X}$ (which is taken to be the same as for $\bar{c} \to \mu^-\bar{\nu}_\mu\mathrm{X}$).

Then

$$\frac{\mathrm{d}^2\sigma}{\mathrm{d}x\mathrm{d}y}(\nu \mathrm{N_0} \to \mu^-\mu^+\mathrm{X})_c - \frac{\mathrm{d}^2\sigma}{\mathrm{d}x\mathrm{d}y}(\bar{\nu} \mathrm{N_0} \to \mu^+\mu^-\mathrm{X})_c$$

$$= \frac{G^2 s\xi_c B_c}{2\pi}|V_{cd}|^2[u_V(\xi_c) + d_V(\xi_c)] \tag{18.2.7}$$

Since the valence quark distributions are well measured and their normalization is fixed, (18.2.5) yields a value for $|V_{cd}|^2 B_c$. The branching fraction is known from studies of ν_μ interactions in emulsions:

$$B_c = 0.110 \pm 0.009 \tag{18.2.8}$$

so that $|V_{cd}|$ can be found. One obtains

$$|V_{cd}| = 0.220 \pm 0.016. \tag{18.2.9}$$

[The above description is somewhat idealized. For details the reader should consult CCFR (1990b).]

Using unitarity,

$$|V_{cs}|^2 = 1 - |V_{cb}|^2 - |V_{cd}|^2 \tag{18.2.10}$$

so that from (18.1.19) and (18.2.7):

$$|V_{cs}| = 0.974 \pm 0.004 \quad \text{[unitarity]}. \tag{18.2.11}$$

Recall that in the old Cabibbo theory one had $V_{ud} = V_{cs} = \cos\theta_C$. The values given in (18.2.9) and (18.1.5) indicate that in the u, d, s, c sector the KM generalization is remarkably close to the Cabibbo description.

Unitarity now allows one to calculate

$$|V_{td}| = \left[1 - |V_{ud}|^2 - |V_{cd}|^2\right]^{1/2}$$

and

$$|V_{ts}| = \left[1 - |V_{us}|^2 - |V_{cs}|^2\right]^{1/2}. \tag{18.2.12}$$

We obtain

$$|V_{td}| = 0.046 \pm 0.09 \quad \text{[unitarity]} \tag{18.2.13}$$

and

$$|V_{ts}| = 0.054 \pm 0.08 \quad \text{[unitarity]} \tag{18.2.14}$$

which, while showing that the matrix elements are very small, do not pin them down very accurately.

The matrix elements V_{td} and V_{ts} play a crucial rôle in B^0–$\bar{B}^0$ mixing as discussed in Section 19.4. Ultimately their values will be pinned down by experiments in this field; provided, of course, that the SM *is* capable of explaining the data.

18.3 Summary

We summarize below our present knowledge of the *magnitude* of the KM matrix elements. Entries marked † have been derived on the basis of unitarity. [See eqn (9.2.6) for notation.]

$$|V_{ij}| = \begin{pmatrix} 0.9744 \pm 0.0010 & 0.220 \pm 0.002 & \approx 0.005 \\ 0.220 \pm 0.016 & 0.974 \pm 0.004^{\dagger} & 0.049 \pm 0.006 \\ 0.046 \pm 0.09^{\dagger} & 0.054 \pm 0.08^{\dagger} & 0.995 \pm 0.009^{\dagger} \end{pmatrix}. \quad (18.3.1)$$

The phases of the V_{ij} are discussed in Chapter 19.

19
Mixing and CP violation

Mixing and CP violation effects were first discovered in the strange meson K^0–$\bar{K}^0$ system. To date this is the only system in which CP violation has been observed and one of the most exciting questions at present is whether analogous effects will show up in the charm meson D^0–$\bar{D}^0$ and bottom meson B^0–$\bar{B}^0$ systems. Equally challenging is the question as to whether the 'natural' CP violation arising from the KM matrix can adequately explain the observations.

19.1 General phenomenology of mixing and CP violation

We shall first set up a general formalism for mixing and CP violation, utilizing hypothetical pseudo-scalar mesons P^0 and $\bar{P}^0$. Everything in the general formalism applies equally well to $K^0, \bar{K}^0$, $D^0, \bar{D}^0$ and $B^0, \bar{B}^0$. Questions specific to the particular mesons will be discussed thereafter.

To simplify the notation we shall often use symbols like $P_0(t)$ or $\bar{P}_0(t)$ as a shorthand for the wave-functions associated with the state vectors $|P^0\rangle$, $|\bar{P}^0\rangle$.

19.1.1 General formalism for mixing

Consider $|P^0\rangle$ and $|\bar{P}^0\rangle$ which are eigenstates of the strong interaction field-theoretic Hamiltonian H_s which conserves strangeness, charm and bottom. $|P^0\rangle$ and $|\bar{P}^0\rangle$ are assumed to differ in one or more of the above quantum numbers.

But the electroweak Hamiltonian H_w does not conserve strangeness, charm or bottom and thus transitions $\langle \bar{P}^0|\hat{T}_w|P^0\rangle$ are possible, with the consequence that $|P^0\rangle$ and $|\bar{P}^0\rangle$ are not eigenstates of the total Hamiltonian $H = H_s + H_w$. Thus they are not stationary states or simply decaying states, and do not have a simple $\exp[-iEt/\hbar]$ time dependence.

Consider the complete state vector $|\Psi(t)\rangle$ which has evolved from a $|P^0\rangle$ or $|\bar{P}^0\rangle$ state at say $t = 0$. We are interested in the projection of $|\Psi(t)\rangle$ (let us call it $|\psi(t)\rangle$) onto the subspace spanned by $|P^0\rangle$ and $|\bar{P}^0\rangle$. Thus we put

$$|\psi(t)\rangle = P_0(t)|P^0\rangle + \bar{P}_0(t)|\bar{P}^0\rangle. \tag{19.1.1}$$

It can be shown [see, for example, Marshak, Riazuddin and Ryan (1969)] that the time dependence of the 'wave functions' $P_0(t), \bar{P}_0(t)$ is controlled by a coupled Schrödinger equation,

$$\mathrm{i}\hbar\frac{\mathrm{d}}{\mathrm{d}t}\begin{pmatrix} P_0(t) \\ \bar{P}_0(t) \end{pmatrix} = \begin{pmatrix} H_{11} & H_{12} \\ H_{21} & H_{22} \end{pmatrix}\begin{pmatrix} P_0(t) \\ \bar{P}_0(t) \end{pmatrix} \tag{19.1.2}$$

where, as will be shown in Section 19.3, the H_{ij} are matrix elements of the weak interaction transition operator $\hat{T}_w$ defined in Appendix 5. One has

$$\begin{aligned} H_{11} &= \langle P^0|\hat{T}_w|P^0\rangle \qquad H_{22} = \langle \bar{P}^0|\hat{T}_w|\bar{P}^0\rangle \\ H_{12} &= \langle P^0|\hat{T}_w|\bar{P}^0\rangle \qquad H_{21} = \langle \bar{P}^0|\hat{T}_w|P^0\rangle \end{aligned} \tag{19.1.3}$$

where the states are normalized to *unity*. The matrix $\boldsymbol{H}$ is an 'effective Hamiltonian' which is not, in general, hermitian, i.e. H_{11} and H_{22} are not in general real, nor is $H_{12} = H_{21}^*$. If, as usual, we assume CPT invariance, then, as we show in Appendix 6, one has $H_{11} = H_{22} \equiv H$. CPT gives no information about H_{12} or H_{21}.

It turns out that one can split each matrix element, in a physically meaningful way, into two parts, as will be explained in Section 19.3, in the form:

$$H = M - \mathrm{i}\Gamma/2, \quad H_{12} = M_{12} - \mathrm{i}\Gamma_{12}/2, \quad H_{21} = M_{12}^* - \mathrm{i}\Gamma_{12}^*/2 \tag{19.1.4}$$

where M and Γ are real. Then (19.1.2) becomes

$$\mathrm{i}\hbar\frac{\mathrm{d}}{\mathrm{d}t}\begin{pmatrix} P_0(t) \\ \bar{P}_0(t) \end{pmatrix} = \begin{pmatrix} M - \mathrm{i}\Gamma/2 & M_{12} - \mathrm{i}\Gamma_{12}/2 \\ M_{12}^* - \mathrm{i}\Gamma_{12}^*/2 & M - \mathrm{i}\Gamma/2 \end{pmatrix}\begin{pmatrix} P_0(t) \\ \bar{P}_0(t) \end{pmatrix}. \tag{19.1.5}$$

Note that despite superficial appearances, one cannot calculate H_{21} from H_{12} just on the basis of (19.1.4) since M_{12} and Γ_{12}, being complex, are *not* determined uniquely by the value of H_{12}.

Equation (19.1.5) can be diagonalized by using certain linear combinations of $|P^0\rangle$ and $|\bar{P}^0\rangle$ as basis states. These combinations can be written in the form

$$p|P^0\rangle \pm q|\bar{P}^0\rangle. \tag{19.1.6}$$

From the process of diagonalization, one finds only a condition relating q to p, namely,

$$\frac{q}{p} = \left(\frac{H_{21}}{H_{12}}\right)^{1/2} = \left(\frac{M_{12}^* - \mathrm{i}\Gamma_{12}^*/2}{M_{12} - \mathrm{i}\Gamma_{12}/2}\right)^{1/2} \tag{19.1.7}$$

which determines q/p up to a sign. Let

$$H_{12} = |H_{12}|\mathrm{e}^{\mathrm{i}\phi_{12}} \qquad H_{21} = |H_{21}|\mathrm{e}^{\mathrm{i}\phi_{21}} \tag{19.1.8}$$

and let us write

$$\frac{q}{p} = \left|\frac{H_{21}}{H_{12}}\right|^{1/2} \mathrm{e}^{\mathrm{i}(\phi_{21}-\phi_{12})/2}. \tag{19.1.9}$$

The reason for the above phase choice is explained later. It is in accord with recent papers on B^0–$\bar{B}^0$ physics.

Regarding the overall phase of the linear combinations (19.1.6) which diagonalize (19.1.5), these are completely free since we are solving a homogeneous equation. *By convention* we define

$$|P_\pm^0\rangle \equiv \frac{1}{\sqrt{|p|^2 + |q|^2}} \left\{p|P^0\rangle \mp q|\bar{P}_0\rangle\right\}. \tag{19.1.10}$$

The reason for this rather odd notation is that with the sign convention for q/p mentioned previously, it will turn out that $|P_\pm^0\rangle$ have CP eigenvalues ± 1 *if* CP is conserved. Substituting into (19.1.5) one finds that the time evolution of these states is now simple:

$$|P_\pm^0, t\rangle = P_\pm(t)|P_\pm^0\rangle \tag{19.1.11}$$

where

$$P_\pm(t) = \mathrm{e}^{-\Gamma_\pm t/2}\mathrm{e}^{-\mathrm{i}m_\pm t} \tag{19.1.12}$$

with

$$m_\pm = M \pm \mathrm{Re}\,[H_{12}H_{21}]^{1/2} \tag{19.1.13}$$

$$\Gamma_\pm = \Gamma \mp 2\mathrm{Im}\,[H_{12}H_{21}]^{1/2}. \tag{19.1.14}$$

If we define

$$\Delta m \equiv m_- - m_+ \qquad \Delta\Gamma \equiv \Gamma_+ - \Gamma_- \tag{19.1.15}$$

then we have from (19.1.13,14)

$$\Delta m + \mathrm{i}\Delta\Gamma/2 = -2(H_{12}H_{21})^{1/2} = -2|H_{12}H_{21}|^{1/2}\mathrm{e}^{\mathrm{i}(\phi_{12}+\phi_{21})/2} \tag{19.1.16}$$

where we have used (19.1.8 and 9) to evaluate the phase of the square root.

Consider now the time evolution of the states $|P^0\rangle$ or $|\bar{P}^0\rangle$ produced in a strong interaction at $t = 0$. There are two forms of the result which will be useful later. Solving for $|P^0\rangle$, $|\bar{P}^0\rangle$ from (19.1.10) and using (19.1.12), the states at some later time t have become, respectively,

$$
\begin{aligned}
|\psi(t)\rangle_{P^0} &= \frac{\sqrt{|p|^2+|q|^2}}{2p}\left\{P_+(t)|P_+^0\rangle + P_-(t)|P_-^0\rangle\right\} \\
|\psi(t)\rangle_{\bar{P}^0} &= \frac{\sqrt{|p|^2+|q|^2}}{2q}\left\{P_-(t)|P_-^0\rangle - P_+(t)|P_+^0\rangle\right\}.
\end{aligned}
\tag{19.1.17}
$$

This form will be useful for studying CP violation in the K^0–$\bar{K}^0$ system.

If we now substitute (19.1.10) into (19.1.17) we obtain an alternative form, particularly useful for B^0–$\bar{B}^0$ mixing,

$$
\begin{aligned}
|\psi(t)\rangle_{P^0} &= f_+(t)|P^0\rangle - \left(\frac{q}{p}\right) f_-(t)|\bar{P}^0\rangle \\
|\psi(t)\rangle_{\bar{P}^0} &= f_+(t)|\bar{P}^0\rangle - \left(\frac{p}{q}\right) f_-(t)|P^0\rangle
\end{aligned}
\tag{19.1.18}
$$

where

$$
f_\pm(t) = \tfrac{1}{2}\left[\mathrm{e}^{-im_+t}\mathrm{e}^{-\Gamma_+t/2} \pm \mathrm{e}^{-im_-t}\mathrm{e}^{-\Gamma_-t/2}\right]. \tag{19.1.19}
$$

We see that a state which at $t=0$ was, say, a pure $|P^0\rangle$ is, at later times, a mixture of $|P^0\rangle$ and $|\bar{P}^0\rangle$. The probabilities of finding the various states at time t, starting with either $|P^0\rangle$ or $|\bar{P}^0\rangle$ at $t=0$, are then:

$$
\left.
\begin{aligned}
\mathcal{P}(P^0 \to P^0; t) &= |f_+(t)|^2 \\
\mathcal{P}(P^0 \to \bar{P}^0; t) &= \left|\tfrac{q}{p}\right|^2 |f_-(t)|^2
\end{aligned}
\right\}
\tag{19.1.20}
$$

and

$$
\left.
\begin{aligned}
\mathcal{P}(\bar{P}^0 \to \bar{P}^0; t) &= |f_+(t)|^2 \\
\mathcal{P}(\bar{P}^0 \to P^0; t) &= \left|\tfrac{p}{q}\right|^2 |f_-(t)|^2.
\end{aligned}
\right\}
\tag{19.1.21}
$$

It should be stressed that the above phenomenon of strangeness or charm or bottom mixing has, *a priori*, nothing to do with CP violation, and would occur even if CP were absolutely conserved, as will become clear later.

19.1.2 General formalism for CP violation

Let us now turn to the question of CP invariance. Under C a particle is changed into its antiparticle, so that given that we are dealing with pseudoscalar mesons, for a particle at rest it is natural to take $\mathcal{CP}|P^0\rangle = -|\bar{P}^0\rangle$

and $\mathcal{CP}|\bar{P}^0\rangle = -|P^0\rangle$. But, as with all discrete symmetries, there is some freedom of choice in the phase used in defining the transformation (Feinberg and Weinberg, 1959). Of course the physics cannot change but the formalism does depend upon the phase. In pre-quark days, when P^0 and $\bar{P}^0$ (in fact $K^0, \bar{K}^0$) were considered as fundamental fields other phase conventions were useful. But in the SM where everything is expressed in terms of quark fields whose transformations under C and P are the conventional ones (see Appendix A1.2) the above CP convention is necessary. Thus we take

$$\mathcal{CP}|P^0\rangle = -|\bar{P}^0\rangle, \qquad \mathcal{CP}|\bar{P}^0\rangle = -|P^0\rangle. \tag{19.1.22}$$

The transition amplitudes for $P^0 \to \bar{P}^0$ and $\bar{P}^0 \to P^0$ are

$$\left.\begin{aligned} \langle\bar{P}^0|\hat{T}_w|P^0\rangle &= H_{21} = M_{12}^* - \mathrm{i}\Gamma_{12}^*/2 \\ \langle P^0|\hat{T}_w|\bar{P}^0\rangle &= H_{12} = M_{12} - \mathrm{i}\Gamma_{12}/2. \end{aligned}\right\} \tag{19.1.23}$$

If CP were conserved we would have by (19.1.22) and by the definitions of M_{12} and Γ_{12} given in (19.3.5 and 6)

$$H_{12} = H_{21} \quad \text{or} \quad M_{12} = M_{12}^*, \Gamma_{12} = \Gamma_{12}^* \qquad \text{[CP conserved]} \tag{19.1.24}$$

implying M_{12} and Γ_{12} are real, or, in particular,

$$\mathrm{Im}(M_{12}^*\Gamma_{12}) = 0 \qquad \text{[CP conserved]}. \tag{19.1.25}$$

Conversely, if $\mathrm{Im}(M_{12}^*\Gamma_{12}) \neq 0$ then there is some CP violation in the dynamics that determines the eigenstates $|P_\pm^0\rangle$. This conclusion is actually independent of any phase convention. Thus we have

$$\mathrm{Im}(M_{12}^*\Gamma_{12}) \neq 0 \qquad \Rightarrow \text{CP violation in } \boldsymbol{H}. \tag{19.1.26}$$

With our phase convention, from (19.1.8, 24) one has, when CP is conserved,

$$\frac{q}{p} = 1 \quad \text{[CP conserved]}. \tag{19.1.27}$$

In an arbitrary phase convention an unambiguous consequence of CP violation is

$$\left|\frac{q}{p}\right| \neq 1 \quad \Leftrightarrow \quad \text{CP violation}. \tag{19.1.28}$$

It is easy to check that this is consistent with (19.1.26) since one finds after some algebra that

$$\frac{|p|^2 - |q|^2}{|p|^2 + |q|^2} = \frac{\mathrm{Im}(M_{12}^*\Gamma_{12})}{|M_{12}|^2 + |\Gamma_{12}/2|^2 + \frac{1}{4}[(\Delta m)^2 + (\Delta\Gamma/2)^2]}. \tag{19.1.29}$$

It is interesting to compute the overlap $\langle P^0_+|P^0_-\rangle$. One finds from (19.1.10)

$$\langle P^0_+|P^0_-\rangle = \frac{|p|^2 - |q|^2}{|p|^2 + |q|^2} \tag{19.1.30}$$

so that $|P^0_+\rangle, |P^0_-\rangle$ are orthogonal only if $\boldsymbol{H}$ conserves CP.

Note finally that if CP is conserved then using (19.1.22, 27) in (19.1.10) we have

$$\mathcal{CP}|P^0_\pm\rangle = \pm|P^0_\pm\rangle \qquad \text{[CP conserved]} \tag{19.1.31}$$

as was indicated earlier.

19.1.3 Practical aspects of mixing and CP violation

We have seen that CP conservation implies $|q/p| = 1$. Returning to the mixing formulae (19.1.20,21) we see that mixing can perfectly well take place when CP is conserved, but the probabilities are equal:

$$\mathcal{P}(P^0 \to \bar{P}^0; t) = \mathcal{P}(\bar{P}^0 \to P^0; t). \tag{19.1.32}$$

Any departure from (19.1.32) will thus signal CP violation in $\boldsymbol{H}$.

Let us now consider some of the practical aspects of mixing and/or CP violation. We shall suppose that we can identify $|P^0\rangle, |\bar{P}^0\rangle$ by some characteristic reaction signature. For example, for kaons one might ideally look for $\bar{K}^0 + \mathrm{p} \to \Lambda^0 + \pi^0$, but in practice one might use the $\Delta S = \Delta Q$ rule of semi-leptonic decays which implies

$$K^0 \to \pi^- \mathrm{e}^+ \nu_\mathrm{e}, \qquad \bar{K}^0 \to \pi^+ \mathrm{e}^- \bar{\nu}_\mathrm{e}. \tag{19.1.33}$$

Thus by measuring the time dependence of the production rate of these final states (or, equivalently since the original K^0 or $\bar{K}^0$ is moving in the LAB, the distance dependence of these rates) one can effectively measure the $|f_\pm(t)|^2$ of eqns (19.1.20,21).

Now one has from (19.1.19 and 15)

$$|f_\pm(t)|^2 = \tfrac{1}{4}\left\{\mathrm{e}^{-\Gamma_+ t} + \mathrm{e}^{-\Gamma_- t} \pm 2\mathrm{e}^{-\Gamma t}\cos(\Delta m t)\right\} \tag{19.1.34}$$

where, from (19.1.14), we have used

$$\tfrac{1}{2}(\Gamma_+ + \Gamma_-) = \Gamma. \tag{19.1.35}$$

Hence oscillations in the production rate will be observable provided the period $2\pi/\Delta m$ is not too long compared with the decay time $1/\Gamma$. Thus the condition for observable oscillations is roughly

$$|\Delta m| \geq \Gamma. \tag{19.1.36}$$

Let us now consider the various possibilities for $P^0, \bar{P}^0$.

(a) The K^0-$\bar{K}^0$ system. In the K^0–$\bar{K}^0$ system the dominant decays are

$$K^0_+ \to 2\pi, \qquad K^0_- \to 3\pi$$

so that because of phase space

$$\Gamma_+ \gg \Gamma_- \tag{19.1.37}$$

and the lifetime of $|K^0_+\rangle$ is much shorter than $|K^0_-\rangle$. For this reason these states are usually called (S for short, L for long)

$$|K^0_S\rangle \equiv |K^0_+\rangle \quad \text{and} \quad |K^0_L\rangle \equiv |K^0_-\rangle. \tag{19.1.38}$$

Indeed for the lifetimes one has experimentally

$$\begin{aligned} \tau_S &= (0.8922 \pm 0.0020) \times 10^{-10}\text{s} \\ \tau_L &= (5.15 \pm 0.04) \times 10^{-8}\text{s} \end{aligned} \tag{19.1.39}$$

so that

$$\left.\begin{aligned} &\Gamma_S \approx 600\Gamma_L \approx 10^{10}\text{s}^{-1} \\ &\Gamma \approx \tfrac{1}{2}\Gamma_S. \end{aligned}\right\} \tag{19.1.40}$$

Also, empirically,

$$\Delta m = m_L - m_S = (3.522 \pm 0.016) \times 10^{-12}\text{MeV}/c^2 \tag{19.1.41}$$

so that

$$\Delta m \approx 0.54 \times 10^{10} s^{-1} \approx \tfrac{1}{2}\Gamma_S.$$

Thus (19.1.35) is roughly satisfied empirically, i.e.

$$\Delta m \approx \Gamma \tag{19.1.42}$$

and the observation of oscillations is feasible. Indeed the fantastically small mass difference Δm is derived from measurements of these oscillations.

The fortuitous fact that $\Gamma_S \gg \Gamma_L$ is a great boon experimentally. Any beam of neutral K mesons, either K^0 or $\bar{K}^0$, traversing a vacuum, will, if we wait long enough, become essentially a pure K_L beam according to (19.1.17 and 12). It is then straightforward to get an idea of the size of the CP violation in $\boldsymbol{H}$. For, upon using (19.1.33), and the fact that the $K^0, \bar{K}^0$ rates for these reactions are equal by CPT, we get from the measured asymmetry in the semi-leptonic decays of the K_L, from (19.1.10),

$$\begin{aligned} \delta_\ell &\equiv \frac{\Gamma(K_L \to \pi^-\ell^+\nu_\ell) - \Gamma(K_L \to \pi^+\ell^-\bar{\nu}_\ell)}{\Gamma(K_L \to \pi^-\ell^+\nu_\ell) + \Gamma(K_L \to \pi^+\ell^-\bar{\nu}_\ell)} \\ &= \frac{|p|^2 - |q|^2}{|p|^2 + |q|^2} \approx 3.3 \times 10^{-3}. \end{aligned} \tag{19.1.43}$$

Thus in the kaon system $|q/p|$ is very close to 1.

$$\left|\frac{q}{p}\right|_K \approx 1 - (3.3 \times 10^{-3}) \qquad (19.1.44)$$

(b) The D^0–$\bar{D}^0$ system. The situation in the D^0–$\bar{D}^0$ system is somewhat different. Firstly the $D^0, \bar{D}^0$ decays are Cabibbo favoured (see Section 13.1). Secondly, because of their larger mass, both have many more channels open for their decay. Consequently their lifetime is much shorter than the K's: $\tau(D^0_\pm) \sim 4 \times 10^{-13}s$, and their decay widths are similar $\Gamma_+ \sim \Gamma_-$ so that $\Delta\Gamma/\Gamma \ll 1$. Thus the observation of oscillations does not seem feasible. However one could try to compare the total number of $\bar{D}^0$ found per D^0 produced, and vice versa, i.e. utilize the integrated version of (19.1.20,21). From (19.1.19) one finds

$$\begin{aligned} r &\equiv \frac{\int_0^\infty \mathcal{P}(D^0 \to \bar{D}^0; t)dt}{\int_0^\infty \mathcal{P}(D^0 \to D^0; t)dt} = \left|\frac{q}{p}\right|^2 \frac{x^2+y^2}{2+x^2-y^2} \\ \bar{r} &\equiv \frac{\int_0^\infty \mathcal{P}(\bar{D}^0 \to D^0; t)dt}{\int_0^\infty \mathcal{P}(\bar{D}^0 \to \bar{D}^0; t)dt} = \left|\frac{p}{q}\right|^2 \frac{x^2+y^2}{2+x^2-y^2} \end{aligned} \qquad (19.1.45)$$

where

$$x = -\Delta m/\Gamma, \qquad y = \Delta\Gamma/2\Gamma. \qquad (19.1.46)$$

These simple results show clearly what is needed for the largest mixing effects. Either $y^2 \approx 1$, i.e. $|\Delta\Gamma| \approx 2\Gamma$, and/or $x^2 \geq 1$, i.e. $|\Delta m| \geq \Gamma$, will ensure that both r and $\bar{r}$ are not small for given $|q/p|$. Neither of these conditions is met in the D^0–$\bar{D}^0$ system and the data given in Section 13.2.3 [see (13.2.16)] correspond to $r < 0.0037$. (A more detailed explanation of the dynamics responsible for this is given later.)

(c) The B^0–$\bar{B}^0$ system. The situation in the B^0–$\bar{B}^0$ case is much more encouraging, and mixing effects have indeed been seen. Here one expects small y^2 but quite large x^2. We assume that $B\bar{B}$ are produced *in pairs* at $t = 0$, and bearing in mind (19.1.42), try to count the number N of events having respectively BB, $\bar{B}\bar{B}$ and $\bar{B}B$ pairs in the final state. A non-zero value of $[N(BB)+N(\bar{B}\bar{B})]/N(\bar{B}B)$ will thus signal mixing. The complete identification of a B or $\bar{B}$ is not easy, so one relies upon the semi-leptonic decays

$$B^0 \to \ell^+\nu X \qquad \text{whereas} \qquad \bar{B}^0 \to \ell^-\bar{\nu}X$$

and thus counts events with $\ell^-\ell^-, \ell^+\ell^+$ and $\ell^+\ell^-$ in the final state. However, there is a danger of picking up leptons from D decay. Hence one restricts attention to high energy leptons which could only have come from B decay. Of course it is still necessary to make many background subtractions.

Let R be the fraction of events with pairs of B^0s or pairs of $\bar{B}^0$s, so that

$$R \equiv \frac{N(B^0B^0) + N(\bar{B}^0\bar{B}^0)}{N(B^0B^0) + N(\bar{B}^0\bar{B}^0) + N(B^0\bar{B}^0)}. \tag{19.1.47}$$

If the only sources of the leptons are the $B^0, \bar{B}^0$, we have

$$R = \frac{N(\ell^+\ell^+) + N(\ell^-\ell^-)}{N(\ell^+\ell^+) + N(\ell^-\ell^-) + N(\ell^+\ell^-)}. \tag{19.1.48}$$

The relationship between R defined in (19.1.47) and $r, \bar{r}$ defined as in (19.1.45), but with $B^0, \bar{B}^0$ replacing $D^0, \bar{D}^0$, depends upon the correlation between the initially produced B^0 and $\bar{B}^0$. Two cases are of particular interest:

1. *Incoherent production and decay*: If the B^0 and $\bar{B}^0$ are produced in a high multiplicity final state it is reasonable to assume them to be uncorrelated. Then any decay probability for the $B^0\bar{B}^0$ pair is taken to be built up from the product of the probabilities for the B^0 and $\bar{B}^0$ to decay. For example, one takes

$$N(B^0\bar{B}^0) \propto \mathcal{P}(B^0 \to B^0)\mathcal{P}(\bar{B}^0 \to \bar{B}^0) + \mathcal{P}(B^0 \to \bar{B}^0)\mathcal{P}(\bar{B}^0 \to B^0)$$

etc. Then one finds from the definition of $r, \bar{r}$ in (19.1.45)

$$R = \frac{r + \bar{r}}{1 + r + \bar{r} + r\bar{r}}. \tag{19.1.49}$$

In the absence of CP violation one has $r = \bar{r}$, thus

$$R = \frac{2r}{(1+r)^2} \qquad \text{[CP conserved]}. \tag{19.1.50}$$

2. *$B^0, \bar{B}^0$ from $\Upsilon(4S)$ decay*: One of the best sources of Bs is the decay of $\Upsilon(4S)$. Because its mass is less than $2m(B_s^0)$ it can only decay into $B_d^0\bar{B}_d^0$ pairs. The production via a virtual photon: $e^+e^- \to$'γ'$\to \Upsilon(4S) \to B_d^0\bar{B}_d^0$ implies that the Bs are in an eigenstate of odd charge conjugation, $C = -1$. Thus at time t the two-particle state, in its CM, is, in a notation similar to that used in Section 19.1.1, given by

$$|\phi(t)\rangle^{C=-1}_{B^0\bar{B}^0} = \frac{1}{\sqrt{2}}\left\{|\psi(t)\rangle_{B^0(\boldsymbol{p})}|\psi(t)\rangle_{\bar{B}^0(-\boldsymbol{p})} - |\psi(t)\rangle_{\bar{B}^0(\boldsymbol{p})}|\psi(t)\rangle_{B^0(-\boldsymbol{p})}\right\}. \tag{19.1.51}$$

Upon substituting (19.1.18,19) in (19.1.51) one finds a very simple time dependence for $|\phi(t)\rangle$, namely

$$|\phi(t)\rangle^{C=-1}_{B^0\bar{B}^0} = \mathrm{e}^{-\mathrm{i}Mt}\mathrm{e}^{-\Gamma t}\frac{1}{\sqrt{2}}\left\{|B^0(\boldsymbol{p})\bar{B}^0(-\boldsymbol{p})\rangle - |B^0(-\boldsymbol{p})\bar{B}^0(\boldsymbol{p})\rangle\right\}. \tag{19.1.52}$$

Suppose, for example, we wish to evaluate the probability of finding $\ell^+\ell^+$ in the final state. To begin with we consider the amplitude $\mathcal{A}_1$ to produce $(\ell^+\mathrm{X})$ with momentum $\boldsymbol{p}$ at time t_1 followed by the amplitude $\mathcal{A}_2$ to produce $(\ell^+\mathrm{X})$ with momentum $-\boldsymbol{p}$ at time t_2. Given that only B^0 can decay into ℓ^+, $\mathcal{A}_1$ will be proportional to the amplitude to find $|B^0(\boldsymbol{p})\bar{B}^0(-\boldsymbol{p})\rangle$ in $|\phi(t)\rangle^{C=-1}_{B^0\bar{B}^0}$ at time t_1, i.e. to

$$\frac{1}{\sqrt{2}}\mathrm{e}^{-\mathrm{i}Mt_1}\mathrm{e}^{-\Gamma t_1}.$$

This must be multiplied by the amplitude $\mathcal{A}_2$ to find $|B^0(-\boldsymbol{p})\rangle$ at time t_2 in the state which has evolved from being $|\bar{B}^0(-\boldsymbol{p})\rangle$ at $t = t_1$. The latter is given by (19.1.18) with t replaced by $t_2 - t_1$. Thus the amplitude $\mathcal{A}_2$ is proportional to

$$\left(\frac{p}{q}\right) f_-(t_2 - t_1).$$

Finally, since in both $\mathcal{A}_1$ and $\mathcal{A}_2$ we have the semi-leptonic decay $B^0 \to \ell^+\mathrm{X}$, we must multiply by the amplitudes $A(B^0 \to \ell^+\mathrm{X})A(B^0 \to \ell^+\mathrm{X})$.

Similar arguments apply to the other charge combinations.

Taking the modulus squared to get the rate, integrating over t_1, t_2 and assuming CP invariance of the semi-leptonic decays one finds eventually

$$R = \frac{r + \bar{r}}{2 + r + \bar{r}}. \tag{19.1.53}$$

Note the difference between this and (19.1.49).

In the absence of CP violation $\bar{r} = r$, so that

$$R = \frac{r}{1 + r} \qquad \text{[CP conserved]} \tag{19.1.54}$$

to be compared with (19.1.50).

There is one other complication that one has to worry about. Because the $B^0, \bar{B}^0$ are identified only by their semi-leptonic decay, by counting the leptons we are also counting the leptons from the B^+B^- produced in the $\Upsilon(4S)$ decay.

Let f_{00} and $f_{+-}(f_{00}+f_{+-}=1)$ be the fractions of $B_d^0\bar{B}_d^0$ and B^+B^- produced in $\Upsilon(4S)$ decay. Then, assuming CP conservation in the decays,

$$\left.\begin{aligned} N(\ell^-\ell^-) &= N(\bar{B}^0\bar{B}^0)(BR_0)^2 \\ N(\ell^+\ell^+) &= N(B^0B^0)(BR_0)^2 \\ N(\ell^+\ell^-) &= N(B^0\bar{B}^0)(BR_0)^2 + N(B^+B^-)(BR_+)^2 \end{aligned}\right\} \tag{19.1.55}$$

where

$$BR_+ = BR(B^\pm \to \ell^\pm \mathrm{X}), \qquad BR_0 = BR(B^0 \to \ell^+ \mathrm{X}).$$

Thus

$$N(\ell^+\ell^-) = N(B^0\bar{B}^0)(BR_0)^2\left[1+\frac{f_{+-}}{f_{00}}\left(\frac{BR_+}{BR_0}\right)^2\right]. \tag{19.1.56}$$

Hence the physically interesting R (19.1.47) is in this case given by

$$R = \frac{N(\ell^+\ell^+)+N(\ell^-\ell^-)}{N(\ell^+\ell^+)+N(\ell^-\ell^-)+N(\ell^+\ell^-)\left[1+\frac{f_{+-}}{f_{00}}\left(\frac{BR_+}{BR_0}\right)^2\right]}, \tag{19.1.57}$$

instead of by (19.1.48). It is likely that $B^\pm$ and $(B^0, \bar{B}^0)$ have the same semi-leptonic *widths* (from the spectator diagrams) but not the same branching ratios since different non-spectator hadronic decays are possible in the two cases. Also there is no unanimous agreement on the values of f_{+-} and f_{00}, and in present analyses it is usually *assumed* that $f_{+-} \approx f_{00} \approx 0.5$. We see that the extraction of R from the data is not completely unambiguous.

Finally we note that experimental results are sometimes presented in terms of a quantity called χ rather than r, and related to it by:

$$\chi \equiv \frac{N(B^0 \to \bar{B}^0)}{N(B^0 \to B^0) + N(B^0 \to \bar{B}^0)} = \frac{r}{1+r}$$

and

$$\bar{\chi} \equiv \frac{N(\bar{B}^0 \to B^0)}{N(\bar{B}^0 \to \bar{B}^0) + N(\bar{B}^0 \to B^0)} = \frac{\bar{r}}{1+\bar{r}}. \tag{19.1.58}$$

Also note that if χ is measured by counting leptons, and if there is a mixture of B_d^0 and B_s^0 produced with fractions f_d and f_s $(f_d+f_s=1)$, then what is measured is actually

$$\chi_{\mathrm{meas}} = \frac{f_d BR(B_d^0 \to \ell^+\mathrm{X})\chi_d + f_s BR(B_s^0 \to \ell^+\mathrm{X})\chi_s}{f_d BR(B_d^0 \to \ell^+\mathrm{X}) + f_s BR(B_s^0 \to \ell^+\mathrm{X})}. \tag{19.1.59}$$

If we *assume* equal semi-leptonic branching ratios for B_s^0 and B_d^0 this simplifies to

$$\chi_{\mathrm{meas}} = f_d\chi_d + f_s\chi_s \tag{19.1.60}$$

which is the formula which has been used in present analyses (see Section 13.5).

The confrontation between theory and experiment for the B^0–$\bar{B}^0$ system will be discussed in Section 19.4.

19.2 Detailed phenomenology of CP violation in the K^0–$\bar{K}^0$ system

We consider now in detail the theoretical description and the experimental information on the CP violating phenomena seen in the K^0–$\bar{K}^0$ system.

19.2.1 Formalism and summary of data

Up to now CP violation has only been detected in the K^0–$\bar{K}^0$ system. The most dramatic manifestation is in the 2π decay of K_L. A $(\pi^+\pi^-)$ or $(\pi^0\pi^0)$ pair with $J = 0$ must have $CP = +1$. Thus if CP were conserved in the weak hamiltonian the amplitude for $K_L \rightarrow 2\pi$ would be zero by (19.1.31).

Experimentally the K_L decay is observed and one defines, as a measure of the CP violation,

$$\begin{aligned} \eta_{+-} &\equiv \frac{A(K_L \rightarrow \pi^+\pi^-)}{A(K_S \rightarrow \pi^+\pi^-)} \\ \eta_{00} &\equiv \frac{A(K_L \rightarrow \pi^0\pi^0)}{A(K_S \rightarrow \pi^0\pi^0)}. \end{aligned} \tag{19.2.1}$$

Experimentally $|\eta_{+-}| \approx |\eta_{00}| \approx 2 \times 10^{-3}$ so that the CP violation is very small. Our first aim is to relate these quantities to the basic parameters that occur in (19.1.5).

The pion pair can be in a state of isospin $I = 0$ or 2. Define, for the transition amplitudes,

$$\begin{aligned} \langle(\pi\pi)_{I=0}|\hat{T}_w|K^0\rangle &= a_0 \mathrm{e}^{\mathrm{i}\delta_0} \\ \langle(\pi\pi)_{I=2}|\hat{T}_w|K^0\rangle &= a_2 \mathrm{e}^{\mathrm{i}\delta_2} \end{aligned} \tag{19.2.2}$$

where $\delta_{0,2}$ are the ***strong interaction*** $\pi\pi$ s-wave phase shifts. Their presence arises because we are calculating a transition to a strongly interacting state (Watson, 1954). Intuitively the result can be understood as follows. Consider the reaction taking place in two steps: first the weak interactions turn the K^0 into a $\pi\pi$ pair in some very small region of space; second the

$\pi\pi$ separate while interacting strongly with each other:

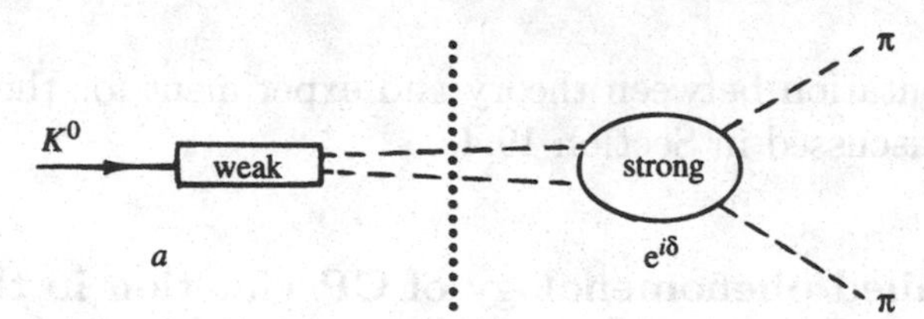

The first reaction has amplitude a. To see why the second has amplitude $e^{i\delta}$ recall that in a *scattering process*

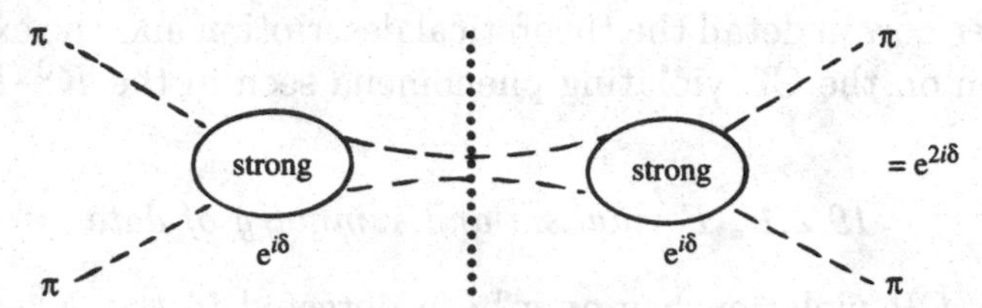

each partial wave has amplitude $e^{2i\delta}$, half the total phase building up as the particles approach each other and the other half as they separate.

Now using CPT invariance one can show that

$$\begin{aligned}\langle(\pi\pi)_{I=0}|\hat{T}_w|\bar{K}^0\rangle &= -a_0^* e^{i\delta_0} \\ \langle(\pi\pi)_{I=2}|\hat{T}_w|\bar{K}^0\rangle &= -a_2^* e^{i\delta_2}.\end{aligned} \qquad (19.2.3)$$

(See e.g. Marshak, Riazuddin and Ryan, 1969. This equation is given incorrectly in many textbooks and review articles.)

If, in addition, CP were conserved *in the decay*, use of (19.1.22) in (19.2.2 and 3) would yield

$$a_0 = a_0^*, \qquad a_2 = a_2^* \qquad \text{[CP conserved]}. \qquad (19.2.4)$$

If we put

$$a_0 = |a_0|e^{i\theta_0}, \qquad a_2 = |a_2|e^{i\theta_2} \qquad (19.2.5)$$

then CP conservation implies via (19.2.4)

$$\theta_0 = 0 \quad \text{or} \quad \pi, \qquad \theta_2 = 0 \quad \text{or} \quad \pi \qquad (19.2.6)$$

i.e. CP conservation implies

$$\theta_2 - \theta_0 = 0 \quad \text{or} \quad \pm\pi. \qquad (19.2.7)$$

It is important to realize that we cannot test CP conservation by using just *one* of the relations in (19.2.6). The reason is that by a change of CP phase convention one can actually accommodate any value of θ_0 or θ_2 separately but not both. But if (19.2.7) fails that is independent of any CP phase convention and CP is genuinely violated in the decay amplitude.

Thus CP violation *in the decay amplitude* is signalled by a relative phase between a_0 and a_2 which differs from 0 or π or $-\pi$.

Returning to the $K_{L,S}$ decays, we now have via (19.1.10)

$$\begin{aligned} a_0^{S,L} &\equiv \langle(\pi\pi)_{I=0}|\hat{T}_w|K_{S,L}\rangle = \frac{\mathrm{e}^{\mathrm{i}\delta_0}}{\sqrt{|p|^2+|q|^2}}\{pa_0 \pm qa_0^*\} \\ a_2^{S,L} &\equiv \langle(\pi\pi)_{I=2}|\hat{T}_w|K_{S,L}\rangle = \frac{\mathrm{e}^{\mathrm{i}\delta_2}}{\sqrt{|p|^2+|q|^2}}\{pa_2 \pm qa_2^*\}. \end{aligned} \tag{19.2.8}$$

Further, the isotopic spin decomposition of the two pion states is

$$\begin{aligned} |\pi^0\pi^0\rangle &= \tfrac{1}{\sqrt{3}}|(\pi\pi)_{I=0}\rangle - \sqrt{\tfrac{2}{3}}|(\pi\pi)_{I=2}\rangle \\ \tfrac{1}{\sqrt{2}}\left(|\pi^+\pi^-\rangle + |\pi^-\pi^+\rangle\right) &= \sqrt{\tfrac{2}{3}}|(\pi\pi)_{I=0}\rangle + \tfrac{1}{\sqrt{3}}|(\pi\pi)_{I=2}\rangle \end{aligned} \tag{19.2.9}$$

where the charged pion state in the second equation is the correctly symmetrized state that should be used in (19.2.1). Then one obtains for the measured ratios (19.2.1),

$$\begin{aligned} \eta_{+-} &= \frac{\sqrt{2}a_0^L + a_2^L}{\sqrt{2}a_0^S + a_2^S} \\ \eta_{00} &= \frac{a_0^L - \sqrt{2}a_2^L}{a_0^S - \sqrt{2}a_2^S}. \end{aligned} \tag{19.2.10}$$

Using now (19.2.5) one finds, after some algebra,

$$\eta_{+-} = \frac{(1-\eta)[1+w\cos(\theta_2-\theta_0)] + \mathrm{i}(1+\eta)w\sin(\theta_2-\theta_0)}{(1+\eta)[1+w\cos(\theta_2-\theta_0)] + \mathrm{i}(1-\eta)w\sin(\theta_2-\theta_0)} \tag{19.2.11}$$

where

$$\eta \equiv \frac{q}{p}\mathrm{e}^{-2\mathrm{i}\theta_0} \tag{19.2.12}$$

and

$$w \equiv \frac{1}{\sqrt{2}}\left|\frac{a_2}{a_0}\right|\mathrm{e}^{\mathrm{i}(\delta_2-\delta_0)}. \tag{19.2.13}$$

Also one finds

$$\eta_{00} = \frac{(1-\eta)[1-2w\cos(\theta_2-\theta_0)] - 2\mathrm{i}(1+\eta)w\sin(\theta_2-\theta_0)}{(1+\eta)[1-2w\cos(\theta_2-\theta_0)] - 2\mathrm{i}(1-\eta)w\sin(\theta_2-\theta_0)}. \tag{19.2.14}$$

We see immediately that if CP is conserved in the *decay amplitudes*, so that from (19.2.7), $\sin(\theta_2-\theta_0)=0$, one will have

$$\eta_{+-} = \eta_{00} \qquad \text{[CP conserved in decay].} \tag{19.2.15}$$

Recalling (19.1.27), namely that $q/p = 1$ if CP is conserved, and also, in this case, that $2\theta_0 = 0$ or 2π, implying $\eta = 1$, we see from (19.2.12) that the small violation of CP will correspond to $\eta \approx 1$. In addition, it is an empirical fact (the so-called $\Delta I = \frac{1}{2}$ rule for non-leptonic K decay) that

$$|w| \approx 1/25 \ll 1. \tag{19.2.16}$$

Thus the second term in the denominators of η_{+-}, η_{00} is totally negligible, and we may take

$$\begin{aligned} \eta_{+-} &\approx \left(\frac{1-\eta}{1+\eta}\right) + \frac{iw\sin(\theta_2-\theta_0)}{1+w\cos(\theta_2-\theta_0)} \\ \eta_{00} &\approx \left(\frac{1-\eta}{1+\eta}\right) - \frac{2iw\sin(\theta_2-\theta_0)}{1-2w\cos(\theta_2-\theta_0)}. \end{aligned} \tag{19.2.17}$$

Finally we write these in the standard form

$$\begin{aligned} \eta_{+-} &= \epsilon + \frac{\epsilon'}{1+\omega/\sqrt{2}} \\ \eta_{00} &= \epsilon - \frac{2\epsilon'}{1-\sqrt{2}\omega} \end{aligned} \tag{19.2.18}$$

where we have introduced the conventional notation*

$$\begin{aligned} \epsilon' &= \frac{\mathrm{i}}{\sqrt{2}} \mathrm{e}^{\mathrm{i}(\delta_2-\delta_0)} \left|\frac{a_2}{a_0}\right| \sin(\theta_2-\theta_0) \\ \omega &= \mathrm{e}^{\mathrm{i}(\delta_2-\delta_0)} \left|\frac{a_2}{a_0}\right| \cos(\theta_2-\theta_0) \end{aligned} \tag{19.2.19}$$

and we have defined

$$\epsilon \equiv \frac{1-\eta}{1+\eta} = \frac{1-(q/p)\mathrm{e}^{-2\mathrm{i}\theta_0}}{1+(q/p)\mathrm{e}^{-2\mathrm{i}\theta_0}}. \tag{19.2.20}$$

[Note that because $|\omega|$ is so small ω is often ignored in the denominator of the formulae (19.2.18).]

Clearly, since η_{+-} and η_{00} are experimental quantities, both ϵ and ϵ' must be independent of any phase convention. But the *relationship* between ϵ and q/p *is* convention dependent.

Many different kinds of experiment have been performed on the K^0–$\bar{K}^0$ system, often of great ingenuity. For example to compare the amplitudes

* The reader is warned that there are at least three other definitions of ω in the literature.

for $K_L \to f$ and $K_S \to f$, for any state f, one has from (19.1.17) that the rates of production of f at time t, starting with either a K^0 or a $\bar{K}^0$ at $t = 0$, are

$$R^f_{K^0 \text{ or } \bar{K}^0}(t) = \left\{ \begin{array}{c} (1+|q/p|^2) \\ \text{or} \\ (1+|p/q|^2) \end{array} \right\} \frac{|A^f_S|^2}{4} \left\{ e^{-t/\tau_S} + |\eta_f|^2 e^{-t/\tau_L} \pm \right.$$
$$\left. \pm |\eta_f| \exp\left[-\frac{t}{2}\left(\frac{1}{\tau_S}+\frac{1}{\tau_L}\right)\right] \cos(\Delta m t + \phi_f) \right\} \tag{19.2.21}$$

where $A^f_{S,L}$ are the amplitudes for $K_{S,L} \to f$ and, analogous to (19.2.1), $\eta_f = A(K_L \to f)/A(K_S \to f) \equiv |\eta_f| e^{i\phi_f}$. Study of the time dependence in (19.2.21) yields information on both the magnitude and phase of η_f.

For access to the literature about the many experiments on the K^0–$\bar{K}^0$ system, the reader should consult Table 1 of Carosi *et al.* (1990). The latter paper gives the most recent and precise information on η_{+-} and η_{00}. One has

$$\begin{aligned} \eta_{+-} &= (2.272 \pm 0.021) \times 10^{-3} e^{i(46.9\pm 2.2)^0} \\ \eta_{00} &= (2.249 \pm 0.027) \times 10^{-3} e^{i(47.1\pm 2.8)^0} \end{aligned} \tag{19.2.22}$$

showing that η_{+-} and η_{00} are very nearly equal, both in magnitude and phase, so that ϵ' in (19.2.18) must be exceedingly tiny, and $|\epsilon'| \ll |\epsilon|$. We can thus safely write (19.2.18) in the form

$$\begin{aligned} \eta_{+-} &\approx \epsilon + \epsilon' \\ \eta_{00} &\approx \epsilon - 2\epsilon' \end{aligned} \tag{19.2.23}$$

so that

$$|\epsilon| \approx \tfrac{1}{3}|(2\eta_{+-} + \eta_{00})| = (2.26 \pm 0.02) \times 10^{-3}. \tag{19.2.24}$$

The interesting parameter ϵ', which will only be non-zero if there is CP violation in the decay amplitudes, is in principle given by

$$\epsilon' \approx \tfrac{1}{3}(\eta_{+-} - \eta_{00}) \tag{19.2.25}$$

but because of the almost exact equality of η_{+-} and η_{00}, this is not a practical option for obtaining a significant value for ϵ'. Instead, present approaches are based upon the measurement of

$$\left|\frac{\eta_{00}}{\eta_{+-}}\right|^2 \approx \left|\frac{\epsilon - 2\epsilon'}{\epsilon + \epsilon'}\right|^2 \simeq 1 - 6\mathrm{Re}(\epsilon'/\epsilon). \tag{19.2.26}$$

The LHS of (19.2.26) can be obtained to great precision from the very accurate measurement of the ratios $\Gamma(K_L \to \pi^0\pi^0)/\Gamma(K_L \to \pi^+\pi^-)$ and

$\Gamma(K_S \to \pi^0\pi^0)/\Gamma(K_S \to \pi^+\pi^-)$, in terms of which

$$\left|\frac{\eta_{00}}{\eta_{+-}}\right|^2 = \frac{\Gamma(K_L \to \pi^0\pi^0)/\Gamma(K_L \to \pi^+\pi^-)}{\Gamma(K_S \to \pi^0\pi^0)/\Gamma(K_S \to \pi^+\pi^-)}. \tag{19.2.27}$$

The first positive evidence for CP violation in the decay amplitude came from the NA31 collaboration at CERN (NA31, 1988) who found

$$\text{Re}\,(\epsilon'/\epsilon) = (3.3 \pm 1.1) \times 10^{-3}. \tag{19.2.28}$$

In this experiment the K_S and the K_L propagate and decay in a vacuum tube surrounded by detectors. Because of the vast difference in lifetimes the K_L decays occur over a long (48 m) stretch of the vacuum pipe whereas the K_S all decay very close to the production target. As a beautiful refinement to eliminate errors due to variations with distance of the detector efficiency, the K_S are produced from a movable target which can be set in 1.2 m steps along the 48 m K_L decay region. Thereby the K_S decays are monitored by essentially the same parts of the detectors which monitor the K_L decays.

Unfortunately, preliminary results from the E731 collaboration at Fermilab (E731, 1990) yielded

$$\text{Re}\,(\epsilon'/\epsilon) = (-0.5 \pm 1.5) \times 10^{-3} \tag{19.2.29}$$

which is consistent with zero.

More recently at the EPS and Lepton-Photon Conference (1991) the above groups presented slightly modified results with better statistics.

$$\begin{aligned} \text{NA31}: \quad & \text{Re}(\epsilon'/\epsilon) = (2.3 \pm 0.7) \times 10^{-3} \\ \text{E731}: \quad & \text{Re}(\epsilon'/\epsilon) = (0.60 \pm 0.69) \times 10^{-3}. \end{aligned} \tag{19.2.30}$$

It is difficult to assess which result is to be trusted. The Fermilab experiment is somewhat more complicated in that it involves the regeneration of K_S from K_L via a thick target, but we are unaware of any definitive argument in favour of one or the other. We shall use both values in the theoretical analysis of CP violation.

Let us now consider the relationship between the phenomenological parameters ϵ, ϵ' and the fundamental quantities in the effective Hamiltonian (19.1.5).

19.2.2 Relation between phenomenological parameters and the CP-violating Hamiltonian

In a purely phenomenological approach, since the phase of K^0 relative to the non-strange sector is not fixed, one is free to choose one of the phases in (19.2.5) arbitrarily. An early pre-quark convention, due to Wu and Yang (1964), was to choose $\theta_0 = 0$ and to define $\mathcal{CP}|K^0\rangle = |\bar{K}^0\rangle$ in contrast to (19.1.22). In this convention (19.2.20) becomes

$$\epsilon = \frac{p-q}{p+q} \qquad \text{[Wu–Yang convention]}. \tag{19.2.31}$$

However this is not a convenient CP convention in the SM. Indeed using the KM matrix with the parametrization given in (9.2.7), a_0 will turn out to be complex. Moreover the charged current part of the SM lagrangian (9.3.8 and 10) is CP invariant if $\delta = 0$ in the KM matrix, when we use the natural CP convention [see (19.1.22)]

$$\mathcal{CP}|K^0\rangle = -|\bar{K}^0\rangle, \qquad \mathcal{CP}|\bar{K}^0\rangle = -|K^0\rangle. \tag{19.2.32}$$

Until quite recently it was believed that a_2 is essentially real in the SM, so that the 'opposite' convention to the Wu–Yang one, i.e. $\theta_2 = 0$, was often used—the so-called *quark phase convention.* But it is now claimed that higher order contributions (to be discussed in Section 19.3) yield a non-zero value of θ_2, though it is still safe to take $|\sin\theta_2| \ll |\sin\theta_0|$. For these reasons we continue, throughout this chapter, *to use the CP convention (19.2.32).* The values of $\theta_0, \theta_2, q/p$ will emerge from dynamical calculations and the general formulae (19.2.19 and 20) must be used for comparing with the experimentally determined parameters η_{+-}, η_{00}.

We have seen that ϵ and ϵ' are extremely small so that η in (19.2.20) is very close to 1. Thus we may safely take

$$\epsilon = \frac{1-\eta}{1+\eta} = \frac{1-\eta^2}{(1+\eta)^2} \simeq \frac{1-\eta^2}{4}. \tag{19.2.33}$$

Substituting for η from (19.2.12) and using (19.1.7)

$$\epsilon \simeq \frac{\mathrm{e}^{2\mathrm{i}\theta_0}H_{12} - \mathrm{e}^{-2\mathrm{i}\theta_0}H_{21}}{4\mathrm{e}^{2\mathrm{i}\theta_0}H_{12}}. \tag{19.2.34}$$

Moreover, because η is so close to 1, also from (19.1.7)

$$\sqrt{H_{12}} \approx \sqrt{H_{21}}\mathrm{e}^{-2\mathrm{i}\theta_0} \tag{19.2.35}$$

so that we can write in the denominator of (19.2.34)

$$\mathrm{e}^{2\mathrm{i}\theta_0}H_{12} = \mathrm{e}^{2\mathrm{i}\theta_0}\sqrt{H_{12}}\sqrt{H_{12}} \simeq [H_{12}H_{21}]^{1/2}. \tag{19.2.36}$$

Thus

$$\epsilon \simeq \frac{\mathrm{e}^{2\mathrm{i}\theta_0}H_{12} - \mathrm{e}^{-2\mathrm{i}\theta_0}H_{21}}{4[H_{12}H_{21}]^{1/2}}. \tag{19.2.37}$$

Finally using (19.1.4) and (19.1.16) we have

$$\epsilon \simeq \frac{\mathrm{i}\mathrm{Im}(\mathrm{e}^{2\mathrm{i}\theta_0}M_{12}) + \frac{1}{2}\mathrm{Im}(\mathrm{e}^{2\mathrm{i}\theta_0}\Gamma_{12})}{\Delta m + \mathrm{i}\Delta\Gamma/2} \tag{19.2.38}$$

where for the K^0–$\bar{K}^0$ system

$$\Delta m = m_L - m_S, \qquad \Delta\Gamma = \Gamma_S - \Gamma_L. \tag{19.2.39}$$

We shall now show that the formulae for the interesting physical parameters Δm, ϵ and ϵ' simplify when we feed in certain empirical information. The chain of argument will be rather long as a consequence of our avoiding any specific and restrictive phase convention for either θ_0 or θ_2. Thus the reader may if she wishes safely proceed to the results (19.2.59, 60 and 61).

Firstly, when we come to the dynamical calculation of the parameters in (19.2.38) in terms of the standard model lagrangian and the KM matrix, we shall find that

$$\mathrm{Im}(\mathrm{e}^{2\mathrm{i}\theta_0}\Gamma_{12}) \approx 0 \tag{19.2.40}$$

so that

$$\epsilon \simeq \frac{\mathrm{i}\mathrm{Im}(\mathrm{e}^{2\mathrm{i}\theta_0}M_{12})}{\Delta m + \mathrm{i}\Delta\Gamma/2}. \tag{19.2.41}$$

Moreover, empirically [see (19.1.39,40)]

$$\Delta\Gamma \approx \Gamma_S \approx 2\Delta m \tag{19.2.42}$$

so that

$$\epsilon \simeq \frac{\mathrm{Im}(\mathrm{e}^{2\mathrm{i}\theta_0}M_{12})}{\sqrt{2}\Delta m}\mathrm{e}^{\mathrm{i}\pi/4}. \tag{19.2.43}$$

Further, because of (19.2.36) we have from (19.1.13)

$$\begin{aligned}
\Delta m &= m_- - m_+ = -2\mathrm{Re}[H_{12}H_{21}]^{1/2} \\
&\approx 2\mathrm{Re}(\mathrm{e}^{2\mathrm{i}\theta_0}H_{12}) \\
&= 2\left\{\mathrm{Re}(\mathrm{e}^{2\mathrm{i}\theta_0}M_{12}) + \tfrac{1}{2}\mathrm{Im}(\mathrm{e}^{2\mathrm{i}\theta_0}\Gamma_{12})\right\} \\
&\simeq 2\mathrm{Re}(\mathrm{e}^{2\mathrm{i}\theta_0}M_{12})
\end{aligned} \tag{19.2.44}$$

by (19.2.40).

Now, from (19.1.24) M_{12} would be real if CP were conserved. Given the small CP violation, we thus expect

$$|\mathrm{Re}M_{12}| \gg |\mathrm{Im}M_{12}| \tag{19.2.45}$$

and also, from the discussion following (19.2.15), that

$$\mathrm{e}^{2\mathrm{i}\theta_0} \approx 1. \tag{19.2.46}$$

As a consequence,

$$\begin{aligned}
\mathrm{Re}(\mathrm{e}^{2\mathrm{i}\theta_0}M_{12}) &= \cos 2\theta_0 \mathrm{Re}M_{12} - \sin 2\theta_0 \mathrm{Im}M_{12} \\
&\approx \mathrm{Re}M_{12}
\end{aligned} \tag{19.2.47}$$

to a very good approximation, so that

$$\Delta m \approx 2\mathrm{Re}M_{12}. \tag{19.2.48}$$

In (19.2.43) one has

$$\mathrm{Im}(\mathrm{e}^{2\mathrm{i}\theta_0}M_{12}) = \cos 2\theta_0 \mathrm{Im}M_{12} + \sin 2\theta_0 \mathrm{Re}M_{12} \tag{19.2.49}$$

and we shall argue presently that it is safe to take

$$\left|\frac{\mathrm{Im}M_{12}}{\mathrm{Re}M_{12}}\right| \gg \tan 2\theta_0 \tag{19.2.50}$$

so that we may approximate (19.2.43) as

$$\epsilon \approx \frac{\cos 2\theta_0 \mathrm{Im}M_{12}}{\sqrt{2}\Delta m}\mathrm{e}^{\mathrm{i}\pi/4} \approx \frac{1}{2\sqrt{2}}\left(\frac{\mathrm{Im}M_{12}}{\mathrm{Re}M_{12}}\right)\mathrm{e}^{\mathrm{i}\pi/4} \tag{19.2.51}$$

and thus, finally,

$$|\epsilon| \approx \frac{1}{2\sqrt{2}}\left|\frac{\mathrm{Im}M_{12}}{\mathrm{Re}M_{12}}\right|. \tag{19.2.52}$$

Furthermore, we shall find that as expected from (19.2.6)

$$\theta_2 \approx 0 \text{ or } \pi; \quad \Rightarrow \quad |\cos\theta_2| \approx 1. \tag{19.2.53}$$

Taking, as already discussed, $|\sin\theta_2| \ll |\sin\theta_0|$ we have from (19.2.19)

$$\epsilon' \approx -\frac{1}{\sqrt{2}}\left|\frac{a_2}{a_0}\right| \sin\theta_0 \mathrm{e}^{\mathrm{i}\pi/4} \tag{19.2.54}$$

where we have used the empirical fact that

$$\delta_2 - \delta_0 \approx -\pi/4. \tag{19.2.55}$$

Then ϵ'/ϵ is approximately real, and therefore [see (19.2.28, 29)]

$$\epsilon'/\epsilon \approx \mathrm{Re}\,(\epsilon'/\epsilon) \approx 10^{-3} \tag{19.2.56}$$

implying

$$|\epsilon'| \approx 10^{-6}. \tag{19.2.57}$$

Putting $|a_2|/|a_0| \approx 1/20$ in (19.2.54) yields

$$|\sin\theta_0| \approx 3 \times 10^{-5} \tag{19.2.58}$$

consistent with the expectation (19.2.6) $\theta_0 \approx 0$ or π.

Then we can justify (19.2.50) since

$$\begin{aligned}\frac{|\mathrm{Im}M_{12}|}{|\mathrm{Re}M_{12}|} &\approx 2\sqrt{2}|\epsilon| \approx 6 \times 10^{-3} \\ &\gg |\tan 2\theta_0| \approx 6 \times 10^{-5}.\end{aligned}$$

To summarize, use of the general theoretical expressions combined with the empirical data leads us to the following simplified formulae for the dynamical calculation of $\Delta m, \epsilon$ and ϵ':

$$\Delta m \approx 2\mathrm{Re}M_{12} \tag{19.2.59}$$

$$\epsilon \approx \frac{\mathrm{Im}M_{12}}{\sqrt{2}\Delta m}\mathrm{e}^{\mathrm{i}\pi/4} \tag{19.2.60}$$

$$\epsilon' \approx \frac{1}{\sqrt{2}\,20}\sin(\theta_2-\theta_0)\mathrm{e}^{\mathrm{i}\pi/4}. \tag{19.2.61}$$

19.3 Dynamics of mixing and CP violation

19.3.1 Connection with the SM (weak) Hamiltonian

In the phenomenological description of mixing and CP violation the key ingredients are the off-diagonal elements in the 'effective' Hamiltonian $\boldsymbol{H}$ of (19.1.2), as can be seen from (19.1.7). In this section we relate M_{12} and Γ_{12} to the weak interaction Hamiltonian and thus relate ϵ to the KM matrix elements.

For the purposes of illustration let us consider the K^0–$\bar{K}^0$ system and transitions $K^0 \leftrightarrow \bar{K}^0$ which involve $\Delta S = \pm 2$. At the quark level the transitions are

$$d\bar{s} \leftrightarrow s\bar{d}. \tag{19.3.1}$$

The weak interaction transition operator $\hat{T}_w$ is given in terms of the SM Hamiltonian H_w by the perturbative expansion eqn (A5.6).

Now H_w can produce transitions with $\Delta S = \pm 1$ but in doing so will turn a down-type quark into an up-type quark [see (9.3.8,10)]. Thus the lowest order for which the transition (19.3.1) is possible is fourth order in H_w. Thus, in lowest order (H_s is the strong interaction Hamiltonian),

$$\begin{aligned} H_{12} &\equiv \langle K^0|\hat{T}_w|\bar{K}^0\rangle = \langle K^0|\hat{T}_w^{(4)}|\bar{K}^0\rangle \\ &= \langle K^0|H_w\frac{1}{m_K-H_s+\mathrm{i}\epsilon}H_w\frac{1}{m_K-H_s+\mathrm{i}\epsilon}H_w\frac{1}{m_K-H_s+\mathrm{i}\epsilon}H_w|\bar{K}^0\rangle \end{aligned} \tag{19.3.2}$$

where we have taken $K^0, \bar{K}^0$ at rest and thus put $E = m_K$ = mass of K^0 and $\bar{K}^0$. Inserting a complete set of physical states $|n\rangle$, we may write this as

$$H_{12} = \sum_n \frac{\langle K^0|\hat{T}_w^{(2)}|n\rangle\langle n|\hat{T}_w^{(2)}|\bar{K}^0\rangle}{m_K - E_n + \mathrm{i}\epsilon}. \tag{19.3.3}$$

Utilizing eqn (A5.8) we can put

$$H_{12} = M_{12} - \mathrm{i}\Gamma_{12}/2 \tag{19.3.4}$$

where

$$M_{12} \equiv \mathcal{P}\sum_n \frac{\langle K^0|\hat{T}_w^{(2)}|n\rangle\langle n|\hat{T}_w^{(2)}|\bar{K}^0\rangle}{m_K - E_n + \mathrm{i}\epsilon} \tag{19.3.5}$$

$$\Gamma_{12} \equiv 2\pi\sum_n \langle K^0|\hat{T}_w^{(2)}|n\rangle\langle n|\hat{T}_w^{(2)}|\bar{K}^0\rangle\delta(E_n - m_K). \tag{19.3.6}$$

These are referred to as the *dispersive* and *absorptive* parts of H_{12}, respectively. Note that contributions to Γ_{12} can only come from states $|n\rangle$ into which a K^0 or $\bar{K}^0$ can really decay, e.g. $2\pi, 3\pi, \pi e\nu$ etc., and the sum over states is then dominated by the large decay rate into 2π with $I = 0$. Thus a good estimate of Γ_{12} is

$$\Gamma_{12} \approx 2\pi\langle K^0|\hat{T}_w^{(2)}|(\pi\pi)_{I=0}\rangle\langle(\pi\pi)_{I=0}|\hat{T}_w^{(2)}|\bar{K}^0\rangle\rho_{2\pi} \tag{19.3.7}$$

where $\rho_{2\pi}$ is the 2π phase space factor in K-decay.

We now wish to show that the factor $\mathrm{e}^{2\mathrm{i}\theta_0}\Gamma_{12}$ occurring in (19.2.38) is essentially real.

Since the Hamiltonian is hermitian, we have

$$\begin{aligned}\langle K^0|\hat{T}_w^{(2)}|\pi\pi\rangle &= \langle K^0|H_w\frac{1}{m_K - H_s + \mathrm{i}\epsilon}H_w|\pi\pi\rangle \\ &= \langle\pi\pi|H_w\frac{1}{m_K - H_s - \mathrm{i}\epsilon}H_w|K^0\rangle^*.\end{aligned} \tag{19.3.8}$$

Now H_w acting once produces a $W^\pm$ so that any intermediate states in $\hat{T}_w^{(2)}$ in (19.3.8) must contain a $W^\pm$ and therefore cannot satisfy energy conservation and the denominator cannot equal zero. Thus we may put $\epsilon = 0$ in (19.3.8) so that $\hat{T}_w^{(2)}$ behaves here as if it were hermitian†, yielding

$$\langle K^0|\hat{T}_w^{(2)}|\pi\pi\rangle = \langle\pi\pi|\hat{T}_w^{(2)}|K^0\rangle^*. \tag{19.3.9}$$

Then, from (19.2.2,3)

$$\begin{aligned}\Gamma_{12} &\approx 2\pi\langle(\pi\pi)_{I=0}|\hat{T}_w^{(2)}|K^0\rangle^*\langle(\pi\pi)_{I=0}|\hat{T}_w^{(2)}|\bar{K}^0\rangle\rho_{2\pi} \\ &= -2\pi a_0^{*2}\,\rho_{2\pi} = -2\pi|a_0|^2\mathrm{e}^{-2\mathrm{i}\theta_0}\rho_{2\pi}.\end{aligned} \tag{19.3.10}$$

† This fact may be used to show that it is M_{12}^* and Γ_{12}^* that appear in the expression (19.1.4) for H_{21}.

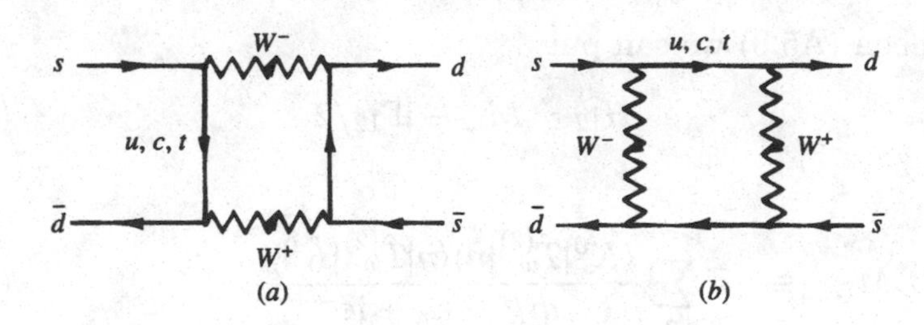

Fig. 19.1. Lowest order Feynman diagrams yielding $d\bar{s} \leftrightarrow s\bar{d}$ transitions.

Hence, indeed, as claimed earlier, in (19.2.40),

$$e^{2i\theta_0}\Gamma_{12} \approx \text{real} \tag{19.3.11}$$

and the expressions (19.2.59,60) for Δm and ϵ are justified.

We turn therefore to the calculation of M_{12}.

Now we actually have two expressions for M_{12}. One of them, given in (19.3.5), involves genuine physical intermediate states $|n\rangle$. On the other hand, from eqn (A5.9) for states normalized to unity in a volume V, we have that

$$\langle K^0|\hat{T}_w|\bar{K}^0\rangle = \frac{i}{2m_K}\mathcal{M} \tag{19.3.12}$$

where $\mathcal{M}$ is the Feynman amplitude for $\bar{K}^0 \to K^0$, so that

$$M_{12} = \frac{1}{2m_K}[\text{dispersive part of } (i\mathcal{M})] \equiv \frac{1}{2m_K}\mathcal{M}^D. \tag{19.3.13}$$

The two expressions (19.3.5 and 13) *ought* to be equivalent but they are not in any straightforward way because of the complications of confinement in QCD. Firstly, $\mathcal{M}$ in perturbation theory really describes the transitions $\bar{d}s \to d\bar{s}$ rather than $\bar{K}^0 \to K^0$; the relevant Feynman diagrams are shown in Fig. 19.1. Secondly we see that the intermediate states in Fig. 19.1 (*b*) are quark–antiquark states rather than physical states.

Thus the two expressions for M_{12} cannot possibly agree. The resolution of this 'paradox' is straightforward. If the *strong* QCD interactions between the quark and antiquark were included they would bind them into the expected physical states. But this is a highly non-perturbative process, so the question remains as to how much trust we should put in the Feynman amplitude calculation.

Ironically, the earliest calculation of the K_L–K_S mass difference Δm (Gaillard and Lee, 1974) using the box-diagrams of Fig. 19.1, ignored this question, yet led, as discussed in Section 9.4, to a sensible result provided there existed a charm quark with a mass not too different from m_p. It is now felt that this was a fortuitous coincidence! We shall not further discuss the calculation of Δm, but turn to the CP parameters ϵ and ϵ'.

19.3.2 Estimate for ϵ in the SM

Now, usually, the dispersive part of a box diagram $\mathcal{M}^D$ is real, and this would indeed be so if the KM matrix elements V_{ij} were real. But if they are not we shall have a non-zero imaginary part $\text{Im}\mathcal{M}^D$, and hence via (19.2.60) the CP parameter $\epsilon \neq 0$. This implies that the expression for ϵ will be proportional to the imaginary parts of the V_{ij}, and hence, from (9.2.7), one of the two quark lines in Fig. 19.1 has to be either charm or a top quark. Thus ϵ receives contributions only from heavy mass intermediate states and it seems reasonable therefore to calculate it from the box diagrams.

As explained in Appendix 4, if we neglect the initial and final quark momenta compared with the heavy propagator masses, we can extract from the Feynman amplitudes $\mathcal{M}(\text{box})$ an equivalent operator $\hat{\mathcal{M}}(\text{box})$ which plays the rôle of a local effective Hamiltonian. We then have via (19.2.60) and (19.3.13)

$$\epsilon = \frac{e^{i\pi/4}}{2\sqrt{2}\Delta m m_K}\text{Im}\langle K^0|\hat{\mathcal{M}}^D(\text{box})|\bar{K}^0\rangle \qquad (19.3.14)$$

where here the $|K^0\rangle, |\bar{K}^0\rangle$ states are normalized relativistically [see eqn (A4.1)].

The original calculations are due to Inami and Lin (1981), but we follow the more modern notation of Buchalla, Buras and Harlander (1990).

Consider the Feynman amplitude for the box diagram Fig. 19.1 (*a*) with just c and t intermediate quarks, for quarks and antiquarks all of the same colour (say, l). Neglecting the external momenta, evaluating the Feynman amplitude and applying the substitution eqn (A4.5) explained in Appendix 4 leads to an operator

$$\hat{O} = 2\mathcal{F}[\bar{d}_l\gamma^\mu(1-\gamma_5)s_l][\bar{d}_l\gamma^\mu(1-\gamma_5)s_l] \qquad (19.3.15)$$

where $\mathcal{F}$, a function of the masses and KM matrix elements, will be given presently.

If we write each field operator in terms of creation and annihilation operators and multiply out we find that $\hat{O}$ contains four terms which contribute to our reaction $s\bar{d} \to \bar{d}s$, two corresponding to Fig. 19.1 (*a*) and two to Fig. 19.1 (*b*). Thus the correct operator is one half of (19.3.15), i.e.

$$\hat{\mathcal{M}}^D(\text{box}) = \mathcal{F}[\bar{d}_l\gamma^\mu(1-\gamma_5)s_l][\bar{d}_l\gamma^\mu(1-\gamma_5)s_l]. \qquad (19.3.16)$$

However, the colour structure requires some thought.

Bearing in mind the sum over colours in the coupling of bosons to quarks, the above must have come equally from the colour structures

$$\left[\sum_i \bar{d}_i\gamma^\mu(1-\gamma_5)s_i\right]\left[\sum_j \bar{d}_j\gamma^\mu(1-\gamma_5)s_j\right]$$

and

$$\sum_{i,j}\left[\bar{d}_i\gamma^\mu(1-\gamma_5)s_j\right]\left[\bar{d}_j\gamma^\mu(1-\gamma_5)s_i\right].$$

Thus we should replace the matrix element of (19.3.16) by

$$\begin{aligned}\langle K^0|\hat{\mathcal{M}}^D(\text{box})|\bar{K}^0\rangle &= \mathcal{F}\frac{1}{2}\Bigg\{\langle K^0|\left[\sum_i \bar{d}_i\gamma^\mu(1-\gamma_5)s_i\right] \\ &\qquad\times\left[\sum_j \bar{d}_j\gamma^\mu(1-\gamma_5)s_j\right]|\bar{K}^0\rangle \\ &\qquad+\langle K^0|\sum_{i,j}\left[\bar{d}_i\gamma^\mu(1-\gamma_5)s_j\right] \\ &\qquad\times\left[\bar{d}_j\gamma^\mu(1-\gamma_5)s_i\right]|\bar{K}^0\rangle\Bigg\} \\ &\equiv \mathcal{F}X_K. \end{aligned}\tag{19.3.17}$$

The function $\mathcal{F}$ is given by

$$\mathcal{F} = \frac{G^2M_W^2}{16\pi^2}\left\{S(z_c)\lambda_c^2 + 2S(z_c,z_t)(\lambda_c\lambda_t) + S(z_t)\lambda_t^2\right\}\tag{19.3.18}$$

where

$$\begin{aligned}\lambda_c &= V_{cd}^*V_{cs} \qquad \lambda_t = V_{td}^*V_{ts} \\ z_c &= m_c^2/M_W^2 \quad\; z_t = m_t^2/M_W^2\end{aligned}\tag{19.3.19}$$

and the $S(z)$ are simple functions of z,

$$S(z) = z\left[\frac{1}{4}+\frac{9}{4(1-z)}-\frac{3}{2(1-z)^2}\right]+\frac{3}{2}\left(\frac{z}{z-1}\right)^3\ln z\tag{19.3.20}$$

and

$$\begin{aligned}S(z_c,z_t) &= z_cz_t\Bigg\{\left[\left(\frac{1}{4}+\frac{3}{2(1-z_t)}-\frac{3}{4(1-z_t)^2}\right)\frac{\ln z_t}{z_t-z_c}\right] \\ &\qquad+[z_t\leftrightarrow z_c]-\frac{3}{4}\frac{1}{(1-z_t)(1-z_c)}\Bigg\}.\end{aligned}\tag{19.3.21}$$

$m_t(\mathrm{GeV}/c^2)$	$S(z_t)$	$S(z_c, z_t)$
97	1.0	2.4×10^{-3}
127	1.6	2.5×10^{-3}
151	2.1	2.6×10^{-3}

Table 19.1.

When QCD corrections are taken into account their effect is approximately:

$$\begin{aligned} S(z_c) &\rightarrow 0.85\ S(z_c) \\ S(z_c, z_t) &\rightarrow 0.36\ S(z_c, z_t) \\ S(z_t) &\rightarrow 0.62\ S(z_t) \end{aligned} \qquad (19.3.22)$$

These values should be regarded as reasonable estimates only. They depend slightly on the QCD scale factor Λ_{QCD} and on the heavy quark masses.

For the calculation of ϵ we require the imaginary part of (19.3.17) which entails replacing $\mathcal{F}$ by its imaginary part. This, in turn, implies using $\mathrm{Im}(\lambda_c^2), \mathrm{Im}(\lambda_c\lambda_t)$ and $\mathrm{Im}(\lambda_t^2)$ in (19.3.18).

We now try to estimate the two terms $\mathrm{Im}\mathcal{F}$ and the hadronic matrix element X_K which occur in (19.3.17).

For the KM matrix elements, using (9.2.7), and bearing in mind the magnitude of the matrix elements as summarized in (9.2.6), one finds:

$$\begin{aligned} 2\mathrm{Im}(\lambda_c\lambda_t) &\approx -\mathrm{Im}\lambda_c^2 \approx (2s_{23}s_{13}\sin\delta)\cdot s_{12} \\ \mathrm{Im}\lambda_t^2 &\approx (2s_{23}s_{13}\sin\delta)\cdot s_{23}(s_{23}s_{12} - s_{13}\cos\delta) \end{aligned} \qquad (19.3.23)$$

where we have used the fact that all the c_{ij} are very nearly equal to unity. Thus

$$\begin{aligned} \mathrm{Im}\mathcal{F} \approx\ & \frac{G^2M_W^2}{8\pi^2}s_{23}s_{13}\sin\delta\Big\{s_{12}[0.36S(z_c, z_t) - 0.85S(z_c)] \\ & + 0.62s_{23}(s_{23}s_{12} - s_{13}\cos\delta)S(z_t)\Big\}. \end{aligned} \qquad (19.3.24)$$

For the function S, since $m_c^2 \ll M_W^2$, one has

$$S(z_c) \approx z_c = 3 \times 10^{-4}.$$

In much of the earlier work on this subject it was assumed also that $m_t^2 \ll M_W^2$ and approximations were made in estimating $S(z_t)$. With the value $m_t = 127^{+24}_{-30}$ given in (7.8.2) one should use the exact formulae. Some relevant values are given in Table 19.1.

Since the functions don't change dramatically, let us get a feeling for the size of $\mathrm{Im}\mathcal{F}$ by taking $m_t = 127\ \mathrm{GeV}/c^2$, $s_{12} = 0.22$, $s_{23} \approx 0.049$, $s_{13} \approx 0.005$ (the latter values follow from the discussion in Sections 18.1 and 18.2). One finds

$$\mathrm{Im}\mathcal{F} \approx 1.2 \times 10^{-21} \sin\delta(1 - 0.34\cos\delta)(\mathrm{MeV})^{-2}. \tag{19.3.25}$$

We turn now to the hadronic matrix element X_K in (19.3.17).

We can insert a complete sum over intermediate physical states between the operators in (19.3.17). To the extent that the operators can be considered as free fields (i.e. neglecting strong interactions) and the K^0 and $\bar{K}^0$ considered as just made up of $\bar{s}d$ and $s\bar{d}$, the only state that will contribute is the vacuum state. Inserting the vacuum state should thus give us some idea of the magnitude of X_K.

Some care is required since when the fields are multiplied out in terms of creation and annihilation operators, there are four terms that contribute equally.

The colour aspect is handled using the fact that

$$\langle K^0|\bar{d}_i\gamma^\mu(1-\gamma_5)s_j|0\rangle = \delta_{ij}\langle K^0|\bar{d}_i\gamma^\mu(1-\gamma_5)s_i|0\rangle \tag{19.3.26}$$

where the matrix element on the RHS is actually independent of the colour i.

The result is, after carrying out the sums over colour in (19.3.17),

$$\mathrm{X}_K = \tfrac{1}{2} \times 4 \times \tfrac{4}{3}\langle K^0|\bar{d}\gamma^\mu(1-\gamma_5)s|0\rangle\langle 0|\bar{d}\gamma_\mu(1-\gamma_5)s|\bar{K}^0\rangle \tag{19.3.27}$$

where the shorthand notation

$$\bar{d}\gamma^\mu(1-\gamma_5)s \equiv \sum_i \bar{d}_i\gamma^\mu(1-\gamma_5)s_i \tag{19.3.28}$$

is used.

By an isospin rotation we can relate these matrix elements to the matrix elements that occur in $K^- \to \ell\bar{\nu}$:

$$\begin{aligned}\langle 0|\bar{d}\gamma^\mu(1-\gamma_5)s|\bar{K}^0(q)\rangle &= \langle 0|\bar{u}\gamma^\mu(1-\gamma_5)s|K^-(q)\rangle \\ &= if_K q^\mu.\end{aligned} \tag{19.3.29}$$

The last step is the analogue of (13.2.5), and, from experiment, $f_K \simeq$ 160 MeV. Thus we estimate

$$\mathrm{X}_K \approx \tfrac{8}{3} f_K^2 m_K^2. \tag{19.3.30}$$

Attempts to improve this approximation are notoriously difficult (see Section 5.3 of Paschos and Türke, 1988). One parametrizes

$$\mathrm{X}_K = \tfrac{8}{3} f_K^2 m_K^2 B_K \tag{19.3.31}$$

B_K	Minimum m_t(GeV/c^2)
0.6	90
0.75	75
0.9	65

Table 19.2.

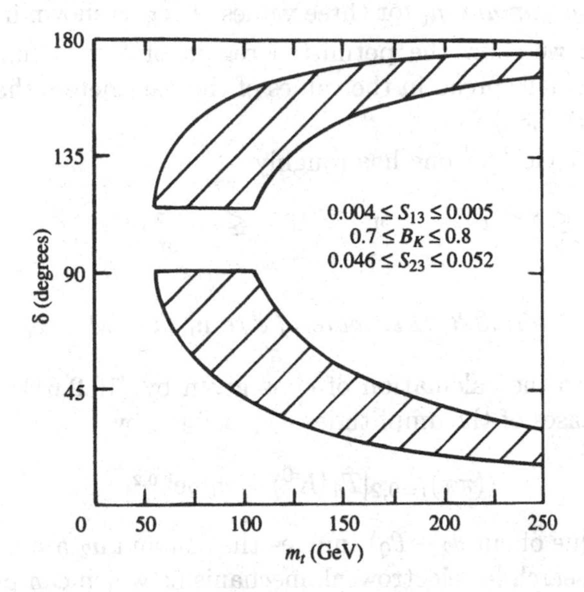

Fig. 19.2. Permitted region of δ *vs* m_t resulting from fitting ϵ_{Expt} and allowing for a small range of the parameter $\theta_{23}, \theta_{13}, B_K$. (Adapted from Buchalla, Buras and Harlander, 1990.)

and tries to estimate B_K in various ways (bag model, QCD sum rules, lattice gauge theory etc.). It is now believed that

$$0.6 \leq B_K \leq 0.9. \tag{19.3.32}$$

Inserting (19.3.31 and 25) into (19.3.17) and then into (19.3.14) yields for our theoretical value of ϵ:

$$\epsilon_{\text{Th}} = 4.1 \times 10^{-3} B_K \sin\delta(1 - 0.34\cos\delta)\mathrm{e}^{\mathrm{i}\pi/4} \tag{19.3.33}$$

to be compared with the empirical value that follows from (19.2.22)

$$\epsilon_{\text{Expt}} = (2.264 \pm 0.023) \times 10^{-3}\mathrm{e}^{\mathrm{i}(47\pm2.5)^\circ}. \tag{19.3.34}$$

Hence the calculated phase is perfectly compatible with the measured phase and the calculated magnitude is fine provided B_K is not too small. Assuming the latter, we see that the natural CP violation in the SM can explain the value of ϵ. Moreover, equating ϵ_{Th} and ϵ_{Expt} will fix δ somewhere in the range $0 < \delta < \pi$, up to a twofold ambiguity.

We have presented the calculation for $m_t = 127$ GeV/c^2. For smaller m_t the coefficient in (19.3.33) decreases and since the maximum possible value of $\sin\delta(1 - 0.34\cos\delta)$ is just slightly larger than unity we see that agreement between theory and experiment is only possible for sufficiently large m_t. The *minimum* m_t for three values of B_K is shown in Table 19.2.

In Fig. 19.2 we show the permitted region of δ as a function of m_t allowing for a small spread in the values of the parameters that were used in obtaining (19.3.33).

For $m_t = 127$ GeV/c^2 one has roughly

$$30^\circ \le \delta \le 70^\circ \quad \text{or} \quad 135^\circ \le \delta \le 173^\circ. \tag{19.3.35}$$

19.3.3 Estimate of ϵ'/ϵ in the SM

We turn now to the calculation of ϵ' as given by (19.2.61). Recall that $\theta_{0,2}$ are the phases of the amplitudes $a_{0,2}$ defined by

$$\langle(\pi\pi)_{I=0,2}|\hat{T}_w|K^0\rangle = a_{0,2}\mathrm{e}^{\mathrm{i}\delta_{0,2}}. \tag{19.3.36}$$

A non-zero value of $\sin(\theta_2 - \theta_0)$ implies that a_0 and a_2 are not relatively real. We thus search for electroweak mechanisms which can produce non-real amplitudes in $K^0 \to \pi\pi$. (It is the strong interaction, i.e. the interaction of the two pions with each other, *after* production that is responsible for the phases $\mathrm{e}^{\mathrm{i}\delta_{0,2}}$.)

We show below examples of the various classes of Feynman diagrams that contribute to $K^0 \to \pi^+\pi^-$, interpreted as $\bar{s}d \to (u\bar{d}) + (d\bar{u})$. Similar diagrams contribute to $\pi^0\pi^0$. The relevant KM matrix elements are indicated. The possible isospin states of the pion pair are shown, as is the phase of the Feynman amplitude. Brief explanations are given after each diagram.

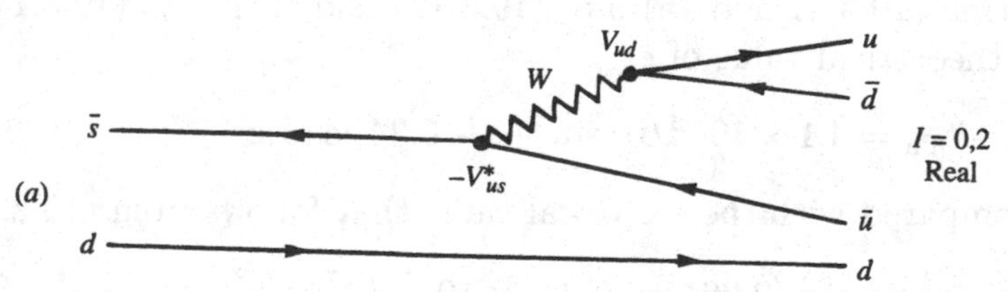

Diagram (a) is the naive ‘spectator’ diagram. Since the electroweak interactions do not conserve isospin the upper $(u\bar{d}\bar{u})$ system can have $I = \frac{1}{2}$ or $\frac{3}{2}$ so that both $I = 0$ and 2 are possible. The amplitude is real because both V_{ud} and V_{us} are real.

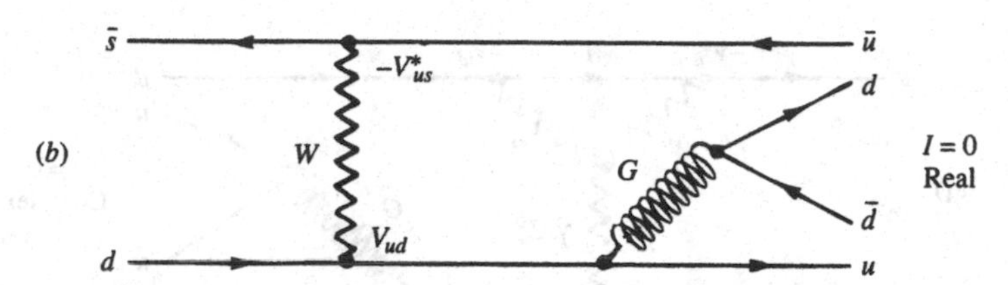

In (b), known as the 'exchange' diagram, a gluon is utilized to produce the $d\bar{d}$ pair which must therefore have $I = 0$. Since $u\bar{u}$ can only have $I = 0$ or 1 and since $I = 1$ for the $\pi\pi$ system is forbidden by Bose statistics, only $I = 0$ is possible. Reality follows as in (a).

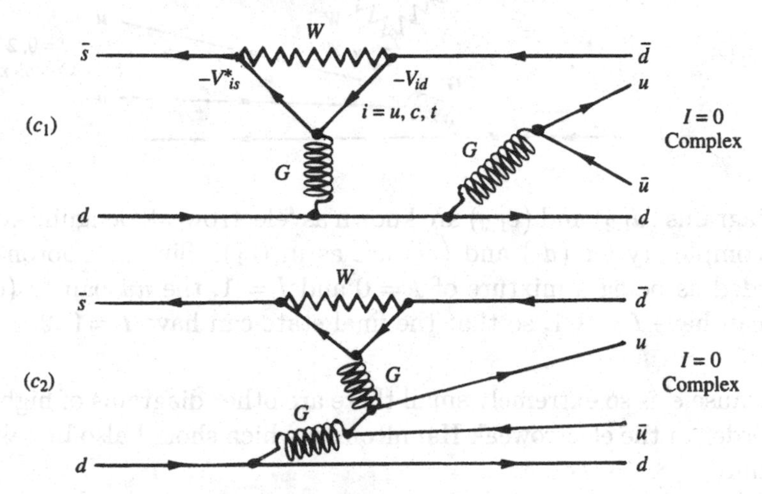

These are QCD 'penguin' diagrams and give $I = 0$ as in (b). The amplitude is complex because V_{cs}, V_{cd}, V_{ts} and V_{td} are complex.

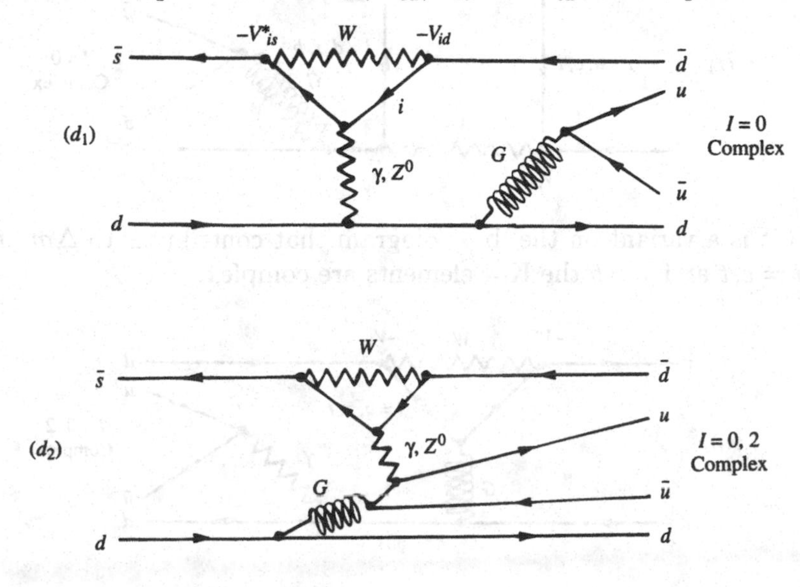

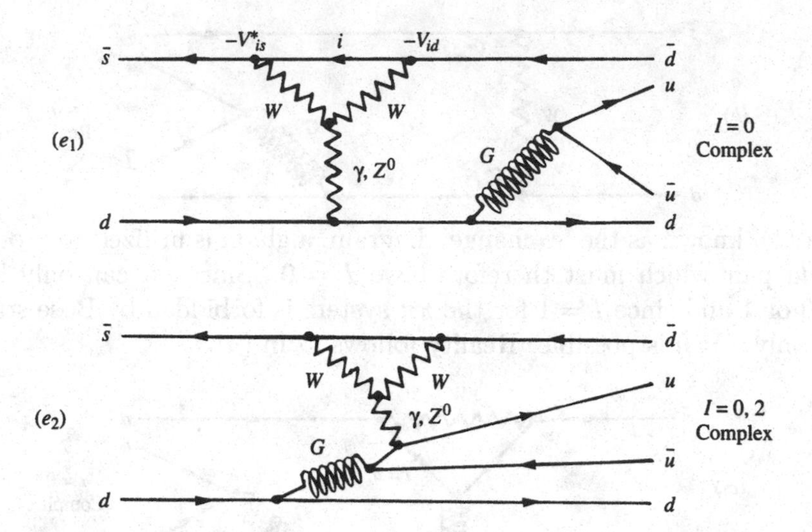

Diagrams $(d_{1,2})$ and $(e_{1,2})$ are known as 'electroweak penguins'. Isospin and complexity for (d_1) and (e_1) are as in (c_1). Since a photon can be regarded as being a mixture of $I = 0$ and $I = 1$, the $u\bar{u}$ pair in (d_2) and (e_2) can have $I = 0, 1$, so that the final state can have $I = 0, 2$.

Because ϵ' is so extremely small there are other diagrams of higher than 2nd order in the electroweak Hamiltonian which should also be taken into account:

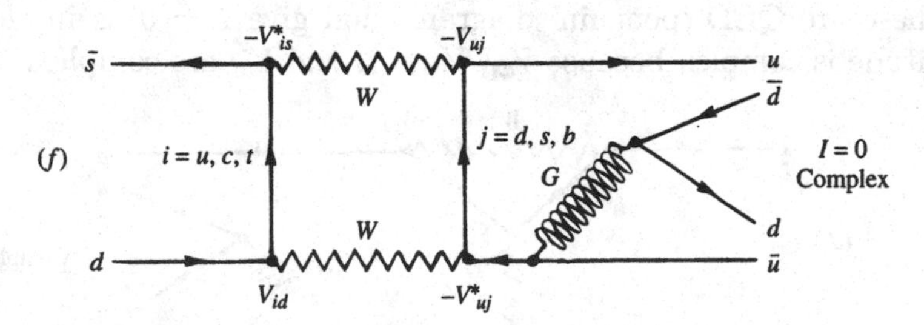

This is a variant on the 'box' diagram that contributes to Δm and ϵ. For $i = c, t$ and $j = b$ the KM elements are complex.

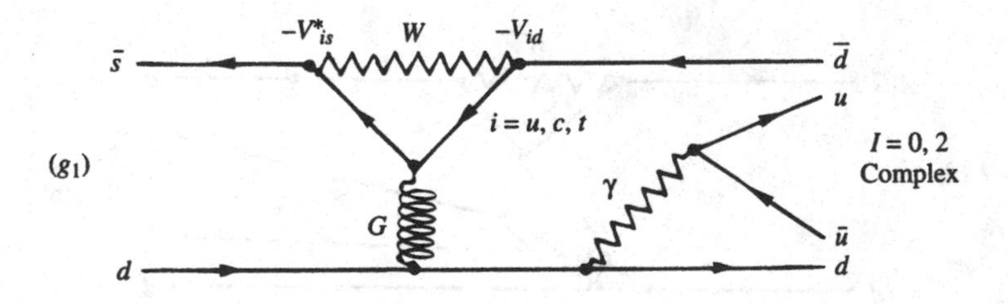

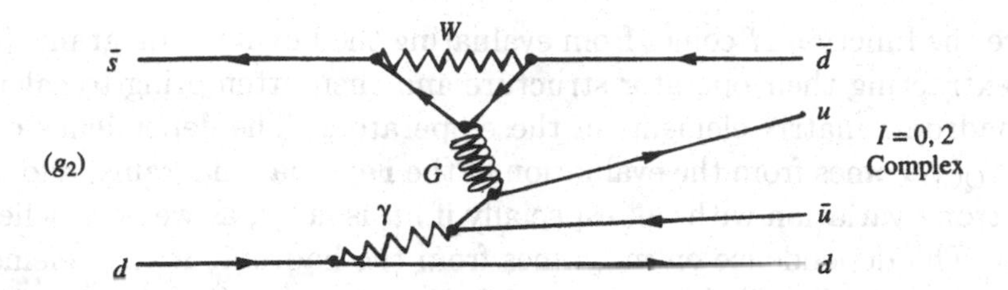

These are variants of the QCD penguins (c_1) and (c_2). Because a photon can be considered as being a mixture of $I = 0$ and 1, one gets contributions to both $I = 0, 2$.

Finally there is the question of the small breaking of isospin invariance amongst the hadrons, which gives rise to π^0–η–η' mixing. As a consequence any amplitude for $K^0 \to \pi^0\pi^0$ will also get contributions from the following diagrams:

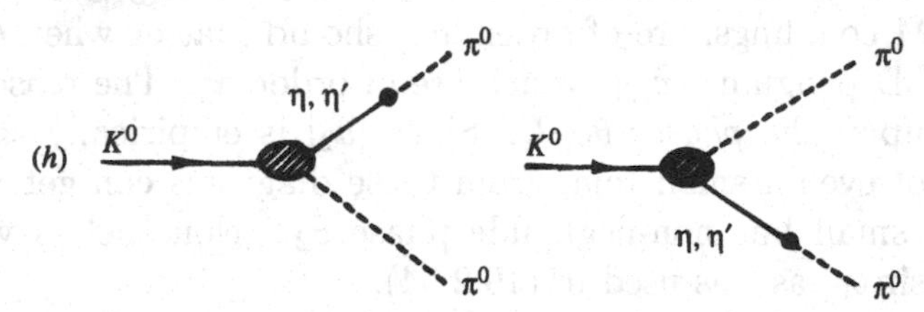

Now the amplitudes $A(K^0 \to \pi^0\eta)$ and $A(K^0 \to \pi^0\eta')$ both get imaginary parts from the QCD diagrams (c) and this will, via the diagrams (h), induce an imaginary part in $A(K^0 \to \pi^0\pi^0)$ and thus in both $I = 0$ and 2 of the $K^0 \to 2\pi$ amplitude. That the effect occurs in both $I = 0, 2$ can be seen from the inverse of (19.2.9).

Manifestly the detailed calculations are horrendously complicated. The technique is to avoid guessing wave-functions for the quarks in the hadrons by using the Feynman diagrams to obtain an effective Hamiltonian for $K \to 2\pi$ (as was done for $K^0 \leftrightarrow \bar{K}^0$), and then trying to estimate the *hadronic* matrix elements of these operators. We shall only quote the results. Access to the theoretical literature can be obtained from Buchalla, Buras and Harlander (1991).

Upon looking at the KM matrix elements involved in the diagrams (c) to (h), it is easy to see that the only complex factors are proportional to λ_c^* and λ_t^* [see (19.3.19)]. The imaginary parts are thus proportional to the factor $s_{23}s_{13}\sin\delta$ which occurred in the imaginary parts (19.3.23) of the box diagrams.

Since ϵ' and ϵ have essentially the same phase, it is convenient to consider the ratio ϵ'/ϵ. Feeding in the experimental value of ϵ (19.3.34) and taking the values of the s_{ij} used in obtaining (19.3.25), one can write

$$\epsilon'/\epsilon \approx 1.56 \times 10^{-2} \sin\delta H(m_t, m_s, \Lambda_{\text{QCD}}) \tag{19.3.37}$$

where the function H comes from evaluating the Feynman diagrams (c) to (h), extracting their operator structure and then attempting to calculate the hadronic matrix elements of these operators. The dependence on m_t and Λ_{QCD} comes from the evaluation of the Feynman diagrams, and there is a strong variation with m_t, especially if m_t is large, as we now believe it to be. The dependence on m_s comes from the hadronic matrix elements. Here m_s is the so-called 'strange quark current mass' (not its constituent mass) which is supposed to be in the region of 150 MeV/c^2.

We consider, in outline, some of the contributions to $H(m_t)$. Originally, when it was believed that $m_t \ll M_{\text{W}}$ only the QCD penguin (c_1) was taken into account. It turns out that for large m_t the importance of the electroweak penguins (d), (e) and (g) grows and their net contribution has opposite sign to the QCD penguin, thus reducing the value of ϵ'/ϵ. It may seem strange that diagrams like (d_2), ($e_{1,2}$) and ($g_{1,2}$) which, aside from the KM couplings, are of order $\alpha\alpha_s$ should matter when compared with the QCD penguins ($c_{1,2}$) which are of order α_s^2. The reason is that ϵ' depends upon the *phases* θ_0, θ_2. Since $|a_2|$ is empirically $\ll |a_0|$ the occurrence of even a small Ima_2 from these diagrams can get magnified into a very small but non-negligible phase θ_2. Nonetheless we expect $|\sin\theta_2| \ll |\sin\theta_0|$ as was used in (19.2.54).

It is found that H decreases strongly as m_t increases, reaching zero for m_t in the region of 220 GeV/c^2. H also depends upon m_s and roughly $H \propto 1/m_s^2$. Finally H increases slowly as Λ_{QCD} grows.

Returning to (19.3.37), if one now (i) takes the range of allowed values of δ which followed from the fit to ϵ_{Expt} (see Fig. 19.2) and (ii) permits Λ_{QCD} to vary between 0.1 GeV and 0.3 GeV, one obtains the graphs of ϵ'/ϵ *vs* m_s for various fixed values of m_t shown in Fig. 19.3.

19.3.4 Summary on ϵ and ϵ' in the K^0–$\bar{K}^0$ system

We see that for the preferred value of $m_t \approx 127$ GeV/c^2 the SM is not compatible with the NA31 measurement for any acceptable value of m_s. The situation gets worse as m_t increases, and only for m_t in the region of or smaller than 90 GeV/c^2 is there the possibility of agreement. The E731 data on the other hand are compatible with the SM for $m_t \approx 127$ GeV/c^2.

We end this discussion with two comments:

1. It is hard to believe that the calculations of individual hadronic matrix elements are accurate to better than say 20%–30%. In calculating ϵ'/ϵ there is considerable cancellation amongst these so that the answer may be quite inaccurate. But even allowing for this it only seems barely possible to get agreement between the SM and NA31 results and then only for a value of m_s smaller than is usually contemplated.

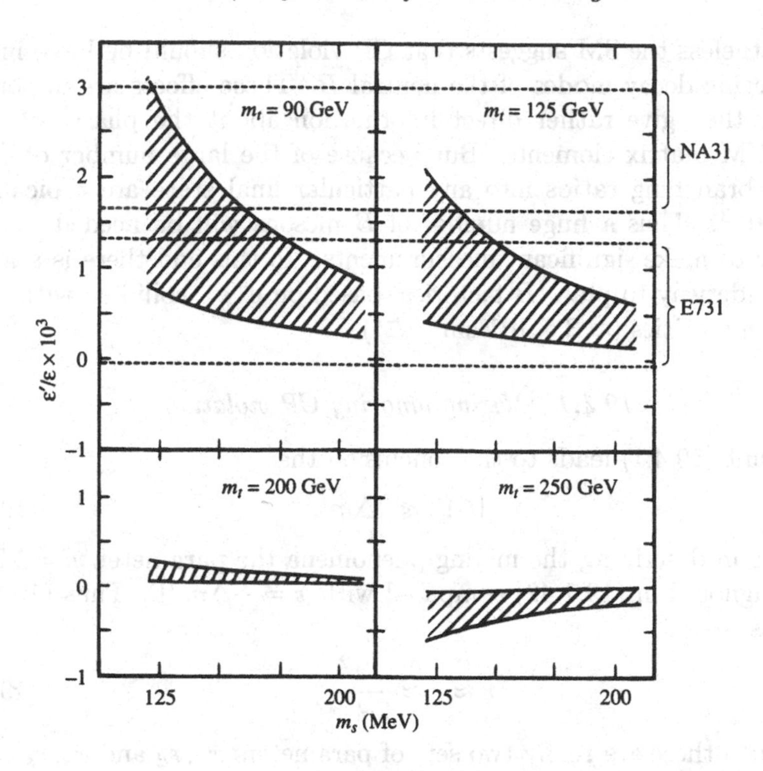

Fig. 19.3. Calculated values of $(\epsilon'/\epsilon) \times 10^3$ for fixed m_t as a function of the strange quark current mass m_s with δ determined from fitting ϵ_{Expt} and $0.1\ \mathrm{GeV} < \Lambda_{\mathrm{QCD}} < 0.3\ \mathrm{GeV}$ (adapted from Buchalla, Buras and Harlander, 1990).

2. The NA31 group has taken further data and it will be of the greatest interest to see if their central value moves down towards the E731 result.

Clearly a resolution of the experimental situation is of the greatest importance.

19.4 Dynamics of B^0–$\bar{B}^0$ mixing

Direct CP violation effects are expected to be small in B^0–$\bar{B}^0$ mixing. This follows basically from estimates (see Section 19.4.2) which suggest

$$|\Gamma_{12}| \ll |M_{12}| \tag{19.4.1}$$

and imply via (19.1.7) that to a very good approximation

$$\left|\frac{q}{p}\right| = 1. \tag{19.4.2}$$

Nonetheless the SM suggests that CP violation should be large in certain specific decay modes of the neutral B. These effects are important because they give rather direct information about the phases of some of the KM matrix elements. But because of the large number of decay modes, branching ratios into any particular final state are typically of order 10^{-3}. Thus a huge number of B mesons will be needed ($\geq 10^8$) in order to make significant measurements. To this end there is a major effort underway to plan 'B factories' based on e^+e^- colliders with enormous luminosities ($\approx 3 \times 10^{33}\mathrm{cm}^{-2}\mathrm{s}^{-1}$).

19.4.1 Mixing ignoring CP violation

The result (19.4.1) leads to the conclusion that

$$|\Delta\Gamma| \ll |\Delta m|. \tag{19.4.3}$$

Hence in describing the mixing phenomena the parameter $y = \Delta\Gamma/2\Gamma$ can be ignored in (19.1.42) compared with $x = -\Delta m/\Gamma$. Thus (19.1.42) becomes

$$r \approx \bar{r} \approx \frac{x^2}{2+x^2}. \tag{19.4.4}$$

(Of course there are really two sets of parameters x_d, r_d and x_s, r_s corresponding to the bottom mesons B_d^0 and B_s^0.)

The analysis exactly parallels the K^0–$\bar{K}^0$ system, simplified by the assumption of CP conservation. Thus corresponding to (19.2.44) and (19.3.13) we will have

$$x = -\frac{1}{\Gamma_B m_B}\mathrm{Re}\mathcal{M}^D \tag{19.4.5}$$

where $\mathcal{M}$ now comes from the box diagrams for $\bar{B}^0 \to B^0$ (interpreted as $\bar{d}b \to d\bar{b}$ for $\bar{B}_d^0 \to B_d^0$ or $\bar{s}b \to s\bar{b}$ for $\bar{B}_s^0 \to B_s^0$) analogous to those in Fig. 19.1.

Because of the relative sizes of the $S(z)$ functions and the combination of KM matrix elements involved, far and away the dominant contribution comes from the heavy t quark and one obtains analogously to (19.3.17, 18 and 31)

$$\begin{aligned} x_d &= \tau_{B_d}\frac{G^2M_W^2}{6\pi^2}m_{B_d}(B_{B_d}f_{B_d}^2)\eta_{\mathrm{QCD}}S(z_t)\mathrm{Re}\left[V_{td}^{*2}V_{tb}^2\right] \\ x_s &= \tau_{B_s}\frac{G^2M_W^2}{6\pi^2}m_{B_s}(B_{B_s}f_{B_s}^2)\eta_{\mathrm{QCD}}S(z_t)\mathrm{Re}\left[V_{ts}^{*2}V_{tb}^2\right] \end{aligned} \tag{19.4.6}$$

where $\eta_{\mathrm{QCD}} \approx 0.84$ is a QCD correction factor and $S(z)$ is given in (19.3.20). The constants B_B, f_B (analogous to B_K and f_K) are related to the hadronic matrix elements of the box operators and to the purely leptonic B decay rate respectively. It is expected that $B_B \approx 1$ but f_B is not known. Generally it is assumed that 150 MeV $\leq f_B \leq$ 200 MeV.

Regarding the KM matrix elements, from (9.2.7 and 6) we may take $V_{tb}^2 = 1$. From (9.2.7) and the values of the s_{ij} given above eqn (19.3.25), we may also take

$$\mathrm{Re}\,(V_{td}^*)^2 \approx s_{12}s_{23}(s_{12}s_{23} - 2s_{13}\cos\delta) \tag{19.4.7}$$

$$\approx (1-\cos\delta)\times 10^{-4} \tag{19.4.8}$$

whose value is clearly sensitive to the sign of $\cos\delta$, i.e. to which quadrant δ lies in.

To estimate x_d let us take $m_B = 5.28$ GeV/c^2, $\tau_B = 1.5\times 10^{-12}$s, $m_t = 127$ GeV/c^2 so that $S(z_t) \simeq 1.6$. Normalizing to the central values $B_B = 1$, $f_B = 175$ MeV yields

$$x_d^{\mathrm{Theory}} \approx 0.56\left(\frac{B_B}{1}\right)\left(\frac{f_B}{175\ \mathrm{MeV}}\right)^2 (1-\cos\delta). \tag{19.4.9}$$

Experimentally, by studying B mesons produced in $\Upsilon(4S)$ decays, one can be sure that one is dealing only with B_d^0–$\bar{B}_d^0$. The ARGUS (1987b) and CLEO (1989) collaborations found consistent evidence for mixing, with a weighted average

$$\chi_d = 0.16 \pm 0.04 \tag{19.4.10}$$

for the parameter $\chi = r/(1+r)$ given in (19.1.55). This leads to

$$x_d^{\mathrm{Expt}} = 0.69 \pm 0.13. \tag{19.4.11}$$

From (19.4.9) we see that if we use the central values for B and f_B we require $\cos\delta < 0$ to obtain agreement with (19.4.11). For these values (19.4.11) implies approximately

$$90^\circ \leq \delta \leq 116^\circ \tag{19.4.12}$$

which however is not consistent with the allowed range of δ from fitting ϵ using $m_t = 127\,\mathrm{GeV}/c^2$ (see Fig. 19.2). Clearly a small adjustment to the value of $B_B f_B^2$ will permit consistency with the fit to ϵ. Once these parameters are better known B_d^0–$\bar{B}_d^0$ mixing will remove the two-fold ambiguity in the knowledge of δ.

The information on B_s^0–$\bar{B}_s^0$ mixing is problematic since it comes from an analysis of lepton pairs from an undifferentiated mixture of B mesons and baryons produced at the Z^0 at LEP (see Section 13.5), so the measured parameter is

$$\chi_{\mathrm{meas}} = \mathrm{f}_d\chi_d + \mathrm{f}_s\chi_s \tag{19.4.13}$$

where f_d and f_s are the fractions of leptons present in the sample from B_d^0 and B_s^0. Based upon a model for heavy quark production it is usually assumed that

$$f_d \approx 0.375, \qquad f_s \approx 0.15. \tag{19.4.14}$$

The experimental values for χ_{meas} are:

$$\begin{aligned} \text{ALEPH (1991)} &: \quad 0.132^{+0.027}_{-0.026} \\ \text{L3 (1991)} &: \quad 0.178^{+0.049}_{-0.040}. \end{aligned} \tag{19.4.15}$$

If one takes the values for $f_{d,s}$ seriously and uses the value of χ_d (19.4.10) one obtains

$$\text{ALEPH}: \quad \chi_s = 0.46 \pm 0.27 \qquad \text{or L3}: \quad 0.79^{+0.47}_{-0.34}. \tag{19.4.16}$$

Of course χ_s cannot be bigger than 0.5 so it is perhaps best to interpret these as indicating simply that χ_s is large, i.e. close to its maximum allowed value.

If now in (19.4.6) we assume that the lifetime and B, f_B parameters are essentially the same for B_d^0 and B_s^0 we see that

$$x_s \approx \frac{\text{Re}(V_{ts}^*)^2}{\text{Re}(V_{td}^*)^2} x_d. \tag{19.4.17}$$

To a good approximation we have

$$\begin{aligned} \text{Re}(V_{ts}^*)^2 &\approx (s_{23})^2 \\ &\approx 2.5 \times 10^{-3} \end{aligned} \tag{19.4.18}$$

yielding, via (19.4.8),

$$x_s \approx \left(\frac{25}{1 - \cos\delta}\right) x_d. \tag{19.4.19}$$

Thus there is no difficulty in producing a value of x_s large enough to make χ_s close to 0.5. But because of the uncertainty in the values of $f_{d,s}$ it does not seem worthwhile to pursue the comparison any further at present.

With the eventual building of B meson 'factories' it will become possible to study the B^0–$\bar{B}^0$ system in the same sort of detail as has been achieved for the K^0–$\bar{K}^0$ system. For the moment it seems that the SM can account perfectly well for what has thus far been measured in the B^0–$\bar{B}^0$ system.

19.4.2 CP violation in the B^0–$\bar{B}^0$ system

As in the D^0–$\bar{D}^0$ system (Section 19.1.3b) we expect $\Gamma_+ \approx \Gamma_-$ so that $\Delta\Gamma \ll \Gamma$. Thus we have no analogue of the useful feature that a K^0 or $\bar{K}^0$

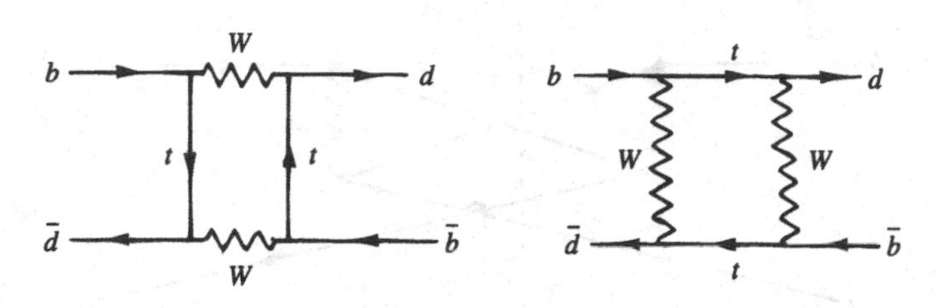

Fig. 19.4. Feynman diagrams for $\bar{d}b \to d\bar{b}$.

beam eventually becomes a pure K_L beam. On the other hand because of the strong dominance of particular diagrams the relationship between the phenomenological parameters and the fundamental theoretical ones is very direct.

Thus the dispersive part M_{12} of the transition $\langle B_d^0|\hat{T}_W|\bar{B}_d^0\rangle$ is dominated by the t quark Feynman diagrams in Fig. 19.4 (in contrast to Fig. 19.1 for the K^0 case). These have a common factor $(V_{tb}^* V_{td})^2$. But the dispersive part of the transition $\langle \bar{B}_d^0|\hat{T}_W|B_d^0\rangle$ is given by almost identical diagrams, the only difference being that the common factor is now $(V_{tb}V_{td}^*)^2$. Hence when we take the ratio q/p using (19.4.1) in (19.1.7) and bearing in mind (19.1.3) everything else cancels out and we obtain the remarkably simple result

$$\left(\frac{q}{p}\right)_{B_d} = \frac{V_{tb}^* V_{td}}{V_{tb} V_{td}^*} = \mathrm{e}^{2\mathrm{i}\phi_{td}} \tag{19.4.20}$$

where we have used the fact that V_{tb} is real and where ϕ_{td} is the phase of the KM matrix element V_{td}.

Similarly for B_s^0 we get

$$\left(\frac{q}{p}\right)_{B_s} = \frac{V_{tb}^* V_{ts}}{V_{tb} V_{ts}^*} = \mathrm{e}^{2\mathrm{i}\phi_{ts}}. \tag{19.4.21}$$

The argument for neglecting the absorptive part Γ_{12} compared with M_{12} in the formula for q/p is based upon the following. Γ_{12} gets contributions from genuine physical intermediate states in the transition $B^0 \to \bar{B}^0$. Consider $B_d^0 \to \bar{B}_d^0$. At the quark level this will be dominated by the Cabibbo favoured decays $b \to c\bar{c}d$ as shown in Fig. 19.5 where the dashed line reminds us that only the absorptive part of the diagram involving real physical intermediate states is relevant for Γ_{12}. But by twisting the diagram topologically one recognizes inside it just a typical box diagram for $b\bar{d} \to \bar{b}d$. Comparison with the diagrams of Fig. 19.4 leads to the result

$$\frac{|\Gamma_{12}|}{|M_{12}|} \approx \frac{m_b^2}{m_t^2} \ll 1.$$

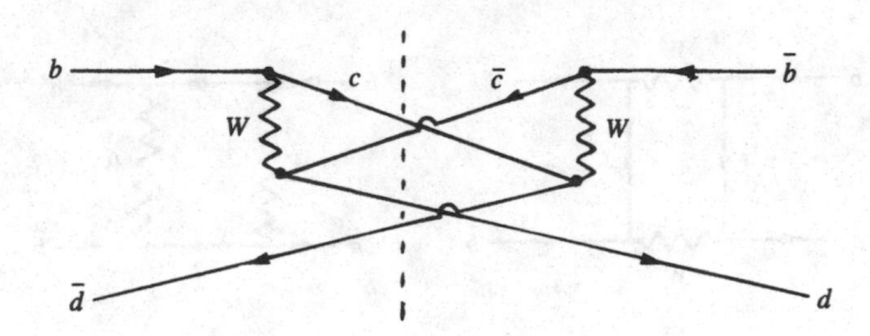

Fig. 19.5. Feynman diagram contributing to Γ_{12}.

For B_s^0 the ratio is even smaller since one needs $B \to c\bar{c}s$ which is suppressed by $\sin\theta_C$.

Let us now consider how one might measure CP violation effects. Given that $\Delta\Gamma \ll \Gamma$ eqns (19.1.19) simplify to

$$\begin{aligned} f_+(t) &= e^{-\Gamma t/2} e^{-iMt} \cos\left(\frac{\Delta m t}{2}\right) \\ f_-(t) &= i e^{-\Gamma t/2} e^{-iMt} \sin\left(\frac{\Delta m t}{2}\right) \end{aligned} \tag{19.4.22}$$

where [see (19.1.13)] M is the average mass

$$M = \tfrac{1}{2}(m_+ + m_-). \tag{19.4.23}$$

Thus (19.1.18) becomes

$$\begin{aligned} |\psi(t)\rangle_{B^0} &= e^{-\Gamma t/2} e^{-iMt} \left\{ \cos\left(\frac{\Delta m t}{2}\right)|B^0\rangle - i\frac{q}{p}\sin\left(\frac{\Delta m t}{2}\right)|\bar{B}^0\rangle \right\} \\ |\psi(t)\rangle_{\bar{B}^0} &= e^{-\Gamma t/2} e^{-iMt} \left\{ \cos\left(\frac{\Delta m t}{2}\right)|\bar{B}^0\rangle - i\frac{p}{q}\sin\left(\frac{\Delta m t}{2}\right)|B^0\rangle \right\}. \end{aligned} \tag{19.4.24}$$

We shall give just one illustration of the use of (19.4.24) and will restrict ourself to the simplest case where the interference effects are maximal. It is possible, as we shall explain presently, to find a decay mode of B^0 and $\bar{B}^0$ into a final state $|f_{CP}\rangle$, *which is an eigenstate of CP*, such that the *rates* $|B^0\rangle \to |f_{CP}\rangle$ and $|\bar{B}^0\rangle \to |f_{CP}\rangle$ are equal. Let

$$\begin{aligned} A_f &\equiv \langle f_{CP}|\hat{T}_W|B^0\rangle \\ \bar{A}_f &\equiv \langle f_{CP}|\hat{T}_W|\bar{B}^0\rangle \end{aligned} \tag{19.4.25}$$

with

$$|A_f|^2 = |\bar{A}_f|^2 \tag{19.4.26}$$

but with the amplitudes differing by a phase.

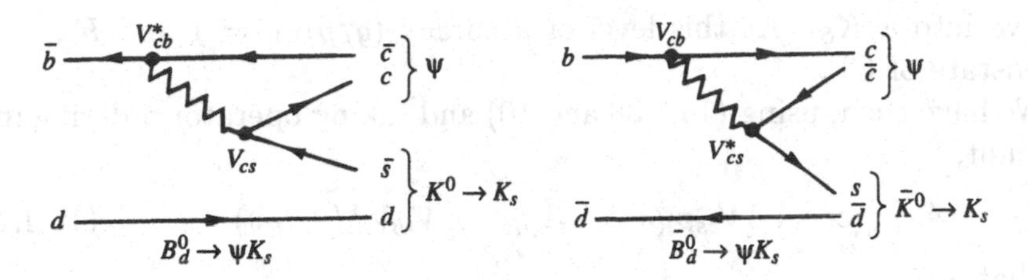

Fig. 19.6. Quark diagrams for B^0_d and $\bar{B}^0_d \to \psi K_S$.

Then, using (19.4.24) and (19.4.2), the rate for producing $|f_{\text{CP}}\rangle$ at a time t after production of a B^0 or $\bar{B}^0$ is proportional to

$$|\langle f_{\text{CP}}|\psi(t)\rangle_{B^0/\bar{B}^0}|^2 = \mathrm{e}^{-\Gamma t}|A_f|^2\left\{1 \pm \text{Im}\left(\frac{q\bar{A}_f}{pA_f}\right)\sin(\Delta m t)\right\} \qquad (19.4.27)$$

allowing therefore, in principle, a measurement of Δm and $\text{Im}(q\bar{A}_f/pA_f)$.

There is a non-trivial technical problem associated with the fact that we are using a final state $|f_{\text{CP}}\rangle$ into which both B^0 and $\bar{B}^0$ decay equally. Since any beam we prepare is a mixture of B^0 and $\bar{B}^0$ and since we do not have any phenomenon similar to the $K^0, \bar{K}^0$ beam becoming pure K_L, we cannot tell whether $|f_{\text{CP}}\rangle$ arose from the decay of $|\psi(t)\rangle_{B^0}$ or $|\psi(t)\rangle_{\bar{B}^0}$. Thus one needs independent information on the flavour of the decaying neutral B meson; this is called *flavour tagging*. It can be achieved either (i) by finding a B^0–$\bar{B}^0$ production reaction like $\mathrm{e}^+\mathrm{e}^- \to b\bar{b}$ jets where the forward-backward asymmetry allows one to identify which neutral B is involved or (ii) by tagging the B^0 or $\bar{B}^0$ which does not decay into $|f_{\text{CP}}\rangle$ by means of some decay specific to one or the other, e.g. $B^0 \to \ell^+\nu\text{X}$, $\bar{B}^0 \to \ell^-\nu\text{X}$. This is one more reason why one requires a huge number of Bs to get significant results.

In general the connection between the decay amplitudes A_f and $\bar{A}_f$ depends upon detailed dynamics. But in some cases, where the decay is dominated by one Feynman diagram at the quark level, all the dynamics cancel out and the ratio $\bar{A}_f/A_f$ depends on ratios of KM matrix elements and their complex conjugates. From the structure of the KM matrix (9.2.7) and our knowledge of the angles involved the only elements which could have large phases are V_{ub} and V_{td}. To good accuracy we may treat all the other KM elements as real in the following discussion. We shall consider one example:

$$B^0_d \to \psi K_S \qquad \text{and} \qquad \bar{B}^0_d \to \psi K_S \qquad (19.4.28)$$

where the decay proceeds via the spectator diagram shown in Fig. 19.6. The process is really sequential: first a K^0 or $\bar{K}^0$ is produced which then evolve into a K_S. At this level of accuracy $(q/p)_{K^0} = 1$ and K_S is an eigenstate of CP.

We have then, using (19.1.38 and 10) and taking operator ordering into account,

$$A_{\psi K_S} \propto -V_{cb}^* V_{cs} p_{K^0} \qquad \bar{A}_{\psi K_S} \propto V_{cb} V_{cs}^* (-q_{K^0}) \tag{19.4.29}$$

so that

$$\frac{\bar{A}_{\psi K_S}}{A_{\psi K_S}} \approx 1. \tag{19.4.30}$$

Hence, for the measured parameter, in B_d^0–$\bar{B}_d^0$, we have, from (19.4.20) and (19.4.30)

$$\text{Im}\,(q\bar{A}_{\psi K_S} \,/\, pA_{\psi K_S}) = \sin 2\phi_{td}. \tag{19.4.31}$$

Thus a measurement of the oscillatory term in (19.4.27) would yield direct information on the phase of V_{td}. From our present fragmentary knowledge of the KM matrix we have only the wide range

$$0.08 \leq \sin 2\phi_{td} \leq 1. \tag{19.4.32}$$

Many other reactions have been studied theoretically and a rich experimental programme will be possible if B factories are ever constructed.

For access to the literature, see Nir and Quinn, 1992. The reader is warned that many different notational conventions exist and that some early writers used $B_d^0 = b\bar{d}$ instead of $\bar{b}d$.

20
Regularization, renormalization and introduction to the renormalization group

Our eventual aim is a simple and intelligible discussion of QCD and its applications to deep inelastic scattering, the Drell–Yan process and the cross-section for $e^+e^- \to$ hadrons etc. To achieve this, in this chapter we present a heuristic treatment of the ideas of regularization, renormalization and the powerful renormalization group technique. Then we introduce the concept of scaling and asymptotic freedom, initially for the case of scalar particles. The renormalization group results derived for scalar particles will be extended to the realistic case in Chapter 21, and applications follow in Chapter 22. All technical elements are relegated to the Appendix to this chapter (Section 20.10) where we explain the technique of dimensional regularization.

20.1 Introduction

Most discussions of renormalization theory, in textbooks on field theory, are impenetrable and bogged down in technical complexities. Previously this did not matter very much. The important thing was to *know* that the theory was renormalizable, and the calculation of a few important effects could be left to a handful of experts. But the realization of the importance of the renormalization group in perturbative QCD and its constant use in all calculations of hard hadronic processes has meant that it is essential to understand at least the main ideas about schemes of regularization and renormalization. Otherwise labels like 'MS' and '$\overline{\text{MS}}$' which abound in the literature, not just in esoteric theory papers but also in experimental reports, are meaningless.

We shall present a simple and non-technical explanation of the ideas of renormalization, of the existence of different renormalization schemes and of how the freedom to choose one particular scheme can be put to advantage in the framework of the renormalization group. We try to clarify too the rôle and meaning of the parameters, such as masses and coupling constants, that appear in a field theory Lagrangian, and which, perhaps contrary to intuition, do not have the simple obvious meaning that they would seem to have. Our presentation will be largely pedagogical and somewhat qualitative though we go through one example in detail in Section 20.7. Some of the missing technical details can be found in the Appendix to this chapter (which we shall simply refer to as 'the Appendix'), but it is not necessary to read the Appendix in order to follow the general development of the chapter. For a more technical yet pedagogical presentation of the mathematics, see Ryder (1985). For a more rigorous and general approach, see Collins (1984).

Throughout this chapter, we talk, for simplicity, in terms of a scalar field theory and usually illustrate the concepts by reference to ϕ^4 field theory. All the general concepts, however, carry over into QED and QCD.

20.2 Parameters and physical observables in a field theory

A field theory is specified by giving the Lagrangian density as a function of the field operators and their derivatives. A very simple example is the so-called ϕ^4 theory describing a self-interacting electrically neutral scalar field, in which the Lagrangian density is

$$\mathcal{L}(x,t) = \underbrace{\tfrac{1}{2}[\partial_\mu \phi_B(x)]^2 - \tfrac{1}{2} m_B^2 \phi_B^2(x)}_{\mathcal{L}^0} - \underbrace{\frac{g_B}{4!}\phi_B^4(x)}_{\mathcal{L}'} . \qquad (20.2.1)$$

$\mathcal{L}^0$ is the kinetic part of $\mathcal{L}$, and $\mathcal{L}'$ describes the self-interaction. In (20.2.1) ϕ_B, m_B and g_B are referred to as the bare field, bare mass and bare coupling constant (note that it is dimensionless, see Section 1.1), for reasons that will be explained shortly, and

$$\partial_\mu \phi \equiv \frac{\partial}{\partial x^\mu}\phi(x).$$

We also use the shorthand notation $(\partial_\mu \phi)^2$ for

$$\sum_{\mu\nu} g^{\mu\nu}\partial_\mu\phi\partial_\nu\phi \equiv \sum_\mu \partial_\mu\phi\partial^\mu\phi.$$

The overall numerical factor in (20.2.1) is irrelevant, but the sign and size of the coefficient of $m_B^2\phi_B^2$ *relative* to $(\partial_\mu\phi_B)^2$ is chosen to ensure that for a free field without interaction, i.e. if $g_B \equiv 0$, ϕ_B will obey the correct Klein–Gordon field equation. Indeed the Euler–Lagrange equation

$$\partial_\mu \frac{\partial \mathcal{L}}{\partial(\partial_\mu\phi)} = \frac{\partial \mathcal{L}}{\partial \phi} \qquad (20.2.2)$$

becomes just

$$(\Box + m_{\rm B}^2)\phi_{\rm B}(x) = 0 \tag{20.2.3}$$

with

$$\Box \equiv \frac{\partial^2}{\partial x^{02}} - \nabla^2,$$

as desired when $g_{\rm B} = 0$.

The quantum element is introduced by regarding $\phi(x)$ as an operator analogous to the generalized position operators $\hat{q}_j$ in ordinary quantum mechanics, and defining $\pi(x)$, the 'conjugate momentum' to ϕ,

$$\pi(x) \equiv \frac{\partial \mathcal{L}}{\partial \dot{\phi}}; \quad \text{where} \quad \dot{\phi} = \frac{\partial \phi}{\partial x^0} \tag{20.2.4}$$

in complete analogy to the definition of $\hat{p}_j$ in ordinary quantum mechanics, and then demanding that ϕ and π satisfy canonical commutation relations *when their time variables are equal*

$$[\phi(\boldsymbol{x}, t), \phi(\boldsymbol{x}', t)] = 0, \qquad [\pi(\boldsymbol{x}, t), \pi(\boldsymbol{x}', t)] = 0, \tag{20.2.5}$$

$$[\pi(\boldsymbol{x}, t), \phi(\boldsymbol{x}', t)] = -\mathrm{i}\delta^3(\boldsymbol{x} - \boldsymbol{x}'), \tag{20.2.6}$$

the latter being the continuum analogue of $[\hat{p}_j, \hat{q}_k] = -\mathrm{i}\delta_{jk}(\hbar = 1)$. The above 'canonical quantization' with its emphasis on the time derivative suffers from being not manifestly covariant. An alternative approach, the path integral formalism, deals only with classical (non-operator) fields and is manifestly covariant, but involves hair-raising mathematical manipulations. It is well described in Taylor (1976) and Ryder (1985).

Once we are given the Lagrangian we can, in principle, calculate any physical observable, such as a decay rate, a scattering cross-section, or the physical mass of a particle. Clearly our result will depend upon the parameters $m_{\rm B}$ and $g_{\rm B}$ in $\mathcal{L}$. Let us for definiteness refer to the calculation of the S-matrix element for some process $|\mathrm{i}\rangle \to |\mathrm{f}\rangle$. We shall obtain a result of the form

$$S = S(p_1, p_2, \ldots; m_{\rm B}, g_{\rm B}), \tag{20.2.7}$$

where the p_i are momenta of particles involved in the reaction. Generally of course we cannot compute S exactly and must approach it via a perturbative calculation. Usually we take $\mathcal{L}^0$ as the unperturbed Lagrangian and $\mathcal{L}'$ as the perturbation, so that S is given as a power series in $g_{\rm B}$, and the actual evaluation of the terms in the series is simplest using Feynman diagram techniques (Bjorken and Drell, 1964, 1965).

There is, however, one other parameter on which the results of our computation will depend. Equations (20.2.6, 4 and 1) imply that $[\dot{\phi}(\boldsymbol{x}, t), \phi(\boldsymbol{x}', t)] = -\mathrm{i}\delta^3(\boldsymbol{x} - \boldsymbol{x}')$ and this is constantly utilized in the perturbative calculation, albeit in a somewhat hidden way. But we could well have normalized the fields differently by demanding that

$$[\dot{\phi}(\boldsymbol{x},t),\phi(\boldsymbol{x}',t)] = -\frac{\mathrm{i}}{Z_\phi} i\delta^3(\boldsymbol{x}-\boldsymbol{x}'), \tag{20.2.8}$$

where Z_ϕ is some arbitrary number; and the S calculated by perturbation theory would then depend also on the parameter Z_ϕ. The label B on ϕ simply means that we have made the choice $Z_\phi = 1$, that is, we are working with the *bare* field. Let us for the moment continue to utilize ϕ_{B} and therefore suppress the functional dependence of Z_ϕ in (20.2.7).

The parameters $m_{\mathrm{B}}, g_{\mathrm{B}}$ despite their intuitive rôle have no immediate physical significance. The mass m of the physical quanta of the ϕ field is given by the value of p^2 at which the single particle propagator has a pole. To zeroth order in g_{B} this is at $p^2 = m_{\mathrm{B}}^2$, but higher order terms corresponding to diagrams like

will shift the pole to the point $p^2 = m^2$, which depends upon m_{B} and g_{B}. If the perturbative corrections were really small we would have $m \approx m_{\mathrm{B}}$ so that it would be meaningful to regard m_{B} as an approximation to the physical mass. But in practice the corrections are not only large, they are infinite! This brings us to the subject of renomalization, whereby the apparently infinite results are rendered both finite and meaningful.

20.3 The idea of renormalization

If the parameters $m_{\mathrm{B}}, g_{\mathrm{B}}$ and $Z_{\phi_{\mathrm{B}}}$ are regarded as *fixed numbers*, then it is found that in the evaluation of many S-matrix elements by perturbation theory the integrals involved in certain Feynman diagrams diverge. For example in QED the electron self-energy or propagator correction

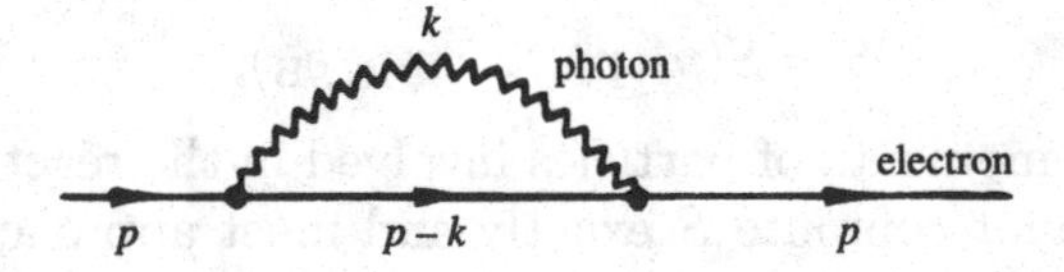

contains an integral of the general form

$$e_{\mathrm{B}}^2 \int \frac{\mathrm{d}^4k}{k^2[(p-k)^2 - m_{\mathrm{B}}^2]}. \tag{20.3.1}$$

For very large k, since p is fixed, this looks like $e_{\mathrm{B}}^2 \int \mathrm{d}^4k/k^4$ and 'diverges logarithmically', i.e. if we pretend that k is a Euclidean vector with $|k|^2 =$

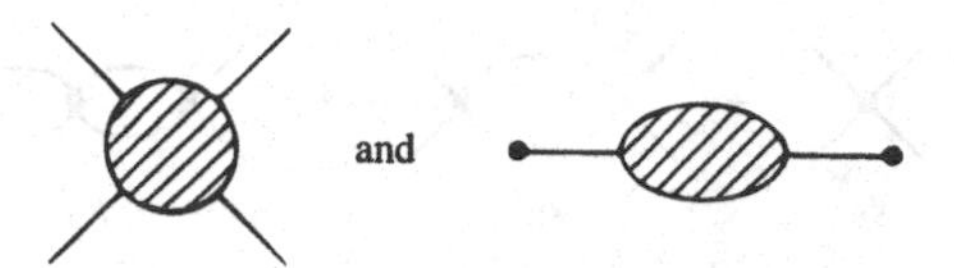

Fig. 20.1. Complete 4-point vertex and complete propagator in ϕ^4 theory.

$k_x^2 + k_y^2 + k_z^2 + k_4^2, k_4 \equiv ik_0$, and if we cut off the integral at $|k| = \lambda$, then the result is proportional to $e_{\rm B}^2 \ln \lambda$.

In general terms, if we cut off all divergent integrals at λ we shall find that our calculated S-matrix elements (call them $S_{\rm B}$) depend upon λ:

$$S_{\rm B} = S_{\rm B}(p_i; m_{\rm B}, g_{\rm B}, Z_{\phi_{\rm B}} = 1, \lambda) \tag{20.3.2}$$

and for many of them

$$\lim_{\lambda \to \infty} S_{\rm B}(p_i; m_{\rm B}, g_{\rm B}, Z_{\phi_{\rm B}} = 1, \lambda) = \infty, \tag{20.3.3}$$

giving rise to nonsensical results.

The renormalization scheme which renders results finite is based upon the brilliant idea of allowing the parameters $m_{\rm B}$, $g_{\rm B}$ *to depend upon* λ and to introduce a new field ϕ for which $Z_\phi = Z_\phi(\lambda) \neq 1$ and to try to adjust the λ dependence so as to cancel out the infinities as $\lambda \to \infty$. We could imagine computing a few $S_{\rm B}$, comparing with experiment and deducing how $m_{\rm B}(\lambda)$ etc. should depend upon λ in order to agree with the finite experimental results. (Of course we are assuming that the theory is a correct one and does describe nature.) In our present example we have available the choice of just three functions $m_{\rm B}(\lambda)$, $g_{\rm B}(\lambda)$ and $Z_\phi(\lambda)$, whereas infinitely many S-matrix elements are found to diverge. So it is far from evident that once we have adjusted our functions, all these infinities will be removed. Indeed the miracle does not occur in general, and it is only in a limited number of field theories, including our present example, that the scheme works. The actual proof that a theory is renomalizable is very complicated (see, for example, Chapter 19 of Bjorken and Drell (1965)) and will not be entered into here. Roughly speaking the scheme works when the number of independently divergent terms in the 'fundamentally' divergent matrix elements is the same as or smaller than the number of functions of λ that we are able to choose freely. In our present example of ϕ^4 field theory the fundamentally divergent matrix elements are shown in Fig. 20.1. Some contributions to these (and their perturbative order) are shown in Fig. 20.2. The complete single-particle propagator is linearly divergent and contains two kinds of divergent terms, one proportional to p^2 and the other independent of p.

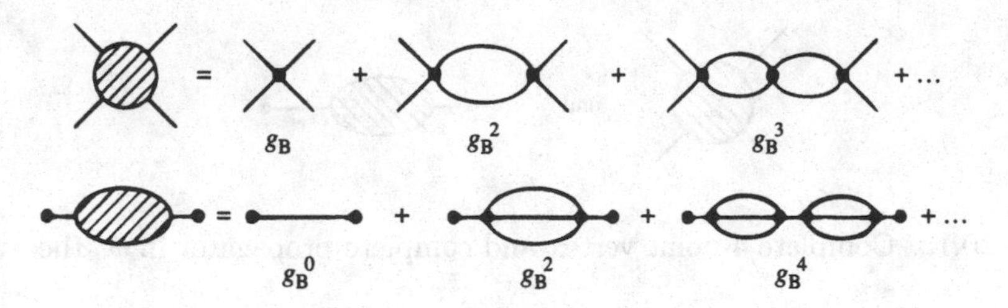

Fig. 20.2. Some contributions to the diagrams of Fig. 20.1.

Since our free functions do not depend on momenta we need two of them $Z_\phi(\lambda)$ and $m_B(\lambda)$ in order to cancel these divergences.

All other matrix elements that diverge can be shown to do so only because they contain within themselves one or more of the above divergent diagrams. Thus by neutralizing this fundamental set we in effect neutralize all the matrix elements.

In practice one makes a multiplicative change of variables from m_B, g_B, ϕ_B to m, g, ϕ as follows

$$\left.\begin{aligned} \phi_B &= Z_\phi^{1/2}(\lambda)\phi \\ g_B &= Z_\phi^{-2}(\lambda)Z_g(\lambda)g \\ m_B^2 &= Z_\phi^{-1}(\lambda)Z_m(\lambda)m^2 \end{aligned}\right\} \qquad (20.3.4)$$

and tries to adjust the dimensionless functions $Z(\lambda)$ so that g, m are finite and independent of λ as $\lambda \to \infty$. This is the only requirement on the $Z(\lambda)$. They are otherwise quite arbitrary and may be chosen for convenience. Usually they are given as a power series in g of the generic form

$$Z(\lambda) = 1 + ga_1(\lambda) + g^2a_2(\lambda) + \dots \qquad (20.3.5)$$

and the coefficient functions $a_j(\lambda)$ contain the necessary divergence as $\lambda \to \infty$. (Note that for some $Z(\lambda)$ only even or odd powers of g will appear.) It is assumed that at fixed λ these series can be manipulated as if they were convergent series. Only at the *very end* does one let $\lambda \to \infty$.

We can regard (20.3.4) as a transformation in the Lagrangian from the *bare* quantities ϕ_B, g_B, m_B to new *renormalized* quantities, the field ϕ and the parameters g and m. Substituting into (20.2.1) gives

$$\begin{aligned} \mathcal{L} &= \tfrac{1}{2}(\partial_\mu\phi_B)^2 - \tfrac{1}{2}m_B^2\phi_B^2 - \frac{g_B}{4!}\phi_B^4 \\ &= \tfrac{1}{2}Z_\phi(\partial_\mu\phi)^2 - \tfrac{1}{2}Z_\phi^{-1}Z_m m^2 Z_\phi\phi^2 - \frac{Z_\phi^{-2}Z_g g Z_\phi^2\phi^4}{4!} \\ &= \tfrac{1}{2}Z_\phi(\partial_\mu\phi)^2 - \tfrac{1}{2}Z_m m^2\phi^2 - \frac{Z_g g}{4!}\phi^4. \end{aligned} \qquad (20.3.6)$$

Diagram	$\phi_{\rm B}$ scheme	ϕ scheme
Undressed propagator (p)	$\dfrac{\mathrm{i}}{p^2 - m_{\rm B}^2}$	$\dfrac{\mathrm{i}Z_\phi^{-1}}{p^2 - m_{\rm B}^2} = \dfrac{\mathrm{i}}{Z_\phi p^2 - Z_m m^2}$
vertex	$-\mathrm{i}g_{\rm B}$	$-\mathrm{i}Z_g g$
external on mass-shell line in S-matrix	1	$\dfrac{1}{\sqrt{Z_\phi}}$

Table 20.1. Feynman rules in $\phi_{\rm B}$ and ϕ schemes.

It is instructive to consider how one would calculate Feynman diagrams based on the two different forms of $\mathcal{L}$; let us call them the '$\phi_{\rm B}$ scheme' and the 'ϕ scheme' (see Table 20.1).

Using these it is a simple exercise to take an arbitrary Feynman diagram for any S-matrix element and to show that one gets the same answer in both schemes, i.e. $S_{\rm fi}(\phi_{\rm B}$ scheme$) = S_{\rm fi}(\phi$ scheme$)$, but the functions are of course different functions of their respective variables. Notice that this is true without our specifying *what functions* the $Z(\lambda)$ are! All we have used is the structure of the transformation in (20.3.4). Thus there are many schemes, i.e. many different changes of variable, all of which give the same answer for the S-matrix. Note that when we come to discuss the so-called dimensional regularization schemes, the equation for $g_{\rm B}$ in (20.3.4) will be slightly modified for dimensional reasons [see eqn (20.4.8)].

It should be noted that very often in discussions of renormalization one rewrites (20.3.6) in the form

$$\mathcal{L} = \mathcal{L}^0_{(\phi)} + \mathcal{L}'_{(\phi)} \tag{20.3.7}$$

with

$$\mathcal{L}^0_{(\phi)} = \tfrac{1}{2}(\partial_\mu\phi)^2 - \tfrac{1}{2}m^2\phi^2 \tag{20.3.8}$$

and

$$\mathcal{L}'_{(\phi)} = -\frac{g}{4!}\phi^4 - \delta m^2\frac{\phi^2}{2} + \tfrac{1}{2}(Z_\phi - 1)(\partial_\mu\phi)^2 - \frac{\delta g}{4!}\phi^4 \tag{20.3.9}$$

where

$$\begin{aligned} \delta m^2 &= (Z_m - 1)m^2 & (20.3.10)\\ \delta g &= (Z_g - 1)g. & (20.3.11)\end{aligned}$$

The point of this is that one then does perturbation theory based upon the free Lagrangian $\mathcal{L}^0_{(\phi)}$ (20.3.8) with $\mathcal{L}'_{(\phi)}$ being the new interaction part of the Lagrangian. The extra terms in the latter [compared with $\mathcal{L}'$ in (20.2.1)] are called *counter-terms*, and are adjusted to cancel the divergences, order by order, as they arise in perturbation theory.

However, this approach is not necessary for a discussion of renormalization, and we shall continue our treatment based on $\mathcal{L}$ as given in (20.3.6). This does not mean that our treatment avoids perturbation theory. At some point we have to utilize the fact that the $Z(\lambda)$ are given as power series in g.

If now we choose our $Z(\lambda)$ to eliminate the infinities, does this lead to a unique renormalized scheme? Clearly not, since we can trivially go to another scheme, call it the 'ϕ' scheme', by mimicking the transformation in (20.3.4)

$$\left.\begin{aligned} g &= z_{\phi'}^{-2} z_{g'} g', \\ m^2 &= z_{\phi'}^{-1} z_{m'} m'^2, \\ \phi &= z_{\phi'}^{1/2} \phi', \end{aligned}\right\} \qquad (20.3.12)$$

but now using *finite* $z_{\phi'}, z_{g'}, z_{m'}$ so that the results, finite in the ϕ scheme, remain finite in the ϕ' scheme.

We could, of course, have gone directly from ϕ_B to ϕ':

$$\left.\begin{aligned} g_B &= Z_{\phi'}^{-2} Z_{g'} g', \\ m_B^2 &= Z_{\phi'}^{-1} Z_{m'} m'^2, \\ \phi_B &= Z_{\phi'}^{1/2} \phi'. \end{aligned}\right\} \qquad (20.3.13)$$

From (20.3.13,12 and 4) it is easy to see for example that:

$$z_{\phi'} = \frac{Z_{\phi'}(\lambda)}{Z_{\phi}(\lambda)}. \qquad (20.3.14)$$

Putting

$$\begin{aligned} z_{\phi'} &= 1 + ag' + \dots \\ Z_{\phi'} &= 1 + b'(\lambda)g' + \dots \\ Z_{\phi} &= 1 + b(\lambda)g + \dots, \end{aligned}$$

and, using $g' = g$ to lowest order, one has from (20.3.14)

$$1 + ag' + \dots = 1 + [b'(\lambda) - b(\lambda)]g' + \dots$$

i.e.

$$b'(\lambda) - b(\lambda) = a. \tag{20.3.15}$$

Remembering that $b'(\lambda)$ and $b(\lambda)$ are infinite as $\lambda \to \infty$ whereas a is finite, we learn that the difference between the ϕ and ϕ' schemes arises from the *finite* parts of the $Z(\lambda)$. In other words, demanding a particular behaviour as $\lambda \to \infty$ does not fix the $Z(\lambda)$ uniquely. The coefficient functions, while having the same limit as $\lambda \to \infty$, can still differ by an arbitrary finite amount.

This has two consequences. On the one hand, in order to have a definite scheme to be able to calculate with, one must specify the $Z(\lambda)$ more precisely. On the other hand, we can make use of the fact that the S-matrix is invariant under the set of transformations (20.3.12)—the 'multiplicative renormalization group'—and actually derive some practical consequences. One of the most useful prescriptions is the so-called *on-shell* (or *physical*) scheme, which we shall label 'R'. One chooses m in (20.3.4) to be the *physical* mass m_R and demands that the complete propagator has a simple pole at $p^2 = m_R^2$ and that the normalization is such that

complete propagator = [diagram, momentum p] $\overset{p^2 \to m_R^2}{\longrightarrow} \dfrac{i}{p^2 - m_R^2}$. (20.3.16)

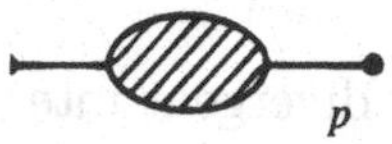

As we shall soon see, these two conditions uniquely fix $Z_{\phi R}$ and Z_{m_R}. Z_{g_R} is fixed by demanding

complete vertex = [diagram, momenta p_1, p_2, p_3, p_4] $= -ig_R$ (20.3.17)

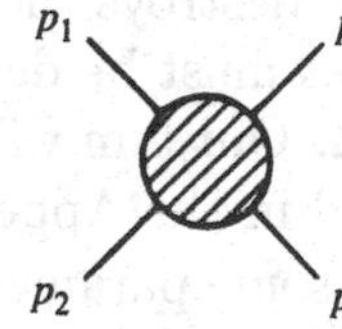

at some specified value of the momenta corresponding to the particles being on-shell, say $p_j^2 = m_R^2$, $p_i \cdot p_j = -m_R^2/3$ $(i \neq j)$. All S-matrix elements are then given as functions of the two parameters m_R and g_R. The value of m_R is known, since it is the measured physical mass. The value of g_R has to be found by comparing theory and experiment for some process. Generally this is a non-trivial undertaking, since the theoretical result is never calculated exactly, i.e. to all orders in g_R. Thus comparing say a second order calculation with experiment may yield a certain best fit value for g_R, whereas comparison between experiment and a better calculation, say to fourth order, may yield a different best fit value of g_R. This difficulty happens to be absent in the best known of all field theories, QED, for there one can prove rigorously that the cross-section for Compton scattering in the long wavelength limit is *exactly* the classical

Thomson expression, i.e.

$$\frac{\mathrm{d}\sigma}{\mathrm{d}\Omega} \xrightarrow{k\to 0} \frac{\alpha^2}{m^2}(\epsilon.\epsilon')^2, \tag{20.3.18}$$

where $m \equiv m_{\mathrm{R}}$ is the physical electron mass, ϵ, ϵ' are the photon polarization vectors and $\alpha \equiv e_{\mathrm{R}}^2/4\pi$ is the fine structure constant. It is for this reason particularly that it is useful to use the on-shell scheme in QED.

In other theories, and in QCD in particular, the analogue of (20.3.18) is not known, or may not exist, so various other prescriptions are used to fix the $Z(\lambda)$ uniquely. But different people choose different prescriptions, so their results will look different, and this can be a source of confusion, especially if notation is careless. For example the same physical observable may be calculated to be $1 + g^2/\pi + g^4/\pi^2$ and $1 + g^2/\pi + 1.7g^4/\pi^4$ by different authors. Of course their variables g are not the same and there is no contradiction! In QED, on the other hand, everybody expresses their results in terms of *the* physical electron mass and *the* fine structure constant α, so there is no confusion.

20.4 Choice of cut-off procedure—regularization

In the above, when faced with divergent integrals like (20.3.1), we introduced a cut-off λ so that we could carry out various manipulations with finite quantities and only at the very end let $\lambda \to \infty$. Literally cutting off an integral, by integrating up to $|k| = \lambda$ instead of infinity, is not useful for multiple integrals, and destroys the Poincaré invariance of the theory, and alternative procedures must be devised. These procedures are called methods of *regularization*. Our aim will be to explain the essential ideas. More details can be found in the Appendix to this chapter.

All methods introduce some parameter, λ (or ϵ), such that the original infinite integral is recovered with $\lambda \to \infty$ (or when $\epsilon \to 0$), but such that the integral is finite for finite λ (or for non-zero ϵ).

An old favourite is the Pauli–Villars (1949) scheme in which each propagator is replaced, as follows:

$$\frac{1}{(p-k)^2 - m_{\mathrm{B}}^2} \to \frac{1}{(p-k)^2 - m_{\mathrm{B}}^2} - \frac{1}{(p-k)^2 - \lambda^2}. \tag{20.4.1}$$

Clearly for $\lambda \to \infty$ we are back to the original propagator, but for *fixed* λ, as $k \to \infty$ the new propagator decreases like $1/k^4$ compared with the $1/k^2$ of the original. Thus an integral like that encountered in (20.3.1) becomes convergent with λ held fixed.

One of the most important criteria in choosing a regularization scheme concerns the symmetries of the theory. It often happens that the possibility of rendering a theory finite depends crucially on cancellations that are due to the symmetries of the theory. It is then important to ensure that the regularization procedure respects the symmetries.

A relatively new and very powerful approach that respects gauge symmetries is the 'dimensional regularization' method ('tHooft and Veltman, 1972). The internal momenta over which one has to integrate in a Feynman diagram are taken to have d components, i.e. we pretend we are working in a d-dimensional space. After certain formal manipulations it is possible to interpret the result as holding for arbitrary complex d. For small enough d the result is finite, and the divergences that we originally had when $d = 4$ now show up as singularities when we continue analytically in d up to $d = 4$. These singularities (poles) can be eliminated by allowing the parameters of the theory to depend on d. In fact if one defines $\epsilon = \frac{1}{2}(4 - d)$ then letting $\epsilon \to 0$ is quite analogous to letting $\lambda \to \infty$ in the older methods, but there are advantages due to the fact that ϵ is dimensionless whereas λ has dimensions of mass.

The idea of dimensional regularization stems from the following observation. Suppose for example that we lived in a two-dimensional world. Then in the integral (20.3.1), which was divergent in our four-dimensional world, we would have had for the physical momentum $p = (p_0, p_x)$ and for the Feynman integrated loop momentum $k = (k_0, k_x)$, with $\mathrm{d}^2k = k\,\mathrm{d}k\,\mathrm{d}\theta$ in polar coordinates. For large k the integrand would have looked like $\mathrm{d}^2k/k^4 \sim \mathrm{d}k/k^3$ which yields a convergent integral. So the idea is to temporarily work in d dimensions, with $d < 4$, carry out the renormalization, and then let $d \to 4$ at the end. For integer $d > 4$ it is trivial to see how to extend the definition of a Feynman integral. For example for $d = 6$ we could put for the physical momenta $\tilde{p} = (p_0, p_1, p_2, p_3, 0, 0)$ and for the integrated loop momenta $\tilde{k} = (k_0, k_1, k_2, k_3, k_4, k_5)$ and integrate over d^6k. But for $d < 4$ this trick won't work. We would lose information about some of the components of p. So for $d < 4$ something more subtle is needed and it is not really a continuation in the trivial sense of just working in fewer dimensions. Nonetheless a continuation is possible and is designed to agree with the above naive generalization whenever $d \geq 4$. For this reason one talks loosely of continuing in the number of dimensions.

Now consider a *convergent* Feynman integral

$$F^{\text{conv.}}(p) = \int \frac{\mathrm{d}^4k}{(2\pi)^4} f(p,k) \tag{20.4.2}$$

where $f(p,k)$ is some product of propagators.

In the Appendix to this chapter we outline how one may define (i.e. give a formula for) a function $F_{(d)}(p)$ for arbitrary d, not necessarily integer or even real, which is *formally* written as the 'd dimensional Feynman integral'

$$F_{(d)}(p) = \int \frac{\mathrm{d}^d k}{(2\pi)^d} f(p,k) \tag{20.4.3}$$

and such that at $d = 4$ we recover the original Feynman integral, i.e.

$$F_{(4)}(p) = F^{\mathrm{conv.}}(p). \tag{20.4.4}$$

Now consider some *divergent* Feynman integral, say one which is logarithmically divergent like (20.3.1)

$$F^{\mathrm{div.}}(p) = \int \frac{\mathrm{d}^4 k}{(2\pi)^4} f(p,k). \tag{20.4.5}$$

It turns out that for $d < 4$ in the above case (and in the case of a more divergent integral, for d sufficiently small) the formula for $F_{(d)}(p)$ makes perfectly good sense. Moreover it can be continued analytically to larger d, and, not surprisingly, something striking happens at $d = 4$ as a souvenir of the fact that (20.4.5) diverged. One finds that $F_{(d)}$ has a pole in the d-plane at $d = 4$. In practice the pole appears in a specific form. One finds

$$F_{(d)}(p) = (4\pi)^{2-d/2}\Gamma(2 - d/2)G_d(p) \tag{20.4.6}$$

where $G_d(p)$ is well behaved at $d = 4$ and the pole at $d = 4$ is hidden in the Γ-function. [Recall that $\Gamma(z) = 1/z$ as $z \to 0$.] But so long as we keep $d < 4$ and small enough we are dealing with finite quantities. Thus for a divergent Feynman integral we define the *regularized* integral by (20.4.3) with d small enough. The renormalization is carried out keeping $d < 4$ and at the very end we led $\epsilon \equiv 2 - d/2 \to 0$.

It should be clear that *any* definition of $F_{(d)}(p)$ could be altered by multiplying it by a function of d which equals one at $d = 4$. If we think of the above as an actual field theory in d dimensions, then we must keep the action of the d dimensional theory dimensionless (in natural units) and this will imply that the coupling constant in d dimensions picks up some extra dimensions of mass to some power. This will be achieved in the next section by multiplying g, which in the theories of interest is dimensionless in four dimensions, by a suitable power of μ^{4-d}, where μ is an *arbitrary* mass parameter. Thus the typical Feynman integral above will appear multiplied by some power of μ^{4-d}. It is easy to see (see Section 1.1) that one needs the following changes in going from four to d dimensions:

$$\left.\begin{array}{lll} \text{scalar } \phi^4 \text{ theory}: & & g \to \mu^{4-d} g \\ \text{QED}: & & e \to \mu^{2-d/2} e \\ \text{QCD}: & & g \to \mu^{2-d/2} g \end{array}\right\} \tag{20.4.7}$$

where the e and g on the RHS are still dimensionless. It is then convenient to alter the equation for g_B in (20.3.4) for ϕ^4 theory, to

$$g_B = Z_\phi^{-2} Z_g \mu^{2\epsilon} g_{[\mu]} \tag{20.4.8}$$

with analogous changes for QED amd QCD, so that the $Z(\epsilon)$ factors remain dimensionless. The vertex in the ϕ scheme given in Table 20.1 will now become

$$\times \quad = \quad -i\mu^{2\epsilon} Z_g g_{[\mu]}. \tag{20.4.9}$$

vertex

The subscript label $[\mu]$ reminds us that the definition of the renormalized g depends explicitly on our choice of the arbitrary mass parameter μ.

Note that when dealing with fermions it is necessary to extend the definition of the Dirac matrices γ^μ to d dimensions. In general this is not a problem, the only exception being γ_5 whose properties, when extended simplistically to d dimensions, lead to the occurrence of non-analytic functions of d. Thus if we ask of $\gamma_5^{(d)}$ that it anticommutes with all γ^μ, $\mu = 0, 1 \ldots, d-1$, we find e.g. $\mathrm{Tr}(\gamma_5^{(d)}) = 0$ *except* at $d = 0$, $\mathrm{Tr}(\gamma_5^{(d)}\gamma_\mu\gamma_\nu) = 0$ *except* at $d = 2$, etc. [see Section 13.2 of Collins (1984)]. Thus the typical traces that occur in calculations are non-analytic (strictly, non-meromorphic) functions of d. 'tHooft and Veltman (1972) avoid this dilemma by requiring $\gamma_5^{(d)}$ to anticommute with γ^μ for $\mu = 0, 1, 2, 3$, but to *commute* with γ^μ for $\mu \geq 4$. There is thus a loss of Lorentz invariance in the d-dimensional Minkowski space, but that turns out to be innocuous. [Other approaches exist. See, for example, Chanowitz, Furman and Hinchcliffe (1979).]

> Recall that the infamous triangle anomaly discussed in Section 9.5.3 was linked to the occurrence of axial, i.e. $\gamma_5\gamma^\mu$ type, currents. The bizarre feature, that the result is dependent upon how the Feynman diagram is labelled, occurs when one uses Pauli–Villars regularization. When dimensional regularization is used the result is also ambiguous; this time it depends upon how one manipulates γ_5 inside the traces that occur.

Finally we mention a totally different approach to regularization which is useful in non-perturbative studies. Since large momenta correspond to short distances, the ultraviolet divergences that arise in the Feynman integrals can be shown to be linked to the singular behaviour of products of field operators like $\phi(x)\phi(y)$ when $x - y \to 0$. In the *lattice regularization*, space-time is discretized into elementary hypercubes of side 'a'. As long as $a \neq 0$ there are no ultraviolet divergences and the theory with $a \neq 0$ is regularized. Clearly by working with cubes we lose rotational invariance, and care has to be exercised in passing ultimately to the limit $a \to 0$. This will be briefly discussed in Section 27.2.

20.5 Choice of renormalization scheme

We present the following in terms of a scalar field $\phi(x)$ but the whole approach applies equally well to QED and QCD.

It was remarked above that *S-matrix* elements do not depend upon the 'scheme' used in their calculation. In a field theory it is also important to deal with two other kinds of amplitudes, the n-leg momentum space Green's functions $G^{(n)}$ and the amputated Green's functions $\Gamma^{(n)}$. The $G^{(n)}$ in coordinate space correspond to the vacuum expectation value of a time-ordered product of n fields $\phi(x)$, i.e. $\langle 0|T[\phi(x_1)\phi(x_2)\dots\phi(x_n)]|0\rangle$. In momentum space they are represented by Feynman diagrams with n external legs, as shown, and with a single particle propagator appearing for each leg.

$G^{(n)} =$

Because of translational invariance $G^{(n)}$ depends on only $(n-1)$ momenta and can be defined as the Fourier transform with respect to $x_1, x_2, \dots, x_{n-1}$ of the time-ordered product with x_n put equal to zero.

The $\Gamma^{(n)}$ are just the $G^{(n)}$ without the propagators for the external legs—hence the nomenclature 'amputated'. They are depicted thus.

$\Gamma^{(n)} =$

Clearly $G^{(n)}_{\mathrm{B}}$ and $G^{(n)}_{\phi}$ computed in the ϕ_{B} and the ϕ schemes will, via (20.3.4), be related by

$$\begin{aligned} G^{(n)}_{\phi} &= \text{FT of } \langle 0|T(\phi\dots)|0\rangle = \text{FT of } Z_{\phi}^{-n/2}\langle 0|T(\phi_{\mathrm{B}}\dots)|0\rangle \\ &= Z_{\phi}^{-n/2} G^{(n)}_{\mathrm{B}}, \end{aligned} \tag{20.5.1}$$

where FT $\Rightarrow$ Fourier transform.

Since $\Gamma^{(n)}$ is obtained from $G^{(n)}$ by dividing by a product of n single particle propagators, each of the form $\langle 0|T(\phi\phi)|0\rangle$, the analogue of (20.5.1) will be

$$\Gamma_\phi^{(n)} = Z_\phi^{n/2}\Gamma_{\rm B}^{(n)}. \tag{20.5.2}$$

We shall now examine the consequences of the freedom we have in fixing the finite parts of the $Z(\lambda)$ functions. Giving a prescription for the finite parts constitutes a choice of a *renormalization scheme.* In (20.3.16 and 17) we did this by choosing the point $p^2 = m_{\rm R}^2$, where $m_{\rm R}$ is the *physical* mass, and demanding a specific form for the complete propagator and complete vertex at this point (the on-shell scheme). But the choice of $p^2 = m_{\rm R}^2$, though convenient, was arbitrary. We are free to choose any point $p^2 = -\mu^2$ provided we don't choose a point where the amplitudes possess singularities. The unphysical (Euclidean) region $p^2 < 0$ is safe.

Each choice of μ fixes a different scheme, so that m, g and ϕ should carry a label μ which is usually left out in the literature, i.e. m_μ, g_μ, ϕ_μ.

But there is an even greater freedom. The above method of demanding a specific form for the complete propagator and vertex at some point in momentum space is just one method of specifying the renormalization scheme. We shall call it the *momentum point subtraction* (MPS) scheme. Later we shall deal with more modern and powerful schemes which are specifically linked to *dimensional regularization* of which two, the *minimal subtraction* (MS) and *modified minimal subtraction* ($\overline{\rm MS}$) schemes, are very popular.

This implies that our renormalized m, g, and ϕ really ought to have two labels (!)—one to indicate the scheme, the other to indicate the dependence on the arbitrary mass parameter μ. To avoid an unholy printing mess we shall leave out these labels unless they are absolutely essential.

In the next two sections we outline the general ideas. A concrete example is given in Section 20.7.

20.5.1 The momentum point subtraction (MPS) scheme

Analogously to (20.3.17) for the complete vertex $\Gamma^{(4)}$, we can demand that

$$\Gamma^{(4)}(p_1,p_2,p_3,p_4) \equiv -\mathrm{i}g_\mu \quad \text{at } p_j^2 = -\mu^2, \quad p_i\cdot p_j = \mu^2/3, \quad i \neq j. \tag{20.5.3}$$

Let us call these values of the momentum $p_j^0, j = 1, \ldots, 4$.

We cannot ask the complete propagator $G^{(2)}$ to have a pole at $p^2 = -\mu^2$ but we can for example take

$$G^{(2)}(p) \to \frac{\mathrm{i}}{p^2 - m_\mu^2} \quad \text{for} \quad p^2 \to -\mu^2. \tag{20.5.4}$$

Note that (20.5.4) does *not* imply that $G^{(2)}(p)$ has a pole at $p^2 = m_\mu^2$, so m_μ^2 is *not* the physical mass. The mass μ, which also has nothing to do with the mass of the physical particles, is referred to as the 'renormalization point' or the 'subtraction point'. It is the point at which the infinities are subtracted out, as will become clear shortly. Note that the *on-shell* scheme is just a special case of the MPS schemes in which one takes $\mu^2 = -m_R^2$, m_R being the physical mass. As mentioned we are suppressing the label 'MPS' which should be attached to g_μ, m_μ and the $Z(\lambda)$.

To see that μ^2 is the point at which the infinities are subtracted out, consider the calculation of the propagator $G^{(2)}(p)$ in an MPS scheme. It is given diagrammatically by

$G^{(2)}(p) \equiv$

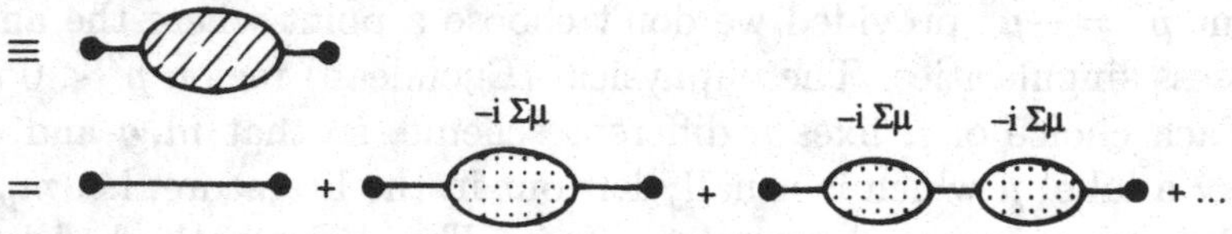

$=$ ... $+$ ... $+$ ... $+ \ldots$

-i Σμ -i Σμ -i Σμ

where the 'blobs'

$-\mathrm{i}\Sigma\mu \equiv$

are 'one-particle irreducible', i.e. cannot be split into two pieces by cutting one single internal line. Σ is called the *self-energy*. The series for $G^{(2)}(p)$ is a geometric one and sums to

$$G^{(2)}(p) = \frac{\mathrm{i}}{Z_\phi p^2 - Z_m m^2 - \Sigma(p^2, m^2)}. \tag{20.5.5}$$

All divergent loop integrals reside in $\Sigma(p^2, m^2)$ which is then regulated using the cut-off λ, but for visual clarity we do not indicate that the Σ etc. depend upon λ. Expanding Σ about the point $p^2 = -\mu^2$

$$G^{(2)}(p) = \frac{\mathrm{i}}{Z_\phi p^2 - Z_m m^2 - \left[\Sigma(-\mu^2) + (p^2 + \mu^2)\Sigma'(-\mu^2) + \tilde{\Sigma}(p^2)\right]}, \tag{20.5.6}$$

where $\Sigma' = \partial\Sigma/\partial p^2$, and $\tilde{\Sigma}(p^2)$ is the rest of the Taylor series for $\Sigma(p^2)$, and vanishes like $(p^2 + \mu^2)^2$ as $p^2 \to -\mu^2$. To satisfy (20.5.4) we now choose in the MPS scheme

$$\left.\begin{aligned} Z_{\phi_\mu} &= 1 + \Sigma'(-\mu^2, m_\mu^2) \\ m_\mu^2(Z_{m_\mu} - 1) &= -\mu^2(Z_{\phi_\mu} - 1) - \Sigma(-\mu^2, m_\mu^2) \end{aligned}\right\} \tag{20.5.7}$$

which fixes Z_{ϕ_μ} and Z_{m_μ} in terms of the quantities Σ and Σ', which are both infinite as $\lambda \to \infty$. The crucial point is that one can show that

$$\tilde{\Sigma}_{(\mu)}^{\text{MPS}}(p^2, m_\mu^2) \equiv \Sigma(p^2, m_\mu^2) - \Sigma(-\mu^2, m_\mu^2) - (p^2+\mu^2)\Sigma'(-\mu^2, m_\mu^2) \tag{20.5.8}$$

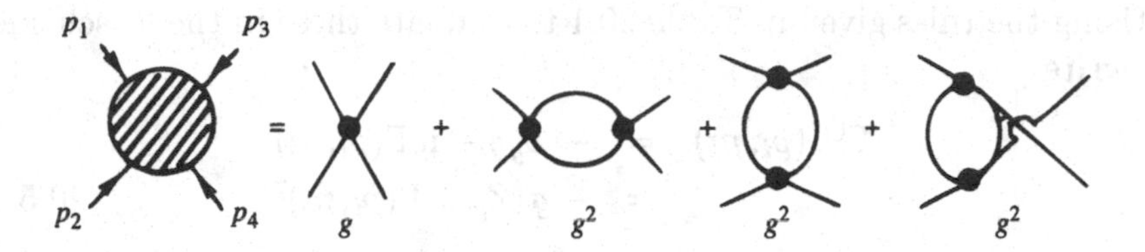

Fig. 20.3. Contributions to the complete vertex in ϕ^4 field theory.

is *finite* as $\lambda \to \infty$, i.e. subtracting the first two terms in the Taylor expansion about the point $p^2 = -\mu^2$ has eliminated the infinite parts of $\Sigma(p^2)$. Using (20.5.7) and (20.5.8) in (20.5.5) gives

$$G^{(2)}(p) = \frac{\mathrm{i}}{p^2 - m_\mu^2 - \tilde{\Sigma}_{(\mu)}(p^2, m_\mu^2)} \tag{20.5.9}$$

which is finite as $\lambda \to \infty$. We see that the point $p^2 = -\mu^2$ is the point at which the infinite subtraction is made which renders the results finite. $G^{(2)}(p)$ and $\tilde{\Sigma}_{(\mu)}(p^2)$ are referred to as the renormalized propagator and self-energy, respectively in the MPS scheme, with μ as renormalization point. We stress again that m_μ is not the physical mass. If we calculated $\tilde{\Sigma}_{(\mu)}$ to some order of perturbation theory and then found where $G^{(2)}(p)$ had a pole we would obtain a relation between m_μ and the physical mass m_{R}, valid to that order of perturbation theory. In QCD, since quarks do not occur as physical particles, we expect that the *exact* quark $G^{(2)}(p)$ does not have any poles.

There is a subtle point involved in the practical use of (20.5.9). In our calculation using the 'ϕ schemes' $\Sigma(p^2, m^2)$ will involve Z_ϕ, Z_m and Z_g according to Table 20.1. But $\Sigma(p^2, m^2)$ is explicitly proportional to g^2. (Of course it contains terms with higher powers of g as well.) Thus if we work only to order g^2 we may replace all the Zs by 1 in $\Sigma(p^2, m^2)$. So $\tilde{\Sigma}_{(\mu)}(p^2, m^2)$, to order g^2, will be of the form

$$\tilde{\Sigma}_{(\mu)}(p^2, m_\mu^2) = g_\mu^2 \times (\text{function of } p^2, \mu^2, m_\mu^2). \tag{20.5.10}$$

To order g^2 it will be as if $\tilde{\Sigma}_{(\mu)}(p^2, m_\mu^2)$ were calculated using the Feynman rules of Table 20.1 with $Z_\phi = Z_m = Z_g = 1$.

To get results to higher order, say g^n, the $Z(\lambda)$ used inside $\Sigma(p^2, m^2)$ need only be correct to order g^{n-2}.

Consider now the fundamental complete vertices or amputated Green's functions of the theory. The number of external lines involved will depend upon the particular theory. For simplicity let us discuss ϕ^4 field theory. The fundamental vertex is then the 4-point function $\Gamma^{(4)}(p_i)$. Some contributions and their perturbative order are shown in Fig. 20.3.

Using the rules given in Table 20.1 to evaluate these in the 'ϕ schemes', we write

$$\begin{aligned} \Gamma^{(4)}(p_i, m) &= -\mathrm{i}Z_g g - \mathrm{i}g\Gamma(p_i, m) \\ &= -\mathrm{i}g\left[Z_g + \Gamma(p_i, m)\right] \end{aligned} \tag{20.5.11}$$

where $\Gamma(p_i, m)$ is explicitly proportional to g, since we have extracted one factor $(-\mathrm{i}g)$ for later convenience. Of course $\Gamma(p_i, m)$ also contains terms of higher order than g.

Now the Feynman integral in $\Gamma(p_i, m)$ can be shown to diverge logarithmically and is regulated using the cut-off λ. Expanding $\Gamma(p_i, m)$ about the renormalization point $p_j = p_j^0$ where $p_j^{0\,2} = -\mu^2$, $p_i^0 \cdot p_j^0 = \mu^2/3$,

$$\Gamma(p_i, m) = \Gamma(p_i^0, m) + \tilde{\Gamma}_{(\mu)}(p_i, m), \tag{20.5.12}$$

it is found that $\tilde{\Gamma}_{(\mu)}(p_i, m)$ is finite as $\lambda \to \infty$. We then have from (20.5.11)

$$\Gamma^{(4)}(p_i, m) = -\mathrm{i}g\left\{Z_g + \Gamma(p_i^0, m) + \tilde{\Gamma}_{(\mu)}(p_i, m)\right\}, \tag{20.5.13}$$

where

$$\tilde{\Gamma}_{(\mu)}(p_i, m) = 0 \text{ at the renormalization point } p_i^2 = -\mu^2, p_i \cdot p_j = \mu^2/3. \tag{20.5.14}$$

The MPS scheme is chosen by demanding that (20.5.3) holds, i.e.

$$-\mathrm{i}g_\mu = -\mathrm{i}g_\mu\left\{Z_{g_\mu} + \Gamma(p_i^0, m_\mu)\right\} \qquad \left(p_i^{0\,2} = -\mu^2\right) \tag{20.5.15}$$

so that the infinite Z_{g_μ} is given by

$$Z_{g_\mu} = 1 - \Gamma(p_i^0, m_\mu). \tag{20.5.16}$$

Again we see that the renormalization point is the point at which the infinity is subtracted out. The finite, renormalized vertex in the MPS scheme is then given by

$$\Gamma^{(4)}(p_i, m_\mu) = -\mathrm{i}g_\mu - \mathrm{i}g_\mu\tilde{\Gamma}_{(\mu)}(p_i, m_\mu). \tag{20.5.17}$$

As with the self-energy, $\tilde{\Gamma}_{(\mu)}$ is to order g_μ calculated with $Z_\phi = Z_g = Z_m = 1$ in it. Thus

$$\tilde{\Gamma}_{(\mu)}(p_i, m_\mu) = g_\mu \times (\text{function of } p_i, \mu^2, m_\mu). \tag{20.5.18}$$

If $\Gamma(p_i, m)$ is to be calculated to higher order, say g^n, the $Z(\lambda)$ used need only be correct to order g^{n-1}.

By doing the above derivations in reverse it is easy to see that an equivalent way to specify the renormalization scheme defined by (20.5.3 and 4) is just to specify *what* infinite piece is to be subtracted from the self-energy Σ and the vertex Γ, i.e. *defining* the renormalized self-energy and vertex in the MPS scheme with p_i^0 ($p_i^{0\,2} = -\mu^2, p_i^0 \cdot p_j^0 = \mu^2/3$) as subtraction point, by

$$\begin{aligned}\tilde{\Sigma}_{(\mu)}^{\mathrm{MPS}}(p^2, m_\mu^2) &\equiv \lim_{\lambda\to\infty}\Big\{\Sigma(p^2, m_\mu^2) - \Sigma(-\mu^2, m_\mu^2) \\ &\qquad -(p^2+\mu^2)\Sigma'(-\mu^2, m_\mu^2)\Big\} \quad (20.5.19)\\ \tilde{\Gamma}_{(\mu)}^{\mathrm{MPS}}(p_i, m_\mu) &\equiv \lim_{\lambda\to\infty}\Big\{\Gamma(p_i, m_\mu) - \Gamma(p_i^0, m_\mu)\Big\}. \quad (20.5.20)\end{aligned}$$

Then choosing the $Z(\lambda)$ to cancel the infinite pieces will lead to (20.5.3 and 4). For renormalization schemes which are *not* based on a Taylor-type expansion about some value of the external momenta, the above is the most convenient way to define the scheme. The subtraction schemes based on dimensional regularization to which we now turn are of this type.

20.5.2 *Renormalization schemes specifically linked to dimensional regularization (DR)*

As mentioned in Section 20.4, in *dimensional regularization* the divergence of a Feynman integral in four dimensions shows up as a pole when $F_{(d)}$ is continued to $d = 4$. The pole occurs in a Γ-function, as will be seen in the examples in Section 20.7. For the 4-point vertex in ϕ^4 theory, in analogy to (20.5.11) we write (we are suppressing the label 'DR')

$$\Gamma^{(4)}(p_i, m) = -ig_{[\mu]}\mu^{2\epsilon}\Big\{Z_{g_{[\mu]}} + \Gamma_\epsilon(p_i, m, \mu)\Big\} \quad (20.5.21)$$

which is similar to (20.5.11), bearing in mind that we must replace $g \to \mu^{2\epsilon}g$ [see eqn (20.4.7)]. It follows that $\Gamma_\epsilon(p_i, m, \mu)$ is proportional to $\mu^{2\epsilon}g$.

The expression for $\Gamma_\epsilon(p_i, m, \mu)$, coming from the diagrams in Fig. 20.3, is logarithmically divergent. One then finds a result of the form

$$\Gamma_\epsilon(p_i, m, \mu) = (4\pi)^\epsilon\Gamma(\epsilon)G_{4-2\epsilon}(p_i, m, \mu). \quad (20.5.22)$$

One factor of $\mu^{2\epsilon}$ has been absorbed into $G_{4-2\epsilon}$ to make it dimensionless. Most importantly, $G_{4-2\epsilon}$ is finite at $\epsilon = 0$.

For $\epsilon \to 0$ we use

$$\Gamma(\epsilon) = 1/\epsilon - \gamma_E + 0(\epsilon) \quad (20.5.23)$$

where $\gamma_E = 0.5772\ldots$ is the Euler–Mascheroni constant. We also expand

$$\begin{aligned}(4\pi)^\epsilon &= \mathrm{e}^{\epsilon\ln(4\pi)}\\ &= 1 + \epsilon\ln(4\pi) + 0(\epsilon^2) \quad (20.5.24)\end{aligned}$$

and

$$G_{4-2\epsilon} = G_4 - 2\epsilon G_4' + 0(\epsilon^2) \quad (20.5.25)$$

where $G' = \partial G/\partial d$.

Then (20.5.22) becomes, as $\epsilon \to 0$,

$$\Gamma_\epsilon(p_i, m, \mu) = [1/\epsilon - \gamma_E + \ln 4\pi]\, G_4 - 2G_4'(p_i, m, \mu) + 0(\epsilon). \quad (20.5.26)$$

The renormalized $\tilde{\Gamma}$ is defined by subtracting a term which is singular as $\epsilon \to 0$, but different prescriptions exist for what precisely one subtracts. In some scheme, label it 's', let us split $\Gamma_\epsilon(p_i, m, \mu)$ into a finite piece $\tilde{\Gamma}_s$ and a piece Γ_s^∞ which becomes infinite when $\epsilon \to 0$, i.e.

$$\Gamma_\epsilon(p_i, m, \mu) = \Gamma_s^\infty + \tilde{\Gamma}_s(p_i, m, \mu). \quad (20.5.27)$$

Substituting in (20.5.21) we now fix the particular scheme by demanding that, for $\epsilon = 0$,

$$\Gamma^{(4)}(p_i, m, \mu) = -ig_s \left\{1 + \tilde{\Gamma}_s(p_i, m, \mu)\right\} \quad (20.5.28)$$

where we have suppressed the μ-dependence of g_s. Thus in the 's scheme'

$$Z_{gs} = 1 - \Gamma_s^\infty. \quad (20.5.29)$$

There are two popular schemes. In the *minimal subtraction* (MS) scheme ('tHooft, 1973) we define

$$\Gamma_{\mathrm{MS}}^\infty \equiv \tfrac{1}{\epsilon} G_4 \quad (20.5.30)$$

i.e. we subtract out just the pole and its residue. However, since the combination $-\gamma_E + \ln 4\pi$ inevitably appears, it is convenient to get rid of it. So in the *modified minimal subtraction* ($\overline{\mathrm{MS}}$) scheme (Bardeen, Buras, Duke and Muta, 1978) we define

$$\Gamma_{\overline{\mathrm{MS}}}^\infty \equiv (1/\epsilon - \gamma_E + \ln 4\pi)\, G_4. \quad (20.5.31)$$

It should be clear that in all the above schemes the renormalization factors $Z(\epsilon)$ will be different. An example will be given in Section 20.7. Consequently the g_μ defined in (20.5.3) and the $g_{[\mu]}$ defined above in each scheme are all different. For the dimensional regularization schemes the $g_{[\mu]}$ should, strictly, carry a label MS or $\overline{\mathrm{MS}}$.

Applying the DR method of regularization and renormalization to the self-energy $\Sigma(p)$ in ϕ^4 theory one finds results of the form:

$$\Sigma_\epsilon(p^2, m^2) = A_\epsilon p^2 - B_\epsilon m^2 \quad (20.5.32)$$

where both A_ϵ and B_ϵ have infinite parts as $\epsilon \to 0$. In our generic 's scheme' one isolates the infinite parts (call them A_s^∞ and B_s^∞) and then has

$$\Sigma(p^2, m^2) = A_s^\infty p^2 - B_s^\infty m^2 + \tilde{\Sigma}_s(p^2, m^2, \mu^2) \quad (20.5.33)$$

where $\tilde{\Sigma}_s$ is finite. The factors Z_{ϕ_s} and Z_{m_s} are then chosen so that

$G^{(2)}(p)$ has the form given in (20.5.9). This yields, via (20.5.5),

$$\begin{aligned} Z_{\phi_s} &= 1 + A_s^\infty \\ Z_{m_s} &= 1 + B_s^\infty . \end{aligned} \tag{20.5.34}$$

The MS and $\overline{\text{MS}}$ prescriptions are specified by choosing A_s^∞ and B_s^∞ analogously either to (20.5.30) or (20.5.31) respectively.

For QED and QCD essentially the same procedure is carried out. Of course the analysis is complicated by the γ-algebra. Details can be found in Sections 9.6 to 9.9 of Ryder (1985).

20.6 The renormalization group

For concreteness let us focus on an MPS type renormalization scheme to begin with. It will prove convenient to switch from the coupling constant to the analogue of the fine structure constant, i.e. to utilize

$$\alpha_{\text{B}} \equiv g_{\text{B}}^2/4\pi \qquad \text{and} \qquad \alpha_\mu \equiv g_\mu^2/4\pi \tag{20.6.1}$$

instead of g_{B} and g_μ. (This is in keeping with recent papers on the subject.)

Consider now the relationship betweeen $\Gamma_{\text{B}}^{(n)}$ and $\Gamma^{(n)}$ calculated in the ϕ_μ scheme, which we shall label $\Gamma_{(\mu)}^{(n)}$. Putting in all the arguments, we write (20.5.2) in the form

$$\Gamma_{\text{B}}^{(n)}(p_1 \dots p_n; m_{\text{B}}, \alpha_{\text{B}}, \lambda) = Z_{\phi_\mu}^{-n/2}\left(\frac{\lambda}{\mu}, \frac{m_\mu}{\mu}, \alpha_\mu\right) \Gamma_{(\mu)}^{(n)}(p_1 \dots p_n; m_\mu, \alpha_\mu), \tag{20.6.2}$$

where we have used the fact that Z_{ϕ_μ} is dimensionless to write it in terms of ratios of dimensional parameters.

In the above we are dealing with a cut-off parameter λ. But the following discussion, with minor modifications, holds also in the dimensional regularization scheme with parameter ϵ.

The LHS of (20.6.2) is independent of μ. If, therefore, we take the derivative of (20.6.2) with respect to μ keeping $m_{\text{B}}, \alpha_{\text{B}}$ and λ fixed, we obtain

$$\begin{aligned} 0 &= -\frac{n}{2} Z_{\phi_\mu}^{-n/2-1} \Gamma_{(\mu)}^{(n)} \frac{\mathrm{d}}{\mathrm{d}\mu} Z_{\phi_\mu} + Z_{\phi_\mu}^{-n/2} \frac{\mathrm{d}}{\mathrm{d}\mu} \Gamma_{(\mu)}^{(n)} \\ &= Z_{\phi_\mu}^{-n/2} \left[\frac{-n}{2Z_{\phi_\mu}} \Gamma_{(\mu)}^{(n)} \frac{\mathrm{d}}{\mathrm{d}\mu} Z_{\phi_\mu} + \left(\frac{\partial}{\partial \mu} + \frac{\mathrm{d}m_\mu}{\mathrm{d}\mu} \frac{\partial}{\partial m_\mu} + \frac{\mathrm{d}\alpha_\mu}{\mathrm{d}\mu} \frac{\partial}{\partial \alpha_\mu} \right) \Gamma_{(\mu)}^{(n)} \right] . \end{aligned} \tag{20.6.3}$$

Cancelling the $Z_{\phi_\mu}^{-n/2}$ factor, and multiplying by μ for later convenience, we get the *renormalization group equation* [the original ideas stem from

Stueckelberg and Peterman (1953) and Gell-Mann and Low (1954)]:

$$\left(\mu\frac{\partial}{\partial\mu}+2\beta\frac{\partial}{\partial\alpha_\mu}-n\gamma\right)\Gamma^{(n)}_{(\mu)}=-\mu\left(\frac{\mathrm{d}m_\mu}{\mathrm{d}\mu}\right)\frac{\partial}{\partial m_\mu}\Gamma^{(n)}_{(\mu)},\tag{20.6.4}$$

where β and γ are short for

$$\beta^{\mathrm{MPS}}(\alpha_\mu)\equiv\frac{\mu}{2}\frac{\mathrm{d}\alpha_\mu}{\mathrm{d}\mu}=\frac{\alpha_{\mathrm{B}}\mu}{2}\frac{\mathrm{d}}{\mathrm{d}\mu}\left[Z^2_{\phi_\mu}Z^{-1}_{g_\mu}\right]^2\tag{20.6.5}$$

in which we have used (20.3.4), and

$$\gamma^{\mathrm{MPS}}(\alpha_\mu)\equiv\frac{\mu}{2Z_{\phi_\mu}}\frac{\mathrm{d}}{\mathrm{d}\mu}Z_{\phi_\mu}=\frac{\mu}{2}\frac{\mathrm{d}}{\mathrm{d}\mu}\left(\ln Z_{\phi_\mu}\right).\tag{20.6.6}$$

[Note that often $(\mu/2)\mathrm{d}/\mathrm{d}\mu$ is written $\mathrm{d}/\mathrm{d}\ln(\mu^2)$ in the literature.]

Equation (20.6.4), and the functions occurring in it, are derived at fixed λ. At the end, however, it is understood that the limit $\lambda\to\infty$ is taken.

In the above we started with a theory with non-zero mass m_{B}. In that case it is actually an unnecessary luxury to have two parameters with the dimensions of mass. Indeed it complicates matters; because while one can show that when $\lambda\to\infty$ the functions β and γ are finite they could, being dimensionless, still depend upon the ratio m_μ/μ as well as upon α_μ.

Thus we may as well use the point $p^2=-m^2_\mu$ as the renormalization point, i.e. take $m_\mu=\mu$. In that case we cannot keep m_{B} fixed as we vary μ and we get the analogue of (20.6.4), the Callan–Symanzik equation,

$$\left(\mu\frac{\partial}{\partial\mu}+2\beta\frac{\partial}{\partial\alpha_\mu}-n\gamma\right)\Gamma^{(n)}_{(\mu)}=Z^{n/2}_{\phi_\mu}\left(\mu\frac{\mathrm{d}m_{\mathrm{B}}}{\mathrm{d}\mu}\right)\frac{\partial}{\partial m_{\mathrm{B}}}\Gamma^{(n)}_{\mathrm{B}}.\tag{20.6.7}$$

This is particularly useful at large momenta, i.e. at momenta $p'_j=\eta p_j$ when $\eta\to\infty$, with p_j an arbitrary fixed set of momentum values. For then it can be shown that the RHS of (20.6.7) can be neglected—it provides corrections of order η^{-1}. So

$$\left(\mu\frac{\partial}{\partial\mu}+2\beta\frac{\partial}{\partial\alpha_\mu}-n\gamma\right)\Gamma^{(n)}_{(\mu)}(\eta p_j;\alpha_\mu,\mu)=0;\quad \eta\to\infty.\tag{20.6.8}$$

In QCD, or in any theory with massless particles, we start with a bare Lagrangian that has no mass parameter m_{B}. The renormalization involves a mass parameter μ which is then the only mass parameter and there is no RHS to (20.6.4), i.e. in a massless theory one has for all values of momentum:

$$\left(\mu\frac{\partial}{\partial\mu}+2\beta\frac{\partial}{\partial\alpha_\mu}-n\gamma\right)\Gamma^{(n)}_{(\mu)}(p_j;\alpha_\mu,\mu)=0.\tag{20.6.9}$$

[Note that the analogue of this for $G^{(n)}_{(\mu)}$ is obtained by simply replacing $n\gamma$ by $-n\gamma$, as can be seen by comparing (20.5.1 and 2).]

In eqns (20.6.7 and 8), since there are only two mass parameters, λ and μ, *after* the limit $\lambda \to \infty$ is taken there is no way β and γ can depend explicitly on μ. Thus we have $\beta = \beta(\alpha_\mu), \gamma = \gamma(\alpha_\mu)$ only.

In dimensional renormalization (DR) schemes the analogue of (20.6.5) is, via (20.4.8)

$$\beta^{\mathrm{DR}}(\alpha_\mu) \equiv \frac{\mu}{2}\frac{\mathrm{d}\alpha_\mu}{\mathrm{d}\mu} = \frac{\alpha_{\mathrm{B}}\mu}{2}\frac{\mathrm{d}}{\mathrm{d}\mu}\left[Z_{\phi_\mu}^2 Z_{g_\mu}^{-1}\mu^{-2\epsilon}\right]^2. \qquad (20.6.10)$$

It is understood that at the end the limit $\epsilon \to 0$ is taken. In these schemes ϵ is dimensionless, there is only one mass parameter μ, so that the Z factors, being dimensionless, cannot depend upon μ. Such a scheme is called a *mass-independent* renormalization scheme. Although β depends upon μ for $\epsilon \neq 0$ in the limit $\epsilon \to 0$ one has $\beta = \beta(\alpha), \gamma = \gamma(\alpha)$ only.

In the theory of most interest to us, QCD, the quarks do not possess a mass in the conventional sense, since they do not exist as free particles. Nonetheless, as is discussed in Section 21.3, they are each characterized by a mass parameter in the effective QCD Lagragian. Thus there are several different mass parameters in the Lagrangian, one for each quark. For u, d and s the values of the mass parameters are very small on the scale of a typical high energy (multi-GeV) experiment and it is conventional to treat these quarks as massless. This may be dangerous for the charm quark and is quite incorrect for bottom and top.

At this point we simply wish to note two important results:

1. Even in the presence of several mass parameters in the Lagrangian, β and γ in the MS and $\overline{\mathrm{MS}}$ renormalization schemes remain functions of α only, i.e. they are independent of the mass parameters. (This is not true in momentum point subtraction schemes.)

2. The renormalization group equation (20.6.9) will hold for momentum scales much larger than the largest renormalized mass parameter in the Lagrangian.

For simplicity, we shall only discuss the massless version of the renormalization group equation in this chapter. Later, when we deal with QCD in Sections 21.5, 6 and 7 we shall learn that it is possible to make allowance for the wide range of quark mass parameters and we shall then see how to use the renormalization group techniques below the hopelessly high scale $p^2 \gg m_t^2$!

The equations (20.6.8) and (20.6.9) are useful because they are *exact* consequences of the theory. They will not generally be satisfied by a $\Gamma^{(n)}$ calculated to a given order in perturbation theory, and they can be used to 'improve' the results of a perturbative calculation. They are also remarkable in that they hold for all the functions $\Gamma^{(n)}$, yet β and γ

are fixed functions that are independent of which Green's function one is studying. To actually compute β and γ to a given order in perturbation theory one must go back to the definitions (20.6.5 or 10) and (20.6.6) and calculate the relevant self-energy or vertex diagrams involved in specifying the $Z(\lambda)$ or the $Z(\epsilon)$.

In order to have a somewhat more concrete picture of the above, we shall illustrate it with an example from ϕ^4 theory in the next section.

20.7 A concrete example of different renormalization schemes

Consider, for simplicity, the scalar ϕ^4 theory. From a straightforward calculation of $\Gamma(p_i)$ of (20.5.11) arising from the order g^2 diagrams in Fig. 20.3 (see the Appendix to this chapter, Section 20.10.4) using Pauli–Villars (PV) regularization one finds that at the subtraction point $p_i = p_i^0$ ($p_i^{0\,2} = -\mu^2, p_i^0 \cdot p_j^0 = \mu^2/3$)

$$\begin{aligned}\Gamma^{\mathrm{PV}}(p_i = p_i^0; m_\mu) \;=\; \frac{3g}{32\pi^2}\Bigg\{&-1 + \ln(\mu^2/\lambda^2) + \ln(m_\mu^2/\mu^2) + \\ &+ \sqrt{1+3m_\mu^2/\mu^2}\,\ln\left(\frac{\sqrt{1+3m_\mu^2/\mu^2}+1}{\sqrt{1+3m_\mu^2/\mu^2}-1}\right)\Bigg\}\end{aligned} \tag{20.7.1}$$

so that, from (20.5.16), to order g, in the momentum point subtraction scheme

$$Z_{g_\mu}^{\mathrm{MPS}} = 1 - \frac{3g}{32\pi^2}\left\{\ln(\mu^2/\lambda^2) + (\text{function of } m_\mu/\mu)\right\}. \tag{20.7.2}$$

If the Lagrangian is written with its operators in *normal order* form, then the self-energy $\Sigma(p)$ is of order g^2 from diagrams like

Thus to order g, from (20.5.7), $Z_{\phi_\mu} = 1$.

To order g^2 then (20.6.5) leads to

$$\begin{aligned}\beta^{\mathrm{MPS}} &= \frac{\alpha_{\mathrm{B}}}{2}\lim_{\lambda\to\infty}\mu\frac{\mathrm{d}}{\mathrm{d}\mu}\left\{1+\frac{3g}{32\pi^2}\left[\ln(\mu^2/\lambda^2) + (\text{function of } m_\mu/\mu)\right]\right\}^2 \\ &= \frac{\alpha_{\mathrm{B}}}{2}\left\{\frac{3g}{8\pi^2} + \frac{3g}{16\pi^2}\mu\frac{\mathrm{d}}{\mathrm{d}\mu}\left[\text{function of } m_\mu/\mu\right]\right\} + 0(g^4) \\ &= \frac{3}{8}\left(\frac{\alpha}{\pi}\right)^{3/2}\left\{1+\frac{\mu}{2}\frac{\mathrm{d}}{\mathrm{d}\mu}\left[\text{function of } m_\mu/\mu\right]\right\} + 0(\alpha^2)\end{aligned} \tag{20.7.3}$$

where we have used the fact that $g_B = g + 0(g^2)$. We have also interpreted the inverse of Z_g in a perturbative sense, i.e. $[1 - a(\lambda)g]^{-1} = 1 + a(\lambda)g + 0(g^2)$. At fixed λ this is justified for small enough g. It is always assumed that this can be done.

Equation (20.7.3) shows (a) that β is finite as $\lambda \to \infty$, and (b) that if we use a scheme with $\mu = m_\mu$ then

$$\beta^{\text{MPS}}(\alpha) = \frac{3}{8}\left(\frac{\alpha}{\pi}\right)^{3/2} + 0(\alpha^2) \quad [\mu = m_\mu \text{ scheme}] \tag{20.7.4}$$

is mass-independent.

In a dimensional regularization (DR) scheme the G_d which occurs in (20.5.22) is (see Appendix, Section 20.10.4)

$$\begin{aligned} G_d(p_i; m, \mu) &= -\frac{g}{32\pi^2}\int_0^1 \mathrm{d}z \left[\frac{\mu^2}{m^2 - z(1-z)s}\right]^{2-d/2} \\ &\quad + (s \to t) + (s \to u) \end{aligned} \tag{20.7.5}$$

where $s = (p_1 + p_2)^2, t = (p_1 + p_3)^2, u = (p_1 + p_4)^2$.

Thus

$$G_4 = -3g/32\pi^2 \tag{20.7.6}$$

and

$$\begin{aligned} \left.\frac{\partial G_d}{\partial d}\right|_{d=4} \equiv G_4' &= \frac{g}{64\pi^2}\int_0^1 \mathrm{d}z \ln\left[\frac{\mu^2}{m^2 - z(1-z)s}\right] \\ &\quad + (s \to t) + (s \to u). \end{aligned} \tag{20.7.7}$$

Comparing with (20.5.26) we have for Γ_ϵ

$$\begin{aligned} \Gamma_\epsilon^{\text{DR}}(p_i; m, \mu) &= -\frac{3g}{32\pi^2}\left[\frac{1}{\epsilon} - \gamma_\epsilon + \ln 4\pi\right] \\ &\quad - 2G_4'(p_i; m, \mu). \end{aligned} \tag{20.7.8}$$

In the MS scheme we take, according to (20.5.30)

$$\Gamma_{\text{MS}}^{\infty} = -\frac{3g}{32\pi^2}\cdot\frac{1}{\epsilon} \tag{20.7.9}$$

so that by (20.5.29)

$$Z_g^{\text{MS}} = 1 + \frac{3g}{32\pi^2}\cdot\frac{1}{\epsilon} \tag{20.7.10}$$

and via (20.5.27)

$$\tilde{\Gamma}_{\text{MS}}(p_i; m.\mu) = -\frac{3g}{32\pi^2}\left[\ln 4\pi - \gamma_E\right] - 2G_4'(p_i; m, \mu). \tag{20.7.11}$$

In the $\overline{\text{MS}}$ scheme we take, following (20.5.31)

$$\Gamma^{\infty}_{\overline{\text{MS}}} = -\frac{3g}{32\pi^2}\left[\frac{1}{\epsilon} - \gamma_E + \ln 4\pi\right] \tag{20.7.12}$$

so that from (20.5.29)

$$Z_g^{\overline{\text{MS}}} = 1 + \frac{3g}{32\pi^2}\left[\frac{1}{\epsilon} - \gamma_E + \ln 4\pi\right] \tag{20.7.13}$$

and from (20.5.27)

$$\tilde{\Gamma}_{\overline{\text{MS}}}(p_i; m, \mu) = -2G_4'(p_i; m, \mu). \tag{20.7.14}$$

In both cases, for β we have from (20.6.10), using $Z_\phi = 1$ to our accuracy

$$\begin{aligned} \beta^s(g_s) &= \frac{\alpha_{\text{B}}}{2}\lim_{\epsilon\to 0} Z_{g_s}^{-1}\mu\frac{\mathrm{d}}{\mathrm{d}\mu}(\mu^{-4\epsilon}) \\ &= -2\alpha_{\text{B}}\lim_{\epsilon\to 0}\left\{\epsilon Z_{g_s}^{-1}\mu^{-4\epsilon}\right\}. \end{aligned} \tag{20.7.15}$$

First expanding the quantities in parenthesis in powers of g_s and then taking the limit $\epsilon \to 0$ we see that $\beta^s(\alpha)$ is finite, and we obtain

$$\beta^{\text{MS}}(\alpha) = \beta^{\overline{\text{MS}}}(\alpha) = \frac{3}{8}\left(\frac{\alpha}{\pi}\right)^{3/2} + 0(\alpha^2). \tag{20.7.16}$$

We see that both are independent of the masses and that both have the same functional form as $\beta^{\text{MPS}}(\alpha)$ in the $\mu = m_\mu$ scheme. That the three functions are the same, to this order, is not an accident, as is explained in the next section. Of course the numerical value of these functions will differ since each must be used with the coupling constant appropriate to the renormalization scheme being utilized.

It is interesting to compare the above βs with the analogous functions in QED and QCD. One has

$$\begin{aligned} \text{QED}: \quad \beta(\alpha) &= \frac{1}{3\pi}\alpha^2 + 0(\alpha^3) \\ \text{QCD (with 3 flavours)}: \quad \beta(\alpha) &= -\frac{9}{4\pi}\alpha^2 + 0(\alpha^3) \end{aligned} \tag{20.7.17}$$

The fact that β^{QCD} is negative at small α will play a crucial rôle in determining the behaviour of reactions at large momentum transfer and will lead to the property of *asymptotic freedom.*

20.8 Consequences of the renormalization group equation

As explained we shall here stick to the massless version of the renormalization group equation (20.6.9). β and γ no longer depend explicitly upon μ, but their form does depend upon the regularization and renormalization scheme which we shall label 's'. To cast (20.6.9) into its most useful form we consider the mass dimension of $\Gamma^{(n)}$ and thereby relate the dependence on μ to the dependence on momentum. Since with $\hbar = c = 1, \phi(x)$ has dimension $[\mathrm{M}]^1$, it is clear that $\langle 0|T(\phi(x_1)\ldots\phi(x_n))|0\rangle$ has mass dimension n and therefore after taking the Fourier transform with respect to $(n-1)$ variables

$$[G^{(n)}] = [\mathrm{M}]^{n-4(n-1)} = [\mathrm{M}]^{4-3n} \tag{20.8.1}$$

and, similarly,

$$[\Gamma^{(n)}] = [\mathrm{M}]^{4-n}. \tag{20.8.2}$$

Since α_μ is dimensionless the function $\Gamma^{(n)}(\eta p_j; \alpha_\mu, \mu)$ must depend upon ηp_j and μ in such a way that

$$\left(\eta\frac{\partial}{\partial\eta} + \mu\frac{\partial}{\partial\mu}\right)\Gamma^{(n)} = (4-n)\Gamma^{(n)}. \tag{20.8.3}$$

If this is obscure the reader should write down an arbitrary function of ηp_j and μ with the correct mass dimension and check that (20.8.3) holds. This is just Euler's theorem on homogeneous functions.

Eliminating the $\mu(\partial/\partial\mu)\Gamma^{(n)}$ term between (20.6.9) and (20.8.3) yields the more useful result

$$\left(\eta\frac{\partial}{\partial\eta} - 2\beta_s\frac{\partial}{\partial\alpha_\mu} + (n-4) + n\gamma_s\right)\Gamma_s^{(n)}(\eta p_j; \alpha_\mu, \mu) = 0, \tag{20.8.4}$$

which relates the dependence on momentum to the dependence on α_μ.

In the following we shall drop the labels s and μ for notational simplicity.

The solution to (20.8.4) is found by introducing a new function

$$\bar{\alpha}(\alpha, t) \equiv \bar{g}^2(g, t)/4\pi \tag{20.8.5}$$

[where $\bar{g}(g,t)$ is sometimes oxymoronically referred to as the 'running coupling *constant*'] defined implicitly by

$$t \equiv 2\ln\eta = \int_\alpha^{\bar{\alpha}(\alpha,t)} \frac{\mathrm{d}x}{\beta(x)} \tag{20.8.6}$$

with

$$\bar{\alpha}(\alpha, t=0) = \alpha \tag{20.8.7}$$

where, at this stage, α is an *arbitrary constant.*

It follows by differentiating (20.8.6) with respect to η or t that

$$\eta\frac{\partial\bar{\alpha}(\alpha,t)}{\partial\eta} = 2\beta[\bar{\alpha}(\alpha,t)] \tag{20.8.8}$$

or

$$\frac{\partial\bar{\alpha}(\alpha,t)}{\partial t} = \beta[\bar{\alpha}(\alpha,t)]. \tag{20.8.9}$$

We have stressed that in any of the renormalization schemes discussed above, the coupling α_μ depends upon a mass parameter μ. From now on we shall emphasize this by writing it as a function of μ. Conventionally one writes it as $\alpha(\mu^2)$ and calls it the coupling *defined at mass-scale* μ.

In (20.8.6) let us consider two values of the mass-scale, μ_0, fixed, and μ, varying. Let us choose

$$t = \ln(\mu^2/\mu_0^2) \tag{20.8.10}$$

and let us also choose for the arbitrary constant

$$\alpha = \alpha(\mu_0^2).$$

Then (20.8.9) reads

$$\frac{\partial\bar{\alpha}[\alpha(\mu_0^2), \ln(\mu^2/\mu_0^2)]}{\partial\ln(\mu^2)} = \beta\left[\bar{\alpha}\left(\alpha(\mu_0^2), \ln(\mu^2/\mu_0^2)\right)\right]. \tag{20.8.11}$$

Thus $\bar{\alpha}\left[\alpha(\mu_0^2), \ln(\mu^2/\mu_0^2)\right]$ satisfies differential equation (20.6.5) which controls the variation of $\alpha(\mu^2)$ with μ. Moreover, by (20.8.7) and our choice for the arbitrary constant, it is equal to $\alpha(\mu_0^2)$ at $\mu = \mu_0$, i.e.

$$\bar{\alpha}\left[\alpha(\mu_0^2), 0\right] = \alpha(\mu_0^2). \tag{20.8.12}$$

Thus $\bar{\alpha}$ is *the* function which gives $\alpha(\mu^2)$ in terms of $\alpha(\mu_0^2)$ as μ varies, i.e.

$$\alpha(\mu^2) = \bar{\alpha}\left[\alpha(\mu_0^2), \ln(\mu^2/\mu_0^2)\right]. \tag{20.8.13}$$

Equivalently, we can write that

$$\left.\begin{aligned} \bar{\alpha}\left[\alpha(\mu^2), t\right] &= \alpha(\mathrm{e}^t\mu^2) \\ \text{or} \quad \alpha(\eta^2\mu^2) &= \bar{\alpha}\left[\alpha(\mu^2), \ln\eta^2\right] \end{aligned}\right\} \tag{20.8.14}$$

Now differentiating (20.8.6) with respect to α we get

$$\frac{\partial\bar{\alpha}(\alpha,t)}{\partial\alpha} = \frac{\beta[\bar{\alpha}(\alpha,t)]}{\beta(\alpha)} \tag{20.8.15}$$

so that, via (20.8.8) (recall that $t \equiv 2\ln\eta$)

$$\left(\eta\frac{\partial}{\partial\eta} - 2\beta(\alpha)\frac{\partial}{\partial\alpha}\right)\bar{\alpha}(\alpha,t) = 0 \tag{20.8.16}$$

from which one finds that for any differentiable function $F[\bar{\alpha}(\alpha,t)]$

$$\left(\eta\frac{\partial}{\partial\eta} - 2\beta(\alpha)\frac{\partial}{\partial\alpha}\right) F[\bar{\alpha}(\alpha,t)] = 0. \tag{20.8.17}$$

It is now easy to see that the solution to (20.8.4) is

$$\begin{aligned}\Gamma^{(n)}[\eta p_i; \alpha(\mu^2), \mu] &= \eta^{(4-n)} \exp\left\{-\frac{n}{2}\int_0^t \gamma\left[\bar{\alpha}\left(\alpha(\mu^2), t'\right)\right] \mathrm{d}t'\right\} \times \\ &\quad \times \Gamma^{(n)}[p_i; \alpha(\eta^2\mu^2), \mu]\end{aligned} \tag{20.8.18}$$

where, we remind the reader once more, $t \equiv 2\ln\eta$.

Using (20.8.9) this can be cast into the form

$$\begin{aligned}\Gamma^{(n)}[\eta p_i; \alpha(\mu^2), \mu] &= \eta^{(4-n)} \exp\left\{-\frac{n}{2}\int_{\alpha(\mu^2)}^{\alpha(\eta^2\mu^2)} \frac{\gamma(x)}{\beta(x)} \mathrm{d}x\right\} \times \\ &\quad \times \Gamma^{(n)}[p_i; \alpha(\eta^2\mu^2), \mu].\end{aligned} \tag{20.8.19}$$

The RHS of (20.8.18) is clearly correct at $\eta = 1$ or $t = 0$. For $\eta \neq 1$ direct differentiation and use of (20.8.17) shows that it satisfies (20.8.4).

The remarkable result (20.8.19) tells us that $\Gamma^{(n)}$ at momentum ηp_j is related to $\Gamma^{(n)}$ at the lower momentum p_j, but evaluated using the coupling defined at the scale $\eta\mu$, i.e. $\alpha(\eta^2\mu^2)$, and multiplied by a factor which is almost, but not quite, $\eta^{(4-n)}$, as will become clear in the next section.

Even more remarkable is the fact that all the η-dependence of $\Gamma^{(n)}(\eta p_j)$ is, aside from the factor $\eta^{(4-n)}$, controlled entirely by the η-dependence of $\alpha(\eta^2\mu^2)$! The dramatic implications of this are spelled out in detail in Section 22.1 in the context of the reaction $e^+e^- \to$ hadrons.

On dimensional grounds, if the ηp_j are so big that masses are irrelevant, we might have guessed that there would be a factor $\eta^{(4-n)}$. The correction term involving γ, as we shall see in a moment, makes $\Gamma^{(n)}$ behave as if its mass dimension was not quite $4-n$. The behaviour of the theory at large momenta is critically dependent on what sort of function $\bar{\alpha}$ is, i.e. on the behaviour of $\beta(\alpha)$. We now digress to study this. [See Gross (1976) for an instructive treatment with more emphasis on the field theoretic details.]

20.9 Scaling and asymptotic freedom

Consider again the defining equation (20.8.6) for $\bar{\alpha}(\alpha,t)$:

$$\int_{\alpha}^{\bar{\alpha}(\alpha,t)} \frac{\mathrm{d}x}{\beta(x)} = 2\ln\eta = t. \tag{20.9.1}$$

We continue to drop all scheme labels s and μ in this section, but it should be remembered that really β is β_s etc. It will

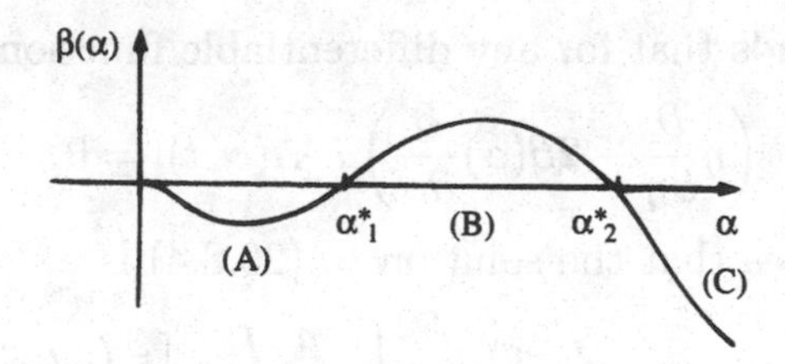

Fig. 20.4. Possible form of $\beta(\alpha)$ *vs* α in QCD.

turn out that the most important results are independent of the scheme.

Since we can certainly take $\eta = 0$ or ∞ the LHS must $\to \pm\infty$ at these values of η. This could happen because $\bar{\alpha}(\alpha, \eta = 0, \infty) \to \infty$, but for the cases of physical interest it occurs because $\beta(\alpha)$ has zeros, called 'fixed points', at $\alpha = 0, \alpha_1^*, \alpha_2^*, \ldots$. For QCD $\beta(\alpha)$ *might* appear as shown in Fig. 20.4.

We don't really know what β looks like for large α, since we calculate it perturbatively, but the small α behaviour is well established. We are interested in the behaviour of $\bar{\alpha}(\alpha, t)$ as $\eta \to \infty$. This will depend on the value of α, as found ultimately from experiment. Suppose α is small and lies in region (A) on the diagram. For large η, $\ln \eta$ is positive. On the other hand, $\beta(\alpha)$ is negative in region (A) and the only way the LHS of (20.9.1) can give a positive answer is if $\bar{\alpha}(\alpha, t)$ lies to the left of α for large η. As η increases the integral has to grow so $\bar{\alpha}(\alpha, t)$ must move further left until finally, as $\eta \to \infty$, $\bar{\alpha}(\alpha, t) \to 0$.

Bearing in mind the role of $\bar{\alpha}$ as an effective coupling [see (20.8.18)] we see that at high energies the theory approaches the behaviour of a free field theory—it is *asymptotically free.*

In QED or in ϕ^4 theory, by contrast, $\beta(\alpha)$ is positive for small α [see (20.7.17 and 16)] implying that $\bar{\alpha}(\alpha, t)$ grows larger as $\eta \to \infty$, perhaps approaching a finite value α_1^*, but possibly growing infinitely large.

A major discovery of the past few years is the proof that non-Abelian gauge theories can be asymptotically free.

We concentate now on QCD, assume that the α in our renormalization scheme lies in region (A), so that $\bar{\alpha} \to 0$ as $\eta \to \infty$, and study the behaviour of (20.8.18) as $\eta \to \infty$. Let us rewrite the term

$$\begin{aligned}\int_0^t \gamma[\bar{\alpha}(\alpha, t')]\,\mathrm{d}t' &= \int_0^t \mathrm{d}t'[\gamma(0) + \gamma(\bar{\alpha}) - \gamma(0)] \\ &= t\gamma(0) + \int_0^t \mathrm{d}t'\,\{\gamma[\bar{\alpha}(\alpha, t')] - \gamma(0)\}\end{aligned}$$

as

$$\equiv 2\gamma(0)\ln\eta + 2r(t). \tag{20.9.2}$$

Putting this into (20.8.18) and writing $\gamma_0 \equiv \gamma(0)$ gives

$$\Gamma^{(n)}(\eta p_j; \alpha(\mu^2), \mu) \overset{\eta\to\infty}{\longrightarrow} \eta^{(4-n-n\gamma_0)} \exp[-nr(t)]\Gamma^{(n)}_{\text{free}}(p_j) \tag{20.9.3}$$

where 'free' means evaluated to zeroth order in perturbation theory, i.e. as in a free field theory.

> Exceptionally, for some $\Gamma^{(n)}$, the free field result will be zero. In that case $\Gamma^{(n)}_{\text{free}}$ is really $\Gamma^{(n)}_{\text{almost free}}$ and should be evaluated to lowest order in perturbation theory that yields a non-zero result.

The precise situation depends upon how fast $\gamma[\bar{\alpha}(\alpha,t)]$ approaches $\gamma(0)$ as η or $t\to\infty$. If the integral giving $r(t)$ converges as $t\to\infty$, then $r(\infty)$ is just a number, and all the η-dependence in (20.9.3) resides in the factor $\eta^{(4-n-n\gamma_0)}$, i.e. the behaviour is power-like or 'scales'. Note, however, that the power of η is *not* what one would naively have expected from the mass dimensions of $\gamma^{(n)}$. There is an 'anomalous' dimension $n\gamma(0)$.

If the integral does not converge as $t\to\infty$, it nevertheless cannot grow as fast as t since its integrand tends to zero as $t'\to\infty$. Then $\exp[-nr(t)]$ cannot behave like $\exp[-nt] = \exp[-2n\ln\eta]$ and thus cannot behave like a power of η. So one will end up with a behaviour $\eta^{(4-n-n\gamma_0)}$ multiplied by terms typically of the form of powers of $\ln\eta$. The latter break the scaling behaviour.

In actual fact, in QCD one has

$$\beta(\alpha) = -b\alpha^2[1 + b'\alpha + \cdots] \tag{20.9.4}$$

and, as we shall see in Chapter 21, there are several different $\gamma(\alpha)$ which typically behave as

$$\gamma(\alpha) = \gamma_0 + \gamma_1\alpha + \cdots \tag{20.9.5}$$

where γ_0 may in some cases be zero.

For a case like (20.9.5) the asymptotic behaviour in (20.9.3) will be controlled by

$$\eta^{(4-n-n\gamma_0)} \exp\left\{-\frac{n}{2}\gamma_1 \int_0^t \bar{\alpha}\left[\alpha(\mu^2), t'\right] dt'\right\}. \tag{20.9.6}$$

Now we can calculate $\bar{\alpha}(\alpha,t)$ from (20.9.1) and (20.9.4), and using just the lowest order result for β, have

$$t = \int_\alpha^{\bar{\alpha}(\alpha,t)} \frac{dx}{-bx^2} = \frac{1}{b}\left[\frac{1}{\bar{\alpha}(\alpha,t)} - \frac{1}{\alpha}\right]$$

so that, to lowest order,

$$\bar{\alpha}(\alpha, t) = \frac{\alpha}{1 + b\alpha t}. \tag{20.9.7}$$

Note that as expected from the general discussion above, $\bar{\alpha}(\alpha, t) \to 0$ as η or $t \to \infty$.

Using (20.9.7) we can evaluate the integral in (20.9.6):

$$\int_0^t \bar{\alpha}(\alpha, t')\, dt' = \int_0^t \frac{\alpha\, dt'}{1 + b\alpha t'} = \frac{1}{b} \ln(1 + b\alpha t) \tag{20.9.8}$$

$$= \frac{1}{b} \ln\left[\alpha / \bar{\alpha}(\alpha, t)\right]. \tag{20.9.9}$$

The η-dependence in (20.9.6) is then

$$\eta^{(4-n-n\gamma_0)} \left[\frac{\bar{\alpha}\left[\alpha(\mu^2), t\right)]}{\alpha(\mu^2)}\right]^{n\gamma_1/2b} = \eta^{(4-n-n\gamma_0)} \left[\frac{\alpha\left(\eta^2\mu^2\right)}{\alpha(\mu^2)}\right]^{n\gamma_1/2b} \tag{20.9.10}$$

To summarize, using the lowest order expressions for $\beta(\alpha)$ and $\gamma(\alpha)$ we have

$$\Gamma^{(n)}\left(\eta p_j; \alpha(\mu^2), \mu\right) = \eta^{(4-n-n\gamma_0)} \left[\frac{\alpha\left(\eta^2\mu^2\right)}{\alpha(\mu^2)}\right]^{n\gamma_1/2b} \times \Gamma^{(n)}\left(p_j; \alpha(\eta^2\mu^2), \mu\right). \tag{20.9.11}$$

We draw attention once again to the remarkable feature that, aside from the factor $\eta^{(4-n-n\gamma_0)}$, all the η-dependence appears through the coupling $\alpha(\eta^2\mu^2)$.

There are two ways one could utilize (20.9.11):

- if one is studying a $\Gamma^{(n)}$ that has a special value at some point $p_j = p_j^0$, one can choose $p_j = p_j^0$, and the $\Gamma^{(n)}$ on the RHS is then exactly known. In this case (20.9.11) gives a result for $\Gamma^{(n)}(\eta p_j^0; \alpha, \mu)$ valid for all positive η and inexact only to the extent that β and γ were approximated by their lowest order forms and that $\bar{\alpha}(\alpha, t)$ is therefore also not exact;

- if $\Gamma^{(n)}$ does not have some special value at $p_j = p_j^0$ one can take arbitrary p_j and use the fact that $\bar{\alpha} \to 0$ as $\eta \to \infty$ to get an expression for the asymptotic behaviour

$$\Gamma^{(n)}(\eta p_j; \alpha(\mu^2), \mu) \overset{\eta\to\infty}{\longrightarrow} \eta^{(4-n-n\gamma_0)} \left[\frac{\alpha(\eta^2\mu^2)}{\alpha(\mu^2)}\right]^{n\gamma_1/2b} \times \Gamma^{(n)}_{\text{free}}(p_j). \tag{20.9.12}$$

where the meaning of $\Gamma^{(n)}_{\text{free}}$ was explained in connection with (20.9.3).

It is not easy to specify precisely the criterion for the validity of (20.9.11) and (20.9.12), but a reasonable requirement is that $\bar{\alpha}$ be small. Ultimately the real test is to compare with a higher order calculation. For many cases in QCD this has been done, and the data sometimes indicate a need to work to higher order in $\bar{\alpha}$. Such higher order calculations are very difficult but much has been achieved in the past few years.

The rôle and mechanism of the renormalization group is illuminated by a practical example, $e^+e^- \to$ hadrons, which is studied in detail in Section 22.1.

In the above we suppressed all labels referring to the renormalization scheme. It is important to note that one can show that the coefficients b and b' in (20.9.4) and γ_0 and γ_1 in (20.9.5) are *independent of the renormalization scheme* for mass-independent schemes. This is not so for the coefficients of higher powers of α. We sketch the proof for $\beta(\alpha)$.

Let α and $\tilde{\alpha}$ be the couplings in two different renormalization schemes in QCD, defined at the same mass scale μ. By definition

$$\beta(\alpha) \equiv \frac{\mu}{2}\frac{\partial\alpha}{\partial\mu} = -b\alpha^2(1 + b'\alpha + \cdots). \tag{20.9.13}$$

Similarly,

$$\tilde{\beta}(\tilde{\alpha}) \equiv \frac{\mu}{2}\frac{\partial\tilde{\alpha}}{\partial\mu} = -\tilde{b}\tilde{\alpha}^2(1 + \tilde{b}'\tilde{\alpha} + \cdots). \tag{20.9.14}$$

But, analogously to (20.3.12) and using the form of the zs given after (20.3.14),

$$\tilde{\alpha} = \alpha(1 + c\alpha + \cdots), \tag{20.9.15}$$

substituting in (20.9.14) yields

$$\tilde{\beta} = -\tilde{b}\alpha^2[1 + (2c + \tilde{b}')\alpha + \cdots]. \tag{20.9.16}$$

However because β and $\tilde{\beta}$ do not depend explicitly upon the mass μ we can write

$$\begin{aligned} \tilde{\beta} &= \frac{\mu}{2}\frac{\partial\tilde{\alpha}}{\partial\alpha}\cdot\frac{\partial\alpha}{\partial\mu} = \beta(\alpha)\frac{\partial\tilde{\alpha}}{\partial\alpha} && (20.9.17)\\ &= \beta(\alpha)[1 + 2c\alpha + \cdots] \quad \text{by (20.9.14)} \\ &= -b\alpha^2[1 + (2c + b')\alpha + \cdots] \quad \text{by (20.9.13)}. && (20.9.18) \end{aligned}$$

Comparing (20.9.16 and 18) we see that

$$\tilde{b} = b \qquad \text{and} \qquad \tilde{b}' = b' \tag{20.9.19}$$

as promised.

We wish now to look at the detailed experimental consequences of the above arguments in the specific case of QCD. We therefore turn to take a closer look at QCD and its properties.

20.10 Appendix to Chapter 20

We first outline the general approach of 'tHooft and Veltman (1972) to the definition of a Feynman integral 'in d dimensions' and to its analytic continuation in the variable d. In Section 20.10.1 we give the *definition* of a d dimensional integral when d is not a positive integer. Section 20.10.2, which gives a general discussion of the convergence and analytic continuation involved, is rather complex and is *not needed* for the rest of the Appendix. It may be skipped by the non-theoretically minded reader. In Section 20.10.3 we derive some useful formulae for commonly occurring d dimensional integrals and finally in Section 20.10.4 use these to compute the order g^2 vertex corrections in ϕ^4 theory.

20.10.1 Definition of a d-dimensional integral

Consider a *convergent* integral

$$I(p) = \int \mathrm{d}^4k \; f(p,k) \tag{20.10.1}$$

where $f(p,k)$ is some product of propagators each of the form

$$\frac{1}{(p-k)^2 - m^2} = \frac{1}{p^2 - 2p\cdot k + k^2 - m^2}. \tag{20.10.2}$$

Note that we are assuming $m \neq 0$ to avoid the totally separate problem of *infrared* divergences.

Suppose we make the extension to $d > 4$ dimensions by putting

$$\begin{aligned} \tilde{p} &= (p_0p_1p_2p_3\; 0\, 0 \ldots 0) \text{ with } \tilde{p}^2 = p^2 \\ \tilde{k} &= (k_0k_1k_2k_3K_1K_2\ldots K_{d-4}) \end{aligned} \tag{20.10.3}$$

with $\tilde{k}^2 = k^2 - K^2$ where $k^2 = k_0^2 - \boldsymbol{k}^2$ and $K^2 = K_1^2 + \cdots + K_{d-4}^2$, and suppose further that the d-dimensional integral is convergent, i.e. that

$$\begin{aligned} I_d(p) &\equiv \int \mathrm{d}^dk \; f(p,k;K) \\ &= \int \mathrm{d}^4k \int \mathrm{d}^{d-4}K \; f(p,k;K) \quad (d \text{ integer}, \geq 5) \end{aligned} \tag{20.10.4}$$

converges.

Because of (20.10.3) the propagators will involve

$$(\tilde{p}-\tilde{k})^2 - m^2 = \tilde{p}^2 - 2\tilde{p}\cdot\tilde{k} + \tilde{k}^2 - m^2 = p^2 - 2p\cdot k + k^2 - K^2 - m^2. \tag{20.10.5}$$

Thus $K_1 \ldots K_{d-4}$ only occur as the scalar K^2, i.e. in a rotationally invariant form in the Euclidean $(d-4)$-dimensional space. Hence in that space we can use polar coordinates and carry out all the angular integrations.

In an N-dimensional Euclidean space the volume element is

$$\begin{aligned} d^N K &= K^{N-1}\, dK (\sin\theta_{N-1})^{N-2}\, d\theta_{N-2} (\sin\theta_{N-2})^{N-3}\, d\theta_{N-3} \cdots \\ &\quad \ldots \sin\theta_2\, d\theta_2\, d\theta_1 \end{aligned} \tag{20.10.6}$$

where $0 \le \theta_j \le \pi$ for $j \ge 2, 0 \le \theta_1 \le 2\pi$.

Integration over the 'solid angle' yields

$$2\pi^{N/2} \Big/ \Gamma(N/2). \tag{20.10.7}$$

[Recall that for positive integer n, $\Gamma(n) = (n-1)!$ with $\Gamma(1) = 1$.]

Thus, using $N = d - 4$ yields

$$I_d(p) = \frac{2\pi^{d/2-2}}{\Gamma(d/2-2)} \int d^4k \int_0^\infty K^{d-5} f(p, k; K^2)\, dK \tag{20.10.8}$$

or, changing to K^2 as integration variable,

$$I_d(p) = \frac{\pi^{d/2-2}}{\Gamma(d/2-2)} \int d^4k \int_0^\infty (K^2)^{d/2-3} f(p, k; K^2)\, dK^2. \tag{20.10.9}$$

Note that

$$f(p, k; K^2 = 0) = f(p, k) \tag{20.10.10}$$

the original physical integrand.

The expressions (20.10.8 or 9), though established for integer $d \ge 5$, now make perfectly good sense for d *non-integer or even complex*, provided only that the integral converges.

The d-dimensional integral, for d non-integer, is *defined* by (20.10.9) for all d for which the integral converges. For values of d where the integral diverges see Section 20.10.2.

20.10.2 Questions of convergence and analytic continuation

The expression (20.10.9), established for $d \ge 5$, actually converges even for $d < 5$ provided $d > 4$. However it is ambiguous at $d = 4$ because the integral then diverges (at $K^2 = 0$) and is multiplied by zero [because $\Gamma(z) = 1/z$ as $z \to 0$].

To avoid this difficulty integrate by parts:

$$\begin{aligned} &\frac{1}{\Gamma(d/2-2)} \int_0^\infty (K^2)^{d/2-3} f(K^2)\, dK^2 \\ &= \frac{1}{\Gamma(d/2-1)} \left\{ (K^2)^{d/2-2} f(K^2)\Big|_0^\infty - \int_0^\infty (K^2)^{d/2-2} \left(\frac{\partial f}{\partial K^2}\right) dK^2 \right\} \end{aligned} \tag{20.10.11}$$

where we have used $z\Gamma(z) = \Gamma(z+1)$. Because the original integral converged at ∞, and because $f(p,k;K^2=0)$ is finite, the first term on the RHS will vanish if $d/2-2>0$, i.e. $d>4$.

Thus we have a new expression

$$I_d^{(1)}(p) = \frac{-\pi^{d/2-2}}{\Gamma(d/2-1)}\int \mathrm{d}^4k \int_0^\infty (K^2)^{d/2-2}\left(\frac{\partial f}{\partial K^2}\right)\mathrm{d}K^2 \qquad (20.10.12)$$

which *is equal* to $I_d(p)$ for $d>4$, but which is well behaved at $d=4$. Indeed if we take $d=4$ we get

$$\begin{aligned} I_4^{(1)}(p) &= -\frac{1}{\Gamma(1)}\int \mathrm{d}^4k \int_0^\infty \left(\frac{\partial f}{\partial K^2}\right)\mathrm{d}K^2 \\ &= -\int \mathrm{d}^4k \left\{f(p,k;K^2)\right\}_{K^2=0}^{\infty} \\ &= \int \mathrm{d}^4k f(p,k;0) \qquad (20.10.13) \end{aligned}$$

so that, upon using (20.10.10), $I_4^{(1)}(p)$ equals the original integral $I(p)$.

The expression (20.10.12) makes perfectly good sense down to $d/2-2>-1$, i.e. $d>2$. So the expression $I_d^{(1)}(p)$ gives a continuation of $I_d(p)$ below $d=4$. In a sense we have succeeded in continuing to a smaller number of dimensions than $d=4$!

The process of partial integration can be repeated to yield

$$I_d^{(\ell)}(p) = \frac{(-1)^\ell \pi^{d/2-2}}{\Gamma(d/2+\ell-2)}\int \mathrm{d}^4k \int_0^\infty (K^2)^{d/2+\ell-3}\left[\frac{\partial^\ell f}{(\partial K^2)^\ell}\right]\mathrm{d}K^2 \qquad (20.10.14)$$

and this is well behaved as long as $d/2+\ell-3>-1$, i.e. for $d>4-2\ell$. So, in this fashion, we can continue as far as we like to smaller and smaller values of d (even to negative values!). And $I_d^{(\ell)}(p)$ will coincide with the genuine d-dimensional integral (20.10.4) for any integer value $d=4,5\ldots$ for which it is *convergent*.

All the above assumed an integral which converged. Suppose now that we have an integral which *diverges* in the physical case of four dimensions. As a concrete example, suppose that

$$f(p,k) = \frac{1}{[(p-k)^2+m^2]^2} \to \frac{1}{k^4} \quad \text{as} \quad k\to\infty \qquad (20.10.15)$$

as was the case in (20.3.1), so that the integral diverges logarithmically. But the integral defining $I_d^{(\ell)}(p)$ in (20.10.14) will, in this case, converge provided

$$6+2(d/2+\ell-3)-(2\ell+4)<0 \qquad (20.10.16)$$

where we have used the fact that, via (20.10.15),

$$\frac{\partial^\ell f}{(\partial K^2)^\ell} \sim \frac{1}{K^{2\ell+4}} \quad \text{as} \quad K \to \infty.$$

Thus the integal defining $I_d^{(\ell)}(p)$ will converge provided Re $d < 4$ independent of ℓ.

The *regularized* integral is now *defined* to be $I_d^{(\ell)}(p)$, with $d < 4$. Taking $d < 4$ plays the rôle of a cut-off.

Since, in fact, $I_d^{(\ell)}(p)$ is well defined even for complex d provided Re $d <$ 4, (20.10.14) yields a definition of an analytic function of d which can then be continued to the right of Re $d = 4$.

In many cases the integral for $I_d^{(\ell)}(p)$ can be done explicitly with the result expressed in terms of well-known functions of d. The analytic continuation in d can then be done using their analytic properties. We shall see some examples in the next section.

For an arbitrary integral, 'tHooft and Veltman (1972) have provided an explicit method of continuing in d based upon a very clever use of partial integration. We shall illustrate how this approach works for the case where $f(p,k) \sim 1/k^4$ as $k \to \infty$.

The integral in (20.10.14) is proportional to $\int \mathrm{d}^4k\,\mathrm{d}K \;\; F(p,k;K^2)$, where F has the structure

$$F(p,k;K^2) = \frac{K^N}{[p^2 - 2p\cdot k + k^2 + m^2 - K^2]^\alpha} \equiv \frac{K^N}{D^\alpha} \qquad (20.10.17)$$

with $N = d + 2\ell - 5$ and $\alpha = 2 + \ell$.

For d small enough that the integral converges, we do partial integration with respect to each of k_0, k_1, k_2, k_3 and K *separately*, in the form

$$\int \mathrm{d}k_\mu F = k_\mu F\big|_{\text{limits}} - \int k_\mu \frac{\partial F}{\partial k_\mu}\,\mathrm{d}k_\mu \qquad (20.10.18)$$

and then add the results. The LHS will thus be $5I_d^{(\ell)}(p)$.

On the RHS all the first terms vanish since the integral is convergent. For the second terms, because of the structure of the propagators one is dealing with

$$-\sum_{\mu=0}^{3} k_\mu \frac{\partial}{\partial k_\mu}\left[-\frac{K^N}{D_\alpha}\right] = \frac{2\alpha K^N(k^2 - p\cdot k)}{D^{\alpha+1}} \qquad (20.10.19)$$

and

$$-K\frac{\partial}{\partial K}\left[\frac{K^N}{D_\alpha}\right] = -\frac{2\alpha K^{N+2}}{D^{\alpha+1}} - \frac{NK^N}{D^\alpha}. \qquad (20.10.20)$$

When we add (20.10.19 and 20) we get

$$\begin{aligned}
&2\alpha\frac{K^N}{D^{\alpha+1}}\left(k^2 - p\cdot k - K^2\right) - N\frac{K^N}{D^\alpha} \\
&= 2\alpha\frac{K^N}{D^{\alpha+1}}\left\{(p^2 - 2p\cdot k + k^2 + m^2 - K^2) + (p\cdot k - m^2 - p^2)\right\} - \frac{NK^N}{D^\alpha} \\
&= (2\alpha - N)\frac{K^N}{D^\alpha} + 2\alpha K^N\frac{(p\cdot k - m^2 - p^2)}{D^{\alpha+1}} \\
&= (2\alpha - N)F(p,k;K^2) + 2\alpha K^N\frac{(p\cdot k - m^2 - p^2)}{D^{\alpha+1}}. \qquad (20.10.21)
\end{aligned}$$

We thus have shown that

$$5I_d^{(\ell)}(p) = (2\alpha - N)I_d^{(\ell)}(p) + \hat{I}_d^{(\ell)}(p) \qquad (20.10.22)$$

where the integrand of $\hat{I}_d^{(\ell)}(p)$ is

$$\left[\frac{2\alpha(p\cdot k - m^2 - p^2)}{p^2 - 2p\cdot k + k^2 + m^2 - K^2}\right] \times F(p,k;K^2) \qquad (20.10.23)$$

which for $K \to \infty$ is of the form

$$\frac{1}{K^2} \times F(p,k;K^2) \qquad (20.10.24)$$

so that $\hat{I}_d^{(\ell)}(p)$ will be convergent for d two units larger than was the case in $I_d^{(\ell)}(p)$, i.e. it will converge for $d < 6$.

Substituting for N and α in (20.10.22), we have $(d-4)I_d^{(\ell)}(p) = \hat{I}_d^{(\ell)}(p)$, i.e. for $d < 5$

$$I_d^{(\ell)}(p) = \left(\frac{1}{d-4}\right)\hat{I}_d^{(\ell)}(p). \qquad (20.10.25)$$

Thus we have succeeded in providing an expression for $I_d^{(\ell)}(p)$ valid beyond $d = 4$, and we see that the divergence which was originally present for $d = 4$ now shows up as a pole in $I_d^{(\ell)}(p)$ at $d = 4$.

In the above we dealt with a single integral $\int d^4k$ as would occur in a single loop Feynman diagram. For multiple integrals which occur in multi-loop diagrams see 'tHooft and Veltman (1972). Beware that in that paper the metric tensor is called $\delta_{\mu\nu}$ and the metric is $(-1,1,1,1)$.

A more mathematical approach to d-dimensional integrals can be found in Collins (1984). Interesting properties are listed in Section 4.3 of the latter. Note that the formulae given in that section refer to an *Euclidean* space.

20.10.3 Some useful d-dimensional integrals

We shall present here with brief derivations a few of the most commonly needed d-dimensional integrals.

Consider the integral

$$I_d(p;m;\alpha) \equiv \int d^d k \frac{1}{[(k-p)^2 - m^2]^\alpha} \tag{20.10.26}$$

where α is a *positive integer*. Suppose that α is large enough so that we may use the definition (20.10.8), i.e. such that

$$I_d(p;m;\alpha) \equiv \frac{2\pi^{d/2-2}}{\Gamma(d/2-2)} \int d^4 k \int_0^\infty \frac{K^{d-5}}{[k^2 - 2k\cdot p + p^2 - m^2 - K^2]^\alpha} dK \tag{20.10.27}$$

converges for $d > 4$.

Put, for the moment, $C = -(k^2 - 2k\cdot p + p^2 - m^2)$ so that

$$\mathcal{I} \equiv \int_0^\infty \frac{K^{d-5}\, dK}{[k^2 - 2k\cdot p + p^2 - m^2 - K^2]^\alpha} = (-1)^\alpha \int_0^\infty \frac{K^{d-5}\, dK}{[K^2 + C]^\alpha}. \tag{20.10.28}$$

Using the known formula

$$\int_0^\infty \frac{K^\beta\, dK}{[K^2 + C]^\alpha} = \frac{\Gamma(1/2 + \beta/2)\,\Gamma(\alpha - (1+\beta)/2)}{2\Gamma(\alpha)\; C^{\alpha-(1+\beta)/2}} \tag{20.10.29}$$

with $\frac{1}{2}(1+\beta) = d/2 - 2$ gives

$$\mathcal{I} = (-1)^\alpha \frac{\Gamma(d/2-2)\Gamma(\alpha+2-d/2)}{2\Gamma(\alpha)\; C^{\alpha+2-d/2}}. \tag{20.10.30}$$

Thus, substituting in (20.10.27) and replacing C, gives

$$I_d(p;m;\alpha) = (-1)^\alpha \frac{\pi^{d/2-2}\Gamma(\alpha+2-d/2)}{\Gamma(\alpha)} \int \frac{d^4 k}{[m^2 - (k-p)^2]^{\alpha+2-d/2}}. \tag{20.10.31}$$

The remaining integral is a standard (convergent) 4-dimensional integral in Minkowski space. Change integration variables to $k_\mu - p_\mu$ so that

$$I_d(p;m;\alpha) = (-1)^\alpha \frac{\pi^{d/2-2}\Gamma(\alpha+2-d/2)}{\Gamma(\alpha)} \int \frac{d^4 k}{[m^2 - k^2 - i\epsilon]^{\alpha+2-d/2}} \tag{20.10.32}$$

where we have shown explicitly the $(i\epsilon)$ term that should really be in all propagators. Note that the integral is, in fact, independent of p. (The latter is no surprise. In the original 4-dimensional integral which we assumed was convergent, we could immediately have substituted $k \to k-p$ as integration variables.)

We now rotate the integration contour $-\infty \leq k_0 \leq \infty$ into $-i\infty \leq k_0 \leq i\infty$ in the complex k_0 plane. The contribution from the arcs at infinity vanish so that, for the above,

$$\int_{-\infty}^{\infty} dk_0 = \int_{-i\infty}^{i\infty} dk_0 = i \int_{-\infty}^{\infty} dk_4 \tag{20.10.33}$$

where we have put $k_0 = ik_4$.

Then, putting $N = \alpha + 2 - d/2$,

$$\int \frac{d^4k}{[m^2 - k^2 - i\epsilon]^N} = i \int \frac{d^4k_E}{[k_E^2 + m^2 - i\epsilon]^N} \tag{20.10.34}$$

where k_E is now 'Euclidean', i.e. $d^4k_E = dk_1\, dk_2\, dk_3\, dk_4$, and we have used $k^2 = k_0^2 - \boldsymbol{k}^2 = -k_4^2 - \boldsymbol{k}^2 \equiv -k_E^2$.

Using polar coordinates in the 4-dimensional Euclidean space and carrying out the angular integrals using (20.10.7)

$$\int \frac{d^4k}{[m^2 - k^2 - i\epsilon]^N} = i \frac{2\pi^2}{\Gamma(2)} \int_0^{\infty} \frac{k_E^3\, dk_E}{[k_E^2 + m^2]^N} \tag{20.10.35}$$

which, via (20.10.29),

$$\begin{aligned} &= i\pi^2 \frac{\Gamma(N-2)}{\Gamma(N)(m^2)^{N-2}} \\ &= \frac{i\pi^2}{(\alpha - d/2 + 1)(\alpha - d/2)} \frac{1}{(m^2)^{\alpha - d/2}}. \end{aligned} \tag{20.10.36}$$

Substituting into (20.10.32) and using

$$\Gamma(\alpha - d/2 + 2) = (\alpha - d/2 + 1)(\alpha - d/2)\Gamma(\alpha - d/2)$$

we end up with

$$I_d(p, m; \alpha) \equiv \int \frac{d^dk}{[(k-p)^2 - m^2]^\alpha} = i(-1)^\alpha \pi^{d/2} \frac{\Gamma(\alpha - d/2)}{\Gamma(\alpha)} \cdot \frac{1}{(m^2)^{\alpha - d/2}}, \tag{20.10.37}$$

the result, as mentioned, being independent of p.

In the above we assumed that the integer α was large enough so that we could use the definition (20.10.8) for I_d. If α is too small to allow this, we should, strictly speaking, utilize the more general expression (20.10.14) with ℓ chosen sufficiently large. But doing this one finds, at the end, exactly the same result (20.10.37)! The reason is that (20.10.37), established for large enough α, can be used to analytically continue in α down to smaller integer values. Thus (20.10.37) gives $I_d(m; \alpha)$ for all d and all α.

When one uses a d-dimensional integral to give a meaning to divergent integrals one sometimes finds surprising results.

Let us put $p = 0$ in (20.10.37). Then, taking $\alpha = 0$ and recalling that $\Gamma(0) = \infty$, we find the peculiar result

$$\int \mathrm{d}^d k = 0 \qquad (d \neq \text{even integer}). \tag{20.10.38}$$

But $\Gamma(\alpha)$ is infinite at all negative integers, $-1, -2, \ldots$, so that one can similarly deduce that

$$\int \mathrm{d}^d k (k^2)^n = 0 \qquad n = 1, 2, \ldots \qquad (d \neq \text{even integer}). \tag{20.10.39}$$

These should be taken as a warning that a divergent 'd-dimensional integral' is not always a very intuitive object.

Now in (20.10.37) take $d = 0$. One has

$$\int \frac{\mathrm{d}^0 k}{[k^2 - m^2]^\alpha} = \frac{\mathrm{i}(-1)^\alpha}{(m^2)^\alpha} = \mathrm{i}\left[\frac{1}{(k^2 - m^2)^\alpha}\right]_{k^2=0}. \tag{20.10.40}$$

For $d = -2$ one has

$$\int \frac{\mathrm{d}^{-2} k}{[k^2 - m^2]^\alpha} = \frac{\mathrm{i}\alpha(-1)^\alpha}{\pi(m^2)^{\alpha+1}} = -\frac{\mathrm{i}}{\pi}\frac{\partial}{\partial k^2}\left[\frac{1}{(k^2 - m^2)^\alpha}\right]_{k^2=0} \tag{20.10.41}$$

with obvious generalizations to $d = -4, -6, \ldots$.

For integrals with components of k in the numerator one can deduce the following results by differentiating (20.10.37) with respect to p_μ.

$$\int \mathrm{d}^d k \frac{k^\mu}{[(k-p)^2 - m^2]^\alpha} = \frac{\mathrm{i}(-1)^\alpha \pi^{d/2}\Gamma(\alpha - d/2)}{\Gamma(\alpha)(m^2)^{\alpha - d/2}} \cdot p^\mu \tag{20.10.42}$$

$$\begin{aligned}\int \mathrm{d}^d k \frac{k^\mu k^\nu}{[(k-p)^2 - m^2]^\alpha} &= \frac{\mathrm{i}(-1)^\alpha \pi^{d/2}}{\Gamma(\alpha)(m^2)^{\alpha - d/2}}\Big\{\Gamma(\alpha - d/2)p^\mu p^\nu - \\ &\quad -\frac{\Gamma(\alpha - 1 - d/2)}{2} m^2 g^{\mu\nu}\Big\}.\end{aligned} \tag{20.10.43}$$

20.10.4 Regularization of the 4-point vertex in ϕ^4 theory

The fundamental vertex in ϕ^4 theory is the 4-point amputated Green's function $\Gamma^{(4)}(p_i, m)$ discussed in Section 20.5. We shall demonstrate explicitly how to regularize it to order g^2 in both a momentum point subtraction (MPS) scheme and in schemes, MS and $\overline{\text{MS}}$, based upon dimensional regularization (DR).

(i) Momentum point subtraction (MPS). According to (20.5.11) we write the order g^2 contribution to $\Gamma^{(4)}$ given by the diagrams in Fig. 20.5 as

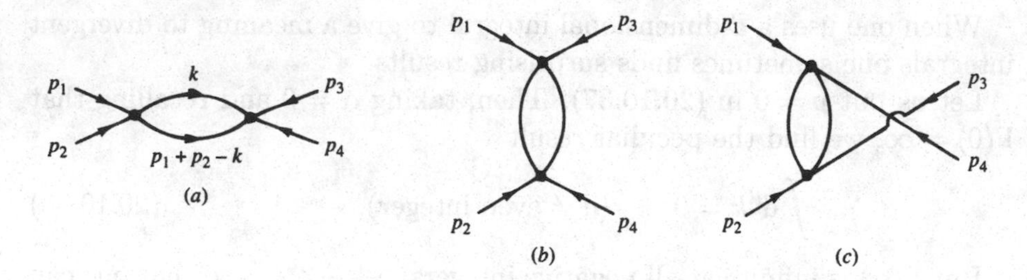

Fig. 20.5. Order g^2 contribution to $\Gamma^{(4)}$ in ϕ^4 theory.

$-\mathrm{i}g\Gamma(p_i, m)$. As explained earlier, to the desired order in g we may put $Z_\phi = Z_m = 1$ in the calculation.

The contribution from diagram (a) is

$$\Gamma_{(a)} = -\frac{\mathrm{i}g}{2}\int \frac{\mathrm{d}^4k}{(2\pi)^4}\cdot\frac{\mathrm{i}}{k^2-m^2+\mathrm{i}\epsilon}\cdot\frac{\mathrm{i}}{(k-p_1-p_2)^2-m^2+\mathrm{i}\epsilon} \tag{20.10.44}$$

where the factor $\frac{1}{2}$ arises because of Bose statistics.

The result, which depends on (p_1+p_2), must be a Lorentz scalar, so can in fact only depend on $s \equiv (p_1+p_2)^2$. Thus we shall write

$$\Gamma_{(a)} = \Gamma(s). \tag{20.10.45}$$

It is then easy to see that the complete result is

$$\Gamma(p_i, m) = \Gamma(s) + \Gamma(t) + \Gamma(u) \tag{20.10.46}$$

in which

$$t = (p_1+p_3)^2 \qquad \text{and} \qquad u = (p_1+p_4)^2 \tag{20.10.47}$$

in the convention of all momenta being incoming.

Consider now the logarithmically divergent integral I that occurs in (20.10.44). Put $P = p_1 + p_2$ and use Pauli–Villars regularization [see (20.4.1)] to replace $1/(k^2-m^2)$ by $1/(k^2-m^2) - 1/(k^2-\lambda^2)$. Then the regularized integral is

$$I^{\mathrm{PV}} \equiv \int \frac{\mathrm{d}^4k}{(2\pi)^4}\frac{1}{(k-P)^2-m^2}\left[\frac{1}{k^2-m^2}-\frac{1}{k^2-\lambda^2}\right]. \tag{20.10.48}$$

We use the famous Feynman trick to combine products of denominators:

$$\frac{1}{D_1D_2D_3\ldots D_N} = (N-1)!\int_0^1 \frac{\mathrm{d}z_1\mathrm{d}z_2\ldots\mathrm{d}z_N\,\delta(1-z_1-z_2\cdots-z_N)}{[z_1D_1+z_2D_2+\cdots+z_ND_N]^N}. \tag{20.10.49}$$

(An even more general form can be obtained by differentiating both sides with respect to any D_j.)

Using the above with $N = 2$

$$
\begin{aligned}
I^{\mathrm{PV}} &= \int_0^1 \mathrm{d}z \int \frac{\mathrm{d}^4k}{(2\pi)^4} \left\{ \frac{1}{[z((k-P)^2 - m^2) + (1-z)(k^2 - m^2)]^2} - \right. \\
&\qquad \left. - \frac{1}{[z((k-P)^2 - m^2) + (1-z)(k^2 - \lambda^2)]^2} \right\} \\
&= \int_0^1 \mathrm{d}z \int \frac{\mathrm{d}^4k}{(2\pi)^4} \left\{ \frac{1}{[k^2 - 2zk \cdot P + zP^2 - m^2]^2} - \right. \\
&\qquad \left. - \frac{1}{[k^2 - 2zk \cdot P + z(P^2 - m^2 + \lambda^2) - \lambda^2]^2} \right\}.
\end{aligned}
$$

Changing integration variable to $k - zP$ we get

$$
\begin{aligned}
I^{\mathrm{PV}} &= \int_0^1 \mathrm{d}z \int \frac{\mathrm{d}^4k}{(2\pi)^4} \left\{ \frac{1}{[k^2 + z(1-z)P^2 - m^2]^2} - \right. \\
&\qquad \left. - \frac{1}{[k^2 + z(1-z)P^2 - (1-z)\lambda^2 - zm^2]^2} \right\}.
\end{aligned}
$$

Since the k-integration is *convergent* we might as well evaluate it using the d-dimensional integration formula (20.10.37) with $\alpha = 2$ and $d \to 4$. (We are *not*, of course, using dimensional *regularization*. I^{PV} is already regularized and convergent. We are just using d-dimensional integration as a quick way to evaluate the integral.)

We get, with $\epsilon = \frac{1}{2}(4 - d)$,

$$
\begin{aligned}
I^{\mathrm{PV}} &= \frac{\mathrm{i}\pi^2}{(2\pi)^4} \int_0^1 \mathrm{d}z \lim_{\epsilon \to 0} \Gamma(\epsilon) \left\{ [m^2 - z(1-z)P^2]^{-\epsilon} - \right. \\
&\qquad \left. - [zm^2 + (1-z)\lambda^2 - z(1-z)P^2]^{-\epsilon} \right\}. \qquad (20.10.50)
\end{aligned}
$$

Using

$$
A^{-\epsilon} = \mathrm{e}^{-\epsilon \ln A} = 1 - \epsilon \ln A + 0(\epsilon^2)
$$

and

$$
\Gamma(\epsilon) = \frac{1}{\epsilon} + 0(1) \qquad \text{as } \epsilon \to 0
$$

we get, recalling that $P^2 = (p_1 + p_2)^2 = s$,

$$
I^{\mathrm{PV}}(s) = \frac{\mathrm{i}}{16\pi^2} \int_0^1 \mathrm{d}z \ln \left\{ \frac{(1-z)\lambda^2 + zm^2 - z(1-z)s}{m^2 - z(1-z)s} \right\} \qquad (20.10.51)
$$

so that, from (20.10.44)

$$
\Gamma^{\mathrm{PV}}(s) = -\frac{g}{32\pi^2} \int_0^1 \mathrm{d}z \ln \left\{ \frac{(1-z)\lambda^2 + zm^2 - z(1-z)s}{m^2 - z(1-z)s} \right\}. \qquad (20.10.52)
$$

Remembering that we are interested in $\lambda \to \infty$, we find after a little algebra that

$$\Gamma^{\text{PV}}(s) = \frac{g}{32\pi^2}\left\{1 + \ln(m^2/\lambda^2) + \int_0^1 \mathrm{d}z \ln[1 - z(1-z)s/m^2]\right\}. \tag{20.10.53}$$

The z integral can be done, and on putting $s = t = u = -4\mu^2/3$ at the symmetric point $p_i = p_i^0$ [see (20.5.3)], and using (20.10.46) we obtain the expression for $\Gamma^{\text{PV}}(p_i = p_i^0; m)$ used in (20.7.1).

(ii) Schemes (MS and $\overline{MS}$) based on dimensional regularization. According to (20.5.20) we write the order g^2 contributions to $\Gamma^{(4)}$ given by the diagrams Fig. 20.5 as $-\mathrm{i}g\mu^{2\epsilon}\Gamma_\epsilon(p_i, m, \mu)$. Bearing in mind (20.4.9), we have with $\epsilon = 2 - d/2$

$$\Gamma_\epsilon(s) = -\frac{\mathrm{i}g\mu^{2\epsilon}}{2}\int \frac{\mathrm{d}^d k}{(2\pi)^d}\frac{\mathrm{i}}{k^2 - m^2}\cdot\frac{\mathrm{i}}{(k-P)^2 - m^2}. \tag{20.10.54}$$

Note that it is conventional to use $(2\pi)^d$ in a d-dimensional Feynman integral. Using (20.10.49) yields

$$\Gamma_\epsilon(s) = \frac{\mathrm{i}g\mu^{2\epsilon}}{2(2\pi)^d}\int_0^1 \mathrm{d}z \int \mathrm{d}^d k \frac{1}{[k^2 + z(1-z)P^2 - m^2]^2}, \tag{20.10.55}$$

which, via (20.10.37), gives

$$\begin{aligned} \Gamma_\epsilon(s) &= \left(\frac{\mathrm{i}g\mu^{2\epsilon}}{2(2\pi)^d}\right)\left\{\mathrm{i}\pi^{d/2}\frac{\Gamma(\epsilon)}{\Gamma(2)}\int_0^1 \frac{\mathrm{d}z}{[m^2 - z(1-z)s]^\epsilon}\right\} \\ &= (4\pi)^\epsilon\Gamma(\epsilon)\left\{-\frac{g}{32\pi^2}\int_0^1\left[\frac{\mu^2}{m^2 - z(1-z)s}\right]^\epsilon \mathrm{d}z\right\}. \end{aligned} \tag{20.10.56}$$

Feeding this into (20.10.46) and comparing with (20.5.21) yields the expression for $G_d(p_i; m, \mu)$ utilized in (20.7.5).

The reader should note that the z integral ultimately appearing in the DR regularized 4-point vertex $\tilde{\Gamma}_{\text{MS}}$ or $\tilde{\Gamma}_{\overline{\text{MS}}}$ [see eqns (20.7.7, 11 and 14)] is of the same form as that in Γ^{PV} in (20.10.53). The same integral, via (20.5.11), therefore also occurs in the MPS renormalized $\tilde{\Gamma}_{(\mu)}$. As an exercise the reader should relate these renormalized vertices to each other.

21

Gauge theories, QCD and the renormalization group

In the previous chapter we studied the general ideas of renormalization of a field theory, in particular, the freedom in the choice of a renormalization scheme and the consequences thereof as embodied in the renormalization group. For simplicity we talked mainly in terms of scalar ϕ^4 theory. But we did illustrate the very important property of asymptotic freedom that emerges when these results are used in QCD.

In this chapter we show how to extend these ideas to the realistic case of gauge theories, and especially to QCD. We begin with a general outline of gauge theories and point out some of their subtleties, highlighting differences between QCD and QED. We then extend the renormalization group results of Chapter 20 to the case of QCD.

21.1 Introduction

In earlier chapters, and in those to follow, we constantly quote QCD corrections to naive quark–parton model estimates in various processes. It is felt at present that QCD is a serious candidate for *the* theory of strong interactions. QCD has many beautiful properties. It is a non-Abelian gauge theory describing the interaction of massless spin $\frac{1}{2}$ objects, the 'quarks', which possess an internal degree of freedom called colour, and a set of massless gauge bosons (vector mesons), the 'gluons' which mediate the force between quarks in much the same way that photons do in QED. Loosely speaking, the quarks come in three colours and the gluons in eight. More precisely, if $q^a(x), a = 1, 2, 3$ and $\mathrm{A}^b_\mu(x)$, $b = 1, \ldots, 8$, are the quark and gluon fields, respectively, then under an $SU(3)$ transformation acting on the colour indices, q and A are defined to transform as the fundamental ($\underset{\sim}{3}$) and the adjoint ($\underset{\sim}{8}$) representations of $SU(3)$ respectively. These $SU(3)$ transformations, acting solely on the colour indices, have nothing at all to do with the usual $SU(3)$ that acts on the flavour labels. For this reason we refer to them as $SU(3)_C$ and $SU(3)_F$ respectively. In what follows it must be understood that the quarks possess a flavour label as well, but it plays no rôle in QCD since the gluons are taken to be flavourless [i.e. to be singlets under $SU(3)_F$] and electrically neutral, so it will not be displayed unless specifically needed.

The theory is *known* to possess the remarkable property of 'asymptotic freedom' and is *supposed* to possess the property of 'colour confinement'. The former implies that for interactions between quarks at very short distances, i.e. for large momentum transfers, the theory looks more and more like a free field theory without interactions—this, ultimately, is the justification for the parton model. The latter means that only 'colourless' objects, i.e. objects which are colour singlets, can be found existing as real physical particles. In other words the forces between two coloured objects grow stronger with distance, so that they can never be separated. This property of confinement is also referred to as 'infrared slavery'. The proof of confinement is still lacking and remains one of the most burning theoretical questions.

All gauge theories are subtle. Non-Abelian gauge theories are very much more so. Our aim in this chapter is to provide an introduction only to the subject, but with special emphasis upon explaining those properties which are peculiar to gauge theories and which consequently are often a source of confusion and bewilderment to the non-expert. Our presentation therefore will be largely pedagogical and somewhat qualitative, and the serious student seeking a deep theoretical understanding of gauge theories is recommended to turn to the more advanced texts of Taylor (1976) or Itzykson and Zuber (1980).

In Chapter 2 we discussed at some length the concept of global and local gauge invariance, and the distinction between Abelian and non-Abelian groups of gauge transformations. There the emphasis was entirely upon the structure and symmetry of the Lagrangian, and, since we were dealing with weak and electromagnetic phenomena, so that lowest order perturbation theory usually sufficed, we did not discuss the deeper dynamical properties of these theories. QCD is a gauge theory of the *strong* interactions so that it is precisely the higher order perturbative or even non-perturbative dynamical effects that will now be important.

To begin with we shall remind ourselves of some of the peculiarities of gauge theories by looking at the best known Abelian theory, QED.

21.2 Gauge theories: QED

It is well known in classical electrodynamics that for a given electric field $\boldsymbol{E}(x)$ and magnetic field $\boldsymbol{B}(x)$, the electromagnetic potentials are not uniquely determined. The theory, expressed in terms of $A_\mu(x)$, is gauge invariant and $\boldsymbol{E}$ and $\boldsymbol{B}$ do not change if $A_\mu(x)$ is replaced by

$$A'_\mu(x) = A_\mu(x) + \partial_\mu \Lambda(x), \tag{21.2.1}$$

where $\Lambda(x)$ is an arbitrary scalar function.

Technically this arises because Maxwell's equations, in terms of the field tensor

$$F_{\mu\nu} = \partial_\mu A_\nu - \partial_\nu A_\mu \tag{21.2.2}$$

(∂_μ is short for $\partial/\partial x^\mu$), are

$$\partial_\mu F^{\mu\nu} = 0 \tag{21.2.3}$$

and $F^{\mu\nu}$ is unaltered by (21.2.1).

It is very convenient to work with the potentials $A_\mu(x)$ rather than the fields $F_{\mu\nu}(x)$, both classically and quantum mechanically, but it is not possible to do so without imposing a 'gauge condition' on the $A_\mu(x)$. A knowledge of the currents will not enable us to solve for the $A_\mu(x)$ since infinitely many $A_\mu(x)$ correspond to the same physics. Mathematically one sees this when one writes Maxwell's equation (21.2.3) in terms of the $A_\mu(x)$:

$$\Box A_\nu - \partial_\nu(\partial_\mu A^\mu) \equiv (g_{\mu\nu}\Box - \partial_\nu\partial_\mu)A^\mu(x) = 0 \tag{21.2.4}$$

or, in the presence of an electromagnetic current density,

$$(g_{\mu\nu}\Box - \partial_\nu\partial_\mu)A^\mu(x) = eJ_\nu^{\text{em}}(x). \tag{21.2.5}$$

[Recall that J_μ^{em} is defined per unit e—see (1.1.1).]

The differential operator on the left has no inverse, so we cannot solve for $A^\mu(x)$ in terms of $J^{\text{em}}(x)$, not even in the classical case!

In classical physics one picks a convenient gauge, e.g. one demands that

$$\partial_\mu A^\mu(x) = 0, \tag{21.2.6}$$

which is known as the Lorentz condition. Then Maxwell's equation (21.2.5) becomes

$$\Box A_\mu(x) = eJ_\mu^{\text{em}}(x), \tag{21.2.7}$$

which can be solved for $A_\mu(x)$, given $J_\mu^{\text{em}}(x)$.

But choosing a gauge raises a subtle question regarding the conservation of the electromagnetic current. We have no doubt that $J_\mu^{\text{em}}(x)$ is conserved. Yet in Section 3.3 we showed that it was the invariance of the Lagrangian under gauge transformations that ultimately was responsible for the conservation of the current. Once we impose (21.2.6) the Lagrangian is no longer invariant under the local gauge transformation [see (2.3.7)]

$$\begin{aligned} \phi_j(x) &\to \mathrm{e}^{-\mathrm{i}q_j\theta(x)}\phi_j(x) \\ \mathrm{A}_\mu(x) &\to \mathrm{A}_\mu(x) + \frac{1}{\mathrm{e}}\frac{\partial\theta(x)}{\partial x^\mu}. \end{aligned} \tag{21.2.8}$$

However, it *is* still invariant under the global transformation with $\theta(x) = \theta =$ constant, and this only was used in Section 2.3 to derive the conservation of the current. The moral of all this is that we must be careful only to use 'gauge-fixing' conditions like (21.2.6) which do not destroy the *global* gauge invariance of the theory.

The constraint (21.2.6) is particularly nice because it is *covariant*; $\mathrm{A}^\mu(x)$ can be a genuine four-vector and satisfy (21.2.6) in every reference frame.

At the quantum level a new problem arises. The gauge invariant Lagrangian which gives rise to the free field Maxwell's equations (21.2.3) or (21.2.4) is

$$\mathcal{L} = -\tfrac{1}{4}F_{\mu\nu}(x)F^{\mu\nu}(x). \tag{21.2.9}$$

If the $\mathrm{A}_\mu(x)$ are considered as 'generalized coordinates' then the canonically conjugate momenta are

$$\pi^\mu(x) \equiv \frac{\partial\mathcal{L}}{\partial(\partial\mathrm{A}_\mu/\partial t)}. \tag{21.2.10}$$

Note π^μ is not really a vector. It is the time component $\pi^{\mu 0}$ of the tensor $\pi^{\mu\nu} \equiv \partial\mathcal{L}/\partial(\partial\mathrm{A}_\mu/\partial x^\nu)$.

From (21.2.9) one finds

$$\pi^0 = 0; \qquad \pi^k(x) = \partial^k\mathrm{A}^0(x) - \partial^0\mathrm{A}^k(x) = F^{k0}(x), \tag{21.2.11}$$

and therefore Maxwell's equations for $F^{\mu\nu}$, for $\nu = 0$, can be written entirely in terms of $\pi^k(x)$:

$$0 = \partial_\mu F^{\mu 0} = \partial_k F^{k0}$$

since $F^{00} \equiv 0$, so

$$\partial_k\pi^k = 0. \tag{21.2.12}$$

(Incidentally, this is just Gauss's law $\boldsymbol{\nabla}\cdot\boldsymbol{E} = 0$.)

Eqn (21.2.12) is a source of trouble in the quantum theory. For the canonical commutation relations one would naturally like to assume the usual form

$$[\pi^j(\boldsymbol{x},t), \mathrm{A}^k(\boldsymbol{x}',t)] = -\mathrm{i}\delta_{jk}\delta^3(\boldsymbol{x}-\boldsymbol{x}'), \tag{21.2.13}$$

which are directly analogous to (20.2.6), but it is easily seen by direct differentiation of (21.2.13) that this contradicts the Maxwell equation (21.2.12).

There are now two possibilities: (i) either alter the commutation relations and hang on to Maxwell's equations for the quantum operators; or (ii) alter Maxwell's equations for the *quantum field operators* $\mathrm{A}_\mu(x)$ but be careful to ensure that expectation values of operators do satisfy Maxwell's equations so as to get agreement in the classical limit.

In both cases it will still be necessary to impose some gauge condition on the $\mathrm{A}_\mu(x)$ and care has to be taken to ensure that this condition itself does not contradict the commutation relations.

21.2.1 Retaining Maxwell's equations for the field operators

An example of the first approach is the method discussed in Chapter 15 of Bjorken and Drell (1965).

The trick is to modify (21.2.13) by replacing

$$\delta_{jk}\delta^3(\boldsymbol{x}-\boldsymbol{x}') \equiv \delta_{jk}\int \frac{\mathrm{d}^3\boldsymbol{k}}{(2\pi)^3}\mathrm{e}^{\mathrm{i}\boldsymbol{k}\cdot(\boldsymbol{x}-\boldsymbol{x}')}$$

by

$$\delta_{jk}^{\mathrm{TRANSVERSE}}(\boldsymbol{x}-\boldsymbol{x}') \equiv \int \frac{\mathrm{d}^3\boldsymbol{k}}{(2\pi)^3}\mathrm{e}^{\mathrm{i}\boldsymbol{k}\cdot(\boldsymbol{x}-\boldsymbol{x}')}\left(\delta_{jk}-\frac{k_jk_k}{\boldsymbol{k}^2}\right) \tag{21.2.14}$$

whose divergence is zero, so (21.2.12 and 13) are now compatible.

One can then show that $\boldsymbol{\nabla}\cdot\mathbf{A}$ commutes with everything and is thus not really an operator. A convenient (non-covariant) gauge is then

$$\boldsymbol{\nabla}\cdot\mathbf{A} = 0 \tag{21.2.15}$$

known as the Coulomb gauge. In this gauge one is effectively only quantizing the transverse oscillations of the field. Since these are the true independent degrees of freedom, corresponding to the fact that the photon can only have helicity ±1, the physics in this approach is clear. But the formalism is messy because $\mathrm{A}_\mu(x)$ is clearly no longer a four-vector if it satisfies (21.2.15) in every reference frame. $\mathrm{A}_0(\boldsymbol{x},t)$ is an operator, but not an independent one. It is given directly in terms of the charge density at the same time t.

$$\mathrm{A}_0(\boldsymbol{x},t) = \frac{e}{4\pi}\int \frac{\psi^\dagger(\boldsymbol{x}',t)\psi(\boldsymbol{x}',t)}{|\boldsymbol{x}-\boldsymbol{x}'|}\mathrm{d}^3\boldsymbol{x}'. \tag{21.2.16}$$

The Feynman rules, i.e. the kind of diagrams and the mathematical expressions corresponding to the diagrams, depend upon the gauge being used.

For example in the Coulomb gauge the bare (transverse) photon propagator is complicated looking:

$$\underset{\gamma_{\text{TRANSVERSE}}}{\overset{k}{\bullet\!\sim\!\sim\!\sim\!\sim\!\bullet}} = \frac{\mathrm{i}}{k^2+\mathrm{i}\epsilon}\left[-g_{\mu\nu} - \frac{k_\mu k_\nu}{(k\cdot\eta)^2-k^2} + \frac{(k\cdot\eta)(k_\mu\eta_\nu+\eta_\mu k_\nu)}{(k\cdot\eta)^2-k^2} - \frac{k^2\eta_\mu\eta_\nu}{(k\cdot\eta)^2-k^2}\right], \tag{21.2.17}$$

where η_μ is a four-vector whose value is (1, 0, 0, 0) in the original reference frame in which the quantization is carried out. It turns out, however, that: (*a*) when coupled to electrons the terms involving k_μ don't contribute, as a result of current conservation and the gauge invariance of the current (see Section 21.3.1d); and (*b*) the Hamiltonian contains terms involving the instantaneous Coulomb interaction between electrons, and there are Feynman graphs corresponding to this,

$$\underset{\text{COULOMB}}{\overset{k}{\bullet\!\sim\!\sim\!\sim\!\sim\!\bullet}} = \frac{\mathrm{i}\eta_\mu\eta_\nu}{(k\cdot\eta)^2-k^2}, \tag{21.2.18}$$

which just cancel the last term in (21.2.17).

The net effect is that in practice one may ignore $\gamma_{\text{TRANSVERSE}}$ and γ_{COLOUMB} and simply use the covariant propagator

$$\overset{k}{\bullet\!\sim\!\sim\!\sim\!\sim\!\bullet} = \frac{-\mathrm{i}g_{\mu\nu}}{k^2+\mathrm{i}\epsilon}, \tag{21.2.19}$$

though some care may be necessary in divergent diagrams.

One lesson to be learnt is that the mathematical expression corresponding to a given Feynman diagram does depend upon the gauge.

It also follows that the result of a single Feynman diagram may not be gauge invariant and it is necessary to combine the results from all Feynman diagrams of a given order in perturbation theory to be sure of getting a gauge invariant answer. A classic example is photoproduction, $\gamma\mathrm{p}\to\pi^+\mathrm{n}$, where, to lowest order in the electromagnetic coupling, all three diagrams:

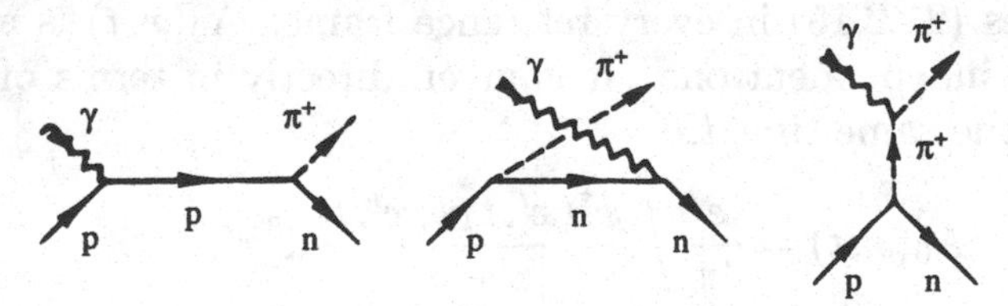

must be added to get a physically acceptable result, i.e. a result for which, in momentum space, replacing the photon polarization vector $\epsilon_\mu(k)$ by $\epsilon_\mu(k)+\lambda k_\mu$, λ arbitrary, does not alter the answer. We shall have more to say on this question in Section 21.3.1d.

In concluding this discussion it should be noted that an interesting question in the case we are examining, where we retain Maxwell's equations for the operators $\mathrm{A}_\mu(x)$, is whether we can choose a *covariant* gauge condition. There are deep theorems that show this to be impossible. [For a sophisticated mathematical discussion of these problems see Strocchi and Wightman (1974).]

21.2.2 Modifying Maxwell's equations for the field operators

Let us now turn to the alternate possibility where we retain the canonical commutation relations but modify Maxwell's equations for the field operators.

A well-known approach, due to Fermi, is to start from the Lagrangian

$$\begin{aligned}\mathcal{L} &= -\tfrac{1}{4}F_{\mu\nu}F^{\mu\nu} - \tfrac{1}{2}(\partial_\mu \mathrm{A}^\mu)^2 \\ &= -\tfrac{1}{2}(\partial_\mu \mathrm{A}^\nu)(\partial^\mu \mathrm{A}_\nu),\end{aligned} \tag{21.2.20}$$

which is no longer fully gauge invariant. (The extra piece in $\mathcal{L}$ is therefore referred to as a 'gauge fixing' term.) The equations of motion in the presence of a current density are

$$\Box \mathrm{A}_\mu(x) = eJ_\mu^{\mathrm{em}}(x) \tag{21.2.21}$$

or

$$\partial^\mu F_{\mu\nu} + \partial_\nu(\partial_\mu \mathrm{A}^\mu) = eJ_\nu^{\mathrm{em}}(x), \tag{21.2.22}$$

which would agree with Maxwell's equations if $\partial_\mu \mathrm{A}^\mu = 0$. Since we cannot demand that Maxwell's equations hold for the operators, a weaker condition, on the admissible physical states of the sytem, is imposed, namely the *subsidiary condition*

$$\langle \text{physical}|\partial_\mu \mathrm{A}^\mu(x)|\text{physical}\rangle = 0, \tag{21.2.23}$$

which ensures that Maxwell's equations hold for the expectation values of the operators.

It is important to realize that to demand (21.2.23) for *all times* is highly non-trivial. We may impose the condition at some time $t = t_0$ but what happens thereafter is controlled by the equations of motion. For a conserved current, $\partial^\mu J_\mu^{\mathrm{em}} = 0$, it is easy to see from (21.2.22) that the object $\partial_\mu \mathrm{A}^\mu(x)$ satisfies a *free field* equation

$$\Box(\partial_\mu \mathrm{A}^\mu(x)) = 0 \tag{21.2.24}$$

so its time variation is known. It is then no problem to ensure (21.2.23) for all times.

There is another rather peculiar point to notice. With the covariant gauge condition (21.2.23), $\mathrm{A}_\mu(x)$ is a true four-vector, and an inevitable consequence is that there exist states in the theory whose norm or length is negative!

A non-rigorous way to see this is to consider the vacuum expectation value of $\mathrm{A}_\mu(x)\mathrm{A}_\nu(y)$ at $x = y$, i.e. $C_{\mu\nu} \equiv \langle 0|\mathrm{A}_\mu(0)\mathrm{A}_\nu(0)|0\rangle$. Such an object is actually highly singular as $x \to y$, but we ignore this in our heuristic discussion. Clearly $C_{\mu\nu}$ is a constant tensor which is independent of any physical vectors. It can thus only be of the form $C_{\mu\nu} = -Cg_{\mu\nu}$. For $\mu = \nu$, $C_{\mu\nu}$ is just the norm of the state $|\mu\rangle \equiv \mathrm{A}_\mu(0)|0\rangle$.

If we assume that the states $|\mu = j\rangle, j = 1, 2, 3$ have positive norm, this implies that $C > 0$. It then follows that the state $|\mu = 0\rangle$ has norm $-C$. When carried through in detail this amounts to saying that a state with a 'time-like photon' has negative norm, i.e. is a ghost. When this happens one talks of an 'indefinite metric' theory.

The Fermi Lagrangian (21.2.20) is really just one of a family in which one takes

$$\mathcal{L} = -\frac{1}{4}F_{\mu\nu}F^{\mu\nu} - \frac{1}{2a}(\partial_\mu \mathrm{A}^\mu)^2 \qquad (21.2.25)$$

with a an arbitrary constant. One finds for the field equations

$$\Box \mathrm{A}_\mu + \frac{1-a}{a}\partial_\mu(\partial_\nu \mathrm{A}^\nu) = eJ_\mu^{\mathrm{em}} \qquad (21.2.26)$$

which, as in (21.2.22), would yield Maxwell's equations for the expectation values if (21.2.23) held.

From (21.2.26) one can deduce that the photon propagator is

$$-\frac{\mathrm{i}}{k^2}\left[g_{\mu\nu} - (1-a)\frac{k_\mu k_\nu}{k^2}\right]. \qquad (21.2.27)$$

Various choices of the value of a correspond to different gauges, and the physics at the end must, of course, be independent of a. Two famous examples in (21.2.27) are $a = 1$, the *Feynman gauge*, and $a = 0$, the *Landau gauge*.

Let us now turn to QCD and see to what extent the previous methods can be taken over.

21.3 Gauge theories: QCD

From the analysis in Section 2.3, the $SU(3)$ non-Abelian, gauge invariant theory for an octet of massless vector gluons interacting with a triplet of massless spin $\frac{1}{2}$ quarks involves:

1. generalized field tensors (i.e. the non-Abelian generalization of $F_{\mu\nu}$)

$$G^a_{\mu\nu} = \partial_\mu \mathrm{A}^a_\nu - \partial_\nu \mathrm{A}^a_\mu + g\, f_{abc} \mathrm{A}^b_\mu \mathrm{A}^c_\nu, \tag{21.3.1}$$

where A^a_μ is the gluon vector potential, the label $a = 1, \dots, 8$ being the octet colour label, and where f_{abc} are the structure constants for $SU(3)$, i.e. the group generators obey

$$[T_a, T_b] = \mathrm{i} f_{abc} T_c. \tag{21.3.2}$$

Note that the structure constants were written c_{ijk} in Section 2.3. Note too that colour indices are, for printing clarity, sometimes written as subscripts, sometimes as superscripts—there is no difference in meaning. The reader is also warned that some QCD papers utilize $(-g)$ where we use g. Our notation is in accordance with electroweak conventions as used in Chapter 2. Since everything calculable in perturbative QCD depends on g^2, this is irrelevant.

2. quark spinor fields ψ_j, where $j = 1, 2, 3$ labels the quark colour. There will be a set of ψ_j for each flavour, but we leave out the flavour label to simplify the notation.

3. covariant derivative operator: symbolically one has the operator

$$\hat{\mathrm{D}}_\mu \equiv \partial_\mu - \mathrm{i} g T_a \mathrm{A}^a_\mu. \tag{21.3.3}$$

When acting on some given field which transforms according to a particular representation of the group one replaces the T_a by the relevant representation matrices. Thus acting on quark fields $\hat{\mathrm{D}}_\mu$ is represented by

$$(\mathbf{D}_\mu)_{ij} = \delta_{ij} \partial_\mu - \mathrm{i} g t^a_{ij} \mathrm{A}^a_\mu \tag{21.3.4}$$

where the $\boldsymbol{t}^a, a = 1, \dots, 8$ are 3×3 Hermitian matrices which for the triplet representation of $SU(3)$ are just one half the Gell-Mann matrices $\boldsymbol{\lambda}^a$.

Acting on the gluon fields the T_a are represented by the structure constants $(T_a)_{bc} \to -\mathrm{i} f_{abc}$, so that $\hat{\mathrm{D}}_\mu$ is represented by

$$(\mathbf{D}_\mu)_{bc} = \delta_{bc} \partial_\mu - g \mathrm{A}^a_\mu f_{abc}. \tag{21.3.5}$$

Note the following useful property of the covariant derivative which is easily derived. For any field ϕ

$$[\hat{\mathrm{D}}_\mu, \hat{\mathrm{D}}_\nu]\phi = -\mathrm{i} g G^a_{\mu\nu} T_a \phi. \tag{21.3.6}$$

The gauge invariant interaction is described by the Lagrangian

$$\mathcal{L} = -\tfrac{1}{4}G^a_{\mu\nu}G^{a\mu\nu} + i\bar{\psi}_i\gamma_\mu(\mathbf{D}^\mu)_{ij}\psi_j, \tag{21.3.7}$$

where really the last term should be a sum of identical terms, one for each flavour, and where we have assumed massless quarks.

It is usually assumed that there are no quark mass terms in the original QCD Lagrangian, so that it is perfectly flavour symmetric and chirally symmetric. The flavour symmetry is presumably spontaneously broken with the quarks acquiring masses from the electroweak Higgs mechanism (Section 9.7) and/or from non-perturbative spontaneous chiral-symmetry breaking effects caused by non-zero vacuum expectation values of $\langle 0|\bar{\psi}_f\psi_f|0\rangle$ (f = flavour; not summed).

Since quarks are not supposed to exist as free physical particles their masses are not masses in the usual sense. The quark mass should be thought of simply as a parameter in the Lagrangian, to be determined in principle from experiment. However, in perturbation theory, a quark propagator has a pole at $p^2 = m^2$, whereas in the exact theory it presumably has no pole at all. So perturbative calculations are only considered reliable in kinematic regions where the momentum transfers, energies etc., are all large compared with m, which can then be neglected. Thus, determination of quark masses must come from non-perturbative studies such as current algebra or QCD sum-rules. (A comprehensive review is given in Gasser and Leutwyler, 1982.) One finds that u and d have masses of a few MeV only ($m_u \sim 4$ MeV/c^2, $m_d \sim 7$ MeV/c^2) and that $m_s \sim$ 125–150 MeV/c^2—these are referred to as 'current quark masses', and should not be confused with the 'constituent quark masses' which were extensively used in previous chapters, especially in discussing hadron spectroscopy.

We shall see later that perturbative QCD can only be applied to 'hard' processes where the energy and momentum scales are sufficiently large so that the running coupling is considerably smaller than 1. At these scales we can ignore m_u, m_d and m_s and it is adequate to utilize the massless Lagrangian (21.3.7). For the 'heavy' quarks t(!), b and perhaps c, one should modify $\mathcal{L}$ to include quark mass terms as will be discussed in Sections 21.6 and 21.7. The theoretically minded reader should consult the illuminating articles by Politzer (1976) and Georgi and Politzer (1976) on the question of the origin of the heavy quark masses.

From (21.3.7) one can, in standard fashion, derive the equations of motion

$$\mathrm{D}^\mu_{ab}G^b_{\mu\nu} = gJ^a_\nu \tag{21.3.8}$$

for the gluon field, where the quark current is

$$J^a_\nu = \bar{\psi}_i\gamma_\nu t^a_{ij}\psi_j \tag{21.3.9}$$

and for the quark fields

$$\partial\!\!\!/\psi_i = \mathrm{i}gt^a_{ij}A\!\!\!/^a\psi_j. \tag{21.3.10}$$

In analogy with QED it is necessary to fix a gauge if one works with the potentials A^a_μ. In contrast to QED the current J^a_μ is not conserved in the usual sense, i.e. $\partial^\mu J^a_\mu \neq 0$. However, from (21.3.8) it can be shown that

$$\mathrm{D}^\mu_{ab}J^b_\mu = 0. \tag{21.3.11}$$

Note, on using (21.3.4), that the interaction between the gluons and the quarks in (21.3.7) is contained in a term of the form

$$-\,\mathrm{i}\bar{\psi}_j\gamma^\mu(\mathrm{i}gt^a_{jk}\mathrm{A}^a_\mu)\psi_k = gJ^a_\mu\mathrm{A}^\mu_a. \tag{21.3.12}$$

21.3.1 Differences between QCD and QED

We here highlight some of the essential differences between the Abelian gauge theory QED and the non-Abelian QCD.

(a) Non-conservation of the quark current. At first sight it may seem puzzling that $J^a_\mu(x)$ in (21.3.8 and 9) is not conserved. After all the theory *is* invariant under the $SU(3)_\mathrm{C}$ group of gauge transformations. The point is that there *are* conserved Noether currents, but they do not coincide with the currents $J^a_\mu(x)$. Indeed the Noether currents $\tilde{J}^a_\mu$ turn out to be

$$\tilde{J}^a_\mu = J^a_\mu + f_{abc}G^b_{\mu\nu}\mathrm{A}^\nu_c. \tag{21.3.13}$$

Interestingly, using (21.3.5) in (21.3.8) yields

$$\partial^\mu G^a_{\mu\nu} = g\tilde{J}^a_\nu \tag{21.3.14}$$

but in this form the current which drives $G^a_{\mu\nu}$ is not just a quark current, but contains $G_{\mu\nu}$ itself! Ultimately this is due to the A^a_μ being themselves 'charged', i.e. having a colour charge in contrast to the photon field which is not electrically charged.

The fact that A^a_μ does not couple to a conserved quark current in the Lagrangian is a major difference from QED and is a cause of some difficulties.

(b) The self-coupling of the gauge bosons (gluons). Perhaps the most important difference from QED is the self-coupling of the gluons. Because of the quadratic term in (21.3.1), the first term in $\mathcal{L}$ in (21.3.7) contains

products of three and four factors of A^a_μ, and these imply diagrams of the type

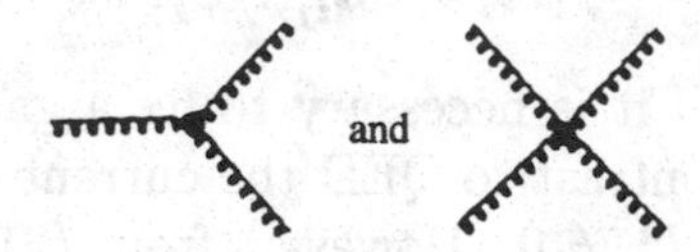

which do not occur for photons. It is in fact these interactions which are responsible for the property of asymptotic freedom in QCD.

(c) Non-gauge invariance of the QCD currents. We have commented on the fact that the quark currents J^a_μ (21.3.9) are not conserved, and mentioned the existence of the conserved Noether currents $\tilde{J}^a_\mu$ (21.3.13). Neither of these currents is invariant under the QCD global gauge transformations (non-Abelian transformations were discussed in Section 2.3)

$$\begin{aligned} \delta\psi_j &= -\mathrm{i}t^b_{jk}\psi_k\theta_b \\ \delta\bar{\psi}_j &= \mathrm{i}\bar{\psi}_k t^b_{kj}\theta_b \\ \delta A^a_\mu &= f_{abc}A^c_\mu\theta_b \\ \delta G^a_{\mu\nu} &= f_{abc}G^c_{\mu\nu}\theta_b. \end{aligned} \tag{21.3.15}$$

One finds for both J^a_μ and $\tilde{J}^a_\mu$

$$\delta J^a_\mu = f_{abc}J^c_\mu\theta_b. \tag{21.3.16}$$

In constrast the electromagnetic current is invariant under the usual Abelian electromagnetic gauge transformations

$$\delta J^{\mathrm{em}}_\mu = 0. \tag{21.3.17}$$

Now it can be shown that the *generators* of any symmetry transformation are the 'charges' associated with the Noether currents, in the following sense. If, for the electromagnetic case, we define the electric charge operator as in (2.3.10)

$$\hat{Q} = \int \mathrm{d}^3x J^{\mathrm{em}}_0(x,t) \tag{21.3.18}$$

and if the change induced in any function $F(x)$ of the field operators by an infinitesimal em gauge transformation is

$$F(x) \to F(x) + \delta F(x) \tag{21.3.19}$$

then one finds that

$$\delta F(x) \propto [\hat{Q}, F(x)] \tag{21.3.20}$$

(see also the discussion in Appendix 3). Since J_μ^{em} is gauge invariant, (21.3.17 and 20) imply that

$$\int \mathrm{d}^3\boldsymbol{x}[J_0^{\text{em}}(\boldsymbol{x},t), J_\mu^{\text{em}}(y)] = 0. \tag{21.3.21}$$

The analogue of (21.3.20) for QCD involves the colour 'charges' $\hat{Q}^a$ associated with the eight Noether currents $\tilde{J}_\mu^a(x)$, via the analogue of (21.3.18), but for the QCD currents, we have for both $J_\mu^b(y)$ and $\tilde{J}_\mu^b(y)$,

$$\int \mathrm{d}^3\boldsymbol{x}[\tilde{J}_0^a(\boldsymbol{x},t), J_\mu^b(y) \text{ or } \tilde{J}_\mu^b(y)] \neq 0 \tag{21.3.22}$$

since the RHS of (21.3.16) is not zero.

We shall see in the next section that this has important implications about the structure of QCD matrix elements.

(d) Concerning $\epsilon^\mu(k) \to \epsilon^\mu(k) + ck^\mu$ in QCD. Consider Compton scattering as a typical QED reaction

$$\gamma(k) + \mathrm{e}(\mathrm{p}) \to \gamma(k') + \mathrm{e}(\mathrm{p}').$$

Its scattering amplitude is of the form

$$(S-1)_{fi} = \epsilon^{\mu^*}(k')M_{\mu\nu}\epsilon^\nu(k) \tag{21.3.23}$$

and it is usually stated that 'as a consequence of gauge invariance' the result must not change if, for any one or more photons, one makes the substitution

$$\epsilon^\mu(k) \to \epsilon^\mu(k) + ck^\mu \tag{21.3.24}$$

where c is an arbitrary number. This implies the relations

$$k'^\mu M_{\mu\nu} = M_{\mu\nu}k^\nu = 0. \tag{21.3.25}$$

Similar results hold for any QED process involving any number of external photons.

It is also frequently stated that the corresponding matrix elements in QCD are *not* invariant under the substitution (21.3.24) for the gluon polarization vectors.

This non-trivial matter deserves further explanation. In QED the invariance under (21.3.24) seems to follow because of gauge invariance from the fact that the theory is invariant under the local gauge transformation

$$\mathrm{A}_\mu(x) \to \mathrm{A}_\mu(x) + \frac{1}{e}\frac{\partial\Lambda(x)}{\partial x^\mu} \tag{21.3.26}$$

and (21.3.24) looks like the Fourier transform of (21.3.26) when $\mathrm{A}_\mu(x)$

describes a plane wave. This is fine in classical electromagnetic theory, but it is incorrect in QED for several reasons. Firstly, the replacement (21.3.24) will only follow from (21.3.26) if $\Lambda(x)$ is not an ordinary function but is chosen to be a scalar quantized field $\hat{\Lambda}(x)$ linear in the photon creation and annihilation operators. Secondly, we had to make a choice of gauge, and once this is done the theory is no longer invariant under *local* gauge transformations.

However, as explained earlier, we always impose the gauge choice in such a way that *global* gauge invariance is preserved. The invariance of the QED matrix elements actually follows from three properties:

1. The global gauge invariance implies that there exists a conserved Noether current.
2. $\mathrm{A}_\mu(x)$ couples to this conserved current in the QED Lagrangian.
3. This current is itself gauge invariant.

The most direct proof that these imply invariance under the substitution (21.3.24) utilizes the LSZ reduction formalism (see, for example, Bjorken and Drell, 1965). Let us return to Compton scattering as a concrete illustration. The amplitude $M_{\mu\nu}$ in (21.3.23) is given by

$$M_{\mu\nu} = (-\mathrm{i}e)^2 \int \mathrm{d}^4x \mathrm{d}^4y \, \mathrm{e}^{\mathrm{i}k'\cdot y - \mathrm{i}k\cdot x} \langle \mathrm{p}'|T\left[J_\mu^{\mathrm{em}}(y) J_\nu^{\mathrm{em}}(x)\right]|\mathrm{p}\rangle \qquad (21.3.27)$$

where T is the time-ordering operator and where as in (21.2.21)

$$\Box \mathrm{A}_\mu(x) = eJ_\mu^{\mathrm{em}}(x). \qquad (21.3.28)$$

We shall now show that the first of (21.3.25) holds.

From (21.3.27) we can write

$$k'^\mu M_{\mu\nu} = \frac{(-\mathrm{i}e)^2}{\mathrm{i}} \int \mathrm{d}^4x \mathrm{d}^4y \left[\frac{\partial}{\partial y_\mu}\mathrm{e}^{\mathrm{i}k'\cdot y}\right] \mathrm{e}^{-\mathrm{i}k\cdot x} \langle \mathrm{p}'|T\left[J_\mu^{\mathrm{em}}(y) J_\nu^{\mathrm{em}}(x)\right]|\mathrm{p}\rangle. \qquad (21.3.29)$$

Integrating by parts and discarding the surface terms

$$k'^\mu M_{\mu\nu} = \mathrm{i}(-\mathrm{i}e)^2 \int \mathrm{d}^4x \mathrm{d}^4y \, \mathrm{e}^{\mathrm{i}k'\cdot y} \mathrm{e}^{-\mathrm{i}k\cdot x} \langle \mathrm{p}'|\frac{\partial}{\partial y_\mu} T\left[J_\mu^{\mathrm{em}}(y) J_\nu^{\mathrm{em}}(x)\right]|\mathrm{p}\rangle. \qquad (21.3.30)$$

The spatial derivatives are uninfluenced by T, i.e.

$$\frac{\partial}{\partial y_i} T\left[J_i^{\mathrm{em}}(y) J_\nu^{\mathrm{em}}(x)\right] = T\left(\frac{\partial J_i^{\mathrm{em}}}{\partial y_i}(y) J_\nu^{\mathrm{em}}(x)\right) \qquad (21.3.31)$$

whereas for the time derivative

$$\begin{aligned}\frac{\partial}{\partial y_0}T\left[J_0^{\rm em}(y)J_\nu^{\rm em}(x)\right] &= \frac{\partial}{\partial y_0}\left[\theta(y_0-x_0)J_0^{\rm em}(y)J_\nu^{\rm em}(x)\right.\\ &\qquad\left.+\,\theta(x_0-y_0)J_\nu^{\rm em}(x)J_0^{\rm em}(y)\right]\\ &= \delta(y_0-x_o)\left[J_0^{\rm em}(y)J_\nu^{\rm em}(x)-J_\nu^{\rm em}(x)J_0^{\rm em}(y)\right]\\ &\qquad+T\left(\frac{\partial J_0^{\rm em}}{\partial y_0}(y)J_\nu^{\rm em}(x)\right).\end{aligned} \tag{21.3.32}$$

Adding (21.3.31 and 32) gives

$$\begin{aligned}\frac{\partial}{\partial y_\mu}T\left[J_\mu^{\rm em}(y)J_\nu^{\rm em}(x)\right] &= \delta(y_0-x_0)\left[J_0^{\rm em}(y),J_\nu^{\rm em}(x)\right]\\ &\qquad+T\left(\frac{\partial J_\mu^{\rm em}}{\partial y_\mu}(y)J_\nu^{\rm em}(x)\right).\end{aligned} \tag{21.3.33}$$

For the conserved electromagnetic current, the last term is zero. The other term is an equal-time commutator, so that taking $J_\mu^{\rm em}$ of the form $\bar\psi\gamma_\mu\psi$ and using the canonical anticommutation relations, one can evaluate it explicitly. One finds

$$\left[J_\mu^{\rm em}(y),J_\nu^{\rm em}(x)\right]_{x_0=y_0}=\bar\psi(x)[\gamma_\mu\gamma_0\gamma_\nu-\gamma_\nu\gamma_0\gamma_\mu]\psi(x)\delta^3(\boldsymbol{x}-\boldsymbol{y}) \tag{21.3.34}$$

whence

$$[J_0^{\rm em}(y),J_\nu^{\rm em}(x)]_{x_0=y_0}=0. \tag{21.3.35}$$

Thus the RHS of (21.3.33) is zero and hence (21.3.25) is proved. The result generalizes to reactions with any number of external photons.

The vanishing of the equal-time commutator in (21.3.34) for $\mu=0$ looks like a miraculous accident, but it really is linked to the fact that $J_\mu^{\rm em}$ is gauge invariant. For (21.3.35) is consistent with the requirement (21.3.21).

In QCD for the 'Compton' reaction on nucleons

$$G^b(k)+N(\mathrm{p})\to G^a(k')+N(\mathrm{p}')$$

the analogue of (21.3.27) and (21.3.28) is

$$M_{\mu\nu}^{ab}=(-\mathrm{i}g)^2\int \mathrm{d}^4x\mathrm{d}^4y\,\mathrm{e}^{\mathrm{i}k'\cdot y}\mathrm{e}^{-\mathrm{i}k\cdot x}\langle\mathrm{p}'|T\left[\hat J_\mu^a(y)\hat J_\nu^b(x)\right]|\mathrm{p}\rangle \tag{21.3.36}$$

where $\hat J_\mu^a(x)$ is the current such that

$$\Box\mathrm{A}_\mu^a(x)=g\hat J_\mu^a(x). \tag{21.3.37}$$

From (21.3.14) and (21.3.1) one finds

$$\hat J_\mu^a(x)=\tilde J_\mu^a(x)-f_{abc}\partial^\nu(\mathrm{A}_\nu^b\mathrm{A}_\mu^c). \tag{21.3.38}$$

The current $\hat{J}^a_\mu$ is conserved because the Noether current $\tilde{J}^a_\mu$ is, and because the last term in (21.3.38) vanishes, by antisymmetry in the colour indices, when acted on by ∂^μ.

However the current $\hat{J}^a_\mu(x)$, like J^a_μ, is not gauge invariant [see (21.3.16)]; nor is the Noether current. So there is no reason for the equal-time commutator of the currents to vanish. Hence the QCD analogue of (21.3.35) is

$$[\hat{J}^a_0(y), \hat{J}^b_\nu(x)]_{x_0=y_0} \neq 0. \tag{21.3.39}$$

Thus the first term on the RHS of the QCD analogue of (21.3.33) will not vanish. Hence

$$k'^\mu M^{ab}_{\mu\nu} \neq 0; \qquad M^{ab}_{\mu\nu} k^\nu \neq 0. \tag{21.3.40}$$

However, it is also possible for example to write for the scattering amplitude, instead of (21.3.23),

$$(S-1)_{fi} = \epsilon^{\mu^*}(k') M^{ab}_\mu \tag{21.3.41}$$

where, for QCD

$$M^{ab}_\mu = (-\mathrm{i}g) \int \mathrm{d}^4y \, \mathrm{e}^{\mathrm{i}k'\cdot y} \langle \mathrm{p}'|\hat{J}^a_\mu(y)|\mathrm{p}; k, b\rangle. \tag{21.3.42}$$

A simple integration by parts then shows that

$$k'^\mu M^{ab}_\mu = 0 \tag{21.3.43}$$

purely as a consequence of the conservation of $\hat{J}^a_\mu$.

Since one has, also

$$M^{ab}_\mu = M^{ab}_{\mu\nu} \epsilon^\nu(k) \tag{21.3.44}$$

(21.3.43) implies [note the difference from (21.3.40)]

$$k'^\mu M^{ab}_{\mu\nu} \epsilon^\nu(k) = 0. \tag{21.3.45}$$

But $\epsilon^\nu(k)$ is, apart from the condition $k.\epsilon(k) = 0$, an arbitrary 4-vector. (Its normalization is clearly irrelevant.) So (21.3.45) must also hold if $\epsilon^\nu(k)$ is replaced by k^ν *provided* $k^2 = 0$.

The general result in QCD is the following. For any number of gluons if the scattering amplitude is written in the form

$$\epsilon^{\mu_1{}^*}(k'_1) \ldots \epsilon^{\mu_n{}^*}(k'_n) M^{a_1\ldots a_n; b_1\ldots b_n}_{\mu_1\ldots\mu_n;\nu_1\ldots\nu_n} \epsilon^{\nu_1}(k_1) \ldots \epsilon^{\nu_m}(k_m) \tag{21.3.46}$$

then one gets zero if *any number* (≥ 1) of the $\epsilon^{\mu_j}(k_j)$ are replaced by $k_j^{\mu_j}$ *provided* that all these k_j, with the exception of *at most one* of them, satisfy $k_j^2 = 0$.

We shall now see that the failure of (21.3.25) to hold in QCD has significant consequences.

(e) Ghosts. An immediate complication follows if one insists on working covariantly with all four A^a_μ for each a. There is an apparent failure of unitarity in the sense that $S^\dagger S \neq 1$ and one has to introduce scalar 'ghost' fields to rectify the matter.

The failure can be seen most simply by putting $S = 1 + \mathrm{i}T$, noting that $S^\dagger S = 1$ implies that for an elastic scattering process $\mathrm{Im}\, T = \frac{1}{2} T^\dagger T$, and this should hold in each order of perturbation theory. Consider $q\bar{q} \to q\bar{q}$ in fourth order via two-gluon exchange. Unitarity requires that

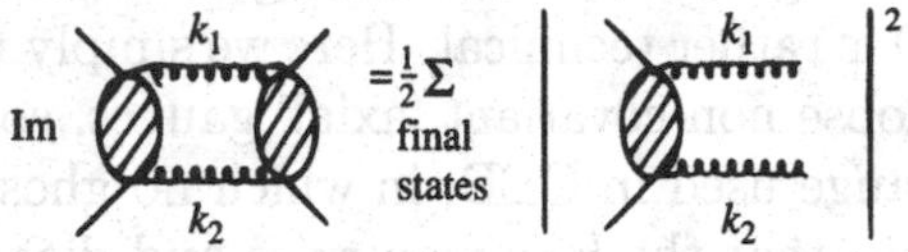

Im [k_1, k_2] $= \frac{1}{2} \sum_{\text{final states}}$ | [k_1, k_2] $|^2$ (21.3.47)

where

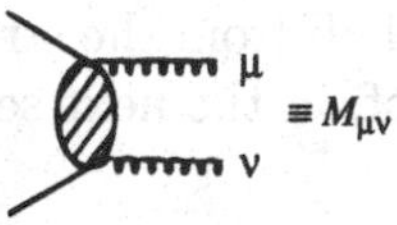

[μ, ν] $\equiv M_{\mu\nu}$

is the $q\bar{q} \to$ two-gluon amplitude in order g^2. If the gluon propagators are of the form $g_{\mu\nu}/(k^2 + \mathrm{i}\epsilon)$ then taking the imaginary part puts the gluons on their mass shell and one has

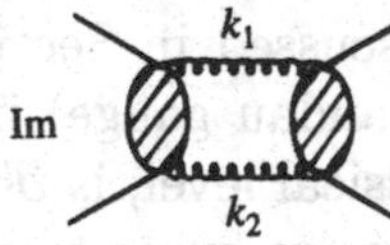

$$\text{Im} \,[k_1, k_2] \propto M^*_{\mu_1\mu_2} g_{\mu_1\nu_1} g_{\mu_2\nu_2} M_{\nu_1\nu_2} = M^*_{\mu\nu} M_{\mu\nu}, \tag{21.3.48}$$

whereas

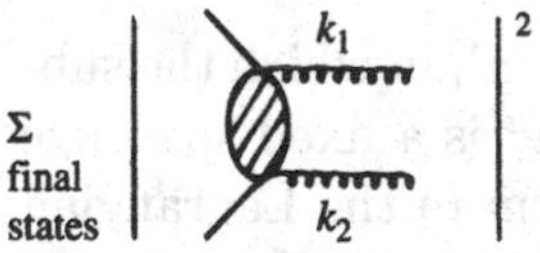

$$\sum_{\substack{\text{final}\\ \text{states}}} \Big| [k_1, k_2] \Big|^2 \propto \sum_{\lambda_1 = \pm 1} \sum_{\lambda_2 = \pm 1} |M_{\mu_1\mu_2} \epsilon_{\mu_1}(k_1, \lambda_1) \epsilon_{\mu_2}(k_2, \lambda_2)|^2$$

$$= M^*_{\mu_1\mu_2} M_{\nu_1\nu_2} \left[\sum_{\lambda_1} \epsilon^*_{\mu_1}(k_1, \lambda_1) \epsilon_{\nu_1}(k_1, \lambda_1) \right] \left[\sum_{\lambda_2} \epsilon^*_{\mu_2}(k_2, \lambda_2) \epsilon_{\nu_2}(k_2, \lambda_2) \right]. \tag{21.3.49}$$

The sums over helicity do not give $g_{\mu_1\nu_1} g_{\mu_2\nu_2}$. Each sum differs from $g_{\mu\nu}$ by terms proportional to k_μ or k_ν and (21.3.49) does not agree with (21.3.48). Note that, in QED, with photons replacing the gluons, the two results *would agree* because terms proportional to k_μ or k_ν vanish because $k_\mu M_{\mu\nu} = k_\nu M_{\mu\nu} = 0$ by (21.3.25). This does not happen in QCD because of (21.3.40).

The reader should beware, as the above implies, that the oft used *trick* of replacing $\sum_\lambda \epsilon_\mu^*(\lambda)\epsilon_\nu(\lambda)$ by $(-g_{\mu\nu})$ in sums over helicity in perturbative QED calculations can generally not be used in QCD except for the *first* gluon dealt with.

In order to restore unitarity one has to introduce extra 'ghost' fields which induce ghost exchange in addition to the gluon exchange in (21.3.47 and 48) and which are designed plainly and simply to make (21.3.49) agree with (21.3.48).

The introduction of ghost fields and the rigorous *derivation* of the Feynman rules for QCD is rather technical. Here we simply note that it is actually possible to choose non-covariant 'axial' gauges, somewhat analogous to the Coulomb gauge used in QED, in which no ghosts are needed, and we stress once again that the Feynman rules and diagrams depend upon the gauge being utilized. Aside from the question of ghosts, the Feynman rules can essentially be read off from the structure of the Lagrangian, and we give them, without proof, in the next section.

21.4 Feynman rules for QCD

We present here a list of the Feynman rules for QCD valid in two classes of gauges:

- the *covariant* gauges labelled by 'a', as discussed in Section 21.2 ($a = 1$ is the Feynman gauge; $a = 0$ the Landau gauge) in which the subsidiary condition, at least at the classical level, is $\partial^\mu \mathrm{A}^c_\mu = 0$ for all values of the colour label c, and the gauge fixing term in the Lagrangian is $-1/2a \sum_c(\partial^\mu \mathrm{A}^c_\mu)^2$.

- an *axial* gauge, one of a family again labelled by 'a', in which the subsidiary condition is $n^\mu \mathrm{A}^c_\mu = 0$ for all c, where n^μ is a fixed spacelike or null 4-vector, and where the gauge fixing term in the Lagrangian is $-1/2a \sum_c(n^\mu \mathrm{A}^c_\mu)^2$.

We allow the quarks to have a mass parameter m which should be put to zero when working with massless quarks.

21.4.1 The propagators

lepton $\xrightarrow{\;p\;}$ $$\frac{\mathrm{i}(\not p + m)}{p^2 - m^2 + \mathrm{i}\epsilon}$$

quark $j \xrightarrow{\;p\;} l$ $$\delta_{jl}\frac{\mathrm{i}(\not p + m)}{p^2 - m^2 + \mathrm{i}\epsilon}$$

In the above the arrow indicates the flow of fermion number and p is the 4-momentum in that direction. (Note: j, l are quark colour labels, b, c gluon and ghost colour labels.)

gluon

b, β ——— k ——— c, γ

$$\delta_{bc}\frac{\mathrm{i}}{k^2+\mathrm{i}\epsilon}\times\begin{cases}\text{Covariant gauges:}\\ \left[-g_{\beta\gamma}+(1-a)\frac{k_\beta k_\gamma}{k^2+\mathrm{i}\epsilon}\right]\\ \\ \text{Axial gauges with } a=0\text{:}\\ \left[-g_{\beta\gamma}+\frac{n_\beta k_\gamma+n_\gamma k_\beta}{n\cdot k}-\frac{n^2 k_\beta k_\gamma}{(n\cdot k)^2}\right]\end{cases}$$

Note that in the above axial gauges the propagator is orthogonal to n^β; and is orthogonal to k^β when $k^2 = 0$.

ghost

b - - - - p - - - - c

$$\delta_{bc}\frac{\mathrm{i}}{p^2+\mathrm{i}\epsilon} \quad \text{(Covariant gauges only)}$$

21.4.2 The vertices

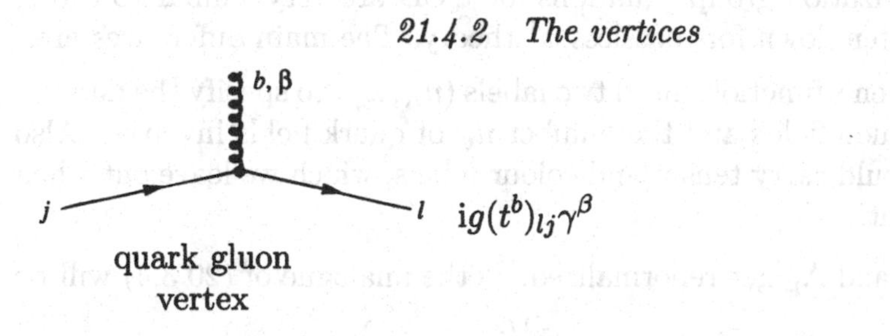

$$\mathrm{i}g(t^b)_{lj}\gamma^\beta$$

a, α ; p ; r ; q ; c, γ ; b, β

triple gluon vertex

$$gf_{abc}\left[g^{\alpha\beta}(p-q)^\gamma + g^{\beta\gamma}(q-r)^\alpha + g^{\gamma\alpha}(r-p)^\beta\right]$$

$$(p+q+r=0)$$

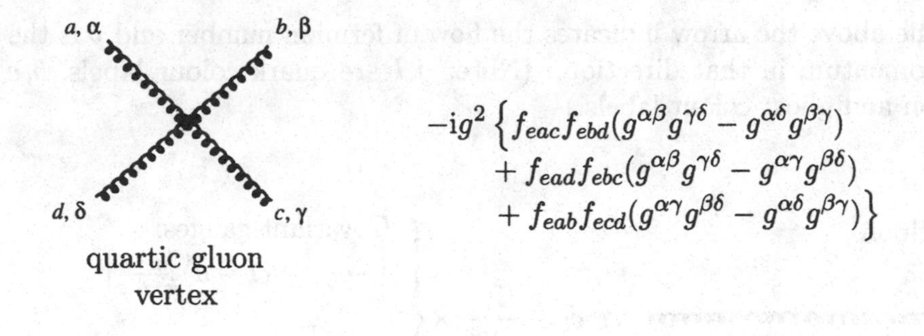

$$-ig^2 \left\{ f_{eac} f_{ebd} (g^{\alpha\beta} g^{\gamma\delta} - g^{\alpha\delta} g^{\beta\gamma}) + f_{ead} f_{ebc} (g^{\alpha\beta} g^{\gamma\delta} - g^{\alpha\gamma} g^{\beta\delta}) + f_{eab} f_{ecd} (g^{\alpha\gamma} g^{\beta\delta} - g^{\alpha\delta} g^{\beta\gamma}) \right\}$$

quartic gluon vertex

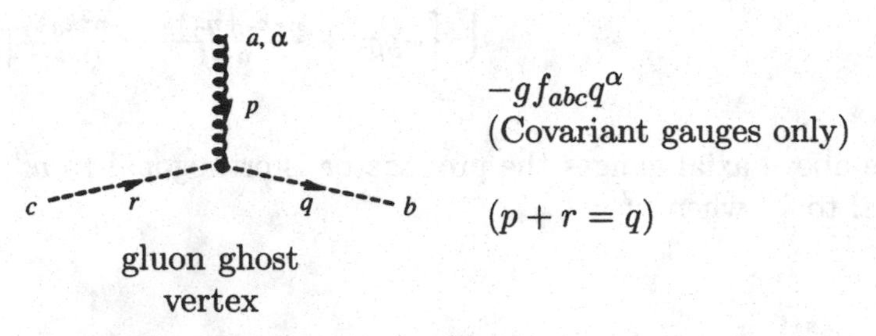

$-g f_{abc} q^{\alpha}$
(Covariant gauges only)

$(p + r = q)$

gluon ghost vertex

Note that the ghosts are scalar fields, but a factor (-1) must be included for each closed loop, as is the case for fermions.

We turn now to the question of renormalization and the renormalization group in QCD.

21.5 The renormalization group for QCD

The renormalization group equations for QCD are very similar to those, (20.6.9), written down for massless ϕ^4 theory. The main differences are:

1. The Green's functions need two labels $(n_{\mathrm{A}}, n_{\psi})$ to specify the number n_{A} of gluon fields and the number n_{ψ} of quark fields involved. Also they should carry tensor and colour labels, which we leave out when irrelevant.

2. Both ψ and A_{μ} get renormalized, so the analogue of (20.3.4) will be

$$\left.\begin{aligned} \psi_{\mathrm{B}} &= Z_{\psi}^{1/2} \psi \\ (\mathrm{A}_{\mathrm{B}}^{a})_{\mu} &= Z_{\mathrm{A}}^{1/2} \mathrm{A}_{\mu}^{a}. \end{aligned}\right\} \qquad (21.5.1)$$

   For the coupling constant renormalization the form depends upon whether we are using a momentum point subtraction scheme (MPS) or a dimensional regularization scheme (DR) [compare (20.3.4) with (20.4.8)]. One takes

$$\begin{aligned} &\text{MPS:} \quad g_{\mathrm{B}} = Z_{\mathrm{A}}^{-1/2} Z_{\psi}^{-1} Z_{g} g_{\mu} \\ &\text{DR:} \quad\;\; g_{\mathrm{B}} = Z_{\mathrm{A}}^{-1/2} Z_{\psi}^{-1} Z_{g} \mu^{\epsilon} g_{[\mu]} \end{aligned} \qquad (21.5.2)$$

Note that in the literature the Zs are sometimes given the following labels:

$$Z_{\mathrm{A}} \to Z_3; \qquad Z_\psi \to Z_2; \qquad Z_g \to Z_1.$$

As in (20.6.5 or 10) β is defined by

$$\beta[\alpha(\mu^2)] = \frac{\mu}{2}\frac{\mathrm{d}}{\mathrm{d}\mu}\alpha(\mu^2). \tag{21.5.3}$$

3. As a consequence the term $n\gamma$ in (20.6.9) is replaced by $n_{\mathrm{A}}\gamma_{\mathrm{A}} + n_\psi\gamma_\psi$ where, analogously to (20.6.6),

$$\gamma^{\mathrm{A}} \equiv \frac{\mu}{2}\frac{\mathrm{d}}{\mathrm{d}\mu}(\ln Z_{\mathrm{A}}) \quad \text{and} \quad \gamma^\psi \equiv \frac{\mu}{2}\frac{\mathrm{d}}{\mathrm{d}\mu}(\ln Z_\psi). \tag{21.5.4}$$

4. In the gauges discussed in Section (21.2.2) the parameter a may also get renormalized, i.e. the initial gauge-fixing parameter, call it a_{B}, may get altered as a consequence of the interactions. In this situation one must introduce the renormalized 'a' via

$$a_{\mathrm{B}} = Z_{\mathrm{A}}a. \tag{21.5.5}$$

In the Landau gauge, where $a_{\mathrm{B}} = 0$, it remains true that $a = 0$ after renormalization. This gauge is simplest since there is no term to express how a responds to a change in μ. *In all that follows we work in the Landau gauge.*

The analogue of (20.6.9) for QCD in the Landau gauge is thus, for *all* p_i,

$$\left(\mu\frac{\partial}{\partial\mu} + 2\beta\frac{\partial}{\partial\alpha} - n_{\mathrm{A}}\gamma^{\mathrm{A}} - n_\psi\gamma^\psi\right)\Gamma^{(n_{\mathrm{A}},n_\psi)}[p;\alpha(\mu^2),\mu] = 0, \tag{21.5.6}$$

where, of course, α is short for α_s, the strong interaction coupling.

Note that if we allow a quark of flavour f to have a non-zero bare mass in the QCD Lagrangian, then analogously to (20.6.4) the LHS of (21.5.6) will contain an additional term $2\gamma_{m_f}\partial/\partial\ln m_f^2$ where $\gamma_{m_f} \equiv \partial\ln m_f^2/\partial\ln\mu^2$, and the Green's functions will be functions of m_f as well.

21.5.1 Specification of the renormalization scheme in QCD

We now outline how one goes about calculating the important functions $\beta, \gamma^{\mathrm{A}}, \gamma^\psi$. First of all we must specify more precisely the renormalization scheme labelled by μ. In analogy with (20.3.16) we demand in an MPS

scheme

$$\stackrel{k^2\to-\mu^2}{\longrightarrow} \frac{\mathrm{i}\left(-g_{\beta\gamma}+\frac{k_\beta k_\gamma}{k^2}\right)}{k^2+\mathrm{i}\epsilon}\delta_{bc} \qquad (21.5.7)$$

and

$$\stackrel{p^2\to-\mu^2}{\longrightarrow} \frac{\mathrm{i}\,\not p}{p^2+\mathrm{i}\epsilon}\delta_{jl}. \qquad (21.5.8)$$

These will be satisfied if the amputated Green's functions satisfy

$$\Gamma^{(2,0)} \stackrel{k^2\to-\mu^2}{\longrightarrow} \mathrm{i}(g_{\mu\nu}k^2-k_\mu k_\nu)\delta_{bc} \qquad (21.5.9)$$

and

$$\Gamma^{(0,2)} \stackrel{p^2\to-\mu^2}{\longrightarrow} -\mathrm{i}\,\not p\,\delta_{jl} \qquad (21.5.10)$$

For the three-gluon vertex

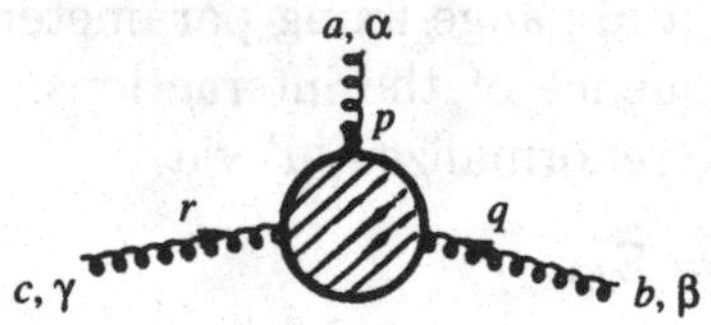

$$\equiv \overset{(3,0)\alpha\beta\gamma}{\Gamma}{}_{abc}\,(p,q,r) \qquad (21.5.11)$$

the tensor structure in the Landau gauge is more complicated than that given for the bare vertex in Section 21.4.2. There are also terms involving three powers of the momenta $p^\alpha q^\beta r^\gamma$ etc. But their coefficient is finite, whereas the coefficient of the tensor structure of the bare vertex diverges. In analogy to (20.3.17) we can demand, for example, that at the symmetric point $p^2=q^2=r^2=-\mu^2$

$$\overset{(3,0)\alpha\beta\gamma}{\Gamma}{}_{abc}\,(p,q,r) \;\to\; gf_{abc}\left[g^{\alpha\beta}(p-q)^\gamma+g^{\beta\gamma}(q-r)^\alpha+g^{\gamma\alpha}(r-p)^\beta\right]$$
$$+\text{ terms cubic in the momenta.} \qquad (21.5.12)$$

Finally, for the gluon–quark vertex,

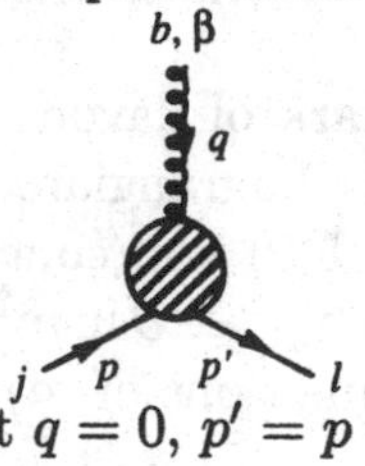

$$\equiv \overset{(1,2)\beta,b}{\Gamma}{}_{lj}\,(p',q,p) \qquad (21.5.13)$$

we can impose that at $q=0$, $p'=p=k$ where $k^2=-\mu^2$

$$\overset{(1,2)\beta,b}{\Gamma}{}_{l,j}\,(k,0,k) \stackrel{k^2\to-\mu^2}{\longrightarrow} \mathrm{i}g\gamma^\beta t^b_{lj}. \qquad (21.5.14)$$

What all this amounts to is that we have demanded the lowest order results to hold at $k^2\to-\mu^2$ in order to fix the normalization of $\Gamma^{(2,0)}$, $\Gamma^{(0,2)}$, $\Gamma^{(3,0)}$ and $\Gamma^{(1,2)}$.

With this renormalization scheme one now computes the lowest order corrections to the gluon propagator (- - - stands for ghost),

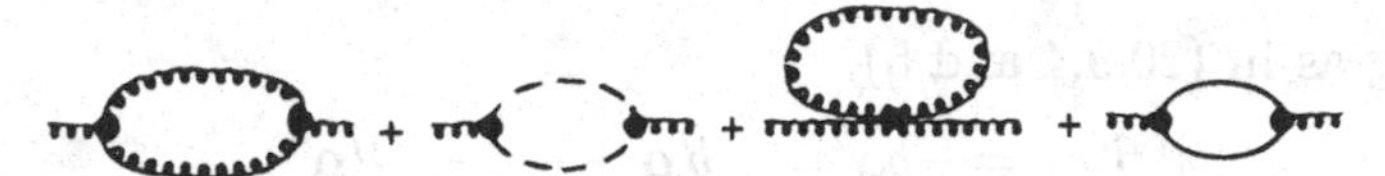

to the quark propagator,

to the three-gluon vertex,

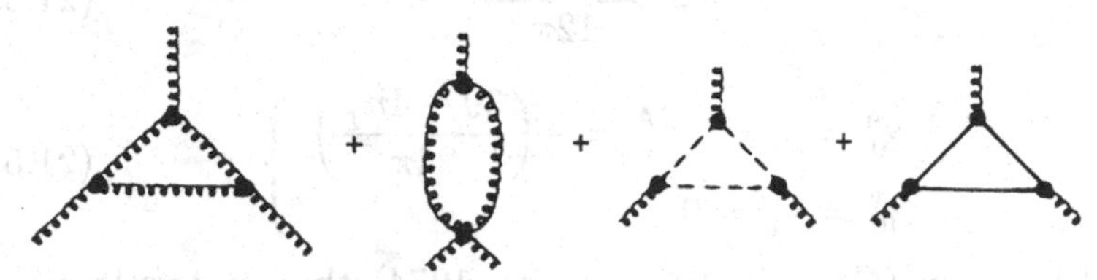

and to the gluon–quark vertex

We do not give the calculations [a comprehensive discussion is given in Gross (1976)], but note that the results need only be found for k^2 close to $-\mu^2$. When the results are substituted into (21.5.6) and the limit $k^2 \to -\mu^2$ taken, one finds

$$\gamma^{\psi} = 0 + O(\alpha^2) \tag{21.5.15}$$

$$\gamma^{\mathrm{A}} = -\frac{\alpha}{4\pi}\left[\frac{13}{6}C_2(G) - \frac{4}{3}T(R)\right] + O(\alpha^2) \tag{21.5.16}$$

$$\beta(\alpha) = -\frac{\alpha^2}{4\pi}\left[\frac{11}{3}C_2(G) - \frac{4}{3}T(R)\right] + O(\alpha^3) \tag{21.5.17}$$

where $C_2(G)$ is a constant that depends on the group involved

$$\delta_{ab}C_2(G) = f_{acd}f_{bcd} \tag{21.5.18}$$

and one has

$$C_2[SU(3)] \equiv C_{\mathrm{A}} = 3. \tag{21.5.19}$$

$T(R)$ is a constant that depends upon the representation that the quarks belong to and the number n_f of quark flavours,

$$\delta_{ab}T(R) \equiv n_f \mathrm{Tr}(t^a t^b) \tag{21.5.20}$$

and for $SU(3)$ and the triplet representation

$$T[SU(3)\,\text{triplet}] = \tfrac{1}{2}n_f.$$

Writing as in (20.9.4 and 5)

$$\beta(\alpha) = -b\alpha^2[1 + b'\alpha + \cdots] + O(\alpha^4) \tag{21.5.21}$$

and

$$\gamma^{\mathrm{A}} = \gamma_0^{\mathrm{A}} + \gamma_1^{\mathrm{A}}\alpha + O(\alpha^2) \tag{21.5.22}$$

we see that for $SU(3)_C$

$$b = \frac{33 - 2n_f}{12\pi} \tag{21.5.23}$$

$$\left.\begin{aligned} \gamma_0^{\mathrm{A}} = 0; \qquad \gamma_1^{\mathrm{A}} &= -\left(\frac{39 - 4n_f}{24\pi}\right) \\ \gamma_0^{\psi} = \gamma_1^{\psi} = 0 & \end{aligned}\right\} \tag{21.5.24}$$

It can be shown (Caswell, 1974; Jones, 1974) that at two-loop order one finds the following expression for b' in (21.5.21)

$$b' = \frac{153 - 19n_f}{2\pi(33 - 2n_f)}. \tag{21.5.25}$$

Because these values were deduced in a massless MPS renormalization scheme, they are, as explained in Section 20.9, the same in the MS and $\overline{\text{MS}}$ schemes.

From (21.5.23) we can see, at last, the justification for the shape of the curve of $\beta(\alpha)$ (Fig. 20.4) at small α. Provided

$$n_f \leq 16 \tag{21.5.26}$$

we have $b > 0$ and $\beta(\alpha)$ will go negative for small positive α. This is, of course, the criterion for asymptotic freedom.

We can also see in (21.5.17) the rôle of the non-Abelian group. For QED, with its Abelian group, all f_{abc} are zero, i.e. $C_2(G) = 0$ and $\beta(\alpha)$ is positive for small positive α, and we do not have asymptotic freedom.

For QCD the running coupling $\bar{\alpha}(\alpha, t)$ in leading approximation will be given by (20.9.7) with b given by (21.5.23).

21.5.2 Consequences of the renormalization group in QCD

Finally in order to understand the high momentum behaviour of QCD we must repeat the steps leading from (20.6.9) to (20.8.4). The mass dimension of the Green's function in QCD is

$$\begin{aligned} [G^{(n_{\mathrm{A}},n_\psi)}] &= [\mathrm{M}]^{n_{\mathrm{A}}+3n_\psi/2-4(n_{\mathrm{A}}+n_\psi-1)} \\ &= [\mathrm{M}]^{4-3n_{\mathrm{A}}-5n_\psi/2} \end{aligned} \tag{21.5.27}$$

and of the amputated Green's functions, is

$$[\Gamma^{(n_\mathrm{A},n_\psi)}] = [\mathrm{M}]^{4-n_\mathrm{A}-3n_\psi/2}. \tag{21.5.28}$$

All results of Section 20.8 will now hold for QCD with the replacement

$$(4-n) \to 4 - n_\mathrm{A} - 3n_\psi/2. \tag{21.5.29}$$

If we leave out inessential labels, the analogue of (20.8.19) will be

$$\Gamma^{(n_\mathrm{A},n_\psi)}[\eta p_i; \alpha(\mu^2), \mu] = \eta^{(4-n_\mathrm{A}-3n_\psi/2)} \times \\ \times \exp\left\{-\tfrac{1}{2}\int_{\alpha(\mu^2)}^{\alpha(\eta^2\mu^2)} \frac{n_\mathrm{A}\gamma^\mathrm{A}(x)+n_\psi\gamma^\psi(x)}{\beta(x)}\,\mathrm{d}x\right\}\Gamma^{(n_\mathrm{A},n_\psi)}[p_i; \alpha(\eta^2\mu^2), \mu]. \tag{21.5.30}$$

To lowest order, bearing in mind (21.5.15) and (21.5.22), we find for the analogue of (20.9.11)

$$\Gamma^{(n_\mathrm{A},n_\psi)}[\eta p_i; \alpha(\mu^2), \mu] = \eta^{(4-n_\mathrm{A}-3n_\psi/2)}\left[\frac{\alpha(\eta^2\mu^2)}{\alpha(\mu^2)}\right]^{n_\mathrm{A}\gamma_1^\mathrm{A}/2b} \times \\ \times \Gamma^{(n_\mathrm{A},n_\psi)}[p_i; \alpha(\eta^2\mu^2), \mu]. \tag{21.5.31}$$

In the above we used a momentum point subtraction to renormalize the theory and $-\mu^2$ was the point at which the infinite subtraction took place. In the MS or $\overline{\mathrm{MS}}$ renormalization schemes (21.5.30) continues to hold, but μ then refers to the mass parameter introduced in (21.5.2).

We must now consider the generalization of (21.5.30) when one allows for non-zero quark masses in the QCD Lagrangian.

21.6 The effect of heavy quarks

Consider a theory in which there are just two quarks of very different mass, a light quark q of mass m and a heavy quark Q of mass M with $M \gg m$. (The more realistic case of several quarks of widely different masses will be dealt with quite analogously.)

We have already remarked, on several occasions, that it is conventional to treat a light quark as massless when studying physics at 'large' momentum scales. But for a dimensionful quantity the epithet 'large' is somewhat meaningless. What is meant is $p^2 \gg m^2$, and what is implied is that m^2 can be neglected compared with p^2 in this region. But although this sounds very intuitive, it is not quite trivial and clearly requires the theory to be well-behaved in the limit $m \to 0$. For QCD this can be justified in the dimensional renormalization schemes and in the general momentum point subtraction schemes, but not in the on-shell scheme.

If we deal with gigantic momentum scales where $p^2 \gg M^2$ then we can neglect both m^2 and M^2 compared with p^2 and once again we may effectively utilize the massless version of the theory, this time with two 'massless' quarks.

The interesting region is the 'in between' one where $m^2 \ll p^2 \ll M^2$. Here we may again treat the light quark as massless, i.e. effectively use $m = 0$, and the real question is what, if any, is the rôle of the heavy quark Q.

Intuitively we would be inclined to feel that Q can play no rôle at scales $p^2 \ll M^2$ since at such scales we do not expect to be able to excite the dynamical degrees of freedom involving Q and we ought therefore to be able to ignore Q completely. If it were not for renormalization this basic intuition would be quite simple to justify. With renormalization it is nonetheless correct in QCD, but the demonstration is not trivial and forms the basis of the *decoupling theorem* of Applequist and Carazzone (1975). The result, for a sequence of quarks $q_1, q_2, \ldots, q_N$ with renormalized mass parameters $m_1 < m_2 < m_3 < \cdots < m_N$, is as follows. If the sequence is such that for some j $m_j^2 \ll m_{j+1}^2$ then for momentum scales p^2 such that $m_j^2 \ll p^2 \ll m_{j+1}^2$ one may:

1. put $m_1 = m_2 = \cdots = m_j = 0$;
2. completely ignore the quarks $q_{j+1}, \ldots, q_N$.

The latter implies that we work with a theory in which the number of quarks is just j, not N.

> This should be compared with what happens in electroweak theory. For example in Section 2.1 we discussed $\nu_e e \to \nu_e e$ via W exchange (Fig. 2.2). In the kinematic domain $q^2 \ll M_W^2$, the W propagator *is* tiny, of order $1/M_W^2$, but we did not therefore throw away the diagram. The reason is that if we had done so the amplitude for our process would have been zero! In QCD on the other hand we are throwing away diagrams whose contribution is down by some power of p^2/M^2 compared with *unity*.

To see heuristically how the decoupling comes about consider the simplest case of just one light and one heavy quark. For momenta $p^2 \ll M^2$ we need only discuss Feynman diagrams in which the external lines are either light quark or gluon lines. Then because there are no flavour changing couplings in QCD there will be no tree graphs involving the heavy quark propagator. The heavy quark will only appear in loops internal to the Feynman diagrams. Since the external momenta all satisfy $p_i^2 \ll M^2$

terms involving the p_i can be dropped from the expressions for the propagators. Thus one is dealing with loop integrals $\int d^4q$ which, after continuation in the q_0 variable, may be treated as Euclidean integrals, typically of the form

$$I = \int \frac{d^4q}{(q^2 + M^2)^n}.$$

We can divide the integration range into three regions: (i) $q^2 \ll M^2$, (ii) $q^2 \approx M^2$, (iii) $q^2 \gg M^2$. Suppose first that the integral is convergent so that $n \geq 3$. Then the contribution from region (i) must be of size $1/M^{2n}$. Region (ii) will give a contribution of size M^4/M^{2n} which is smaller than or of order M^{-2}. Because of the convergence, region (iii) gives a negligible contribution, $\ll M^{4-2n}$. In general, for a convergent integral the contribution from heavy quark loops goes to zero like some power of p^2/M^2 for $M^2 \gg p^2$.

If the integral I is divergent it has to be rendered finite by renormalization. The analysis then depends upon the renormalization scheme used. It is simplest if one uses a momentum point subtraction scheme with μ^2 chosen so that $\mu^2 \ll M^2$. Then it can be argued that the renormalized integral behaves with regard to M essentially as the convergent integral did.

As a consequence of the decoupling theorem we can now understand how to modify (21.5.30) in the presence of non-negligible renormalized quark mass parameters. Let $q_1, q_2, \ldots, q_N$ be the sequence of quarks with mass parameters $m_1 < m_2 < \cdots < m_N$ and suppose that $m_j^2 \ll m_{j+1}^2$. Then for momentum scales in the range

$$m_j^2 \ll (p^2, \eta^2 p^2) \ll m_{j+1}^2 \tag{21.6.1}$$

equation (21.5.30) holds for Green's functions involving n_A gluons and n_ψ 'light' quarks, i.e. quarks belonging to the set $(q_1, q_2, \ldots, q_j)$. The functions $\beta, \gamma^A, \gamma^\psi$ are to be calculated as if the theory contained just j massless quarks. This implies that n_f in (21.5.23, 24 and 25) has to be taken equal to j. To emphasize this we shall write n_f^* to indicate the *number of active flavours*. The relationship between $\alpha(\eta^2\mu^2)$ and $\alpha(\mu^2)$ will of course reflect this change in the interpretation of n_f.

In the realistic QCD case we have the following approximate situation:

Momentum scale	Number of active flavours
$1 < p^2 < (3\ \text{GeV}/c)^2$	3
$(3\ \text{GeV}/c)^2 < p^2 < (10\ \text{GeV}/c)^2$	4
$(10\ \text{GeV}/c)^2 < p^2 < m_t^2$	5
$p^2 > m_t^2$	6

Clearly the behaviour given by (21.5.30) cannot be accurate too close to the boundaries between the regions, since (21.6.1) is then not well satisfied.

We mention finally that it is possible to do better than the above, by making less drastic approximations. One such approach, based on eqn (20.6.4) (Georgi and Politzer, 1976), makes use of running quark masses $m_i(\mu^2)$ and does not discard terms of order $m_i^2(\mu^2)/\mu^2$ for the light quarks nor $\mu^2/m_i^2(\mu^2)$ for the heavy quarks. The price, however, is complexity. The β-function depends on both α and all the mass ratios m_i/μ and the relationship between α defined at scale μ^2 and α defined at some other scale μ_0^2 cannot be written down in analytic form; it has to be dealt with numerically.

21.7 The running coupling in QCD

From (20.8.6) and (20.8.14) the exact relationship between couplings defined at scales $\sqrt{s}$ and μ is given by

$$\int_{\alpha(\mu^2)}^{\alpha(s)} \frac{\mathrm{d}x}{\beta(x)} = \ln(s/\mu^2). \tag{21.7.1}$$

Using the lowest order (LO) form for $\beta(\alpha)$ given by (21.5.21) we have

$$\alpha(s) = \frac{\alpha(\mu^2)}{1 + b(n_f^*)\alpha(\mu^2)\ln(s/\mu^2)} \tag{21.7.2}$$

where, on the basis of Section 21.6, we have written $b(n_f^*)$ to emphasize the dependence upon the number of *active* quark flavours in the kinematic region $\mu^2 \leftrightarrow s$.

Since $\alpha(\mu^2)$ should be a continuous function of μ^2 care must be taken when crossing from one region to another where the number of active flavours changes. Let the number of active flavours be n_f^* in the region $\mu_\mathrm{A}^2 \le \mu^2 \le \mu_\mathrm{B}^2$ and $n_f^* + 1$ in the region $\mu_\mathrm{B}^2 \le \mu^2 \le \mu_\mathrm{C}^2$ and suppose that α has been measured at some scale μ_0 lying in the second region, i.e. $\mu_\mathrm{B}^2 \le \mu_0^2 \le \mu_\mathrm{C}^2$. Then we have

$$\alpha(\mu^2) = \frac{\alpha(\mu_0^2)}{1 + b(n_f^* + 1)\alpha(\mu_0^2)\ln(\mu^2/\mu_0^2)} \quad (\mu_\mathrm{B}^2 \le \mu^2 \le \mu_\mathrm{C}^2). \tag{21.7.3}$$

Using (21.7.3) we can evaluate $\alpha(\mu_\mathrm{B}^2)$. Then for μ^2 in the first region we take

$$\alpha(\mu^2) = \frac{\alpha(\mu_\mathrm{B}^2)}{1 + b(n_f^*)\alpha(\mu_\mathrm{B}^2)\ln(\mu^2/\mu_\mathrm{B}^2)} \quad (\mu_\mathrm{A}^2 \le \mu^2 \le \mu_\mathrm{B}^2). \tag{21.7.4}$$

which ensures the continuity of $\alpha(\mu^2)$ at $\mu^2 = \mu_B^2$. Clearly a similar matching can be made at each boundary between regions of differing n_f^*.

Often equation (21.7.2) is written in the form

$$\alpha(s) = \frac{1}{b\ln(s/\Lambda^2)} \quad \text{(LO)} \tag{21.7.5}$$

where

$$\Lambda^2 = \mu^2 e^{-1/b\alpha(\mu^2)}. \tag{21.7.6}$$

One sees that Λ is a parameter that tells us how big the QCD coupling is. Of course it has to be found from experiment and its value will depend upon the number of active flavours n_f^* used in the expression (21.5.23) for b. Fig. 25.26 in Chapter 25 shows how $\Lambda(5)$ is related to $\Lambda(4)$, on the basis of putting $\alpha^{(4)}(\mu^2) = \alpha^{(5)}(\mu^2)$ at $\mu^2 = 25\ \mathrm{GeV}^2$. Note that the value of $\alpha(s)$ determined by comparing a calculated physical observable with experiment will depend upon the renormalization *scheme* utilized, which implies that Λ will require a label, e.g. MS or $\overline{\mathrm{MS}}$, and will have different values in different schemes. However, in any chosen scheme, Λ is a renormalization group invariant in the sense that it does not depend upon the scale μ. To see this explicitly write (21.7.6) as

$$\ln\Lambda^2 = \ln\mu^2 - \frac{1}{b\,\alpha(\mu^2)}. \tag{21.7.7}$$

Differentiating with respect to $\ln\mu^2$ yields

$$\frac{d\ln\Lambda^2}{d\ln\mu^2} = 1 + \frac{1}{b\,\alpha^2(\mu^2)}\frac{d\alpha(\mu^2)}{d\ln\mu^2}$$

which via (20.6.5 or 10) yields

$$\frac{d\ln\Lambda^2}{d\ln\mu^2} = 1 + \frac{1}{b\,\alpha^2(\mu^2)}\cdot\beta[\alpha(\mu^2)]. \tag{21.7.8}$$

Upon using the lowest order result $\beta(\alpha) = -b\,\alpha^2$ we see that the RHS of (21.7.8) is zero.

Recall that in QED one obtains the fundamental strength of electromagnetic interactions by measuring *the* fine structure constant α_{QED} via, for example, the long wavelength limit of Compton scattering on electrons. As explained in Section 20.3 this actually corresponds to measuring the coupling defined at a scale μ = physical mass of the electron. In pre-electroweak days, given the very slow variation of $\alpha_{\mathrm{QED}}(\mu^2)$ with μ^2 (which follows from the smallness of α_{QED}), and given the relatively small energy scales involved, the concept of a running coupling was virtually unheard of. However in discussing LEP physics at the huge scale M_Z it is helpful to introduce $\alpha_{\mathrm{QED}}(M_Z^2)$ as was done for example in Section 7.3.

For QCD, given that there is no specially favoured scale at which we could define *the* α_{QCD}, it seems, at first sight, sensible to regard Λ as *the fundamental measure* of the strong coupling. In that case we should not rely upon the lowest order form (21.7.5 and 6) for our definition of Λ. An exact definition is given by noticing that $s = \Lambda^2$ corresponds to the point at which $\alpha(s) \to \infty$. Thus we can define Λ by

$$\int_{\alpha(s)}^{\infty} \frac{\mathrm{d}x}{\beta(x)} \equiv \ln(\Lambda^2/s). \tag{21.7.9}$$

It is easily checked that this reproduces (21.7.5) if the lowest order form is used for $\beta(\alpha)$.

In next to leading order (NLO) we have

$$\ln(\Lambda^2/s) = -\int_{\alpha(s)}^{\infty} \frac{\mathrm{d}x}{bx^2(1+b'x)} \tag{21.7.10}$$

where b' is given in (21.5.25). This yields

$$\frac{1}{\alpha(s)} + b' \ln\left[\frac{b'\alpha(s)}{1+b'\alpha(s)}\right] = b\,\ln(s/\Lambda^2) \qquad \text{(NLO)} \tag{21.7.11}$$

from which $\alpha(s)$ can be found numerically for a given value of Λ, or, from which Λ can be found given an experimentally determined value of $\alpha(s)$ at some scale $\sqrt{s}$.

However, we feel that it is dangerous to put so much emphasis on Λ. Firstly its value depends upon n_f^*. Secondly different people use different approximations to (21.7.11) to obtain Λ from a measured $\alpha(s)$ [see eqn (25.5.1)]. Thus it might be better to use as fundamental parameter the value of $\alpha(s)$, measured in $e^+e^- \to$ hadrons at some large enough energy $\sqrt{s}$, say at $s = M_Z^2$. (For a practical assessment see Chapter 25.) The running coupling would then be given at any other scale $\sqrt{s}$ by (21.7.1) in which we take $\mu = M_Z$.

To try to avoid confusion the Particle Data Group have recommended the use of the following approximate solution to (21.7.11)

$$\begin{aligned}\alpha(s) \;=\; & \frac{12\pi}{(33-2n_f^*)\ln(s/\Lambda^2)}\left\{1 - 6\frac{153-19n_f^*}{(33-2n_f^*)^2}\frac{\ln[\ln(s/\Lambda^2)]}{\ln(s/\Lambda^2)}\right\} + \\ & + O\left[\frac{1}{[\ln(s/\Lambda^2)]^3}\right].\end{aligned} \tag{21.7.12}$$

It corresponds to the choice

$$\Lambda^2 = \mu^2 \exp\left\{-\frac{1}{b}\left[\frac{1}{\alpha(\mu^2)} + b'\ln[b\alpha(\mu^2)]\right]\right\} \tag{21.7.13}$$

instead of (21.7.6).

21.7.1 Renormalization scheme dependence of α and Λ

Some of the current literature quotes values of α and Λ in the MS, some in the $\overline{\text{MS}}$ renormalization schemes. It is therefore useful to discuss how values of α, Λ in different schemes are related to each other.

Consider two schemes with couplings α and $\tilde{\alpha}$ related, as in (20.9.15), by

$$\tilde{\alpha} = \alpha(1 + c\alpha + \cdots). \tag{21.7.14}$$

From (21.7.9) we have

$$\int_{\alpha(s)}^{\infty} \frac{dx}{\beta(x)} \equiv \ln(\Lambda^2/s); \qquad \int_{\tilde{\alpha}(s)}^{\infty} \frac{dx}{\tilde{\beta}(x)} \equiv \ln(\tilde{\Lambda}^2/s). \tag{21.7.15}$$

To next to leading order we showed in Section 20.9 that $\tilde{\beta} = \beta$. Thus subtracting the two definitions in (21.7.15) yields

$$\ln(\Lambda^2/\tilde{\Lambda}^2) = \int_{\alpha(s)}^{\tilde{\alpha}(s)} \frac{dx}{-bx^2(1 + b'x)} \qquad \text{(NLO)}. \tag{21.7.16}$$

But the LHS is independent of s, so we may take $s \to \infty$ on the RHS, which implies $\alpha(s), \tilde{\alpha}(s) \to 0$, and the $b'x$ term is then negligible. Hence

$$\ln(\Lambda^2/\tilde{\Lambda}^2) = \frac{1}{b} \lim_{s\to\infty} \left[\frac{1}{\tilde{\alpha}(s)} - \frac{1}{\alpha(s)} \right] = -\frac{c}{b}. \tag{21.7.17}$$

Finally, then,

$$\tilde{\Lambda} = \Lambda e^{c/2b} \qquad \text{(NLO)}. \tag{21.7.18}$$

For the MS and $\overline{\text{MS}}$ schemes it is possible to deduce from (21.5.2,3 and 4) that

$$\alpha^{\overline{\text{MS}}}(\mu^2) = \alpha^{\text{MS}}(\mu^2) + b(\ln 4\pi - \gamma_E)[\alpha^{\text{MS}}(\mu^2)]^2 + \cdots \tag{21.7.19}$$

so that, from (21.7.18)

$$\Lambda^{\overline{\text{MS}}} = \Lambda^{\text{MS}} e^{(\ln 4\pi - \gamma_E)/2}. \tag{21.7.20}$$

21.8 Conclusion

Armed with the Feynman rules and the results of the renormalization group for QCD we shall now turn to some important physical applications of the above formalism. In particular we use the simplest of these, the reaction $e^+e^- \to$ hadrons, to illustrate in detail the remarkable power and implications of the renormalization group, not just in its lowest order form (21.5.31), but also in its more exact form (21.5.30).

22
Applications of the QCD renormalization group

An enormous amount of work has gone into the long and complicated calculations that are needed in order to extract from QCD the kind of detailed predictions that one would wish to compare with experimental data. We shall only be able to present what we believe to be the essential elements of these calculations. For further details and for access to the original literature see Buras (1980) and Altarelli (1982). For more recent work see Ellis and Stirling (1990).

Our first example, $e^+e^- \to$ hadrons, is the reaction *par excellence* to illustrate both the power and subtlety of the renormalization group. We then turn to inclusive deep inelastic lepton–hadron scattering, where a combination of the operator product expansion and renormalization group techniques teach us how to make dynamical improvements to the simple parton model, leading to the slow logarithmic breaking of Bjorken scaling that was mentioned in Chapters 16 and 17. These results have an intuitive interpretation which suggests a more general approach to a QCD improved parton model which is taken up in Chapter 23.

22.1 $e^+e^- \to$ hadrons

Our first application will be to calculate the ratio

$$R \equiv \frac{\sigma(e^+e^- \to \text{hadrons})}{\sigma(e^+e^- \to \mu^+\mu^-)}$$

that was discussed extensively in earlier chapters and for which the parton model gave the result (17.4.6)

$$R = \sum_{\substack{\text{flavours,}\\ \text{colours}}} Q_j^2. \tag{22.1.1}$$

We shall see that QCD gives this result to lowest order, but provides

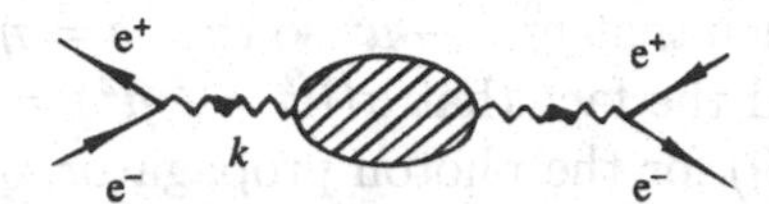

Fig. 22.1. $e^+e^- \to e^+e^-$ via a virtual photon.

a momentum dependent correction term in higher order. We must now consider the combined strong and electromagnetic interactions, but work to lowest order only in the electromagnetic coupling e. According to the optical theorem

$$\sigma_{\mathrm{tot}}(e^+e^-) = F\mathcal{A}(e^+e^- \to e^+e^-), \tag{22.1.2}$$

where $\mathcal{A}$ is the imaginary part of the spin-averaged forward amplitude for $e^+e^- \to e^+e^-$, and F is a flux etc. factor that will turn out to be irrelevant. To lowest order in e^2 the imaginary part comes from the Feynman diagram of Fig. 22.1 where , usually denoted by $\mathrm{i}\pi_{\mu\nu}(k^2)$, with

$$\pi_{\mu\nu}(k^2) = \left(g_{\mu\nu} - \frac{k_\mu k_\nu}{k^2}\right)\pi(k^2) \tag{22.1.3}$$

is 'one-particle irreducible', i.e. the blob cannot be split apart by cutting just one line inside it.

It is clear that if we isolate the contributions to coming from purely hadronic intermediate states [call this $\pi_{\mathrm{h}}(k^2)$] and from a $\mu^+\mu^-$ intermediate state (π_{muons}) then

$$\left.\begin{aligned} \sigma(e^+e^- \to \mathrm{h}) &= F\mathcal{A}_{\mathrm{h}}(e^+e^- \to e^+e^-), \\ \sigma(e^+e^- \to \mu^+\mu^-) &= F\mathcal{A}_{\mathrm{muons}}(e^+e^- \to e^+e^-). \end{aligned}\right\} \tag{22.1.4}$$

In forming the ratio R all kinematic terms cancel out and we are left with

$$R = \frac{\mathrm{Im}\pi_{\mathrm{h}}(k^2)}{\mathrm{Im}\pi_{\mathrm{muons}}(k^2)}. \tag{22.1.5}$$

We are now dealing with the combined theory of QCD and QED, so Green's functions will have an additional label n_{ph} to specify the number of photons involved. Our $\pi_{\mu\nu}(k^2)$ is then really $\Gamma_{\mu\nu}^{(n_{\mathrm{ph}}=2, n_{\mathrm{A}}=0, n_\psi=0)}$ and (21.5.6) should have an extra term $-n_{\mathrm{ph}}\gamma^{\mathrm{ph}}$ in it. Clearly the changes are trivial and we may directly utilize (21.5.30) with $n_{\mathrm{A}} \to n_{\mathrm{A}} + n_{\mathrm{ph}}$. For our particular case we then need $n_{\mathrm{A}} = 0, n_{\mathrm{ph}} = 2, n_\psi = 0$, and (21.5.30) becomes, after cancelling common tensor factors,

$$\pi(s = k^2) = s\ \exp\left\{-\int_{\alpha_s(\mu^2)}^{\alpha_s(-s)} \frac{\gamma^{\mathrm{ph}}(x)}{\beta(x)}\,\mathrm{d}x\right\} \tag{22.1.6}$$

wherein we chose p_i such that $p_i^2 = -\mu^2$ so that $s = \eta^2 p_i^2 = -\mu^2\eta^2$, and is *negative.* We also used the fact that $\pi(k^2 = -\mu^2) = -\mu^2$ in accord with the analogue of (21.5.9) for the photon propagator.

As was stressed in connection with (20.8.19), note the remarkable feature that aside from the initial factor s, all the remaining s-dependence is contained in $\alpha_s(-s)$, the running QCD coupling.

To find γ^{ph} to lowest order in e^2 one has to evaluate the following contributions to $\pi_{\mu\nu}$:

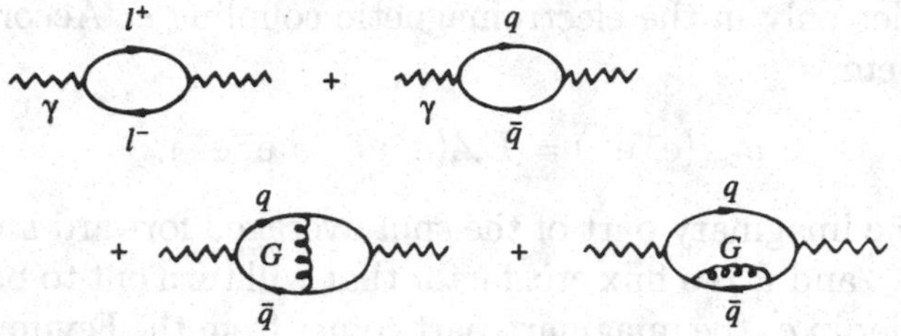

and by substituting in the renormalization group equation identify

$$\gamma^{\text{ph}} = Ae^2 \left[\sum_{\text{leptons}} Q_\ell^2 + \sum_{\text{quarks}} Q_q^2(1 + B\alpha_s) \ldots \right], \tag{22.1.7}$$

where A is a constant that will be irrelevant, and

$$B = \frac{3C_2(R)}{4\pi} \tag{22.1.8}$$

where $C_2(R)$ is analogous to $C_2(G)$ but depends on the quark representation:

$$\delta_{ij} C_2(R) = \sum_a t^a_{ik} t^a_{kj}. \tag{22.1.9}$$

For $SU(3)$ and the fermion, i.e. triplet quark representation,

$$C_2[SU(3) \text{ triplet}] \equiv C_F = \tfrac{4}{3} \tag{22.1.10}$$

so that

$$B[SU(3) \text{ triplet}] = \frac{1}{\pi}. \tag{22.1.11}$$

From (22.1.7) we can identify γ_0^{ph} and γ_1^{ph}, to use in the QCD–QED form of (20.9.11), with $n\gamma_0$ and γ_1 replaced by $2\gamma_0^{\text{ph}}$ and γ_1^{ph} respectively. Since both are proportional to e^2 and therefore small we can use the formula

$$x^\epsilon = \mathrm{e}^{\epsilon \ln x} = 1 + \epsilon \ln x + \ldots \tag{22.1.12}$$

to get, for s *negative*,

$$\pi(s) \approx s\left\{1 - Ae^2\left[\sum_{\substack{\text{leptons}\\\text{and quarks}}} Q_j^2 \ln\left(\frac{s}{-\mu^2}\right)\right.\right.$$

$$\left.\left. -\frac{B}{b}\sum_{\substack{\text{quarks}\\\text{only}}} Q_j^2 \ln\left(\frac{\alpha_s(-s)}{\alpha_s(\mu^2)}\right)\right] + \cdots\right\} \qquad (22.1.13)$$

valid provided s is not so large that $e^2 \ln s$ is large.

The analytic structure of $\pi(s)$ is well understood and it has a discontinuity, i.e. an imaginary part, only for $s > 0$. It is not a trivial matter to extend (22.1.13) to positive s, but if one assumes reasonably non-pathological behaviour one can simply take the form (22.1.13) at large negative s and analytically continue to large positive s, and then take its discontinuity. It is clear from (22.1.13) how to identify π_{h} and π_{muons}, and in forming the ratio (22.1.5) the factor A cancels out, leaving

$$R(s) = \sum_{\text{quarks}} Q_j^2\left[1 + \frac{B}{b\ln(s/\mu^2)} + \cdots\right]$$

$$= \sum_{\text{quarks}} Q_j^2\left[1 + \frac{\alpha_s(s)}{\pi} + \cdots\right] \qquad (22.1.14)$$

wherein we have used (22.1.11) and, via (21.6.2), the lowest order result

$$\alpha_s(s) = \frac{\alpha_s(\mu^2)}{1 + b\alpha_s(\mu^2)\ln(s/\mu^2)} \qquad (22.1.15)$$

$$\simeq \frac{1}{b\ln(s/\mu^2)} \qquad \text{for } s \gg \mu^2. \qquad (22.1.16)$$

Note that the sum over quark flavours in (22.1.14) must include only the 'active' quarks at scale $\sqrt{s}$, i.e. those light enough to be pair produced at CM energy $\sqrt{s}$. Also that b in (22.1.15 and 16) is really $b(n_f^*)$ where n_f^* is the number of active quarks at scale $\sqrt{s}$, as discussed in Section 21.6.

The result (22.1.14) is extremely interesting. Firstly it shows that to zeroth order in the strong coupling we recover the naive parton model result (22.1.1) and it predicts that $R(s)$ will ultimately decrease towards this value as $s \to \infty$. Secondly it allows us to understand a little more clearly the rôle of the renormalization group equation. If we had not used the latter, but simply computed $\pi_{\mu\nu}$ to order g^2 from the diagrams above (22.1.7), we would have ended up with (22.1.14) for R with $\alpha_s(\mu^2)$ instead of $\alpha_s(s)$, i.e.

$$R(s) = \sum_{\text{quarks}} Q_j^2 \left[1 + \frac{\alpha_s(\mu^2)}{\pi}\right]. \tag{22.1.17}$$

(Note that the calculation is non-trivial. It involves a cancellation of infrared singularities between the 3rd and 4th diagrams.) If we recall the form of $\alpha_s(s)$ and expand it as a power series in $\alpha_s(\mu^2)$ we have from (22.1.15)

$$\alpha_s(s) = \alpha_s(\mu^2)\left[1 - b\alpha_s(\mu^2)\ln(s/\mu^2) + (b\alpha_s(\mu^2)\ln(s/\mu^2))^2 + \cdots\right] \tag{22.1.18}$$

so that the renormalization group equation using the lowest order version of $\alpha_s(s)$ has painlessly succeeded in summing a certain part of the perturbation expansion to *all orders.* In effect it picks out the largest terms in each order of perturbation theory, namely those where each factor $\alpha_s(\mu^2)$ is multiplied by $\ln(s/\mu^2)$—this is referred to as the *leading logarithmic approximation.*

There are several other interesting lessons to be learned from this. Consider the calculation of $R(s)$ to order g^4 in *perturbation theory.* (The Feynman diagrams are shown in Fig. 24.7.) This is a difficult exercise, involving both infrared singularities and ultraviolet divergences [see Ellis and Stirling (1990) for details and references]. Because renormalization is involved the result is scheme dependent. In the $\overline{\text{MS}}$ scheme one finds

$$R(s) = \sum_{\text{quarks}} Q_j^2 \left\{1 + \frac{\alpha_s(\mu)}{\pi} + \left[\frac{33 - 2n_f}{12}\ln(s/\mu^2) + \frac{365}{24} - 11\zeta(3) + n_f\left(\frac{2}{3}\zeta(3) - \frac{11}{2}\right)\right]\left(\frac{\alpha_s(\mu^2)}{\pi}\right)^2\right\} \tag{22.1.19}$$

where, strictly speaking, α_s should carry a label '$\overline{\text{MS}}$', and where $\zeta(z)$ is the Riemann zeta-function, $\zeta(3) = 1.2021\ldots$. (The result is now known to order α_s^3—see Section 24.3.3.)

Now even if $\alpha_s(\mu^2)$ is reasonably small we see from (22.1.19) that the higher order term $\ln(s/\mu^2)[\alpha_s(\mu^2)]^2$ will eventually become large for $s \gg \mu^2$, thus invalidating the perturbative expansion. Moreover this trouble reappears in all orders through terms of the form $[\ln(s/\mu^2)]^n[\alpha_s(\mu^2)]^{n+1}$. But from (22.1.19) and (21.5.23) we see that the dangerous term appears in the form

$$\begin{aligned}\frac{\alpha_s(\mu^2)}{\pi}&\left[1 - \frac{33 - 2n_f}{12\pi}\ln(s/\mu^2)\alpha_s(\mu^2)\right]\\ &= \frac{\alpha_s(\mu^2)}{\pi}\left[1 - b\alpha_s(\mu^2)\ln(s/\mu^2)\right]\end{aligned} \tag{22.1.20}$$

which are the first two terms in the expansion of $\alpha_s(s)$ in (22.1.18). This, of course, had to be so, since the general result (22.1.6) assures us that $R(s)$ depends on s *only* through $\alpha_s(s)$!

Now the renormalization scale μ is, in principle, arbitrary. If therefore we choose it equal to $\sqrt{s}$, i.e. $\mu^2 = s$, then we are guaranteed that no other s-dependence will appear which could spoil the perturbative expansion. That is choosing $\mu^2 = s$ and making a perturbative expansion in powers of $\alpha_s(s)$ we are sure that the coefficients in the power series will be s-independent. We thus have a genuine expansion in powers of $\alpha_s(s)$. This can be seen quite explicitly in (22.1.19) upon putting $\mu^2 = s$.

The general lesson to be learned from this is that for a physical quantity with just one large scale, say s, we can safely do perturbation theory in $\alpha_s(s)$ (provided, of course, that s is large enough so that $\alpha_s(s) \ll 1$). But if the physical quantity is more complicated and depends on several *independent* scales, for example sub-energies $s_1, s_2, \ldots$ in a multiparticle or multijet production amplitude, then perturbation theory may not be safe. For example choosing $\mu^2 = s_1^2$ might leave dangerous terms of the form

$$\ln(s_2/\mu^2)\alpha_s(s_1) = \ln(s_2/s_1)\alpha_s(s_1) \tag{22.1.21}$$

in the perturbative series. Only when all the s_i are comparable (which really implies *one* large scale) do these terms become innocuous.

In reactions with several scales it is not obvious what value of μ to choose as the renormalization scale. In an *exact* calculation the result, for some physical observable, must be independent of μ, but in any approximate calculation the result will, of course, vary as μ is changed. There are two popular guesses in use as to how to estimate a 'best' value of μ. Suppose one has a calculation of some physical observable $\sigma(\alpha)$ up to order n, i.e. one has expressions for $\sigma^{(n)}[\alpha(\mu^2)]$ and, manifestly, for $\sigma^{(n-1)}[\alpha(\mu^2)]$. In the first method one searches for a value of μ which minimizes $\sigma^{(n)} - \sigma^{(n-1)}$—this is called the 'fastest apparent convergence' method. In the second, one searches for a value of μ where $\mathrm{d}\sigma^{(n)}[\alpha(\mu^2)]/\mathrm{d}\mu = 0$—this is the 'principle of minimal sensitivity' approach. Both, clearly, are just intuitive guesses, but seem rather reasonable.

Returning to the order α^2 result (22.1.19) and putting $\mu^2 = s$ gives in the $\overline{\mathrm{MS}}$ scheme

$$\begin{aligned} R(s) &= 3 \sum_{\text{flavours}} Q_f^2 \left\{ 1 + \frac{\alpha_s(s)}{\pi} + \left[\frac{365}{24} - 11\zeta(3) \right.\right. \\ &\qquad \left.\left. + n_f \left(\frac{2}{3}\zeta(3) - \frac{11}{2} \right) \right] \left(\frac{\alpha_s(s)}{\pi} \right)^2 \right\}. \end{aligned} \tag{22.1.22}$$

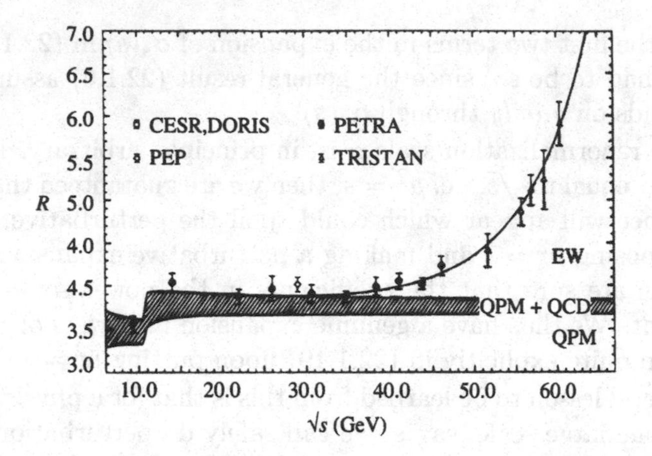

Fig. 22.2. Averaged experimental data on R as function of the centre of mass energy. See text.

The expression (22.1.22) is a smooth function of s except that it has steps whenever $\sqrt{s}$ reaches a threshold for the production of a new quark flavour. It clearly is not valid in the region of these thresholds, nor, obviously, in the region of resonance bumps, though it may give a reasonable description if bumps and dips are averaged over.

The behaviour of $R(s)$ at lower energies has already been discussed in Section 9.5.4, principally in regard to the evidence for colour (see Fig. 9.6). Its behaviour around the charm threshold was shown in Fig. 14.1.

Let us now look at the behaviour in the region $10 \leq \sqrt{s} \leq 60$ GeV, where the number of active flavours is $n_f^* = 5$, since only the top cannot be produced.

Then (22.1.22) becomes

$$R(s) = 3 \sum_{\substack{\text{flavours} \\ u,d,s,c,b}} Q_f^2 \left\{ 1 + \frac{\alpha_s(s)}{\pi} + 1.411 \left(\frac{\alpha_s(s)}{\pi} \right)^2 \right\}. \tag{22.1.23}$$

Note that we should strictly have added a label γ to emphasize that R was calculated on the basis only of virtual photon exchange in Fig. 22.1.

At higher energies Z exchange becomes very important, and dominates as $\sqrt{s}$ approaches M_Z so (22.1.23) must be modified. The modifications are straightforward, very similar to the calculations in Section 8.5, and do not involve the QCD aspects of the calculation. Formulae can be found in d'Agostini, de Boer and Grindhammer (1989), and we show in Fig. 22.2 their comparison of theory with experiment. It can be seen that the agreement is excellent both in the sector where Z exchange is negligible and where it is important. (The electroweak contribution is labelled EW.) It is also clear that the QCD corrections to the quark parton model (QPM) are vital. The upper solid line in Fig. 22.2 is a best fit in which M_Z, θ_W and $\alpha_s[(34 \text{ GeV})^2]$ were left free. The fit yielded

$$M_Z = 89.4 \pm 1.3 \text{ GeV}/c^2, \qquad \sin^2\theta_W = 0.220 \pm 0.02$$

and

$$\alpha_s[(34 \text{ GeV})^2] = 0.142 \pm 0.018. \tag{22.1.24}$$

In Fig. 24.27 of Section 24.13, we summarize various measurements of α_s at different scales. It is seen that the value given in (22.1.24) has a central value somewhat on the high side compared with measurements at a similar scale, but quite compatible given the large error in (22.1.24).

22.2 Deep inelastic lepton scattering

We saw earlier, (15.3.5), that the electromagnetic inelastic nucleon tensor $W^{\alpha\beta}$ is related to the matrix element of a product of current operators:

$$W^{\alpha\beta}(q^2,\nu) = \frac{1}{4\pi}\sum_{\text{spins}}\int \mathrm{d}^4z\, \mathrm{e}^{-\mathrm{i}q\cdot z}\langle P|J^\alpha(0)J^\beta(z)|P\rangle, \tag{22.2.1}$$

where P is the momentum of the target and q is the four-momentum transfer. Using translational invariance we prefer to write (22.2.1) in the more symmetrical form

$$W^{\alpha\beta}(q^2,\nu) = \frac{1}{4\pi}\sum_{\text{spins}}\int \mathrm{d}^4z\, \mathrm{e}^{\mathrm{i}q\cdot z}\langle P|J^\alpha(\tfrac{1}{2}z)J^\beta(-\tfrac{1}{2}z)|P\rangle. \tag{22.2.2}$$

For different processes different currents, i.e. em or weak, will occur, but the basic object will always be of the form of (22.2.2).

As always with QCD we are incapable of dealing directly with the hadronic states since they reflect a fundamentally non-perturbative aspect of the theory. Nonetheless, important information on the ν and q^2 dependence of $W^{\alpha\beta}$ can be obtained by a judicious combination of calculable perturbative effects, principally obtained from the renormalization group analysis, and certain non-calculable constants which have to be obtained from a comparison with experiment.

The classic approach to deep inelastic scattering is via Wilson's *operator product expansion.* [For an intelligible summary and further references to Wilson's work, see de Alfaro *et al.* (1973).] We shall follow this approach initially. Later we shall see how the results can be obtained from a somewhat different point of view which emphasizes the perturbative QCD aspects of the problem.

22.2.1 The operator product expansion

For simplicity let us ignore tensor indices and let us pretend that the currents are scalars, i.e. we consider a product of the type $J(\frac{1}{2}z)J(-\frac{1}{2}z)$. From the discussion of the Appendix to Chapter 16 the behaviour of $W^{\alpha\beta}$ for $Q^2 \equiv -q^2 \to \infty$, $\nu \to \infty$, but Q^2/ν fixed, will be controlled by the behaviour of $\langle P|J(\frac{1}{2}z)J(-\frac{1}{2}z)|P\rangle$ for z near the light cone, i.e. for $z^2 \approx 0$. (This of course does *not* mean that each component z_μ is small.) We thus require the 'light cone behaviour' of the operator product $J(\frac{1}{2}z)J(-\frac{1}{2}z)$.

Such products are known to be singular as $z^2 \to 0$ (as can be seen even for currents made up from free field operators), and Wilson's result is that these singularities can be isolated in ordinary, i.e. non-operator, functions multiplied by *non-singular* operators:

$$J(\tfrac{1}{2}z)J(-\tfrac{1}{2}z) = \sum_{j,N} \tilde{C}_j^N(z^2) z^{\mu_1} z^{\mu_2} \dots z^{\mu_N} \hat{O}^{j,N}_{\mu_1\dots\mu_N}(0). \tag{22.2.3}$$

The operators $\hat{O}^{j,N}_{\mu_1\dots\mu_N}(0)$ are non-singular and are evaluated at $x = \frac{1}{2}[\frac{1}{2}z + (-\frac{1}{2}z)] = 0$. They are chosen to be *symmetric in their tensor indices*, and traceless. They are referred to as 'spin N' operators. The coefficient functions $\tilde{C}_j^N(z^2)$ are singular as $z^2 \to 0$ and, since we are dealing with interacting fields, the $\tilde{C}_j^N$ will also depend upon the strong coupling α_s. The terms with larger N should not be thought of as 'smaller' in (22.2.3) since the z^{μ_i} need not be small.

Let the mass dimensions of J and $\hat{O}^{j,N}$ be d_J and $\hat{d}_{j,N}$ respectively. Then balancing dimensions in (22.2.3) gives

$$[\tilde{C}_j^N] = [\mathrm{M}]^{2d_J+N-\hat{d}_{j,N}}.$$

On naive dimensional grounds, up to logarithmic factors, we would then expect the behaviour

$$\begin{aligned} \tilde{C}_j^N(z^2) &\xrightarrow{z^2\to 0} \left(\frac{1}{z^2}\right)^{\frac{1}{2}(2d_J+N-\hat{d}_{j,N})} \\ &= \left(\frac{1}{z^2}\right)^{\frac{1}{2}(2d_J-\tau_{j,N})}, \end{aligned} \tag{22.2.4}$$

where

$$\begin{aligned} \tau_{j,N} &\equiv \hat{d}_{j,N} - N \\ &= (\text{dimension of } \hat{O}_{j,N}) - (\text{spin of } \hat{O}_{j,N}) \end{aligned} \tag{22.2.5}$$

is called the 'twist' of the operator $\hat{O}_{j,N}$.

We see from (22.2.4) that, the smaller $\tau_{j,N}$ is, the more singular $\tilde{C}_j^N$ will be. The dominant terms as $z^2 \to 0$ will thus be controlled by the *lowest twist operators.* As an illustration we list some simple operators and their properties:

Operator	Mass dimension	Spin	Twist
Scalar field $\phi(x)$	1	0	1
Vector field $A_\mu(x)$	1	1	0
$\bar{\psi}\gamma_\mu\psi$	3	1	2
$\phi^\dagger\partial_{\mu_1}\partial_{\mu_2}\ldots\partial_{\mu_N}\phi$	$N+2$	N	2

In analysing a product of currents the operators $\hat{O}_{j,N}$ will always be at least bilinear in the fields and the lowest twist occurring will be $\tau = 2$.

It is more useful to label the operators by their twist τ, spin N and, if there are several operators with the same (τ, N), by a further label i. In the following we suppress the label i and also the label τ since we are interested in the dominant terms as $z^2 \to 0$ which is controlled by $\tau = 2$ operators. *Thus from now on all operators and their coefficient functions are twist 2.* Then the dominant term in (22.2.3) as $z^2 \to 0$ is

$$J(\tfrac{1}{2}z)J(-\tfrac{1}{2}z) \overset{z^2\to 0}{\approx} \sum_{N=0}^{\infty} \tilde{C}^N(z^2) z^{\mu_1} \ldots z^{\mu_N} \hat{O}_{\mu_1\ldots\mu_N}. \tag{22.2.6}$$

There are now two distinct steps:

- to relate the Fourier transform of $\tilde{C}^N(z^2)$ to the Nth moment of the structure functions $W_j(\nu, q^2)$ defined in (15.3.12), and
- to study the behaviour of these Fourier transforms using the renormalization group equations.

22.2.2 Relating coefficient functions to moments of structure functions

We remarked in Section 15.3 that the $W_j(\nu, q^2)$ looked like the total cross-section for the scattering of virtual γs of 'mass' q^2. We thus expect the W_j to be expressible as the imaginary part of a forward scattering amplitude. Formally, ignoring tensor indices, the spin averaged forward scattering amplitude is given by

$$T(\nu, q^2) = \tfrac{1}{2} \sum_{\text{spins}} \int \mathrm{d}^4 z e^{\mathrm{i}q\cdot z} \langle P | T[J(\tfrac{1}{2}z)J(-\tfrac{1}{2}z)] | P \rangle \tag{22.2.7}$$

where $T[\,]$ implies a time ordered product of currents, and one indeed has

$$W(\nu, q^2) = \frac{1}{2\pi} \mathrm{Im}\; T(\nu, q^2). \tag{22.2.8}$$

Remember that here q^2 is the mass of the scattered virtual γ. It is not the momentum transfer in the scattering. $T(\nu, q^2)$ describes *forward* scattering.

It is simplest to use the Wilson expansion in $T(\nu, q^2)$. When we substitute (22.2.6) into (22.2.7) we shall need the Fourier transform of $z^{\mu_1} \dots z^{\mu_N} \tilde{C}^N(z^2)$, which can only be proportional to $q^{\mu_1} \dots q^{\mu_N}$:

$$\int \mathrm{d}^4 z \mathrm{e}^{\mathrm{i}q\cdot z} z^{\mu_1} \dots z^{\mu_N} \tilde{C}^N(z^2) \equiv q^{\mu_1} \dots q^{\mu_N} \bar{C}^N(q^2).$$

Then

$$T(\nu, q^2) \overset{|q^2|\to\infty}{\approx} \sum_{N=0} \bar{C}^N(q^2) q^{\mu_1} \dots q^{\mu_N} \langle P|\hat{O}_{\mu_1 \dots \mu_N}|P\rangle. \tag{22.2.9}$$

The matrix element on the right is unknown, even if $\hat{O}_{\mu_1 \dots \mu_N}$ is a simple operator, because it depends on the detailed properties of the hadron state $|P\rangle$ of mass m_h, i.e. it would require a knowledge of how the hadron is constructed from quark and gluon fields. However, it can only depend upon the vector P so at least its tensor structure is known. For the *spin averaged* case of interest:

$$\langle P|\hat{O}_{\mu_1 \dots \mu_N}|P\rangle = (P_{\mu_1} P_{\mu_2} \dots P_{\mu_N} - \tfrac{1}{4} P^2 g_{\mu_1 \mu_2} P_{\mu_3 \dots \mu_N} - \dots) O_N, \tag{22.2.10}$$

where O_N is a totally unknown *constant* (it is a sort of reduced hadronic matrix element) and the tensor structure reflects the fact that $\hat{O}_{\mu_1 \dots \mu_N}$ is symmetric and traceless.

When this is substituted into (22.2.9) the contraction with the q^μ factors gives $(q \cdot P)^N = (m_\mathrm{h}\nu)^N$ from the first term in parenthesis, and terms like $\frac{1}{4} P^2 q^2 (P \cdot q)^{N-2} = \frac{1}{4} m_\mathrm{h}^2 q^2 (m_\mathrm{h}\nu)^{N-2}$ from the next factors in parenthesis. Since, in the Bjorken limit, $|q^2| \approx |m_\mathrm{h}\nu|$ we see that all terms other than the first give contributions of order m_h^2/q^2 or more, smaller than the leading one. However, there are N such terms, so our error in neglecting them is of order $N m_\mathrm{h}^2/q^2$. In the Bjorken limit we have the leading behaviour

$$\begin{aligned} T(\nu, q^2) &\overset{\text{Bj. limit}}{\approx} \sum_{N=0} \bar{C}^N(q^2) O_N (m_\mathrm{h}\nu)^N \\ &= 4 \sum_{N=0} C^N(q^2) O_N \omega^N, \end{aligned} \tag{22.2.11}$$

where as usual $\omega = 2m_\mathrm{h}\nu/Q^2$ and

$$C^N(q^2) \equiv \frac{1}{4} \left(-\frac{q^2}{2} \right)^N \bar{C}^N(q^2). \tag{22.2.12}$$

For our scalar case the symmetry of $T(\nu, q^2)$ under $\nu \to -\nu$ implies that only even values of N appear in (22.2.11).

Now it is known that $T(\nu, q^2)$ is an analytic function of ν at fixed q^2 with cuts in the ν plane running from $q^2/2m_\mathrm{h} \to \infty$ and from $-\infty \to -q^2/2m_\mathrm{h}$.

Considered as a function of ω and q^2, T has cuts in the ω plane from $1 \to \infty$ and $-\infty \to -1$. It is then clear that the power series expression (22.2.11) can only be valid for $|\omega| < 1$. We, on the other hand, are interested in $W(\nu, q^2)$ for $\omega \geq 1$. The connection can be made by writing down the dispersion relation for $T(\nu, q^2)$, using it to compute T for small ω in terms of W for $\omega \geq 1$ and then comparing with (22.2.11). The dispersion relation, assuming the need for M subtractions, has the general form, for M odd,

$$T(\omega, q^2) = P_{M-1}(\omega, q^2) + 4\int_1^\infty \frac{\omega^{M+1}}{\omega'^M} \frac{W(\omega', q^2)\mathrm{d}\omega'}{\omega'^2 - \omega^2}, \qquad (22.2.13)$$

where P_{M-1} is a polynomial in ω of degree $M-1$. We have used the fact that $T(-\omega, q^2) = T(\omega, q^2)$ for our scalar currents. An analogous form holds for M even.

Using it to calculate $T(\omega, q^2)$ for $\omega \to 0$, one has

$$\begin{aligned} T(\omega, q^2) &\overset{|\omega|<1}{=} P_{M-1}(\omega, q^2) + 4\omega^{M+1}\int_1^\infty \frac{W(\omega', q^2)\mathrm{d}\omega'}{\omega'^{M+2}[1-(\omega/\omega')^2]} \\ &= P_{M-1}(\omega, q^2) + 4\omega^{M+1}\sum_{n=0}\int_1^\infty \frac{1}{\omega'^{M+2}}\left(\frac{\omega}{\omega'}\right)^{2n} W(\omega', q^2)\mathrm{d}\omega' \\ &= P_{M-1}(\omega, q^2) + 4\sum_{\substack{N=M+1\\ N \text{ even}}}^{\infty} \omega^N \int_1^\infty \frac{W(\omega', q^2)\mathrm{d}\omega'}{\omega'^{N+1}}, \end{aligned} \qquad (22.2.14)$$

which, upon comparison with (22.2.11), yields

$$\int_1^\infty \frac{W(\omega', q^2)\mathrm{d}\omega'}{\omega'^{N+1}} = C^N(q^2)O_N \quad \text{for} \quad \begin{matrix} N \geq M+1 \\ N \text{ even} \end{matrix}. \qquad (22.2.15)$$

Finally we switch to the Bjorken variable $x = 1/\omega$ and find, for M even or odd

$$\int_0^1 x^{N-1}W(x, q^2)\mathrm{d}x \overset{\text{Bj. limit}}{=} C^N(q^2)O_N \quad \text{for} \quad \begin{matrix} N \geq M \\ N \text{ even} \end{matrix}. \qquad (22.2.16)$$

Note that only the moments with N even are related to spin N operators. This does not mean that moments with N odd are zero, but they must be obtained from the N even moments by analytic continuation in N. Some care must be exercised in this process.

If Bjorken scaling held exactly $W(x, q^2)$ would be independent of q^2 implying that $C^N(q^2)$ was independent of q^2. It should be noted that this is what one would expect on naive dimensional grounds, since $[C^N] = [M]^0$ when the $J(z)$ are scalar currents bilinear in the quark fields. It is also the behaviour one would get if one treated the $J(z)$ as built up of *free field* operators. Thus the operator product expansion, in *free field approximation*, reproduces the results of the simple parton model. This is not too surprising—in Section 16.9.2 of the Appendix to Chapter 16 we did basically treat the currents as free field operators! We shall see in a moment how the renormalization group equations modify this behaviour and, in effect, produce a slow q^2 dependent correction to the Bjorken scaling result.

The above can be extended to the realistic case with vector and axial-vector currents, but the analysis is vastly more tiresome because of the algebraic complication of the tensor indices.

The analogue of (22.2.16) is as follows. For any of the deep inelastic reactions ep, en, νp, νn, $\bar{\nu}$p, $\bar{\nu}$n etc. one defines moments of the structure functions $F_k(x, q^2)$ ($k = 1, 2$ for em; $k = 1, 2, 3$ for weak interactions):

$$M_k^{(N)}(q^2) \equiv \int_0^1 x^{N-1-\delta_{k2}} F_k(x, q^2)\, \mathrm{d}x. \qquad (22.2.17)$$

Note that because of the crossing properties of the forward scattering amplitude under $\nu \to -\nu$, and with the definition of the moments given in (22.2.17), one can calculate only the following moments for various physical processes from the operator product expansion:

$$N \text{ even:} \qquad M_1^{(N)\mathrm{em}}, M_2^{(N)\mathrm{em}}, M_1^{(N)\nu+\bar{\nu}}, M_2^{(N)\nu+\bar{\nu}}, M_3^{(N)\nu-\bar{\nu}} \qquad (22.2.18)$$

$$N \text{ odd:} \qquad M_1^{(N)\nu-\bar{\nu}}, M_2^{(N)\nu-\bar{\nu}}, M_3^{(N)\nu+\bar{\nu}}$$

where $M^{\nu\pm\bar{\nu}} \equiv M^{\nu} \pm M^{\bar{\nu}}$.

The value of the moments for other N have to be obtained by analytic continuation in N.

The operators that contribute to $M^{(N)}$ can be divided into two classes according to their *flavour* properties, namely *singlet* (S), i.e. invariant under flavour transformations, and *non-singlet* (NS). The latter are bilinear in the quark fields $\psi_{i,\alpha}$, where i labels colour, and α labels flavour. (Note that the flavour label has never appeared before since we have not needed it up to now.) As an example, for $N = 1$,

$$\hat{O}^{\mu}_{\mathrm{NS},\rho} = \bar{\psi}_{i,\alpha} \left(\frac{\Lambda^{\rho}}{2} \right)_{\alpha\beta} \gamma^{\nu} \psi_{i,\beta}$$

where the matrices Λ^{ρ} are the analogues of Gell-Mann's $SU(3)$ matrices λ^{ρ}, relevant to the flavour symmetry group. In practice one assumes that the only important flavours in the electroweak currents are u, d and s, in which case one replaces Λ^{ρ} above by the Gell-Mann λ^{ρ}.

There are two kinds of singlet operators, those bilinear in the quark fields such as

$$\begin{aligned} \hat{O}^{\mu}_{\psi} &= \bar{\psi}_{i,\alpha} \gamma^{\mu} \psi_{i,\alpha}, \\ \hat{O}^{\mu_1\mu_2}_{\psi} &= \bar{\psi}_{i,\alpha} \gamma^{\mu_1} (\mathbf{D}^{\mu_2})_{ij} \psi_{j,\alpha} \text{ etc.,} \end{aligned}$$

and those bilinear in the gluon fields such as

$$\hat{O}_G^{\mu_1\mu_2} = G^{a\mu_1\nu}G^{a\mu_2\nu},$$
$$\hat{O}_G^{\mu_1\mu_2\mu_3} = G^{a\mu_1\nu}(\mathbf{D}^{\mu_2})_{ab}G^{b\mu_3\nu} \text{ etc.}$$

Every structure function may be broken up into a flavour singlet and non-singlet piece:

$$F_k(x,q^2) = F_k^{\mathrm{S}} + F_k^{\mathrm{NS}}. \qquad (22.2.19)$$

The NS part is easy to isolate since the S part is not sensitive to the flavour content of the target and will cancel out when differences are taken for different targets. Thus

$$(F_{1,2}^{\mathrm{ep}} - F_{1,2}^{\mathrm{en}}), \qquad (F_{1,2,3}^{\nu\mathrm{p}} - F_{1,2,3}^{\nu\mathrm{n}})_{\mathrm{CC\ or\ NC}}$$

are all non-singlet. (Note that this does *not* imply that the sums are singlet—that would only be true if $SU(2)$ were the flavour group, as can be seen from Section 16.4.)

As will become clear later, the singlet case is rather complicated. Let us therefore concentrate upon the non-singlet parts of the scaling functions. It should be noted that it will turn out that although $F_3(x,q^2)$ for a given reaction is not necessarily non-singlet, its behaviour is just like the NS case. *So what follows is valid for the NS parts of F_1 and F_2 and for the whole of F_3.*

In the Bjorken limit one finds for the moments of the scaling functions

$$M_{k,\mathrm{NS}}^{(N)}(q^2) = C_{k,\mathrm{NS}}^{(N)}(q^2)O_N^{\mathrm{NS}}, \qquad k = 1,2,3. \qquad (22.2.20)$$

The unknown constants O_N^{NS} are reduced matrix elements of certain linear combinations, over ρ, of operators like $\hat{O}_{\mathrm{NS},\rho}^{\mu}$ given above. They depend upon the reaction, i.e. upon what matrix elements are being considered. They are defined so as to be dimensionless. Since the scaling functions and their moments are dimensionless, from (22.2.20) also the coefficient functions must be dimensionless. Their value depends upon what product of currents is being considered, but is independent of what matrix elements are taken. As we shall see, their q^2 dependence is entirely specified by the labels NS and N, and is independent of k or ρ.

Finally, recall that (22.2.16) held only for $N \geq M$, where M was the number of subtractions needed in the dispersion relation for T. What is M for the realistic case (22.2.16)? Bearing in mind that ν is effectively the energy in virtual γ–hadron scattering we can use Regge arguments [as we did in the discussion leading to (16.6.4)] to claim that for reactions where the Pomeron can be exchanged we will require two subtractions, whereas one will suffice if Pomeron exchange is forbidden. So, in general, (22.2.20) will hold for $N \geq 1$ or 2 according to which particular physical processes we are studying.

Although the operator product expansion gives direct results only for either the even or odd N moments of certain combinations of scaling functions, we can use this information to get the moments for any N and then undo the combinations, so that ultimately we do have results for any moment of each individual scaling function.

The second step in the analysis is to use the renormalization group equation to learn about the q^2 dependence of the $C^N(q^2)$.

22.2.3 Renormalization group analysis of coefficient functions

Because our interacting field theory is full of infinities, it has to be renormalized, and the renormalization is carried out at some mass scale μ, as discussed extensively in Chapter 20. Equation (22.2.3) must then be thought of as an equation connecting renormalized operators, all renormalized at the same scale μ. The coefficient functions $\tilde{C}_j^N(z^2)$ and consequently the $C^N(q^2)$ in (22.2.11) will thus depend upon μ. The full dependence of the latter is then $C^N(q^2, \alpha(\mu^2), \mu)$, but since they are dimensionless, they must be of the form $C^N[(Q^2/\mu^2), \alpha(\mu^2)]$, where we have introduced $Q^2 = -q^2$ bearing in mind that q^2 is negative in deep inelastic scattering. For simplicity we shall again present the argument for *scalar currents and scalar fields*. One considers generalized 'Green's functions'

$$G_{JJ}^{(n)} \equiv \text{FT of } \langle 0|T[J(\tfrac{1}{2}z)J(-\tfrac{1}{2}z)\phi(x_1)\ldots\phi(x_n)]|0\rangle \qquad (22.2.21)$$

and

$$G_{j,N}^{(n)} \equiv \text{FT of } \langle 0|T[\hat{O}^{j,N}(0)\phi(x_1)\ldots\phi(x_n)]|0\rangle. \qquad (22.2.22)$$

The new operators will be renormalized analogously to (20.3.4) in the form

$$J_{\mathrm{B}} = Z_J J, \qquad \hat{O}_{\mathrm{B}}^{j,N} = Z_{j,N}\hat{O}^{j,N}, \qquad (22.2.23)$$

where the Zs will depend on a cut-off λ or upon the dimension parameter ϵ according to what method of regularization is being used.

The renormalization group equation for $G_{JJ}^{(n)}$ and $G_{j,N}^{(n)}$ will be just like (20.6.9) except that $n\gamma$ will be replaced by $-(n\gamma+4\gamma_J)$ and $-(n\gamma+2\gamma_{j,N})$ respectively, where

$$\gamma_J \equiv \frac{\mu}{2}\frac{\mathrm{d}}{\mathrm{d}\mu}(\ln Z_J), \qquad \gamma_{j,N} \equiv \frac{\mu}{2}\frac{\mathrm{d}}{\mathrm{d}\mu}(\ln Z_{j,N}). \qquad (22.2.24)$$

[If quark masses were being included one would use (20.6.4) rather than (20.6.9).]

We now substitute (22.2.3) into the renormalization group equation for $G_{JJ}^{(n)}$, use the equation satisfied by $G_{j,N}^{(n)}$, and after a lengthy series of manipulations end up with an equation for the coefficient functions C^N:

$$\left(\mu\frac{\partial}{\partial\mu}+2\beta\frac{\partial}{\partial\alpha}+4\gamma_J-2\gamma_N\right)C^N(Q^2/\mu^2,\alpha)=0 \qquad (22.2.25)$$

where now γ_N refers to the twist 2 operator $\hat{O}_{\mu_1\ldots\mu_N}$.

If there are several distinct operators of given twist τ and spin N, they may mix under renormalization in the sense that (22.2.23) is replaced by

$$\hat{O}_{\mathrm{B}}^{\tau,N,j}=Z_{jk}^{(\tau,N)}\hat{O}^{\tau,N,k}.$$

In that case (22.2.25) becomes a matrix equation.

For conserved currents and for the SM weak currents in an anomaly free theory it can be shown (Gross, 1976) that $\gamma_J\equiv 0$. Basically this follows because of the rôle of a conserved current as a generator of symmetry transformations. This rôle has to be preserved under the process of renormalization and leads to $Z_J=1$. Using $\gamma_J=0$ and the fact that the mass dimension of C^N is zero, we find, by steps analogous to those leading from (20.6.9) to (20.8.19), that

$$\begin{aligned}C^N[\eta^2Q_0^2/\mu^2,\alpha(\mu^2)] &= \exp\left\{-\int_{\alpha(\mu^2)}^{\alpha(\eta^2\mu^2)}\frac{\gamma_N(x)}{\beta(x)}\mathrm{d}x\right\}\times\\ &\quad\times C^N[Q_0^2/\mu^2,\alpha(\eta^2\mu^2)], \qquad (22.2.26)\end{aligned}$$

which is the fundamental result concerning the q^2 dependence of C^N. Moreover if we utilize our freedom and choose $\mu^2=Q_0^2$ and put $\eta^2Q_0^2=Q^2$, (22.2.26) becomes (we reinstate the label s on α to emphasize that it is the strong coupling)

$$\begin{aligned}C^N[Q^2/Q_0^2,\alpha_s(Q_0^2)] &= \exp\left\{-\int_{\alpha_s(Q_0^2)}^{\alpha_s(Q^2)}\frac{\gamma_N(x)}{\beta(x)}\mathrm{d}x\right\}\times\\ &\quad\times C^N[1,\alpha_s(Q^2)], \qquad (22.2.27)\end{aligned}$$

showing that with this choice the entire dependence upon q^2 comes through $\alpha_s(Q^2)$.

In the realistic case with vector and axial-vector currents we mentioned that there are two sets of twist 2 flavour singlet operators.

The NS operators get renormalized in a straightforward multiplicative fashion like (22.2.23), whereas the two singlet sets mix under renormalization and the consequent matrix structure will not be dealt with here. It may be found in Buras (1980).

In the analogue of (22.2.25) for the NS case, γ_N^{NS} is independent of the scaling function labelled $k=1,2,3$. It is obtained from a study of the

Green's functions $G^{(0,2)}_{0_{\rm NS}}$ to lowest order in g^2. One finds*

$$\gamma_N^{\rm NS}(\alpha_s) = \gamma_{N,1}^{\rm NS}\alpha_s + O(\alpha_s^2) \tag{22.2.28}$$

with

$$\gamma_{N,1}^{\rm NS} = \frac{C_F}{4\pi}\left[1 - \frac{2}{N(N+1)} + 4\sum_{n=2}^{N}\frac{1}{n}\right] \tag{22.2.29}$$

where C_F was defined in (22.1.9 and 10). Using this in (22.2.27) in which we take Q^2 very large so that $\alpha_s(Q^2) \ll 1$, we get, analogously to (21.5.31), for $C^{(N)}_{k,\rm NS}$

$$C^{(N)}_{k,\rm NS}(Q^2/Q_0^2) \overset{Q^2\to\infty}{=} \left[\frac{\alpha_s(Q^2)}{\alpha_s(Q_0^2)}\right]^{d^N_{\rm NS}} C^{(N)}_{k,\rm NS}(1)_{\rm FREE} \tag{22.2.30}$$

where

$$d^N_{\rm NS} = \gamma_{N,1}^{\rm NS}\ /\ b. \tag{22.2.31}$$

Just as for the case of $e^+e^- \to$ hadrons, b should be evaluated according to the number n_f^* of *active* flavours in the region whose scale is defined by $Q_0^2 \leftrightarrow Q^2$.

For $SU(3)$ of colour one has, on using the lowest order expression (21.5.23) for b, and (22.1.10) for C_F, in (22.2.29)

$$d^N_{\rm NS} = \frac{4}{33-n_f^*}\left[1 - \frac{2}{N(N+1)} + 4\sum_{n=2}^{N}\frac{1}{n}\right]. \tag{22.2.32}$$

Note that $d^N_{\rm NS}$ increases slowly with N.

22.2.4 q^2 dependence of the moments in leading order

To the leading order that we are working with (often called the *leading logarithm approximation*) one requires only the *free field value* of $C^N(q_0^2)$ on the RHS of (22.2.30). A useful way to get this is to write down the free-field version of (22.2.6) and then compute $T_{\mu\nu}(\nu, q^2)$ for virtual γs scattering on *quarks*. In this case the O_N in (22.2.10) can be evaluated and the $C^N_{\rm FREE}(q_0^2)$ identified. One can, in this way, also go further and compute corrections to $C^N_{\rm FREE}(q_0^2)$. Detailed results are given in Buras (1980) for all the interesting physical processes. Here we continue to discuss only the NS case (and F_3) and give a simplified method for understanding the general behaviour that follows from substituting (22.2.30)

* In the older literature $8\pi\gamma_{N,1}^{\rm NS}$ is called $\gamma_N^{\rm NS,0}$.

into (22.2.20). We have then

$$M_{k,\mathrm{NS}}^{(N)}(q^2) = \left[\frac{\alpha_s(Q^2)}{\alpha_s(Q_0^2)}\right]^{d_{\mathrm{NS}}^N} C_{k,\mathrm{NS\ FREE}}^{(N)}(1) O_N^{\mathrm{NS}}, \tag{22.2.33}$$

which implies a slow, logarithmic breaking of perfect Bjorken scaling.

If now we imagine letting $g \to 0$ we must recover the free-field results, i.e. the parton model results. But, as $g \to 0$, $\alpha_s(Q^2)/\alpha_s(Q_0^2) \to 1$. Hence to leading order (22.2.33) must be equivalent to

$$M_{k,\mathrm{NS}}^{(N)}(q^2) = \left[\frac{\alpha_s(Q^2)}{\alpha_s(Q_0^2)}\right]^{d_{\mathrm{NS}}^N} M_{k,\mathrm{NS}}^{(N)}(q_0^2; \text{parton model}), \tag{22.2.34}$$

where the last factor is the moment calculated from the parton model results for the scaling functions in Chapters 15, 16 and 17 determined at some value $q^2 = q_0^2$.

It follows from (22.2.34) and (21.7.5) that, if Q^2 and Q_0^2 are both sufficiently large for the analysis to be valid, then moments at Q^2 and Q_0^2 are related by

$$M_{k,\mathrm{NS}}^{(N)}(q^2) = \left[\frac{\ln(Q^2/\Lambda^2)}{\ln(Q_0^2/\Lambda^2)}\right]^{-d_{\mathrm{NS}}^N} M_{k,\mathrm{NS}}^{(N)}(q_0^2). \tag{22.2.35}$$

It is possible to interpret (22.2.35) in a generalized parton-like language, by allowing the parton distributions $q_i(x)$ of Chapter 16 to depend upon q^2: $q_i(x, q^2)$. Then we use the same formulae as in Chapter 16 to express the $F_k(x, q^2)$ in terms of the $q_i(x, q^2)$, and (22.2.35) can be interpreted as an equation controlling the change with Q^2 of the moments of any non-singlet combination of $q_i(x, q^2)$ such as $u(x, q^2) - d(x, q^2)$, $u(x, q^2) - \bar{u}(x, q^2)$, $u(x, q^2) - \bar{d}(x, q^2)$ etc.

$$q_{i,\mathrm{NS}}^{(N)}(q^2) = \left[\frac{\ln(Q^2/\Lambda^2)}{\ln(Q_0^2/\Lambda^2)}\right]^{-d_{\mathrm{NS}}^N} q_{i,\mathrm{NS}}^{(N)}(q_0^2), \tag{22.2.36}$$

where

$$q_{i,\mathrm{NS}}^{(N)}(q^2) \equiv \int_0^1 x^{N-1} q_{i,\mathrm{NS}}(x, q^2) \mathrm{d}x. \tag{22.2.37}$$

For the singlet case, because of the mixing of the quark and gluon operators, we shall just indicate the form of the result for the moments of a singlet combination of parton distribution functions $q_{\mathrm{S}}(x, q^2)$. The principal new element is that $q_{\mathrm{S}}(x, q^2)$ depends upon both $q_{\mathrm{S}}(x, q_0^2)$ and on the gluon distribution $G(x, q_0^2)$. In Chapters 15 and 16 we never utilized $G(x)$ but commented upon the need for a gluon component to explain the failure of the momentum conservation sum rule (16.7.2). The analogue of (22.2.36) is

$$\begin{aligned} q_{\mathrm{S}}^{(N)}(q^2) &= \left[\frac{\ln(Q^2/\Lambda^2)}{\ln(Q_0^2/\Lambda^2)}\right]^{-d_+^N} q_{\mathrm{S}}^{(N)}(q_0^2) + \\ &+ \left\{\left[\frac{\ln(Q^2/\Lambda^2)}{\ln(Q_0^2/\Lambda^2)}\right]^{-d_-^N} - \left[\frac{\ln(Q^2/\Lambda^2)}{\ln(Q_0^2/\Lambda^2)}\right]^{-d_+^N}\right\} \times \\ &\times \left\{\alpha_N q_{\mathrm{S}}^{(N)}(q_0^2) + \bar{\alpha}_N G^{(N)}(q_0^2)\right\}, \end{aligned} \tag{22.2.38}$$

where $d_\mp^N, \alpha_N, \bar{\alpha}_N$ are constants given in equations (2.77)–(2.87) of Buras (1980).

Let us now study the phenomenological consequences of (22.2.35) or (22.2.36).

1. The Adler sum rules (17.1.2 and 3), the Gross–Llewellyn-Smith rule (17.1.4 and 5), and the Gottfried sum rule (16.3.4) all continue to hold, since they correspond to moments with $N = 1$ and from (22.2.32), $d_{\mathrm{NS}}^{N=1} = 0$.

2. The Callan–Gross relations (16.1.8), (16.4.34, 35) and (16.5.4, 5) continue to hold in the generalized form

$$F_2(x, q^2) = 2xF_1(x, q^2), \tag{22.2.39}$$

for sufficiently large q^2; but recall that the Callan-Gross relation is not valid in regions where quark masses are important. (See Section 16.9.2 of the Appendix to Chapter 16.)

3. Since d_{NS}^N increases with N, the larger N is, the more rapidly does $M^{(N)}(q^2)$ drop as Q^2 increases. But the larger N is, the more sensitively does the moment depend upon $F_k(x, q^2)$ near $x = 1$. So we expect $F_k(x, q^2)$ to drop more rapidly with Q^2 as x gets larger. We shall see later that $F_k(x, q^2)$ is expected to grow with Q^2 at *small* x. Of course the whole parton concept arose from the fact that the measured $F_k(x)$ did not seem to depend upon Q^2. What we now find is that there is indeed a dependence on Q^2, but it is very weak and requires measurements over a huge range of Q^2 before it becomes clearly visible. Moreover the growth at small x and the decrease at large x implies the existence of a region of moderate x where the Q^2 dependence is exceptionally weak. It is for precisely this x range that the original SLAC results were published!

   Now that measurements have been made out to $Q^2 \approx 200\,(\mathrm{GeV}/c)^2$ at Fermilab and CERN, the Q^2 dependence is clearly established (see

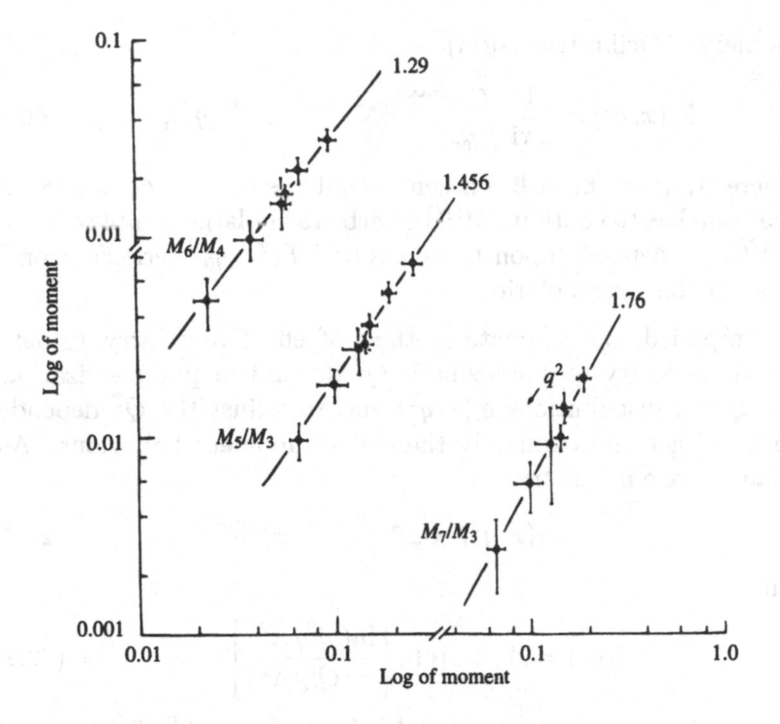

Fig. 22.3. BEBC data for moments of $xF_3(x, q^2)$ showing agreement with (22.2.40). (Data from the Big European Bubble Chamber collaboration.)

Fig. 17.13), though, as discussed in Section 17.1.4, it has taken some time and effort to check the consistency of the various experiments.

4. The logarithms of different NS moments are related to each other:

$$\ln\left[M_{k,\mathrm{NS}}^{(N)}(q^2)\right] = \frac{d_{\mathrm{NS}}^{N}}{d_{\mathrm{NS}}^{N'}} \ln\left[M_{k,\mathrm{NS}}^{(N')}(q^2)\right] + \text{constant.} \qquad (22.2.40)$$

Thus a plot of $\ln[M^{(N)}(q^2)]$ *vs* $\ln[M^{(N')}(q^2)]$ should be a straight line with slope $d_{\mathrm{NS}}^{N}/d_{\mathrm{NS}}^{N'}$. Because of the difficulty of getting accurate data on the moments (because for given Q^2 and given maximum beam energy not all values of x are accessible, e.g. at $Q^2 = 200\ (\mathrm{GeV}/c)^2, E = 250$ GeV, only $x > 0.4$ is attainable) and the insensitivity induced by taking logarithms, the results are perhaps not very convincing. Nevertheless, as shown in Fig. 22.3, the agreement with theory is nice for the moments of $xF_3(x, q^2)$ obtained by the BEBC group at CERN.

5. If $F_k(x, q_0^2)$ is measured as a function of x at some reasonably large Q_0^2, and the moments $M_k^{(N)}(q_0^2)$ are calculated, and via (22.2.35) the moments $M_k^{(N)}(q^2)$, one can then construct $F_k(x, q^2)$ from its

moments (Mellin transform)

$$F_k(x,q^2) = \frac{1}{2\pi \mathrm{i}} \int_{N_0-\mathrm{i}\infty}^{N_0+\mathrm{i}\infty} \mathrm{d}N x^{1-N} M_k^{(N)}(q^2), \tag{22.2.41}$$

where N_0 (real) must be chosen so that the integral converges. Note that one has to continue $M^{(N)}$ to arbitrarily large complex N! Since $M^{(N)}(q^2)$ depends upon the measured $F_k(x, q_0^2)$ serious errors can arise in this extrapolation.

A simplified, approximate method of effectively carrying out the above, is to try to guess simple Q^2 dependent parametrizations of the quark distributions $q_i(x, q^2)$ and to adjust the Q^2 dependence so as to get approximately the correct moment behaviour. As an example one might try

$$u(x,q^2) \propto x^{A(q^2)}(1-x)^{B(q^2)} \tag{22.2.42}$$

with

$$A(q^2) = A_0 + A_1 \ln\left[\frac{\ln(Q^2/\Lambda^2)}{\ln(Q_0^2/\Lambda^2)}\right] \tag{22.2.43}$$

etc., and adjust the constants $A_0, A_1 \ldots$ to roughly fit the required moment behaviour. Details can be found in Buras (1980).

Finally we draw the reader's attention to a refinement of the above development. Nachtmann (1973) has shown how one can define a modified form of moment that exactly projects out the contribution of the twist 2 spin N operators in (22.2.9) without ignoring the secondary terms in (22.2.10). This means that target mass corrections are taken into account. The use of Nachtmann moments is tricky and is discussed in Bitar, Johnson and Wu-Ki Tung (1979). It is only for this type of moment that it would make sense to include the mass dependent corrections discussed in Section 16.9.2.

22.2.5 An interpretation of the Q^2 variation of parton distributions in leading logarithmic approximation

We shall here indicate how the formal results of (22.2.34 and 36) can be given a simple and elegant interpretation in terms of parton distributions which depend upon Q^2. We follow the discussion of Altarelli and Parisi (1977), but all the essential ideas were developed by Gribov and Lipatov (1972).

Let us take some fixed scale μ^2 and put

$$t \equiv \ln(Q^2/\mu^2) \tag{22.2.44}$$

and let us rather loosely write

$$\alpha(Q^2) = \alpha(t), \quad \alpha(\mu^2) = \alpha(t=0), \quad \text{and} \quad M_{k,\mathrm{NS}}^{(N)}(q^2) = M_{k,\mathrm{NS}}^{(N)}(t).$$

According to (21.7.2), in lowest order, we have

$$\frac{\alpha(t=0)}{\alpha(t)} = 1 + b\alpha(t=0)t \tag{22.2.45}$$

from which follows

$$\frac{\mathrm{d}\alpha(t)}{\mathrm{d}t} = -b\alpha^2(t). \tag{22.2.46}$$

Differentiating (22.2.33) and using (22.2.46), one finds that the moments satisfy a differential equation:

$$\begin{aligned} \frac{\mathrm{d}}{\mathrm{d}t} M_{k,\mathrm{NS}}^{(N)}(t) &= -b d_{\mathrm{NS}}^{N} \alpha(t) M_{k,\mathrm{NS}}^{(N)}(t) \\ &= \frac{\alpha(t)}{2\pi} \left(-2\pi\gamma_{N,1}^{\mathrm{NS}}\right) M_{k,\mathrm{NS}}^{(N)}(t), \end{aligned} \tag{22.2.47}$$

where, in the last step, we have used (22.2.31).

Now the convolution theorem for Mellin transforms which states that, if

$$f(x) = \int_x^1 \frac{\mathrm{d}y}{y} h(x/y) g(y), \tag{22.2.48}$$

then the moments of $f(x), g(x), h(x)$ satisfy

$$f^{(N)} = h^{(N)} g^{(N)}, \tag{22.2.49}$$

used in reverse implies that the scaling functions satisfy

$$\frac{\mathrm{d}}{\mathrm{d}t} F^{\mathrm{NS}}(x,t) = \frac{\alpha_s(t)}{2\pi} \int_x^1 \frac{\mathrm{d}y}{y} P(x/y) F^{\mathrm{NS}}(y,t) \tag{22.2.50}$$

provided that $P(x)$ is so chosen that

$$P^{(N)} \equiv \int_0^1 x^{N-1} P(x) \mathrm{d}x = -2\pi\gamma_{N,1}^{\mathrm{NS}}. \tag{22.2.51}$$

Since the Fs are linear combinations of quark distributions with constant coefficients, an identical equation holds with F replaced by $q_{i,\mathrm{NS}}(x, q^2)$ in (22.2.50). Equation (22.2.50) shows how the parton distribution 'evolves' as Q^2 grows. This can be seen more clearly by writing it in the form

$$q_{i,\mathrm{NS}}(x, t+\Delta t) = q_{i,\mathrm{NS}}(x,t) + \left[\frac{\alpha_s(t)}{2\pi} \int_x^1 \frac{\mathrm{d}y}{y} P(x/y) q_{i,\mathrm{NS}}(y,t)\right] \Delta t, \tag{22.2.52}$$

and shows how the partons with given momentum fraction y at t feed the distribution at x and $t + \Delta t$. $P(x/y)$ is called a *splitting function* since

$[a_s(t)/2\pi]P(x/y)\Delta t$ measures the probability change for finding a quark with momentum fraction x inside a quark with momentum fraction y. Note the implication that a quark with momentum fraction y at t feeds quarks with $x < y$ at $t + \Delta t$.

The generalization to arbitrary quarks, antiquarks or gluon distributions is now intuitively clear:

$$\frac{\mathrm{d}q_i(x,t)}{\mathrm{d}t} = \frac{\alpha_s(t)}{2\pi}\int_x^1 \frac{\mathrm{d}y}{y}\left[P_{qq}(x/y)q_i(y,t) + + P_{qG}(x/y)G(y,t)\right] \qquad (22.2.53)$$

$$\frac{\mathrm{d}G(x,t)}{\mathrm{d}t} = \frac{\alpha_s(t)}{2\pi}\int_x^1 \frac{\mathrm{d}y}{y}\left[\sum_j P_{Gq}(x/y)q_j(y,t) + + P_{GG}(x/y)G(y,t)\right]. \qquad (22.2.54)$$

The sum over j is over all quarks *and* all antiquarks.

From their interpretation it is not surprising that the various splitting functions will turn out to correspond, in QCD, to the diagrams below, and are independent of flavour.

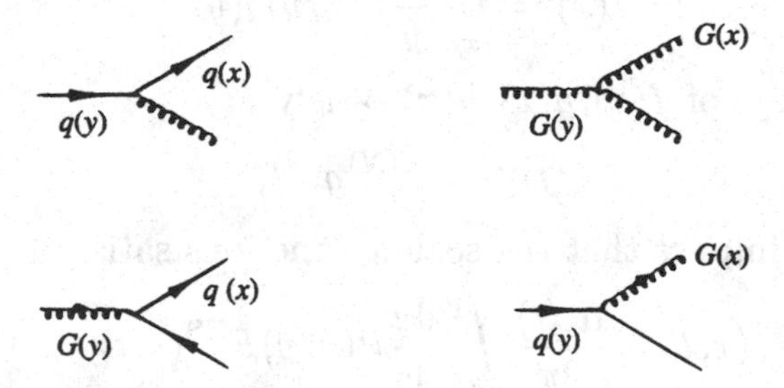

Equation (22.2.53) is completely equivalent to the results of the operator product expansion to the *leading order* and gives a nice heuristic interpretation of them. Moreover, one can utilize this approach in situations where the operator product expansion is not applicable. The explicit form of the various splitting functions will be discussed in Section 23.4. Here we list some general properties of the quark and gluon distributions that can be deduced from equations (22.2.36).

1. The fraction of momentum carried by the valence quarks drops as Q^2 increases. Since total momentum is fixed this implies that the fraction of momentum carried by sea quarks and gluons must increase with Q^2.
2. The average value of x for all the distributions decreases with Q^2.

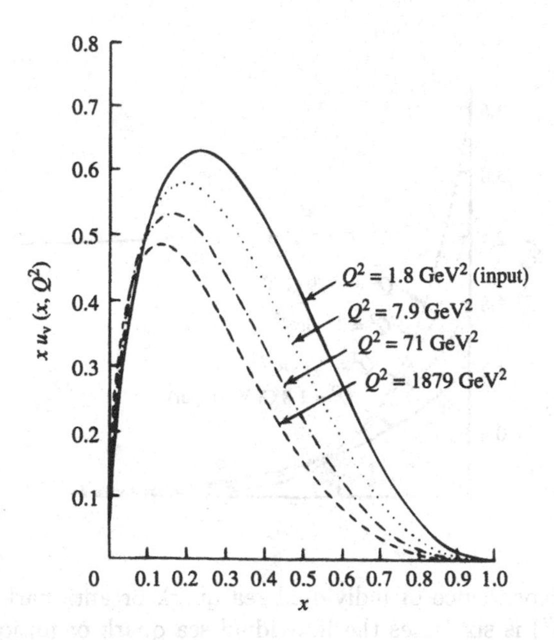

Fig. 22.4. Q^2 dependence of quark distribution functions for the valence u quarks. (From Buras and Gaemers, 1978.)

> For the sea quarks and gluons, this combined with (1) implies that the distributions grow with Q^2 at small x.

These effects are illustrated in Figs. 22.4 and 22.5 taken from a calculation by Buras and Gaemers. These should be compared with the form of some distribution functions deduced from experiment as shown in Fig. 24.21.

22.2.6 q^2 dependence of the moments in higher order

In deriving (22.2.33) we took q^2 so large that $\alpha_s(q^2)$ was small enough to regard the $C_k^{(N)}[\alpha_s(q^2)]$ as essentially given by free field theory. We can improve upon this by using perturbation theory to calculate the $C_k^{(N)}$; and precisely because $\alpha_s(q^2)$ is so small this should be a very reliable procedure.

Let us substitute (22.2.27) into (22.2.20). We shall look only at the simpler case of NS moments, but we shall drop the label NS (and the label s on α_s) for printing clarity. We then have

$$M_k^{(N)}(q^2) = \exp\left\{-\int_{\alpha(Q_0^2)}^{\alpha(Q^2)} \frac{\gamma_N(x)}{\beta(x)}\,\mathrm{d}x\right\} C_k^{(N)}[1,\alpha(Q^2)] \times \times O_N(Q_0) \tag{22.2.55}$$

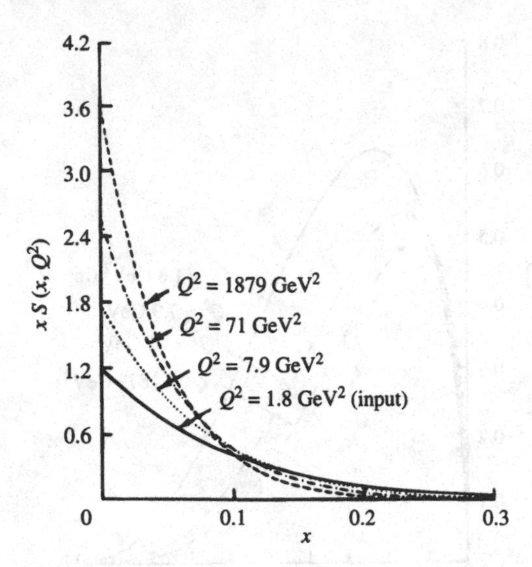

Fig. 22.5. Q^2 dependence of individual sea quark or antiquark distributions. [Note that $S(x, q^2)$ is six times the individual sea quark or antiquark distributions.] (From Buras and Gaemers, 1978.)

where we have indicated that the hadronic matrix elements O_N involve operators all renormalized at the scale $\mu = Q_0$.

If we now go beyond the leading order and take

$$\gamma_N(\alpha) = \gamma_{N,1}\,\alpha + \gamma_{N,2}\,\alpha^2 + O(\alpha^3) \tag{22.2.56}$$

and include the α^3 term in $\beta(\alpha)$ [eqn (21.5.21)] the integral in (22.2.55) is straightforward and gives

$$\exp\left\{-\int_{\alpha(Q_0^2)}^{\alpha(Q^2)} \frac{\gamma_N(x)}{\beta(x)}\mathrm{d}x\right\} = \left[\frac{\alpha(Q^2)}{\alpha(Q_0^2)}\right]^{\gamma_{N,1}/b} \left[\frac{1+b'\alpha(Q^2)}{1+b'\alpha(Q_0^2)}\right]^{\frac{1}{b}\left(\frac{\gamma_{N,2}}{b'}-\gamma_{N,1}\right)} \tag{22.2.57}$$

wherein, for consistency, we should use $\alpha(Q^2)$ calculated to next to leading order, as given in (21.7.12). We then take

$$C_k^{(N)}[1, \alpha(Q^2)] = C_k^{(N)}(1)_{\mathrm{FREE}}\left[1 + C_{k,1}^{(N)}\alpha(Q^2) + \cdots\right] \tag{22.2.58}$$

so that (22.2.55) becomes

$$\begin{aligned} M_k^{(N)}(q^2) \;=\; & \left[\frac{\alpha(Q^2)}{\alpha(Q_0^2)}\right]^{\gamma_{N,1}/b} \left[\frac{1+b'\alpha(Q^2)}{1+b'\alpha(Q_0^2)}\right]^{\frac{1}{b}\left(\frac{\gamma_{N,2}}{b'}-\gamma_{N,1}\right)} \times \\ & \times \left[1 + C_{k,1}^{(N)}\alpha(Q^2) + \cdots\right] C_k^{(N)}(1)_{\mathrm{FREE}} O_N(Q_0). \end{aligned} \tag{22.2.59}$$

Now recall that Q_0 is the scale at which the renormalization is carried out and is arbitrary. Hence the RHS of (22.2.59) cannot depend upon Q_0. It is therefore convenient to group together all terms depending upon Q_0 into one unknown term $\tilde{O}_N$ which is then independent of Q_0:

$$\begin{aligned} M_k^{(N)}(q^2) &= \left[\alpha(Q^2)\right]^{\gamma_{N,1}/b}\left[1+b'\alpha(Q^2)\right]^{\frac{1}{b}\left(\frac{\gamma_{N,2}}{b'}-\gamma_{N,1}\right)} \times \\ &\quad \times\left[1+C_{k,1}^{(N)}\alpha(Q^2)+\cdots\right]C_k^{(N)}(1)_{\mathrm{FREE}}\tilde{O}_N. \end{aligned} \tag{22.2.60}$$

Expanding in $\alpha(Q^2)$ we have finally the next to leading order form for the non-singlet moments:

$$\begin{aligned} M_{k,\mathrm{NS}}^{(N)}(q^2) &= \left[\alpha(Q^2)\right]^{\gamma_{N,1}^{\mathrm{NS}}/b}\left[1+\frac{1}{b}\left(\gamma_{N,2}^{\mathrm{NS}}-b'\gamma_{N,1}^{\mathrm{NS}}+bC_{k,\mathrm{NS},1}^{(N)}\right)\alpha(Q^2)\right]\times \\ &\quad \times C_{k,\mathrm{NS}}^{(N)}(1)_{\mathrm{FREE}}\tilde{O}_N^{\mathrm{NS}}. \end{aligned} \tag{22.2.61}$$

Detailed formulae for the coefficients $\gamma_{N,2}^{\mathrm{NS}}$ and $C_{k,\mathrm{NS},1}^{(N)}$ can be obtained from Buras (1980), as can the analogue of (22.2.61) for the singlet case. We have followed the notation of more recent papers, so one needs to compare equation (7.12, 13 and 14) of Buras with our equations (22.2.56), (21.5.21) and (22.2.58) respectively in order to relate his notation to ours.

It is important to note that the value of the parameters $\gamma_{N,2}$ and $C_{k,1}^{(N)}$ in (22.2.59) depends upon the renormalization scheme. Care must be taken not to combine formulae from different authors using different schemes. (We have already remarked that b and $\gamma_{N,1}$ are scheme independent and b' is invariant within the class of mass independent renormalization schemes.)

Compared with the leading order formulae (22.2.61) differs by terms of order constant$/\ln(Q^2/\Lambda^2)$, and this has the unpleasant consequence that a measurement of Λ *using the leading order formulae* is ambiguous in the sense that we don't know if we are truly measuring Λ or a spurious Λ that is mimicking the higher order term, i.e.

$$\begin{aligned} \frac{1}{\ln(Q^2/\Lambda^2)}\left[1+\frac{c}{\ln(Q^2/\Lambda^2)}\right] &\approx \frac{1}{\ln(Q^2/\Lambda^2)}\frac{1}{1-[c/\ln(Q^2/\Lambda^2)]} \\ &= \frac{1}{\ln(Q^2/\Lambda^2)-c}=\frac{1}{\ln(Q^2/\Lambda^2)-\ln \mathrm{e}^c} \\ &= \frac{1}{\ln(Q^2/\Lambda'^2)} \end{aligned} \tag{22.2.62}$$

where $\Lambda'^2=\mathrm{e}^c\Lambda^2$. In fact the constant c will depend upon N, so we might expect to find that the Q^2 dependence of the moments is best fitted by taking different values of Λ for each N. This indeed seems to happen and the N dependence of Λ_N is consistent with the QCD prediction. The influence of higher order effects upon the value of Λ suggests that leading order fits to different physical processes need not utilize the same value for Λ.

One interesting feature of the higher order formulae is that the Gross–Llewellyn-Smith sum rules (17.1.4 and 5) only hold in the limit $Q^2 \to \infty$. The QCD correction at finite Q^2 was given in (17.1.6). On the other hand there are no QCD corrections to the Adler sum rules (17.1.2 and 3) and corrections to the Gottfried sum rule (16.3.4) and (17.1.8) are negligible. The phenomenology of these sum rules was discussed in Section 17.1.

Also the Callan–Gross relation $F_2 = 2xF_1$ no longer holds, and is replaced by a new relation which we shall discuss in Section 23.8.

In the present, operator product approach, we have next to leading order results for the moments of the structure functions. In Chapter 23 we shall discuss the analogous results for the structure functions themselves.

22.2.7 Conclusion

We have seen in the above how the operator product expansion in conjunction with the renormalization group allows us to make certain improvements on the simple parton model. It has taken into account QCD effects which reflect the fact that the quark–partons are not perfectly free when struck by the hard photon, or vector boson.

Unfortunately there are very few reactions where one can justify the use of the operator product expansion. It is therefore of great importance that one can rephrase all of the above in the language of a very general *QCD-improved parton model.* A first step in this direction was taken in (22.2.5)

In this approach the rôle of the hadronic matrix elements like O_N in (22.2.33), which being non-perturbative are not calculable, is taken over by the equally non-calculable parton distribution functions at some scale Q_0^2. These are blended with a judicious choice of renormalization group improved perturbation theory to provide a very powerful and general tool for analysing large momentum transfer reactions. We turn to study these questions in the next chapter.

23

The parton model in QCD

The operator product expansion plus renormalization group result (22.2.34) tells us how QCD controls the Q^2 variation of the moments of the deep inelastic structure functions. But it does not give us the actual value of the moments, since they depend upon unknown, non-calculable, hadronic matrix elements $O_{N,j}$ of certain operators. In Section 22.2.5 we saw that the moment equations can be replaced by an equation controlling the Q^2 variation of the structure functions themselves, and this could be interpreted [see (22.2.53)] as a Q^2 variation of the parton densities. Again the equation does not give us the actual value of the parton distribution—only their Q^2 evolution is calculable. Thus the rôle of the unknown $O_{N,j}$ in the moment equation is taken by the unknown $q_j(x, Q_0^2)$ in the evolution equation.

It should be clear that all the difficulty stems from the hadrons. They are a non-perturbative manifestation of QCD and the problem is to derive *some* consequences of QCD without being able to handle the genuinely non-perturbative aspect. One is seeking a blend of the perturbative and the non-perturbative and the boundary between them is subtle. If *individual* hadrons are *not* involved, for example, in the totally inclusive reaction $e^+e^- \to$ hadrons, we can use purely perturbative QCD and end up with a genuine calculation of the cross-section to some order in α_s, with no unknown constants or functions appearing. This can be seen in (22.1.22).

In the present chapter we develop a well defined calculational scheme for handling reactions involving individual hadrons—the QCD-improved parton model.

23.1 Partons in a field theoretic context

Aside from the general danger of using a perturbative approach in a strong interaction theory, which is more or less under control in QCD on account of its property of asymptotic freedom, the only reason there is any difficulty in deciding how to calculate a scattering amplitude comes from the fact that the quarks and gluons are not physical particles.

To see the impact of this, consider for pedagogical purposes, old-style strong interaction theory in which the basic fields correspond to quanta like protons, neutrons and pi-mesons, which themselves exist as physical particles. Since the discussion will be heuristic, ignore any questions of spin dependence.

In order to do perturbation theory, i.e. to work out the Feynman rules, we have to decide what to use as the *unperturbed* Lagrangian. Since the bare parameters turn out to be infinite we replace them by the renormalized ones in the unperturbed Lagrangian. But, as we discussed in great detail, there is a huge freedom in the choice of the renormalized parameters. How then do we decide what parameters to use?

Consider the lowest order Feynman diagram for neutron–proton elastic scattering:

$$\mathrm{n}(p_1) + \mathrm{p}(p_2) \rightarrow \mathrm{n}(p_1') + \mathrm{p}(p_2')$$

shown below:

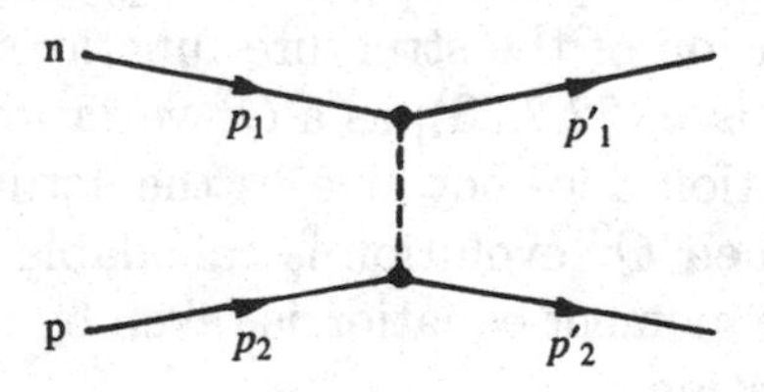

Its momentum transfer dependence is controlled by the free-field pion propagator

$$\frac{1}{(p_1' - p_1)^2 - \mu^2}$$

where μ is the pion mass parameter in the unperturbed Lagrangian $\mathcal{L}_0$. Now it is clear that the angular dependence of the scattering is crucially sensitive to the value of μ, so it would be ridiculous to choose some arbitrary value for it. Of course we choose $\mu = m_\pi$, the mass of the physical pi-meson. Why? Because we know that the *exact* pion propagator has a pole at $p^2 = m_\pi^2$, i.e. is of the form

$$\frac{1}{(p^2 - m_\pi^2)[1 + f(p^2)]}$$

where $f(m_\pi^2) = 0$.

Thus using $\mu = m_\pi$ in the unperturbed Lagrangian is clearly the optimal choice and it ensures that low-order calculations have a credible angular dependence.

What are we to do when the exchanged quantum is a quark, where, because free physical quarks do not exist, we do not expect the exact quark propagator to have a pole? In the simple quark–parton model we pretended that the partons could be regarded as free massless particles, but in a field theory context this, as will be seen, causes difficulties.

We shall now outline heuristically how one arrives at the QCD-improved parton model.

23.1.1 Heuristic reinterpretation of simple Feynman diagrams

Consider first the familiar lowest order (QED) Feynman diagram for $e(\ell)+\mu(p) \to e(\ell')+\mu(p')$ shown below:

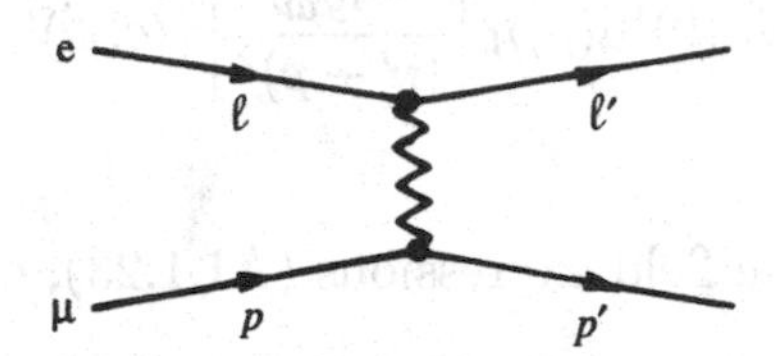

The Feynman amplitude is

$$\mathcal{M} = [\bar{u}(p')(\mathrm{i}e\gamma^{\mu})u(p)]\left[\frac{-\mathrm{i}g_{\mu\nu}}{(p'-p)^2}\right][\bar{u}(\ell')(\mathrm{i}e\gamma^{\nu})u(\ell)]. \tag{23.1.1}$$

The origin of this expression was the 'interaction picture' Dyson expansion (Section A1.3) for the S-operator, which in second order gives

$$\begin{aligned}&\langle e(\ell'),\mu(p')|S^{(2)}|e(\ell),\mu(p)\rangle\\ &= \tfrac{(-\mathrm{i})^2}{2}{}_{\mathrm{I}}\langle e(\ell'),\mu(p')|\textstyle\int d^4x d^4y\, eJ^{\mu}_{\mathrm{I}}(x)\mathrm{i}D^F_{\mu\nu}(x-y)eJ^{\nu}_{\mathrm{I}}(y)|e(\ell),\mu(p)\rangle_{\mathrm{I}}\end{aligned} \tag{23.1.2}$$

where 'I' reminds us that fields and state vectors are in the interaction picture, $J^{\mu}(x)$ is the electromagnetic current and $D^F_{\mu\nu}(x-y)$ is the free-field Feynman photon propagator. We have used the fact that in QED

$$\mathcal{H}_{\mathrm{I}}(x) = eJ^{\mu}_{\mathrm{I}}(x)A^{\mathrm{I}}_{\mu}(x) = -e\bar{\psi}_{\mathrm{I}}(x)\gamma^{\mu}\psi_{\mathrm{I}}(x)A^{\mathrm{I}}_{\mu}(x). \tag{23.1.3}$$

In Feynman gauge

$$D^F_{\mu\nu}(x) = -\frac{g_{\mu\nu}}{(2\pi)^4}\int d^4q\frac{e^{-\mathrm{i}q\cdot x}}{q^2+\mathrm{i}\epsilon}. \tag{23.1.4}$$

To see how (23.1.2) leads to (23.1.1), substitute (23.1.4) and write it in the form (the factor $\frac{1}{2}$ disappears because there are two equivalent ways to take the matrix elements)

$$(-\mathrm{i}e)^2 \int \mathrm{d}^4x \mathrm{d}^4y \int \frac{\mathrm{d}^4q}{(2\pi)^4} \mathrm{e}^{-\mathrm{i}q\cdot(x-y)} \times$$
$$\times\, {}_{\mathrm{I}}\langle\mu(p')|J^\mu_{\mathrm{I}}(x)|\mu(p)\rangle_{\mathrm{I}} \left[\frac{-\mathrm{i}g_{\mu\nu}}{q^2+\mathrm{i}\epsilon}\right] {}_{\mathrm{I}}\langle \mathrm{e}(\ell')|J^\nu_{\mathrm{I}}(y)|\mathrm{e}(\ell)\rangle_{\mathrm{I}}. \quad (23.1.5)$$

Now use the translational property(17.4.3)

$${}_{\mathrm{I}}\langle\mu(p')|J^\mu_{\mathrm{I}}(x)|\mu(p)\rangle_{\mathrm{I}} = \mathrm{e}^{\mathrm{i}(p'-p)\cdot x}\,{}_{\mathrm{I}}\langle\mu(p')|J^\mu_{\mathrm{I}}(0)|\mu(p)\rangle_{\mathrm{I}}. \quad (23.1.6)$$

Then, carrying out the integrations, (23.1.2) becomes

$$\langle \mathrm{e}(\ell'),\mu(p')|S^{(2)}|\mathrm{e}(\ell),\mu(p)\rangle = (2\pi)^4\delta(p'+\ell'-p-\ell)\times$$
$$\times\, {}_{\mathrm{I}}\langle\mu(p')|-\mathrm{i}eJ^\mu_{\mathrm{I}}(0)|\mu(p)\rangle_{\mathrm{I}} \left[\frac{-\mathrm{i}g_{\mu\nu}}{(p'-p)^2}\right] {}_{\mathrm{I}}\langle \mathrm{e}(\ell')|-\mathrm{i}eJ^\nu_{\mathrm{I}}(0)|\mathrm{e}(\ell)\rangle_{\mathrm{I}}.$$
$$(23.1.7)$$

Upon using the free field expressions (A1.1.23), one has

$${}_{\mathrm{I}}\langle\mu(p')|-\mathrm{i}eJ^\mu_{\mathrm{I}}(0)|\mu(p)\rangle_{\mathrm{I}} = \mathrm{i}e\bar{u}(p')\gamma^\mu u(p) \quad (23.1.8)$$

so that as expected from Appendix 2.1,

$$\langle \mathrm{e}(\ell'),\mu(p')|S^{(2)}|\mathrm{e}(\ell),\mu(p)\rangle = (2\pi)^4\delta^4(p'+\ell'-p-\ell)\mathcal{M} \quad (23.1.9)$$

with $\mathcal{M}$ given by (23.1.1). We see therefore that the Feynmann rules expression for $\mathcal{M}$ corresponds to having

$$\mathcal{M} = {}_{\mathrm{I}}\langle\mu(p')|-\mathrm{i}eJ^\mu_{\mathrm{I}}(0)|\mu(p)\rangle_{\mathrm{I}} \left[\frac{-\mathrm{i}g_{\mu\nu}}{(p'-p)^2}\right] {}_{\mathrm{I}}\langle \mathrm{e}(\ell')|-\mathrm{i}eJ^\nu_{\mathrm{I}}(0)|\mathrm{e}(\ell)\rangle_{\mathrm{I}}.$$
$$(23.1.10)$$

Given our knowledge of the photon propagator, this is the most useful expression to use. But imagine now that we are exchanging a quark or a gluon where as discussed in the previous section we don't know the structure of the propagator. In that case it would be better not to display the propagator explicitly. With this motivation let us rewrite (23.1.10) in such a way that the propagator is hidden in one of the matrix elements.

Now the *matrix elements* of an operator are independent of the 'picture', provided states and operators are in the same picture. Thus we may put

$${}_{\mathrm{I}}\langle\mu(p')|J^\mu_{\mathrm{I}}(0)|\mu(p)\rangle_{\mathrm{I}} = \langle\mu(p')|J^\mu_{\mathrm{em}}(0)|\mu(p)\rangle \quad (23.1.11)$$

where on the RHS we have Heisenberg picture states and fields. We have added 'em' to remind us that we are dealing with the electromagnetic current.

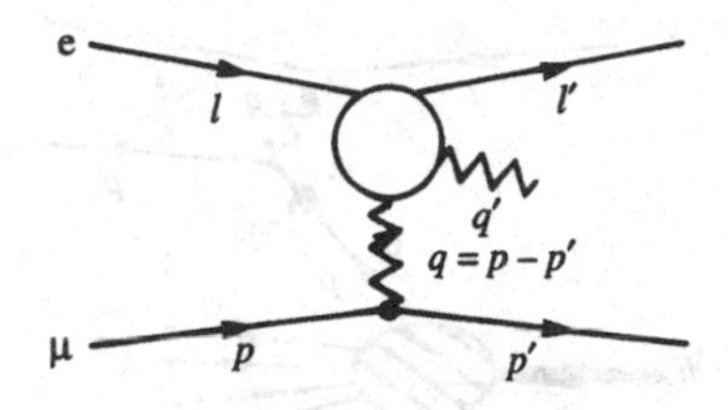

Fig. 23.1. Feynman diagram for $e\mu \to e\gamma\mu$.

Consider now

$$\langle\mu(p')| - ieJ^{\mu}_{\text{em}}(0)|\mu(p)\rangle \left[\frac{-ig_{\mu\nu}}{(p'-p)^2}\right]$$
$$= \int d^4x\, \langle\mu(p')|eJ^{\mu}_{\text{em}}(0)|\mu(p)\rangle e^{i(p'-p)\cdot x} D^F_{\mu\nu}(x) \qquad (23.1.12)$$

where we have used the inverse of (23.1.4).

Now using (23.1.6) the RHS of (23.1.12) becomes

$$\langle\mu(p')| \int d^4x\, eJ^{\mu}_{\text{em}}(x) D^F_{\mu\nu}(x)|\mu(p)\rangle. \qquad (23.1.13)$$

Recall that the Feynman propagator satisfies

$$\Box_x D^F_{\mu\nu}(x-y) = g_{\mu\nu}\delta(x-y) \qquad (23.1.14)$$

so that it is a Green's function for the differential operator $\Box_x$.

Recall also the equation of motion for the interacting photon field $A_\mu(x)$ in Feynman gauge [see (21.2.7)]

$$\Box A_\mu(x) = eJ^{\text{em}}_\mu(x). \qquad (23.1.15)$$

The solution to this, formally, is then

$$A_\mu(x) = A^{\text{Free}}_\mu(x) + e\int d^4y\, D^F_{\mu\nu}(x-y) J^{\nu}_{\text{em}}(y) \qquad (23.1.16)$$

in which only the second term on the RHS is relevant for the interaction with other particles. Hence we may put

$$e\int d^4x\, J^{\mu}_{\text{em}}(x) D^F_{\mu\nu}(x) = A_\nu(0) \qquad (23.1.17)$$

inside the matrix element in (23.1.13).

Finally then we have a new expression for the Feynman amplitude of (23.1.10)

$$\mathcal{M} = \langle\mu(p')|A_\nu(0)|\mu(p)\rangle \times {}_{\text{I}}\langle e(\ell')| - ieJ^{\nu}_{\text{I}}(0)|e(\ell)\rangle_{\text{I}}. \qquad (23.1.18)$$

Consider now the more complicated process shown in Fig. 23.1.

Let $\mathcal{M}_{e\gamma\leftarrow e\gamma}$ be the Feynman amplitude for the *physical* process

$$e(\ell) + \gamma(q) \to e(\ell') + \gamma(q').$$

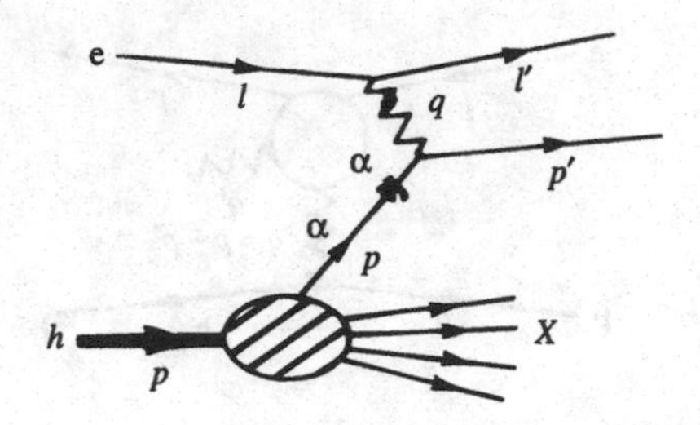

Fig. 23.2. Deep inelastic lepton–hadron scattering in which the virtual photon interacts with a quark.

It is of the form

$$\mathcal{M}_{e\gamma\leftarrow e\gamma} = M^{\nu}(\ell' q'; \ell, q)\epsilon_{\nu}(q) \tag{23.1.19}$$

with $q^2 = 0$.

For the process shown in Fig. 23.1 the analogue of (23.1.18) would be

$$\mathcal{M} = M^{\nu}(\ell', q'; \ell, q)\langle \mu(p')|A_{\nu}(0)|\mu(p)\rangle \tag{23.1.20}$$

where now, of course, $q^2 \neq 0$ in M^{ν}. Diagramatically, we shall depict the first term as

$$\langle \mu(p')|A_{\nu}(0)|\mu(p)\rangle \quad = \tag{23.1.21}$$

p p'

to indicate that no explicit propagator is to be put in, and the second as

$$M^{\nu}(\ell', q'; \ell, q) \quad = \tag{23.1.22}$$

l l' q q'

to indicate that no polarization vector is attached for the virtual photon '$\gamma(q)$'. We now apply our experience gained from QED to QCD.

23.1.2 Application to QCD

Let us now turn to QCD. For a process like deep inelastic scattering off a spin-$\frac{1}{2}$ hadron h where the photon interacts with a *quark* q we shall have the diagram shown in Fig. 23.2 and the corresponding Feynman amplitude

$$\mathcal{M} = M_{\alpha}(\ell', p'; \ell, p)\phi_{\alpha}^{q}(p, \mathrm{X}; P) \tag{23.1.23}$$

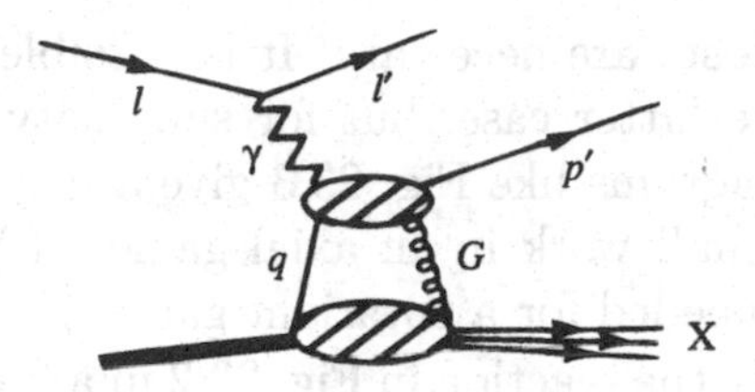

Fig. 23.3. Non-point-like interaction of virtual photon in deep inelastic scattering.

where α is a *spinor* index, in this case, and

$$\phi^q_\alpha(p, \mathrm{X}; P) \equiv \langle \mathrm{X}|\psi_\alpha(0)|P\rangle \tag{23.1.24}$$

with $\psi_\alpha(x)$ the quark field. (We are suppressing any other labels like spin etc. which might be needed to specify the proton state.)

The Feynman amplitude for electron–quark scattering, $\mathrm{e}(\ell) + q(p) \to \mathrm{e}(\ell') + q(p')$, is

$$\mathcal{M}_{\mathrm{e}q\leftarrow \mathrm{e}q} = M_\alpha(\ell', p'; \ell, p)u_\alpha(p) \tag{23.1.25}$$

with $p^2 = 0$. Thus M_α is the Feynman amplitude with the initial spinor (for the incoming quark) removed. [Note that we have suppressed an obvious colour sum in (23.1.23).]

For an antiquark, ϕ^q_α is replaced by

$$\phi^{\bar{q}}_\alpha(\mathrm{X}; P) \equiv \langle \mathrm{X}|\bar{\psi}_\alpha(0)|P\rangle. \tag{23.1.26}$$

Clearly ϕ^q_α and $\phi^{\bar{q}}_\alpha$ can be thought of as the amplitudes to produce a virtual quark or antiquark in the transition $|P\rangle \to |\mathrm{X}\rangle$.

Note that in regarding the diagram in Fig. 23.2 as important, indeed as the dominant contribution to the deep inelastic reaction, we are utilizing the information from Chapter 15, and which led to the invention of the parton model, that only such a point-like interaction of the photon will lead to non-vanishing inelastic form factors as $Q^2 \to \infty$. One might therefore have thought that the interaction depicted in Fig. 23.3, for example, could be considered as the interaction of the virtual photon with an *extended* object, the qG pair, and would not lead to non-vanishing inelastic form factors. But the gluon is chargeless, so the electric charge, with which the photon interacts, is *not* spread out. However, as stressed many times before, in a *gauge* theory the mathematical expression corresponding to a given diagram depends upon the gauge being used. Thus in some gauges, where just the physical transverse degrees of freedom of the gluon can propagate, for example in the family of axial gauges, diagrams of the type depicted in Fig. 23.3 turn out to be unimportant. This is not the case in covariant gauges where all degrees of freedom are treated equally and where ghosts are necessary. It is possible to incorporate the changes needed in the latter case, but for simplicity we shall deal only with gauges where diagrams like Fig. 23.3 give a negligible contribution. Effectively then, we shall work in an axial gauge. (At the end we shall indicate the changes needed for a covariant gauge.)

The cross-section for the reaction in Fig. 23.2 in a frame like the lepton–proton CM, where both lepton and proton have high momentum, is via (A2.1.6) and neglecting the lepton mass

$$d\sigma = \frac{1}{4(\ell \cdot P)} \sum_{X} \int |\mathcal{M}|^2 \frac{d^3\boldsymbol{p}'}{(2\pi)^3 2E_{p'}} \frac{d^3\boldsymbol{\ell}'}{(2\pi)^3 2E_{\ell'}} (2\pi)^4 \delta^4(P_X + p' + \ell' - \ell - P) \tag{23.1.27}$$

where a suitable sum and averaging over spins is implied, and $\sum_X$ is short for a sum over all momenta and spins of the particles making up the state $|X\rangle$.

Let us write $|\mathcal{M}|^2$ in the form

$$|\mathcal{M}|^2 = (M_\alpha \phi^q_\alpha)(M_\beta \phi^q_\beta)^* \tag{23.1.28}$$

and introduce

$$\bar{M}_\beta \equiv M^*_{\beta'} \gamma^0_{\beta'\beta}, \qquad \bar{\phi}^q_\beta = \phi^{q*}_{\beta'} \gamma^0_{\beta'\beta} \tag{23.1.29}$$

so that (23.1.28) becomes

$$|\mathcal{M}|^2 = (\bar{M}_\beta M_\alpha)(\phi^q_\alpha \bar{\phi}^q_\beta). \tag{23.1.30}$$

[Note that if, in (23.1.30) we were to replace $\phi^q_\alpha \to u_\alpha(p)$, $\bar{\phi}^q_\alpha \to \bar{u}_\alpha(p)$, $|\mathcal{M}|^2$ would then be the Feynman matrix element squared for the reaction $e(\ell) + q(p) \to e(\ell') + q(p')$.]

In (23.1.27) let us now introduce

$$\int d^4p\, \delta^4(p' + \ell' - p - \ell) = 1 \tag{23.1.31}$$

to obtain

$$d\sigma = \frac{1}{4(\ell \cdot P)} \int \frac{d^4p}{(2\pi)^4} \left\{ \int (\bar{M}_\beta M_\alpha) dL(p', \ell')(2\pi)^4 \delta^4(p' + \ell' - p - \ell) \right\} \times \\ \times \left\{ \sum_X (\phi^q_\alpha \bar{\phi}^q_\beta)(2\pi)^4 \delta^4(P_X + p - P) \right\} \tag{23.1.32}$$

where we have used the abbreviation

$$dL(p', \ell') \equiv \frac{d^3\boldsymbol{p}'}{(2\pi)^3 2E_{p'}} \frac{d^3\boldsymbol{\ell}'}{(2\pi)^3 2E_{\ell'}} \tag{23.1.33}$$

for the Lorentz invariant phase space factor.

Consider now the last factor in (23.1.32) which we shall call $\Phi_{\alpha\beta}(p,P)$. We have

$$\begin{aligned}\Phi_{\alpha\beta}(p,P) &\equiv \sum_X \phi^q_\alpha \bar{\phi}^q_\beta (2\pi)^4\delta^4(P_X+p-P)\\ &= \sum_X \int d^4y\, e^{iy\cdot(P_X+p-P)}\langle X|\psi_\alpha(0)|P\rangle\langle X|\psi_{\beta'}(0)|P\rangle^* \gamma^0_{\beta'\beta} \quad (23.1.34)\\ &= \sum_X \int d^4y\, e^{iy\cdot(P_X+p-P)}\langle P|\psi^\dagger_{\beta'}(0)\gamma^0_{\beta'\beta}|X\rangle\langle X|\psi_\alpha(0)|P\rangle\\ &= \sum_X \int d^4y\, e^{iy\cdot p}\langle P|\bar{\psi}_\beta(0)|X\rangle\langle X|\psi_\alpha(y)|P\rangle \quad (23.1.35)\end{aligned}$$

where we have used the translation property (17.4.3). Since we now have an unfettered sum over all states $|X\rangle$, completeness yields

$$\Phi_{\alpha\beta}(p,P) = \int d^4y\, e^{iy\cdot p}\langle P|\bar{\psi}_\beta(0)\psi_\alpha(y)|P\rangle. \quad (23.1.36)$$

Thus (23.1.32) becomes

$$\begin{aligned}d\sigma &= \frac{1}{4(\ell\cdot P)}\int \frac{d^4p}{(2\pi)^4}\times\\ &\quad\times\left\{\int(\bar{M}_\beta M_\alpha)dL(p',\ell')(2\pi)^4\delta^4(p'+\ell'-p-\ell)\right\}\Phi_{\alpha\beta}(p,P).\end{aligned} \quad (23.1.37)$$

We shall now see how this leads to a partonic interpretation in field theory.

23.1.3 The parton model in field theory

Now $\Phi_{\alpha\beta}$ is a non-perturbative object reflecting the structure of the hadronic wave-function. We are unable to calculate it, but we assume that it is only large when

$$|p^2| \le m_p^2 \qquad \text{and} \qquad |p\cdot P| \le m_p^2. \quad (23.1.38)$$

To see the implication of the latter constraint, take a reference system such that

$$\left.\begin{aligned}P^\mu &= \left(\sqrt{P^2+m_p^2},0,0,P\right) \approx (P,0,0,P)\\ p^\mu &= (p^0,\mathbf{p}_\perp,p_z)\end{aligned}\right\} \quad (23.1.39)$$

so that

$$p\cdot P \approx P(p^0-p_z) \quad (23.1.40)$$

implying that

$$|p^0 - p_z| \le m_p^2/P. \tag{23.1.41}$$

Let us now define 'light-cone' components

$$p_+ = \frac{1}{\sqrt{2}}(p_0 + p_z), \qquad p_- = \frac{1}{\sqrt{2}}(p_0 - p_z) \tag{23.1.42}$$

so that

$$p^2 = 2p_+p_- - \boldsymbol{p}_\perp^2. \tag{23.1.43}$$

The first constraint in (23.1.38) limits the perpendicular components of $\boldsymbol{p}$ in this frame to be of order m_p^2. In summary, then, we are assuming that the only important regions are

$$|p_-| \le m_p^2/P \qquad \boldsymbol{p}_\perp^2 \le m_p^2 \tag{23.1.44}$$

as $P \to \infty$.

Let us now consider the structure of the 4×4 matrix $\boldsymbol{\Phi}$ whose elements are $\Phi_{\alpha\beta}(p, P)$.

Now any 4×4 matrix can be expanded in terms of the 16 linearly independent matrices $I, \gamma^\mu, \sigma^{\mu\nu}, \gamma^\mu\gamma_5$ and $\mathrm{i}\gamma_5$ in the form

$$\boldsymbol{\Phi} = SI + V_\mu\gamma^\mu + \tfrac{1}{2}T_{\mu\nu}\sigma^{\mu\nu} + A_\mu\gamma^\mu\gamma_5 + P\mathrm{i}\gamma_5 \tag{23.1.45}$$

where $T_{\mu\nu} = -T_{\nu\mu}$. The coefficients are given by

$$\begin{aligned} S &= \tfrac{1}{4}\mathrm{Tr}[\boldsymbol{\Phi}], \qquad V_\mu = \tfrac{1}{4}\mathrm{Tr}[\gamma_\mu\boldsymbol{\Phi}], \qquad \mathrm{i}P = \tfrac{1}{4}\mathrm{Tr}[\gamma_5\boldsymbol{\Phi}] \\ T_{\mu\nu} &= \tfrac{1}{4}\mathrm{Tr}[\sigma_{\mu\nu}\boldsymbol{\Phi}], \qquad A_\mu = \tfrac{1}{4}\mathrm{Tr}[\gamma_5\gamma_\mu\boldsymbol{\Phi}]. \end{aligned} \tag{23.1.46}$$

Now returning to (23.1.37) note from Fig. 23.2 that the matrix $R_{\beta\alpha} \equiv \bar{M}_\beta M_\alpha$ contains a factor $\sum_{\text{spins}} u(p')\bar{u}(p')$ arising from the outgoing quark $q(p')$. Neglecting masses this spin sum just yields $\not{p}'$. It is then easy to see that $\boldsymbol{R}$ is a product of an *odd* number of γ matrices. When, therefore, in (23.1.37) we carry out the summation over α and β

$$\begin{aligned} \sum_{\alpha\beta}(\bar{M}_\beta M_\alpha)(\phi_\alpha^q\bar{\phi}_\beta^q) &= R_{\beta\alpha}\Phi_{\alpha\beta} \\ &= \mathrm{Tr}[\boldsymbol{R\Phi}] \end{aligned} \tag{23.1.47}$$

and we substitute (23.1.45) for $\boldsymbol{\Phi}$, the terms involving $I, \sigma^{\mu\nu}, \gamma_5$ will not contribute because the trace of an odd number of γ matrices is zero.

Thus, effectively, in (23.1.37) we may take

$$\Phi_{\alpha\beta}(p, P) \to (\gamma^\mu)_{\alpha\beta}V_\mu + (\gamma^\mu\gamma_5)_{\alpha\beta}A_\mu \tag{23.1.48}$$

where

$$V_\mu = \tfrac{1}{4}\mathrm{Tr}[\gamma_\mu\boldsymbol{\Phi}] = \tfrac{1}{4}\int \mathrm{d}^4y\, \mathrm{e}^{\mathrm{i}y\cdot p}\langle P|\bar{\psi}(0)\gamma_\mu\psi(y)|P\rangle \tag{23.1.49}$$

and

$$A_\mu = \tfrac{1}{4}\mathrm{Tr}[\gamma_5\gamma_\mu \boldsymbol{\Phi}] = \tfrac{1}{4}\int \mathrm{d}^4y\, \mathrm{e}^{\mathrm{i}y\cdot p}\langle P, S|\bar{\psi}(0)\gamma_5\gamma_\mu\psi(y)|P, S\rangle \qquad (23.1.50)$$

where we have now specified the state of the spin $\frac{1}{2}$ hadron h more precisely in terms of its momentum P^μ and covariant spin (axial) vector S^μ [see eqns (15.6.1 and 2)] and have used the fact, as will be explained presently, that V_μ is independent of S^μ.

Returning now to (23.1.37) we write

$$\mathrm{d}^4p = \mathrm{d}p_+\mathrm{d}p_-\mathrm{d}^2\mathbf{p}_\perp \qquad (23.1.51)$$

and bearing in mind the limited range of p_- and $p_\perp^2$ take the integral over p_- and $p_\perp^2$ through the first factor in parentheses, so that, approximately

$$\begin{aligned} \mathrm{d}\sigma &= \frac{1}{4(\ell\cdot P)}\int\frac{\mathrm{d}p_+}{(2\pi)}\left\{\int(\bar{M}_\beta M_\alpha)\mathrm{d}L(p',\ell')(2\pi)^4\delta^4(p'+\ell'-p-\ell)\right\}\times \\ &\quad\times\int\frac{\mathrm{d}p_-\,\mathrm{d}^2\mathbf{p}_\perp}{(2\pi)^3}\Phi_{\alpha\beta}(p,P). \end{aligned} \qquad (23.1.52)$$

Of course this presupposes that the first factor is a smooth and well behaved function of p^μ.

Using (23.1.36, 49 and 50) the last factor can be written

$$(\gamma^\mu)_{\alpha\beta}\bar{V}_\mu + (\gamma^\mu\gamma_5)_{\alpha\beta}\bar{A}_\mu \qquad (23.1.53)$$

where, on putting $\mathrm{d}^4y = \mathrm{d}y_+\mathrm{d}y_-\mathrm{d}^2\boldsymbol{y}_\perp$,

$$\bar{V}_\mu = \tfrac{1}{4}\int \mathrm{d}y_-\,\mathrm{e}^{\mathrm{i}y_-p_+}\langle P|\bar{\psi}(0)\gamma_\mu\psi(y_-, y_+=0, \boldsymbol{y}_\perp=0)|P\rangle \qquad (23.1.54)$$

and

$$\bar{A}_\mu = \tfrac{1}{4}\int \mathrm{d}y_-\,\mathrm{e}^{\mathrm{i}y_-p_+}\langle P, S|\bar{\psi}(0)\gamma_5\gamma_\mu\psi(y_-, y_+=0, \boldsymbol{y}_\perp=0)|P, S\rangle. \qquad (23.1.55)$$

Now the matrix elements in (23.1.54 and 23.1.55) can, at most, be linear in the axial vector S^μ. Consequently $\bar{V}_\mu$ is independent of S^μ and $\bar{A}_\mu$ is linear in S and thus does not contribute if we average over the spin of the proton, i.e. in the unpolarized case.

Let us consider the unpolarized case where only $\bar{V}_\mu$ contributes. We can write

$$\gamma^\mu\bar{V}_\mu = \gamma_+\bar{V}_- + \gamma_-\bar{V}_+ - \boldsymbol{\gamma}_\perp\cdot\bar{\boldsymbol{V}}_\perp \qquad (23.1.56)$$

where we have introduced

$$\gamma_\pm = \frac{1}{\sqrt{2}}(\gamma_0 \pm \gamma_3), \qquad \bar{V}_\pm = \frac{1}{\sqrt{2}}(\bar{V}_0 \pm \bar{V}_3). \qquad (23.1.57)$$

It is now clear from (23.1.54) that we must have

$$\bar{V}_- = \bar{\boldsymbol{V}}_\perp = 0 \tag{23.1.58}$$

since none of the vectors on which $\bar{V}_\mu$ depends has components in the '$-$' or '$\perp$' directions.

Thus for unpolarized protons, the last term in (23.1.52) becomes simply $\gamma_- \bar{V}_+$. Now note that

$$\begin{aligned} \not{p} &= p_+\gamma_- + p_-\gamma_+ - \boldsymbol{p}_\perp \cdot \boldsymbol{\gamma}_\perp \\ &\approx p_+\gamma_-. \end{aligned}$$

Thus we may write

$$\gamma^- \bar{V}_+ = \frac{1}{p_+} \not{p} \bar{V}_+. \tag{23.1.59}$$

Finally introducing x' via

$$p_+ \equiv x' P_+ \tag{23.1.60}$$

and noting that $(\ell \cdot p) \approx x'(\ell \cdot P)$, eqn (23.1.52) becomes

$$\begin{aligned} \mathrm{d}\sigma = \frac{1}{2} \cdot \frac{1}{4(\ell \cdot p)} \int \mathrm{d}x' \times \\ \times \left\{ \int (\bar{M}_\beta M_\alpha \not{p}_{\alpha\beta}) \mathrm{d}L(p', \ell')(2\pi)^4 \delta^4(p' + \ell' - p - \ell) \right\} f_{q/h}(x') \end{aligned} \tag{23.1.61}$$

where

$$f_{q/h}(x') \equiv \frac{1}{4\pi} \int \mathrm{d}y_-\, \mathrm{e}^{iy_- P_+ x'} \langle h(P)|\bar{\psi}(0)\gamma_+\psi(y_-, y_+ = 0, \boldsymbol{y}_\perp = 0)|h(P)\rangle. \tag{23.1.62}$$

[The appearance of γ_+ is linked to our having chosen $P^\mu \cong (P, 0, 0, P)$. When $P^\mu \approx (P, \boldsymbol{P})$, define the conjugate 4-vector $\tilde{P} = (P, -\boldsymbol{P})$. Then γ_+ is replaced by $(1/\sqrt{2}P)\gamma_\mu \tilde{P}^\mu$.]

But the first factor in (23.1.61) is now just the unpolarized cross-section for the lepto-partonic reaction

$$\mathrm{e}(\ell) + q(p) \to \mathrm{e}(\ell') + q(p')$$

with

$$p \approx x' P. \tag{23.1.63}$$

Thus, using a caret to indicate partonic processes,

$$\begin{aligned} &\mathrm{d}\sigma[\mathrm{e}(\ell) + h(P) \to \mathrm{e}(\ell') + q(p') + \mathrm{X}] \\ &\quad = \int \mathrm{d}x'\, \mathrm{d}\hat{\sigma}_0[\mathrm{e}(\ell) + q(x'P) \to \mathrm{e}(\ell') + q(p')] f_{q/h}(x') \end{aligned} \tag{23.1.64}$$

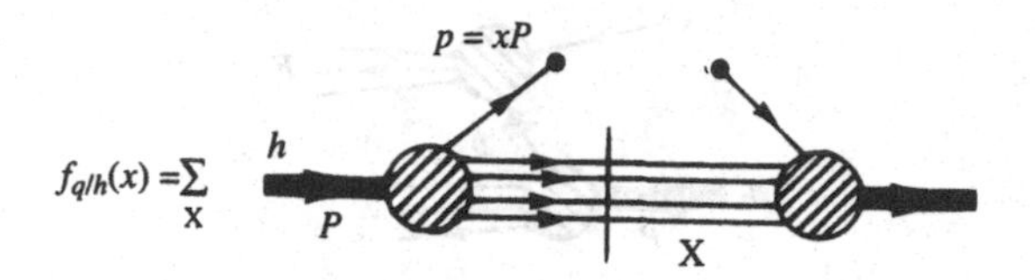

Fig. 23.4. Physical content of $f_{q/h}(x)$.

where we have added the label 0 to indicate that the $eq \to eq$ cross-section has no QCD interactions in it.

We see that $f_{q/h}(x')$ plays exactly the rôle of the quark distribution function, $q_a^h(x')$ which gives the number density of quarks of flavour 'a' in hadron h. It multiplies the $eq \to eq$ cross-section by the number of quarks with momentum fraction x' in the hadron h of momentum P^μ. Of course in the present example the energy momentum conserving δ functions inside $\mathrm{d}\sigma(eq \to eq)$ will produce a term $\delta(x' - x)$, where x is the usual Bjorken variable $Q^2/(2q \cdot P)$, so that only quarks with $x' = x$ participate, and one recovers exactly the simple parton model results.

Equation (23.1.62) provides a precise definition of the quark distribution function in field theoretic language, in terms of matrix elements of quark field operators taken between hadron states of high momentum, when using an axial gauge. Given the definition of $\Phi_{\alpha\beta}$ in (23.1.34), we can depict $f_{q/h}(x')$, aside from the question of spinor labels, as shown in Fig. 23.4 where the vertical line indicates that we are summing over 'physical' states $|\mathrm{X}\rangle$. In QCD these would have to include coloured states.

We do not wish to prolong these rather technical developments any further, so we shall simply mention, without proof, various generalizations of the above. Details and references can be found in Efremov and Radyushkin (1980).

1. Analogous field theoretic definitions can be given for the antiquark and gluon distribution functions in a hadron.

2. For the case of polarized cross-sections analogous expressions to (23.1.62) hold with the spin-dependent parton distributions emerging from the term $\bar{A}_\mu$ in (23.1.53).

3. The cross-section corresponding to the reaction of Fig. 23.5 where A and F are any particles or sets of particles in which there is a large momentum transfer, will have the same structure as (23.1.64), namely

$$\mathrm{d}\sigma[A + h(P) \to F + \mathrm{X}] = \int \mathrm{d}x'\, \mathrm{d}\hat{\sigma}_0[A + q(x'P) \to F] f_{q/h}(x'). \tag{23.1.65}$$

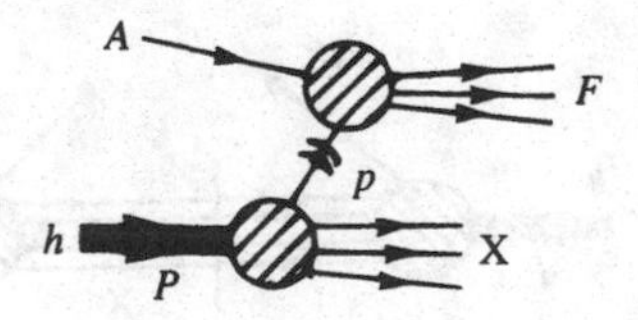

Fig. 23.5. Parton contribution to arbitrary process $Ah \to F\mathrm{X}$.

Recall that in the discussion leading to (23.1.47) the odd number of γ matrices appearing in $R_{\alpha\beta}$ played a crucial rôle. But no matter what reaction $A + q \to F$ is, there will always occur an odd number of γ matrices in the analogue of $R_{\alpha\beta}$ in a gauge theory when fermion masses are ignored. Thus the other pieces of $\boldsymbol{\Phi}$ in (23.1.45) never appear in the QCD parton model for reactions mediated by a single parton.

4. In covariant gauges it can be shown that the multi-gluon exchanges analogous to Fig. 23.3 can be summed into a relatively simple result. One replaces in (23.1.36)

$$\bar{\psi}_\beta(0)\psi_\alpha(y) \to \bar{\psi}_\beta(0) P \exp\left[\mathrm{i}g \int_y^0 t^a A^a_\nu(z) \mathrm{d}z^\nu \right] \psi_\alpha(y) \qquad (23.1.66)$$

where P implies a 'path-ordered' exponential along the straight line from y to 0 and t^a was introduced in (21.3.4).

5. Finally we note that since all field operators have to be renormalized, the matrix elements in (23.1.61 and 62) will depend upon the renormalization scheme and renormalization mass scales.

23.2 QCD corrections to the parton model

In Fig. 23.2 we have a field-theoretic version of the parton model for $\mathrm{e}h \to \mathrm{eX}$. We wish now to consider QCD corrections to this basic diagram. It will turn out that there are non-trivial difficulties in this programme.

23.2.1 Redefinition of $f_{q/h}$

Since all interest will focus on the interaction of the hard virtual photon with the hadron h, we shall no longer show the leptons in the diagrams to follow. Thus we consider

$$\text{'}\gamma\text{'} + h \to \mathrm{X}$$

and the basic diagram for the parton contribution is shown in Fig. 23.6.

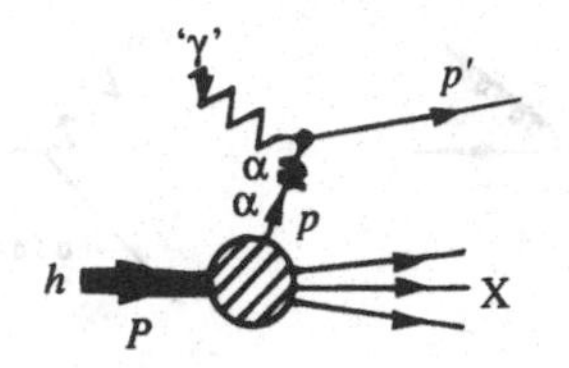

Fig. 23.6. Lowest order parton contribution to 'γ' + $h \to$ X.

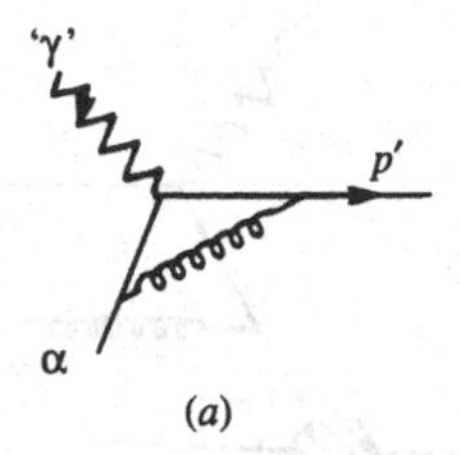

Fig. 23.7.

We now wish to consider QCD corrections to this diagram. Since $\phi^q_\alpha(p, \mathrm{X}; P)$, corresponding to the hadronic vertex, was supposed to be exact, the corrections should only be made to the 'γ'$p \to p'$ vertex. Thus we should consider for example the vertex correction (a) in Fig. 23.7 which, alas, is both ultraviolet and infrared divergent. Now the ultraviolet divergence is taken care of via renormalization, and we are familiar, from QED, with the mechanism that cures the infrared singularity. Namely we make the observation that we can never detect an isolated charged particle. Thus we must consider the possibility that the final state quark and the initial state quark radiate infrared gluons, as shown in diagrams (b) and (c) in Fig. 23.8.

Only when these are included does the sum of the *cross-sections* corresponding to the diagrams (a) and (b) + (c) yield an infrared finite result.

But the diagram (c) poses an immediate problem to us in the parton model. Consider the Feynman diagram in Fig. 23.9.

To give it a partonic interpretation we could introduce the hadronic vertex $\phi^q(p, \mathrm{X}; P)$ of (23.1.23) in either of the two ways shown in Fig. 23.10.

But since 'X' in (23.1.23) is supposed to represent all possible states it clearly must include the state $G + \mathrm{X}'$ shown in Fig. 23.10 (1). Thus diagram (1) ought to be the correct partonic interpretation of Fig. 23.9.

However, this interpretation is not consistent with the crucial assumed properties of $\phi^q(p, \mathrm{X}; P)$, namely that the amplitude is large only for small values of p^2 and $p \cdot P$. For if we take in Fig. 23.10 (1) (recall that P is large),

$$r^\mu \approx (x'P, 0, 0, x'P) \tag{23.2.1}$$

Fig. 23.8. Infrared gluon radiation from final and initial quark.

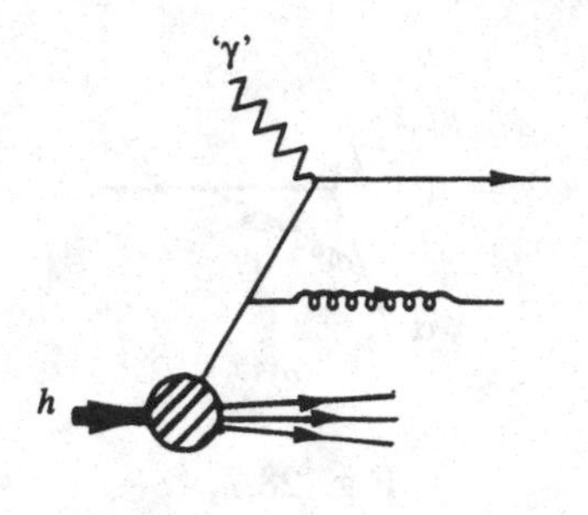

Fig. 23.9.

so that r^2 is small, then we may take

$$k^\mu \approx \left(yx'P + \frac{p_\perp^2}{2yx'P}, \boldsymbol{p}_\perp, yx'P \right) \tag{23.2.2}$$

so that $k^2 \approx 0$, and then to conserve energy and momentum,

$$p^\mu \approx \left((1-y)x'P - \frac{p_\perp^2}{2yx'P}, -\boldsymbol{p}_\perp, (1-y)x'P \right). \tag{23.2.3}$$

But now

$$p^2 \approx \frac{-p_\perp^2}{y} \qquad \text{and} \qquad p \cdot P \approx \frac{-p_\perp^2}{2yx'}. \tag{23.2.4}$$

Hence we can have p^2 and $p \cdot P$ large *without making r^2 large*; so these large values will not be damped out by the propagator involving 4-momentum r.

The resolution to this dilemma is to *redefine* $\phi^q(p, \mathrm{X}; P)$ so that it is *one-particle irreducible* in its quark lines, i.e. it does not contain diagrams which can be split apart into disconnected proper diagrams by cutting just one single quark line. Fig. 23.10 (1) is not one-particle irreducible as can be seen by cutting the line r. Thus Fig. 23.10 (2), and not (1), is the correct partonic interpretation of the Feynman diagram of Fig. 23.9. Of course this redefinition of ϕ^q has no effect on the previous argument leading to (23.1.64 or 65). Only the meaning of the matrix element in (23.1.62) has to be altered to exclude the one-particle reducible diagrams contributing to it.

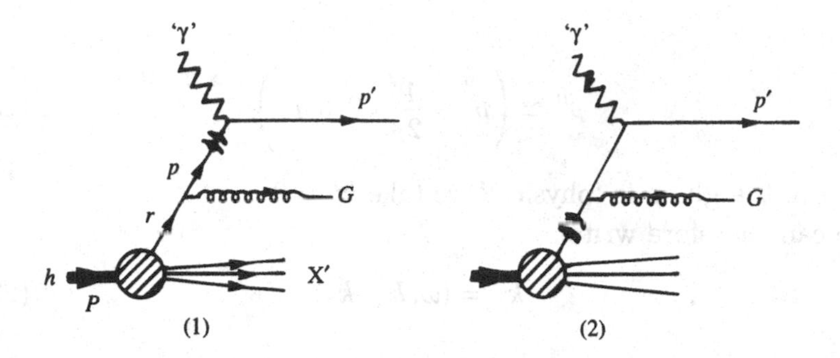

Fig. 23.10. Two possible partonic interpretations of the Feynman diagram of Fig. 23.9.

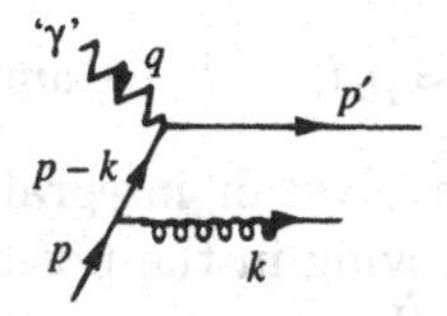

Fig. 23.11. A Feynman diagram which gives rise to an $O(\alpha_s)$ contribution to the cross-section.

With this reinterpretation the generalization of (23.1.64) seems clear:

$$\begin{aligned} &\mathrm{d}\sigma[\mathrm{e}(\ell) + h(P) \to \mathrm{e}(\ell') + \mathrm{X}] \\ &\quad = \sum_{\mathrm{Y}} \int \mathrm{d}x'\, \mathrm{d}\hat{\sigma}[\mathrm{e}(\ell) + q(x'P) \to \mathrm{e}(\ell') + \mathrm{Y}] f_{q/h}(x';\mu) \qquad (23.2.5) \end{aligned}$$

where Y is any allowed partonic final state and the lepto-partonic cross-section $\mathrm{d}\hat{\sigma}(\mathrm{e}q \to \mathrm{eY})$ is calculated, in principle, to all orders in α_s. It will include infinite diagrams requiring renormalization, and this must be done using the same renormalization scheme and renormalization mass scale μ as were used in $f_{q/h}$.

Unfortunately there is still trouble!

23.2.2 Collinear singularities—their physical origin

Consider the contribution of order α_s to the cross-section $`\gamma'(q)+q(p) \to \mathrm{Y}$ arising from the amplitude shown in Fig. 23.11. Here Y consists of a quark plus gluon.

Since the quark $q(p)$ has come from the hadronic vertex we take $(p_z > 0)$

$$p^\mu = \left(\sqrt{p_z^2 + p^2}, 0, 0, p_z\right) \qquad \text{with} \qquad p^2 \approx 0 \qquad (23.2.6)$$

so that

$$p^\mu \simeq \left(p_z + \frac{p^2}{2p_z}, 0, 0, p_z\right) \tag{23.2.7}$$

and since the gluon is 'physical' we take $k^2 = 0$.

We can therefore write

$$k^\mu = (\omega, \boldsymbol{k}_\perp, k_z) \tag{23.2.8}$$

with

$$\omega^2 = \boldsymbol{k}_\perp^2 + k_z^2. \tag{23.2.9}$$

Then

$$p \cdot k \approx p_z(\omega - k_z) + \omega p^2/2p_z. \tag{23.2.10}$$

The cross-section will involve an integral $\int \mathrm{d}^3\boldsymbol{k}$ over the final state gluon. Consider a gluon moving in the positive $0Z$ direction, i.e. in the same sense as the quark p. We can rewrite (23.2.10) as

$$\begin{aligned} p \cdot k \quad &\approx \quad \frac{p_z(\omega^2 - k_z^2)}{\omega + k_z} + \frac{\omega p^2}{2p_z} \\ &= \quad \frac{p_z k_\perp^2}{\omega + k_z} + \frac{\omega p^2}{2p_z}. \end{aligned} \tag{23.2.11}$$

The propagator for the internal quark line of momentum $p - k$ has denominator

$$d = (p-k)^2 - m^2 \tag{23.2.12}$$

where m, which may be zero, is the quark mass parameter being used in the perturbative treatment. Using (23.2.7 and 11) we find that

$$d \approx (p^2 - m^2 - \omega p^2/p_z) - \frac{2p_z k_\perp^2}{\omega + k_z}. \tag{23.2.13}$$

We see that for a gluon *collinear* with quark p, i.e. when $k_\perp^2 = 0$, $d = 0$ if we take $p^2 = m^2 = 0$. As we shall see presently the *cross-section* contains one factor of $k_\perp^2$ in its numerator and the denominator d clearly appears squared. Thus if $p^2 = m^2 = 0$ we have an integral over $\mathrm{d}^2\boldsymbol{k}_\perp$ which involves

$$\int \frac{(k_\perp^2) k_\perp \mathrm{d}k_\perp}{d^2} \propto \frac{1}{2}\int_0 \frac{k_\perp^2 \mathrm{d}k_\perp^2}{(k_\perp^2)^2} \tag{23.2.14}$$

which diverges logarithmically at $k_\perp^2 = 0$.

The maximum permissible value of $k_\perp$ in (23.2.14) can most easily be found by going to the CM of the final state quark and gluon. Then we have

$$\begin{aligned}\hat{s} &\equiv (p+q)^2 \approx 2p\cdot q + q^2 \\ &= \frac{Q^2}{\hat{x}}(1-\hat{x})\end{aligned} \tag{23.2.15}$$

where $\hat{x}$ is the partonic analogue of Bjorken x, i.e.

$$\hat{x} \equiv \frac{Q^2}{2p\cdot q}. \tag{23.2.16}$$

But also

$$\hat{s} = (p'+k)^2 = (p'_0 + \omega)^2_{\mathrm{CM}} \tag{23.2.17}$$

and when all the energy goes into the transverse motion

$$\omega_{\mathrm{CM}} = k_\perp, \qquad p'_{0\mathrm{CM}} \cong k_\perp + \frac{m^2}{2k_\perp} \approx k_\perp \tag{23.2.18}$$

so that (23.2.17 and 15) yield

$$k_\perp^2 \leq \frac{Q^2}{4\hat{x}}(1-\hat{x}). \tag{23.2.19}$$

If now we take the incoming quark p 'on-shell', i.e. take $p^2 = m^2 \neq 0$ in the denominator d in (23.2.13), then the integral in (23.2.14) will contain a term

$$\ln\left[\frac{Q^2(1-\hat{x})/4\hat{x}}{m^2}\right] \simeq \ln\left(\frac{Q^2}{m^2}\right) \tag{23.2.20}$$

showing explicitly the divergence when $m \to 0$.

There are several important observations:

1. If the incoming quark is on-shell then there is a singularity for $m = 0$ arising from gluons moving collinearly to the incoming quark. This is referred to as either a *collinear* or a *mass* singularity.

2. The potentially singular term links the behaviour at small m^2 to the behaviour at large Q^2—the latter is of prime interest in our whole development.

3. Even if $p^2 \neq m^2$ the behaviour for $p^2 \approx 0$ is singular so that the crucial step in the partonic interpretation leading to (23.1.52) would not be justified.

We shall now study these dangerous logarithmic terms (they are called *leading logarithmic terms*) and then show how they can be absorbed into our distribution functions $f_{q/h}(x')$ to provide new, modified Q^2-dependent

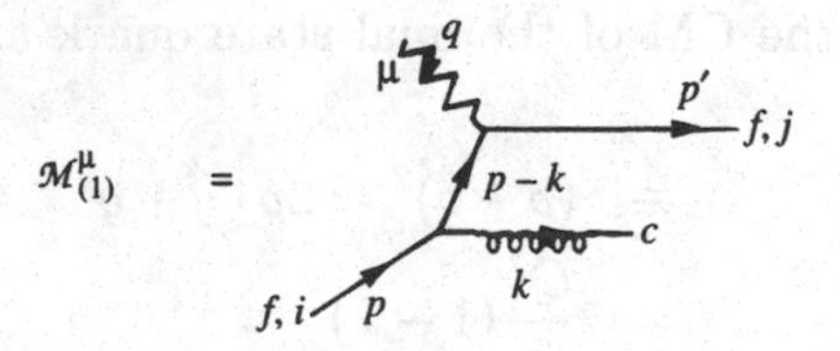

Fig. 23.12. Feynman diagram for 'γ'$(q) + q(p) \to G(k) + q(p')$.

functions $f_{q/h}(x'; Q^2)$ which, of course, are not calculable, but which can be found from experiment, and for which the variation with Q^2 *is* calculable.

23.3 Structure of the leading logarithmic terms

Consider the Feynman amplitude $\mathcal{M}^{\mu}_{(1)}$ shown in Fig. 23.12 which contributes to deep inelastic scattering *on a quark*. In this figure f labels the flavours and i, j, c label the colours of the quarks and gluon. Using the rules of Section 21.4 and after some tidying up

$$\mathcal{M}^{\mu}_{(1)} = \left[\frac{\mathrm{i}egQ_f}{(p-k)^2 - m^2}\right] [\bar{u}(p')\gamma^{\mu}(\not{p} - \not{k} + m)\gamma^{\rho}u(p)\epsilon^*_{\rho}(k)]t^c_{ji} \quad (23.3.1)$$

where Q_f is the quark charge and $\epsilon_{\rho}(k)$ is the gluon polarization vector.

Consider now the contributon of the process Fig. 23.12 to deep inelastic scattering off the quark $q(p)$. The partonic tensor, which is the analogue of the hadronic tensor, is $\hat{W}^{\mu\nu} \equiv W^{\mu\nu}_{\mathrm{em}}$ (quark), in the notation of (15.3.2). The contribution of Fig. 23.12 to it is then

$$\hat{W}^{\mu\nu}_{(1)} = \frac{1}{2e^2} \sum_{\substack{\text{spins}\\ \text{colours}}} \int \mathcal{M}^{\mu}_{(1)} \mathcal{M}^{\nu *}_{(1)} \frac{\mathrm{d}^3\boldsymbol{k}}{(2\pi)^3 2k_0} \frac{\mathrm{d}^3\boldsymbol{p}'}{(2\pi)^3 2p'_0} (2\pi)^3 \delta^4(p' + k - p - q) \quad (23.3.2)$$

where there is a sum over final helicities and colours and an average over the initial quark spin and colour. The label (1) indicates that $\hat{W}^{\mu\nu}_{(1)}$ is first order in α_s.

Neglecting masses, the trace involved, after doing the spin sums, is

$$\begin{aligned} T_{(1)} &\equiv \mathrm{Tr}[\not{p}\gamma^{\sigma}(\not{p} - \not{k})\gamma^{\nu}\ \not{p}'\gamma^{\mu}(\not{p} - \not{k})\gamma^{\rho}] \sum_{\text{hel.}} \epsilon_{\sigma}(k)\epsilon^*_{\rho}(k) \quad (23.3.3)\\ &= \sum_{\substack{\text{gluon}\\ \text{hel.}}} \mathrm{Tr}[\not{p}\ \not{\epsilon}(\not{p} - \not{k})\gamma^{\nu}\ \not{p}'\gamma^{\mu}(\not{p} - \not{k})\ \not{\epsilon}^*] \\ &= \sum \mathrm{Tr}[(-\ \not{\epsilon}\ \not{p} + 2p\cdot\epsilon)(\not{p} - \not{k})\gamma^{\nu}\ \not{p}'\gamma^{\mu}(\not{p} - \not{k})\ \not{\epsilon}^*] \\ &\equiv T_A + T_B \quad (23.3.4) \end{aligned}$$

where, using $\not{p}^2 = 0$,

$$\begin{aligned} T_A &= \sum \mathrm{Tr}[\not\epsilon \not{p} \not{k} \gamma^\nu \not{p}' \gamma^\mu (\not{p} - \not{k}) \not\epsilon^*] \\ &= \mathrm{Tr}[\not{p} \not{k} \gamma^\nu \not{p}' \gamma^\mu (\not{p} - \not{k}) \gamma^\rho S_{\rho\sigma} \gamma^\sigma] \end{aligned} \tag{23.3.5}$$

where

$$\begin{aligned} S_{\rho\sigma} &\equiv \sum_{\text{hel.}} \epsilon_\rho(k) \epsilon^*_\sigma(k) \\ &= -g_{\rho\sigma} + \frac{n_\sigma k_\rho + n_\rho k_\sigma}{n \cdot k} \end{aligned} \tag{23.3.6}$$

and where we are using a null-vector n^μ to specify the gauge (see Section 21.4).

Then

$$\begin{aligned} \gamma^\rho S_{\rho\sigma} \gamma^\sigma &= -g_{\rho\sigma} \gamma^\rho \gamma^\sigma + \frac{1}{n \cdot k} (\not{n} \not{k} + \not{k} \not{n}) \\ &= -4 + 2 = -2. \end{aligned} \tag{23.3.7}$$

Thus T_A becomes

$$\begin{aligned} T_A &= -2\mathrm{Tr}[\not{p} \not{k} \gamma^\nu \not{p}' \gamma^\mu (\not{p} - \not{k})] \\ &= 4p \cdot k \mathrm{Tr}[\not{k} \gamma^\nu \not{p}' \gamma^\mu]. \end{aligned} \tag{23.3.8}$$

Next we look at T_B in (23.3.4):

$$\begin{aligned} T_B &= \sum_{\text{hel.}} 2p \cdot \epsilon \, \mathrm{Tr}[(\not{p} - \not{k}) \gamma^\nu \not{p}' \gamma^\mu (\not{p} - \not{k}) \not\epsilon^*] \\ &= \sum_{\text{hel.}} 2p \cdot \epsilon \, \mathrm{Tr} \{ (\not{p} - \not{k}) \gamma^\nu \not{p}' \gamma^\mu [- \not\epsilon^* (\not{p} - \not{k}) + 2p \cdot \epsilon^*] \} \\ &= \sum_{\text{hel.}} 2p \cdot \epsilon \{ 2p \cdot k \, \mathrm{Tr}[\gamma^\nu \not{p}' \gamma^\mu \not\epsilon^*] \\ &\qquad + 2p \cdot \epsilon^* \, \mathrm{Tr}[(\not{p} - \not{k}) \gamma^\nu \not{p}' \gamma^\mu] \} . \end{aligned} \tag{23.3.9}$$

Now, using (21.3.6), the second term involves

$$\begin{aligned} \sum_{\text{hel.}} p \cdot \epsilon(k) \epsilon^*(k) \cdot p &= p^\rho S_{\rho\sigma} p^\sigma = \\ &= \frac{2(p \cdot n)(k \cdot p)}{n \cdot k}. \end{aligned} \tag{23.3.10}$$

For reasons that will become clear later it will be convenient to define the null vector

$$q'^\mu \equiv q^\mu + \hat{x} p^\mu \tag{23.3.11}$$

and to choose $n^\mu = q'^\mu$. Then

$$p \cdot n = p \cdot q' = p \cdot q, \qquad k \cdot n = k \cdot q + \hat{x} k \cdot p. \tag{23.3.12}$$

Hence we have, for T_B

$$\begin{aligned} T_B &= 4p\cdot k\Big\{\frac{2p\cdot q}{k\cdot q+\hat{x}k\cdot p}\mathrm{Tr}[(\not p-\not k)\gamma^\nu \not p'\gamma^\mu] \\ &\qquad +\sum_{\text{hel.}}(p\cdot\epsilon)\mathrm{Tr}[\gamma^\nu \not p'\gamma^\mu \not\epsilon^*]\Big\}. \end{aligned} \tag{23.3.13}$$

We see that both T_A and T_B have an explicit factor of $p\cdot k$, which, from (23.2.11) is proportional to $k_\perp^2$ when $p^2=0$. This is the origin of the claim in Section 23.2 that the numerator provides one factor of $k_\perp^2$.

Let us now go to the CM to consider the integration over the direction of $\boldsymbol{k}$. We can write

$$\begin{aligned} k^\mu &= (\omega,0,\omega\sin\theta,\omega\cos\theta) \\ p^\mu &\simeq (p,0,0,p) \end{aligned} \tag{23.3.14}$$

so that

$$k^\mu = \frac{\omega}{p}p^\mu + (0,0,\omega\sin\theta,\omega(\cos\theta-1)) \tag{23.3.15}$$

where, in the CM, one finds

$$\frac{\omega}{p} = (1-\hat{x}). \tag{23.3.16}$$

[Recall that $\hat{x}$ is the 'γ'-quark Bjorken variable: $\hat{x}=Q^2/(2q\cdot p)$.] Now, as discussed, we already have one factor

$$p\cdot k = p\omega(1-\cos\theta) \tag{23.3.17}$$

in the numerator, and the denominator in (23.3.1 and 2) will contain

$$\left[\frac{1}{2p\cdot k+m^2}\right]^2 = \frac{1}{4p^2\omega^2}\left[\frac{1}{1-\cos\theta+m^2/2p\omega}\right]^2 \tag{23.3.18}$$

where, in the CM,

$$2p\omega = Q^2/2\hat{x}. \tag{23.3.19}$$

Thus once we have factored out $p\cdot k$ the only remaining terms in the numerator that will yield $\ln Q^2/m^2$ are those which do *not* vanish at $\theta=0$. Hence we may drop the second term in T_B in (23.3.13) since from (23.3.15), $p\cdot\epsilon\to 0$ as $\theta\to 0$. Moreover, via (23.3.15), everywhere that k^μ occurs in these *remaining* terms it can be replaced by

$$k^\mu \to (1-\hat{x})p^\mu. \tag{23.3.20}$$

Thus from (23.3.8, 13 and 4)

$$T_{(1)} = 4p\cdot k\left(\frac{1+\hat{x}^2}{1-\hat{x}}\right)Tr[\not p\gamma^\nu(\not q+\hat{x}\not p)\gamma^\mu]. \tag{23.3.21}$$

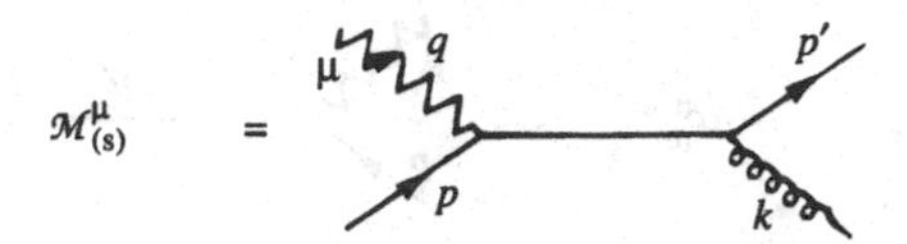

Fig. 23.13. 's-channel' amplitude for 'γ'$(q) + q(p) \to G(k) + q(p')$.

It is important to understand the physical origin of the factor

$$p \cdot k \propto (1 - \cos\theta) \tag{23.3.22}$$

in (23.3.21). The Feynman amplitude $\mathcal{M}^{\mu}_{(1)}$ involves the transition

$$q_{\lambda}(p) \to \text{`}q\text{'}_{\lambda'}(p-k) + G_{\lambda_G}(k)$$

where the labels indicate helicities. As k^{μ} becomes collinear to p^{μ} 'q' becomes a 'real' massless quark, and its helicity λ' will be equal to λ because the helicity of *massless* fermions is conserved for vector (or axial vector) coupling.

But by conservation of angular momentum the collinear amplitude will be zero unless $\lambda = \lambda' - \lambda_G = \lambda - \lambda_G$ which is impossible for a transverse gluon. The vanishing of the amplitude is proportional to

$$\begin{aligned}(\sin\theta/2)^{|(\lambda-\lambda_G)-\lambda|} &= \sin\theta/2 \\ &\propto \sqrt{1-\cos\theta}.\end{aligned} \tag{23.3.23}$$

Since there is a product $\mathcal{M}^{\mu}_{(1)}\mathcal{M}^{\nu*}_{(1)}$ we get the overall factor of $(1-\cos\theta)$ found above.

Now the diagram in Fig. (23.12), (which is referred to as a 't-channel exchange' contribution) is not the only one contributing to $\hat{W}^{\mu\nu}$ in first order in α_s. The 's-channel exchange' diagram Fig. 23.13 also contributes, and should have been added to $\mathcal{M}^{\mu}_{(1)}$ *before* taking the product in $\hat{W}^{\mu\nu}$ in (23.3.2). But the above amplitude is not singular for k^{μ} collinear to p^{μ}, so the term $\mathcal{M}^{\mu}_{s}\mathcal{M}^{\nu*}_{s}$ is irrelevant for our considerations. Moreover the interference term $\mathcal{M}^{\mu}_{s}\mathcal{M}^{\nu*}_{(1)}$ will contain, at worst, a factor $\sqrt{1-\cos\theta}$ in the numerator and just one singular denominator $(1 - \cos\theta + m^2/2p\omega)$. For $m^2 = 0$, the angular integral is then of the form

$$\int_{-1}^{1} \frac{\sqrt{1-z}}{1-z} \mathrm{d}z$$

which is finite. Thus the interference term does not give rise to a collinear mass singularity, and only the diagram of Fig. 23.12 is relevant.

We can now see why the gauge choice $n^{\mu} = q'^{\mu}$ was useful. Any choice of n^{μ} such that $n \cdot k \to 0$ as k^{μ} becomes collinear with p^{μ} would via (23.3.10) produce an extra singular denominator in $T_{(1)}$. But since the overall result

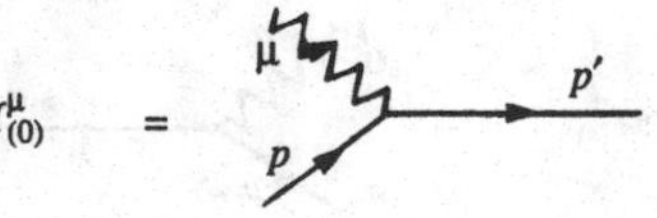

Fig. 23.14. Lowest order diagram for 'γ'$(q) + q(p) \to q(p')$.

must be gauge-invariant this implies that a singular denominator would then appear in $\mathcal{M}^{\mu}_{s}$ and we would not have been able to ignore the diagram in Fig. 23.13.

Finally putting together (23.3.21,18,19) in (23.3.2) and using for the colour average the result [see (22.1.9,10)]

$$\tfrac{1}{3}\mathrm{Tr}[t^c t^c] = C_F \tag{23.3.24}$$

where, for N colours,

$$C_F(N) = \frac{N^2-1}{2N} \tag{23.3.25}$$

and carrying out the integration, the term in $\hat{W}^{\mu\nu}_{(1)}$ which has the collinear singularity $\hat{\hat{W}}^{\mu\nu}_{(1)}$ is

$$\hat{\hat{W}}^{\mu\nu}_{(1)} = \frac{1}{2Q^2}(Q_f)^2 \mathrm{Tr}[\hat{x}\, \not{p}\gamma^{\nu}(\not{q} + \hat{x}\, \not{p})\gamma^{\mu}] \frac{\alpha_s C_F}{2\pi}\left(\frac{1+\hat{x}^2}{1-\hat{x}}\right) \ln Q^2/m^2. \tag{23.3.26}$$

To understand the implications of this result consider now the zero-order (in α_s) process shown in Fig. 23.14. Its contribution to $\hat{W}^{\mu\nu}$ is of course the basic quark–parton contribution and is given by

$$\hat{W}^{\mu\nu}_{(0)} = \frac{1}{2e^2}\sum_{\text{spins}}\int \mathcal{M}^{\mu}_{(0)}\mathcal{M}^{\nu *}_{(0)} \frac{d^3 p'}{(2\pi)^3 2p'_0}(2\pi)^3 \delta^4(p'-q-p) \tag{23.3.27}$$

which will contain the trace

$$\begin{aligned} T_{(0)} &= \mathrm{Tr}[\not{p}\gamma^{\nu}\, \not{p}'\gamma^{\mu}] \\ &= \mathrm{Tr}[\not{p}\gamma^{\nu}(\not{q} + \not{p})\gamma^{\mu}]. \end{aligned} \tag{23.3.28}$$

Handling the energy conserving δ function as in (15.1.8,9):

$$\begin{aligned} \frac{1}{2p'_0}\delta(p'_0 - q_0 - p_0) &= \frac{\hat{x}}{Q^2}\delta(1-\hat{x}) \\ &= \frac{1}{Q^2}\delta(1-\hat{x}). \end{aligned} \tag{23.3.29}$$

Thus we can write

$$\begin{aligned}\hat{W}^{\mu\nu}_{(0)}(q,p) &= \frac{\pi}{Q^2}(Q_f)^2\,\mathrm{Tr}[\not{p}\gamma^\nu(\not{q}+\not{p})\gamma^\mu]\delta(1-\hat{x}) \\ &\equiv \Gamma^{\mu\nu}_{(0)}(q,p)\delta(1-\hat{x}). \end{aligned} \tag{23.3.30}$$

Combining this with (23.3.26) we have the fundamental result for the collinear singular contribution to the quark deep inelastic tensor:

$$\hat{W}^{\mu\nu}_{(0)} + \hat{\hat{W}}^{\mu\nu}_{(1)} = \Gamma^{\mu\nu}_{(0)}(q,\hat{x}p)\left\{\delta(1-\hat{x}) + \frac{\alpha_s C_F}{2\pi}\left(\frac{1+\hat{x}^2}{1-\hat{x}}\right)\ln Q^2/m^2\right\}. \tag{23.3.31}$$

We can write the RHS in the form

$$\begin{aligned}&\int_0^1 \mathrm{d}y\,\Gamma^{\mu\nu}_{(0)}(q,yp)\delta(y-\hat{x})\left\{\delta(1-y) + \frac{\alpha_s C_F}{2\pi}\left(\frac{1+y^2}{1-y}\right)\ln Q^2/m^2\right\} \\ &= \int_0^1 \frac{\mathrm{d}y}{y}\hat{W}^{\mu\nu}_{(0)}(q,yp)\left\{\delta(1-y) + \frac{\alpha_s C_F}{2\pi}\left(\frac{1+y^2}{1-y}\right)\ln Q^2/m^2\right\}\end{aligned} \tag{23.3.32}$$

where we have used the fact which follows from (23.3.30) that

$$\hat{W}^{\mu\nu}_{(0)}(q,yp) = \Gamma^{\mu\nu}_{(0)}(q,yp)\delta(1-\hat{x}/y). \tag{23.3.33}$$

Note that the order α_s correction has a singularity at $\hat{x}=1$. This is an infrared singularity, since it corresponds to $k^\mu = 0$ according to (23.3.15). But we know that there are no infrared singularities when *all* diagrams of order α_s, including the vertex correction (Fig. 23.6), are taken into account. We shall not discuss the technical details. The net result is to replace the function $C_F(1+y^2)/(1-y)$ by $P_{qq}(y)$, the details of which we shall give in the next section.

In the above we have been discussing deep inelastic scattering on the quark $q(p)$. When we consider deep inelastic scattering on a hadron via the analysis of Section 23.1 the quark p will become a virtual quark emitted by the hadron, and m^2 in (23.3.31) will be replaced by p^2. The singular behaviour in p^2 is not physical. It is an artefact of the separation of the perturbative from the non-perturbative.

In the non-singular part of (23.3.31) we can safely put $p = x'P$ and then changing integration variable to $u = x'y$ get

$$\hat{W}^{\mu\nu}_{(0)} + \hat{\hat{W}}^{\mu\nu}_{(1)} = \int_0^1 \frac{\mathrm{d}u}{u}\hat{W}^{\mu\nu}_{(0)}(q,uP)\left\{\delta\left(1-\frac{u}{x'}\right) + \frac{\alpha_s}{2\pi}P_{qq}\left(\frac{u}{x'}\right)\ln\frac{Q^2}{p^2}\right\}. \tag{23.3.34}$$

In Section 23.4 we shall show how to modify the results of Section 23.1 to take account of the singular logarithm in (23.3.34).

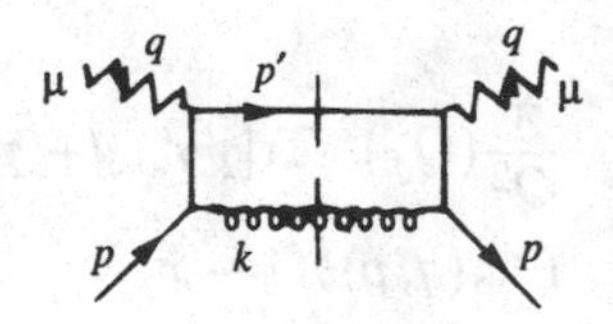

Fig. 23.15.

We end this section by drawing the reader's attention to a slightly different approach to the above which uses Feynman diagrams rather than cross-sections. The essential point, as can easily be checked, is that $\hat{W}^{\mu\nu}_{(1)}$ of (23.3.2) can be obtained from the Feynman 'box' diagram of Fig. 23.15 by putting the propagators marked with a vertical 'slash' on the mass-shell by the substitution

$$\frac{1}{k^2 + \mathrm{i}\epsilon} \to -\mathrm{i}\pi\theta(k_0)\delta(k^2)$$
$$\frac{1}{p'^2 - m^2 + \mathrm{i}\epsilon} \to -\mathrm{i}\pi\theta(p'_0)\delta(p'^2 - m^2). \tag{23.3.35}$$

However this replacement is known to produce the discontinuity of the amplitude considered as a function of complex $\hat{s} = (p + q)^2$. Thus one can study the Feynman amplitude itself and only at the very end take its discontinuity.

By using an axial gauge, where only transverse modes of the gluon field propagate, one finds that only box or ladder-type diagrams are relevant for the leading logarithmic analysis. This is the analogue of our finding above that we could ignore the diagram in Fig. 23.13.

23.4 Q^2-dependent distribution functions

Suppose we are calculating the hadronic DIS tensor $W^{\mu\nu}$. Then the quark $q(p)$ of the previous discussion is itself produced from a hadron of momentum P, so we could repeat all of the analysis leading from (23.1.37) to (23.1.64) with the exception of the step leading to (23.1.52). Now the singular behaviour for small p^2 in (23.3.34) will not allow us to take the integration over p_- and $\boldsymbol{p}_\perp$ through the partonic matrix element and onto $\Phi_{\alpha\beta}$.

However, in deriving (23.3.31) we have been cavalier in the sense that we have integrated from $k_\perp^2 = 0$ to $k_\perp^2 \approx Q^2$, whereas the perturbative calculation can only be trusted for large $k_\perp^2$. To remedy this and at the same time handle the singular p^2 behaviour, we introduce a scale μ, the *factorization scale*, such that

$$Q^2 \gg \mu^2 \geq m_p^2 \tag{23.4.1}$$

and such that we trust the perturbative calculation for $k_\perp^2 > \mu^2$. The expression in parenthesis in (23.3.34) then becomes

$$\delta(1-u/x') + \frac{\alpha_s}{2\pi}P_{qq}(u/x')\ln(Q^2/\mu^2) + \frac{\alpha_s}{2\pi}P_{qq}(u/x')\ln(\mu^2/p^2). \tag{23.4.2}$$

In the above we have used the same symbol μ as was used to specify the renormalization scale. We do so because it is very convenient to work with a renormalization scale equal to the momentum cut-off scale.

We now use a trick very common in renormalization theory to factorize this. If we had had $1+\alpha_s A+\alpha_s B$ we would have been able to claim that, *to order* α_s,

$$1+\alpha_s A+\alpha_s B = (1+\alpha_s A)(1+\alpha_s B). \tag{23.4.3}$$

When a δ function is involved, the factorization requires a convolution. One has, to order α_s with $0 \leq x \leq 1$,

$$\begin{aligned}&\delta(1-x)+\alpha_s A(x)+\alpha_s B(x) = \\ &\int_x^1 \frac{\mathrm{d}y}{y}[\delta(1-y)+\alpha_s A(y)][\delta(1-x/y)+\alpha_s B(x/y)].\end{aligned} \tag{23.4.4}$$

Expression (23.4.2), correct to order α_s, is thus equal to

$$\begin{aligned}&\int \frac{\mathrm{d}y}{y}\left[\delta(1-y)+\frac{\alpha_s}{2\pi}P_{qq}(y)\ln(Q^2/\mu^2)\right]\times \\ &\quad\times\left[\delta(1-u/x'y)+\frac{\alpha_s}{2\pi}P_{qq}(u/x'y)\ln(\mu^2/p^2)\right].\end{aligned} \tag{23.4.5}$$

We now absorb the last factor into our definition of the quark distribution function by keeping it inside the integration over p_- and $\boldsymbol{p}_\perp$ in (23.1.52), i.e. have

$$\int \mathrm{d}p_-\,\mathrm{d}^2\boldsymbol{p}_\perp\left[\delta(1-u/x'y)+\frac{\alpha_s}{2\pi}P_{qq}(u/x'y)\ln(\mu^2/p^2)\right]\Phi_{\alpha\beta}(p,P). \tag{23.4.6}$$

After integrating over x' we are left with

$$\begin{aligned}W^{\mu\nu} \;=\; &\int\frac{\mathrm{d}u}{u}\hat{W}^{\mu\nu}_{(0)}(q,uP)\int\frac{\mathrm{d}y}{y}\left[\delta(1-y)+\frac{\alpha_s}{2\pi}P_{qq}(y)\ln(Q^2/\mu^2)\right]\times \\ &\times f_{q/h}(u/y;\mu^2)\end{aligned} \tag{23.4.7}$$

where $f_{q/h}$ is now a new quark distribution function which in effect includes quarks with $p_\perp^2 \leq \mu^2$ or, via (23.2.4), quarks of 'virtuality' down to $p^2 \approx -\mu^2$.

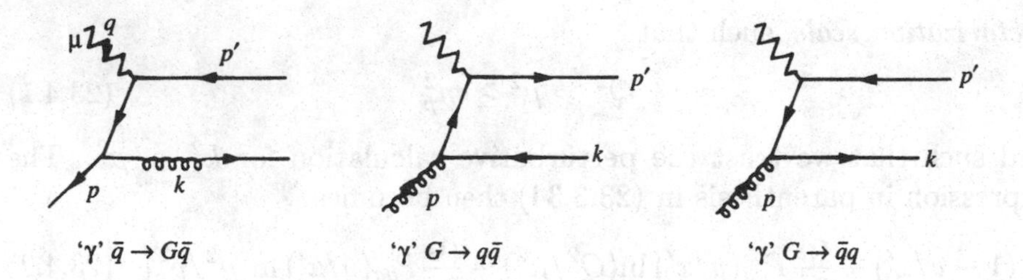

Fig. 23.16. Order α_s contribution to eh → e′X arising from antiquarks and gluons.

Changing integration variables to $x' = u/y$ (23.4.7) becomes

$$W^{\mu\nu} = \int \frac{\mathrm{d}u}{u} \hat{W}^{\mu\nu}_{(0)}(q, uP) \times \\ \times \int_u^1 \frac{\mathrm{d}x'}{x'} \left[\delta(1 - u/x') + \frac{\alpha_s}{2\pi} P_{qq}(u/x') \ln(Q^2/\mu^2) \right] f_{q/h}(x'; \mu^2). \tag{23.4.8}$$

If we consider cross-sections by reinstating the leptonic vertex in an obvious way, we get the QCD corrected version of (23.1.64) in leading logarithmic approximation:

$$\mathrm{d}\sigma[\mathrm{e}(\ell) + h(P) \to \mathrm{e}(\ell') + \mathrm{X}]_{\mathrm{LLA}} = \\ \int_0^1 \mathrm{d}u \, \mathrm{d}\hat{\sigma}_0[\mathrm{e}(\ell) + q(uP) \to \mathrm{e}(\ell') + q(p')] \times \\ \times \int_u^1 \frac{\mathrm{d}x'}{x'} \left[\delta(1 - u/x') + \frac{\alpha_s}{2\pi} P_{qq}(u/x') \ln(Q^2/\mu^2) \right] f_{q/h}(x'; \mu^2). \tag{23.4.9}$$

Note that in interpreting the result of the $\mathrm{d}p_- \, \mathrm{d}^2\mathbf{p}_\perp$ integration in (23.4.6) as the quark distribution function, it is crucial that the function P_{qq} is independent of what reaction we are studying. This is guaranteed by the link between the $\ln Q^2$ growth and the mass singularity. For, as discussed in Section 23.3, the origin of the term $\ln(Q^2/m^2)$ could be traced to kinematics and angular momentum conservation involving the incoming quark $q(p)$ when it radiated the gluon $G(k)$.

In an exactly parallel way we could study the contribution to $\mathrm{e} + h \to \mathrm{e}' + \mathrm{X}$ as initiated by antiquarks and gluons via diagrams like Fig. 23.16. The result (23.4.9) generalizes in an intuitive fashion. As in Section 22.2 we put

$$t \equiv \ln(Q^2/\mu^2) \tag{23.4.10}$$

and then have

$$\begin{aligned}
&\mathrm{d}\sigma[\mathrm{e}(\ell) + h(P) \to \mathrm{e}(\ell') + \mathrm{X}]_{\mathrm{LLA}} = \\
&\qquad \int_0^1 \mathrm{d}u\, \mathrm{d}\hat{\sigma}_0[\mathrm{e}(\ell) + q(uP) \to \mathrm{e}(\ell') + q(p')] \times \\
&\qquad \times \int_u^1 \frac{\mathrm{d}x'}{x'} \left\{ \left[\delta(1 - u/x') + \frac{t\alpha_s}{2\pi} P_{qq}(u/x') \right] q(x', \mu^2) + \right. \\
&\qquad\qquad \left. + \frac{t\alpha_s}{2\pi} P_{qG}(u/x') G(x', \mu^2) \right\} + \\
&\qquad + \int_0^1 \mathrm{d}u\, \mathrm{d}\hat{\sigma}_{(0)}[\mathrm{e}(\ell) + \bar{q}(uP) \to \mathrm{e}(\ell') + \bar{q}(p')] \times \\
&\qquad \times \int_u^1 \frac{\mathrm{d}x'}{x'} \left\{ \left[\delta(1 - u/x') + \frac{t\alpha_s}{2\pi} P_{\bar{q}\bar{q}}(u/x') \right] \bar{q}(x', \mu^2) + \right. \\
&\qquad\qquad \left. + \frac{t\alpha_s}{2\pi} P_{\bar{q}G}(u/x') G(x', \mu^2) \right\}
\end{aligned} \tag{23.4.11}$$

and where we should, of course, sum over flavours. In the above $\bar{q}(x'; \mu^2)$ and $G(x'; \mu^2)$ are the antiquark and gluon distribution functions in the hadron h, defined analogously to (23.4.6).

In fact it turns out that

$$P_{qq} = P_{\bar{q}\bar{q}} \qquad \text{and} \qquad P_{qG} = P_{\bar{q}G} \tag{23.4.12}$$

as a consequence of charge conjugation invariance. Also the Ps are independent of flavour in the above treatment in LLA.

In order to make contact with the operator product expansion of deep inelastic scattering (Section 22.2) let us choose h to be a proton and let us rephrase (23.4.11) in terms of scaling functions rather than cross-sections. We can project $F_{1,2}$ out of $W^{\mu\nu}$ in (23.4.7), in which we should now include an antiquark contribution, using the projection functions given in (16.9.21). To the extent that we ignore masses, the same projection functions acting on $\hat{W}^{\mu\nu}_{(0)}(q, uP)$ will project out the (zeroth order in α_s) *quark* or *antiquark* scaling functions $\hat{F}^{(j)}_{1,2}$, where j refers to both quarks and antiquarks of flavour j. The quark scaling functions are only functions of $Q^2/2q \cdot uP = x/u$. Indeed from (16.1.2) we deduce that for quarks or antiquarks

$$\hat{F}^{(j)}_1(x/u) = \frac{Q_j^2}{2} \delta(1 - x/u) \tag{23.4.13}$$

and

$$\hat{F}^{(j)}_2(x/u) = \frac{2x}{u} \hat{F}^{(j)}_1(x/u). \tag{23.4.14}$$

Carrying out the projection leads to

$$F_i(x,Q^2)_{\mathrm{LLA}} = \sum_j \int_0^1 \frac{\mathrm{d}u}{u} \hat{F}_i^{(j)}(x/u) \int_u^1 \frac{\mathrm{d}x'}{x'} \times$$
$$\left\{ \left[\delta(1-u/x') + \frac{t\alpha_s}{2\pi} P_{qq}(u/x') \right] q_j(x',\mu^2) + \frac{t\alpha_s}{2\pi} P_{qG}(u/x') G(x',\mu^2) \right\} \tag{23.4.15}$$

where the sum over j now includes quarks and antiquarks.

Because of the universality of the terms involving $P_{qq}\ln(Q^2/\mu^2)$ and $P_{qG}\ln(Q^2/\mu^2)$ it is useful to define Q^2-dependent quark and antiquark distribution functions

$$\begin{aligned} q_j(x,Q^2) &= \int_x^1 \frac{\mathrm{d}x'}{x'} \left\{ \left[\delta(1-x/x') + \frac{t\alpha_s}{2\pi} P_{qq}(x/x') \right] q_j(x',\mu^2) + \right. \\ &\quad \left. + \frac{t\alpha_s}{2\pi} P_{qG}(x/x') G(x',\mu^2) \right\}. \end{aligned} \tag{23.4.16}$$

Using these and (23.4.13) in (23.4.15) yields

$$F_1(x,Q^2)_{\mathrm{LLA}} = \sum_j \frac{Q_j^2}{2} q_j(x,Q^2) \tag{23.4.17}$$

$$F_2(x,Q^2)_{\mathrm{LLA}} = 2x\, F_1(x,Q^2) \tag{23.4.18}$$

exactly analogous to the simple parton model results (16.1.6 and 7) with just the replacement

$$q_j(x) \to q_j(x,Q^2). \tag{23.4.19}$$

In summary, in LLA all the simple parton model results continue to hold with the replacement (23.4.19).

In point of fact our definition of $q(x;Q^2)$ in (23.4.16) requires a small refinement. If we take the derivative of (23.4.16) with respect to t we find

$$\begin{aligned} \frac{\mathrm{d}}{\mathrm{d}t} q_j(x,t) &= \frac{\alpha_s}{2\pi} \int_x^1 \frac{\mathrm{d}x'}{x'} \left[P_{qq}(x/x') q_j(x',\mu^2) + \right. \\ &\quad \left. + P_{qG}(x/x') G(x',\mu^2) \right]. \end{aligned} \tag{23.4.20}$$

Correct to this order in α_s we may replace $q_j(x';\mu^2)$, $G(x';\mu^2)$ and α_s by $q_j(x';Q^2)$, $G(x';Q^2)$ and $\alpha_s(Q^2)$ on the RHS respectively. (The implications of this are discussed below.)

Thus we have

$$\begin{aligned} \frac{\mathrm{d}}{\mathrm{d}t} q_j(x,t) &= \frac{\alpha_s(t)}{2\pi} \int_x^1 \frac{\mathrm{d}x'}{x'} \left[P_{qq}(x/x') q_j(x',t) + \right. \\ &\quad \left. + P_{qG}(x/x') G(x',t) \right]. \end{aligned} \tag{23.4.21}$$

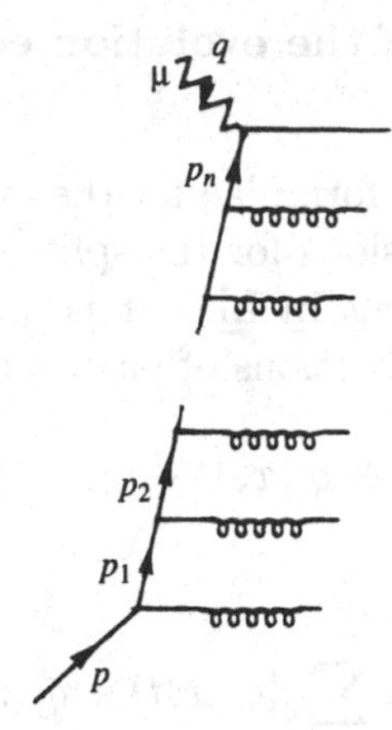

Fig. 23.17. Higher order contributions to 'γ'$q \to q$X.

A similar analysis for processes initiated by a gluon leads to the introduction of a Q^2-dependent gluon distribution function which then satisfies

$$\frac{\mathrm{d}}{\mathrm{d}t}G(x,t) = \frac{\alpha_s(t)}{2\pi}\int_x^1 \frac{\mathrm{d}x'}{x'}\Big[\sum_j P_{Gq}(x/x')q_j(x',t) + + P_{GG}(x/x')G(x',t)\Big] \tag{23.4.22}$$

where j is summed over quarks and antiquarks.

These are just the Altarelli–Parisi evolution equations discussed in Section 22.2 [see (22.2.53)] and just reproduce the leading order QCD renormalization group improved results which there followed from the operator product expansion.

Once the $q_j(x';Q^2)$ are measured at some value of Q^2, the variation with Q^2 is then controlled by the eqns (23.4.21 and 22).

Regarding the change from $\alpha_s(\mu^2)$ to $\alpha_s(Q^2)$ etc. in going from (23.4.20) to (23.4.21), we have already seen [Section 22.1] that use of the running coupling effectively sums the leading logarithmic series $(\alpha_s \ln Q^2)^n$. Thus it is not surprising that one can deduce (23.4.21) by summing the leading logarithmic terms from the diagrams shown in Fig. 23.17.

It is found that the LLA comes from the regions of phase space corresponding to

$$\mu^2 \le p_{1_T}^2 \le p_{2_T}^2 \le \cdots \le p_{n_T}^2 \le Q^2.$$

For a detailed treatment see Paige (1989).

23.5 Summary of the evolution equations in LLA

We now gather together the formulae for the evolution equations in LLA and give the explicit expressions for the splitting functions. Because the latter are flavour independent in LLA it is convenient to define singlet and non-singlet linear combinations of parton distribution functions:

$$\text{Non-singlet: } V_{ij}(x,t) \equiv q_i(x,t) - q_j(x,t), \qquad i \neq j \tag{23.5.1}$$

and

$$\text{Singlet: } \Sigma(x,t) \equiv \sum_{\substack{\text{flavours}\\ j}} [q_j(x,t) + \bar{q}_j(x,t)] \tag{23.5.2}$$

Then the non-singlet evolution, from (23.4.21), is very simple:

$$\frac{\mathrm{d}}{\mathrm{d}t} V_{ij}(x,t) = \frac{\alpha(t)}{2\pi} \int_x^1 \frac{\mathrm{d}y}{y} P_{qq}(x/y) V_{ij}(y,t) \tag{23.5.3}$$

and for the singlet case we have coupled equations:

$$\begin{aligned} \frac{\mathrm{d}}{\mathrm{d}t}\Sigma(x,t) &= \frac{\alpha_s(t)}{2\pi} \int_x^1 \frac{\mathrm{d}y}{y} \left[P_{qq}(x/y)\Sigma(y,t) + 2n_f^* P_{qG}(x/y) G(y,t)\right] \\ \frac{\mathrm{d}}{\mathrm{d}t} G(x,t) &= \frac{\alpha_s(t)}{2\pi} \int_x^1 \frac{\mathrm{d}y}{y} \left[P_{Gq}(x/y)\Sigma(y,t) + P_{GG}(x/y) G(y,t)\right] \end{aligned} \tag{23.5.4}$$

where n_f^* is the number of active flavours.

The above form of the evolution equations seems the most intuitive from a physical point of view. Partons with momentum fraction $y > x$ emit other partons, thereby lose energy and thus feed the distribution at x via the splitting functions $P(x/y)$. But for practical purposes it is simpler to change integration variables from y to $z = x/y$ so that the RHS of the evolution equations is then of the form, for example,

$$\int_x^1 \frac{\mathrm{d}z}{z} P_{qq}(z) V_{ij}(x/z,t).$$

Often the convolution involved is written in the form

$$\begin{aligned} [V \otimes q](x,t) &\equiv \int_x^1 \frac{\mathrm{d}y}{y} V(y) q(x/y,t) = \int_x^1 \frac{\mathrm{d}y}{y} V(x/y) q(y,t) \\ &= [q \otimes V](x,t). \end{aligned} \tag{23.5.5}$$

The splitting functions are:

$$P_{qq}(x) = C_F\left[\frac{1+x^2}{(1-x)_+} + \tfrac{3}{2}\delta(1-x)\right] \tag{23.5.6}$$

$$\begin{aligned} P_{GG}(x) &= 2C_A\left[\frac{x}{(1-x)_+} + \frac{1-x}{x} + x(1-x)\right] + \\ &\quad + \left(\tfrac{11}{6}C_A - \tfrac{2}{3}T\right)\delta(1-x) \end{aligned} \tag{23.5.7}$$

where the '+' on any function implies

$$\int_0^1 f(x)g_+(x)\,\mathrm{d}x \equiv \int_0^1 [f(x)-f(1)]g(x)\,\mathrm{d}x \tag{23.5.8}$$

so that for well behaved $f(x)$ there is no singularity at $x = 1$ when integrating $\int \mathrm{d}x\, f(x)/(1-x)_+$.

Note that the '+' symbol is defined only for an integral between 0 and 1. Thus, for example, for the type of integral that occurs in the evolution equations, one has

$$I(x) \equiv \int_x^1 f(z)g_+(z)\,\mathrm{d}z = \int_0^1 f(z)g_+(z)\,\mathrm{d}z - \int_0^x f(z)g(z)\,\mathrm{d}z$$

where, since z cannot equal 1 the '+' is superfluous in the second integral on the RHS. Thus

$$\begin{aligned} I(x) &= \int_0^1 [f(z)-f(1)]g(z)\,\mathrm{d}z - \int_0^x f(z)g(z)\,\mathrm{d}z \\ &= \int_x^1 [f(z)-f(1)]g(z)\,\mathrm{d}z + \int_0^x [f(z)-f(1)-f(z)]g(z)\,\mathrm{d}z \\ &= \int_x^1 [f(z)-f(1)]g(z)\,\mathrm{d}z - f(1)\int_0^x g(z)\,\mathrm{d}z. \end{aligned} \tag{23.5.9}$$

The group factors C_F, C_A, T were defined in (23.3.24,25) and (21.5.19, 20). For N colours and n_f flavours

$$C_F = (N^2-1)/2N, \qquad C_A = N, \qquad T = n_f/2. \tag{23.5.10}$$

In addition one has

$$P_{qG}(x) = \tfrac{1}{2}[x^2 + (1-x)^2] \tag{23.5.11}$$

and

$$P_{Gq}(x) = C_F\,\frac{1+(1-x)^2}{x}. \tag{23.5.12}$$

The evolution equations and splitting functions for the spin-dependent polarized parton distribution functions are given in Altarelli (1982).

Of course, if we take moments, then the convolution becomes a product. For example

$$\begin{aligned}\frac{\mathrm{d}}{\mathrm{d}t}V_{ij}^{(n)}(t) &\equiv \int_0^1 x^{n-1}\mathrm{d}x\,\frac{\mathrm{d}}{\mathrm{d}t}V_{ij}(x,t)\\ &= \frac{\alpha_s(t)}{2\pi}\int_0^1 x^{n-1}\mathrm{d}x\int_x^1 \frac{\mathrm{d}y}{y}P_{qq}(x/y)V_{ij}(y,t)\\ &= \frac{\alpha_s(t)}{2\pi}\int_0^1 \frac{\mathrm{d}y}{y}V_{ij}(y)\int_0^y x^{n-1}P_{qq}(x/y)\mathrm{d}x\end{aligned}$$

which, on putting $u = x/y$, becomes

$$\begin{aligned}&= \frac{\alpha_s(t)}{2\pi}\int_0^1 y^{n-1}V_{ij}(y,t)\,\mathrm{d}y\int_0^1 u^{n-1}P_{qq}(u)\,\mathrm{d}u\\ &= \frac{\alpha_s(t)}{2\pi}V_{ij}^{(n)}(t)P_{qq}^{(n)}.\end{aligned} \tag{23.5.13}$$

This was the starting point in Section 22.2 for introducing the P_{qq}.

The splitting function moments for $n = 1$ and 2 are of particular significance. One has

$$\int_0^1 \mathrm{d}x\, P_{qq}(x) = 0 \tag{23.5.14}$$

$$\int_0^1 \mathrm{d}x\, x[P_{qq}(x) + P_{Gq}(x)] = 0 \tag{23.5.15}$$

$$\int_0^1 \mathrm{d}x\, x[2n_f^* P_{qG}(x) + P_{GG}(x)] = 0. \tag{23.5.16}$$

These ensure that the Q^2-dependent parton distribution functions still satisfy the conservation laws (16.4.2–4) and that the total 3-momentum of the hadron is carried by its parton constituents, i.e. that

$$\int_0^1 \mathrm{d}x\, x\left\{\sum_{\text{flavours}}\left[q_j(x,Q^2) + \bar{q}_j(x,Q^2)\right] + G(x,Q^2)\right\} = 1 \tag{23.5.17}$$

or

$$\int_0^1 \mathrm{d}x\, x\left[\Sigma(x,Q^2) + G(x,Q^2)\right] = 1. \tag{23.5.18}$$

In terms of moments this simply states that

$$\Sigma^{(2)}(Q^2) + G^{(2)}(Q^2) = 1 \tag{23.5.19}$$

independent of Q^2. It is interesting to see how this emerges from (23.5.4). Taking moments yields

$$\begin{pmatrix}\mathrm{d}\Sigma^{(2)}/\mathrm{d}t\\ \mathrm{d}G^{(2)}/\mathrm{d}t\end{pmatrix} = \frac{\alpha_s(t)}{2\pi}\begin{pmatrix}-4C_F/3 & n_f^*/3\\ 4C_F/3 & -n_f^*/3\end{pmatrix}\begin{pmatrix}\Sigma^{(2)}\\ G^{(2)}\end{pmatrix}. \tag{23.5.20}$$

Diagonalizing these equations, we see that

$$\frac{\mathrm{d}}{\mathrm{d}t}\left(\Sigma^{(2)} + G^{(2)}\right) = 0 \tag{23.5.21}$$

in accord with the Q^2 independence of (23.5.19).

Also, one finds that

$$\frac{\mathrm{d}}{\mathrm{d}t}\left(\Sigma^{(2)} - \frac{n_f^*}{4C_F}G^{(2)}\right) = -\frac{\alpha_s(t)}{6\pi}(4C_F + n_f^*)\left(\Sigma^{(2)} - \frac{n_f^*}{4C_F}G^{(2)}\right) \tag{23.5.22}$$

so that as $Q^2 \to \infty$, using (20.9.9),

$$\Sigma^{(2)}(Q^2) - \frac{n_f}{4C_F}G^{(2)}(Q^2) = \left(\frac{\alpha_s(0)}{\alpha_s(t)}\right)^{d_-^2}\left[\Sigma^{(2)}(\mu^2) - \frac{n_f}{4C_F}G^{(2)}(\mu^2)\right] \tag{23.5.23}$$

where

$$d_-^2 = -\frac{1}{6\pi b}(4C_F + n_f). \tag{23.5.24}$$

Thus the RHS of (23.5.23) $\to 0$ as $Q^2 \to \infty$ and, asymptotically,

$$\Sigma^{(2)}(Q^2) \to \frac{n_f}{4C_F}G^{(2)}(Q^2) \tag{23.5.25}$$

i.e.

$$\frac{\Sigma^{(2)}(Q^2)}{G^{(2)}(Q^2)} \to \frac{n_f}{4C_F} = \frac{Nn_f}{2(N^2-1)} \tag{23.5.26}$$

which, for 3 colours and 3 flavours

$$= \frac{9}{16}. \tag{23.5.27}$$

Combined with (23.5.19) this tells us that the momentum fractions carried by quarks (and antiquarks) and gluons is, asymptotically,

$$\Sigma^{(2)} \to \frac{9}{25}, \qquad G^{(2)} \to \frac{16}{25}. \tag{23.5.28}$$

23.6 Small x behaviour of the Q^2-dependent gluon distribution in LLA

In Section 16.6 we related the behaviour of the quark distribution functions in the limit $x \to 0$ to the Regge trajectories that govern forward 'γ'-proton elastic scattering. Here we ask what the Altarelli–Parisi equations imply for the small x–behaviour. We shall be led to a behaviour which is *not* in accord with the results of Section 16.6. This is not really surprising. The LLA treatment deals with perturbative ladder diagrams whereas small momentum transfer elastic scattering must involve multi-gluon and confinement effects. We shall briefly mention the attempts to deal with this via non-linear Gribov, Levin, Ryskin equations. For access to the literature refer to the excellent summary by Bartels (1991).

Looking at the expressions for the splitting functions (23.5.6, 7, 11 and 12) we see that only P_{GG} and P_{Gq} grow large as $x \to 0$. We shall simplify by ignoring the quark contribution to $\mathrm{d}G/\mathrm{d}t$ in (23.5.4). We approximate P_{GG} by

$$P_{GG}(x) \approx \frac{2C_A}{x} = \frac{2N}{x} \tag{23.6.1}$$

and thus taking moments of (23.5.4) find

$$\frac{\mathrm{d}G^{(n)}}{\mathrm{d}t} = \frac{C_A}{\pi(n-1)}\alpha_s(t)G^{(n)}(t). \tag{23.6.2}$$

Integrating, using (20.9.9) yields

$$G^{(n)}(t) = \exp\left\{\frac{N}{\pi b(n-1)}\ln\left[\frac{\alpha_s(t_0)}{\alpha_s(t)}\right]\right\}G^{(n)}(t_0). \tag{23.6.3}$$

To estimate $G(x,t)$ for $x \approx 0$ we feed this into the inversion formula (22.2.41)

$$\begin{aligned} G(x,t) &= \frac{1}{x}\frac{1}{2\pi\mathrm{i}}\int_{n_0-\mathrm{i}\infty}^{n_0+\mathrm{i}\infty}\mathrm{d}n\, x^{1-n}G^{(n)}(t) \\ &= \frac{1}{x}\frac{1}{2\pi\mathrm{i}}\int \mathrm{d}n\, G^{(n)}(t_0)\mathrm{e}^{f(n)} \end{aligned} \tag{23.6.4}$$

where, via (23.6.3)

$$f(n) \equiv (n-1)\ln(1/x) + \frac{N}{\pi b(n-1)}\ln\left[\alpha_s(t_0)/\alpha_s(t)\right]. \tag{23.6.5}$$

For very small x and for very large Q^2 both terms in $f(n)$ grow large and the integral can be estimated by a saddle-point method: the saddle-point is at

$$n_0 = 1 + \left(\frac{N}{\pi b}\right)^{1/2}\left\{\ln\left[\frac{\alpha_s(t_0)}{\alpha_s(t)}\right] \Big/ \ln\left(\frac{1}{x}\right)\right\}^{1/2} \tag{23.6.6}$$

and one ends up with the estimate

$$G(x,Q^2) \sim \frac{1}{x}\exp\left\{\frac{4N}{\pi b}\ln\left[\frac{\ln(Q^2/\Lambda^2)}{\ln(Q_0^2/\Lambda^2)}\right]\ln\left(\frac{1}{x}\right)\right\}^{1/2} G^{(n_0)}(Q_0^2). \tag{23.6.7}$$

At fixed n_0, i.e. letting $x \to 0$, $t \to \infty$ in such a way as to keep n_0 fixed in (23.6.6), we see that $G(x,t)$ blows up dramatically.

Fig. 23.18. Quark splitting and quark–gluon recombination.

It is customary to discuss small x physics using the variables

$$\xi = \ln\ln(Q^2/\Lambda^2) \tag{23.6.8}$$

and

$$y = \frac{2N}{\pi b}\ln(1/x) \tag{23.6.9}$$

so that

$$G(x, Q^2) \sim \frac{1}{x}\exp\sqrt{2(\xi - \xi_0)y}\; G^{(n_0)}(Q_0^2). \tag{23.6.10}$$

Now recall that for the 'γ'-proton collision

$$s = (P + q)^2 = m_p^2 + Q^2\left(\frac{1}{x} - 1\right) \tag{23.6.11}$$

so that for $x \ll 1$

$$s \approx Q^2/x \tag{23.6.12}$$

and

$$y \propto \ln(s/Q^2). \tag{23.6.13}$$

As a consequence, (23.6.10) will imply a 'γ'-proton cross-section growing faster than any power of $\ln(s/Q^2)$ in contradiction with the Froissart bound $\ln^2(s/Q^2)$. Clearly, therefore, the Altarelli–Parisi evolution equation is inadequate as $x \to 0$. Qualitatively, the reason for the failure can be understood as a saturation effect caused by the growth in the number density of gluons. At some point the density is so great that the gluons cannot be treated as independent of each other. Thus instead of always 'splitting' into more partons, there will be some form of recombination. Fig. 23.18 compares splitting and recombination.

Whereas in the splitting the change in the quark density is linear in the input density, i.e. $\delta q \propto P \otimes q$, the recombination is quadratic in the input densities $\delta q \propto P \otimes qG$. Non-linear evolution equations which take this into account were developed by Gribov, Levin and Ryskin (1982) and are the subject of much interest at present. According to these authors the x–Q^2 plane can be split into three regions as shown in Fig. 23.19. In region A the linear evolution equations hold. In region C highly non-

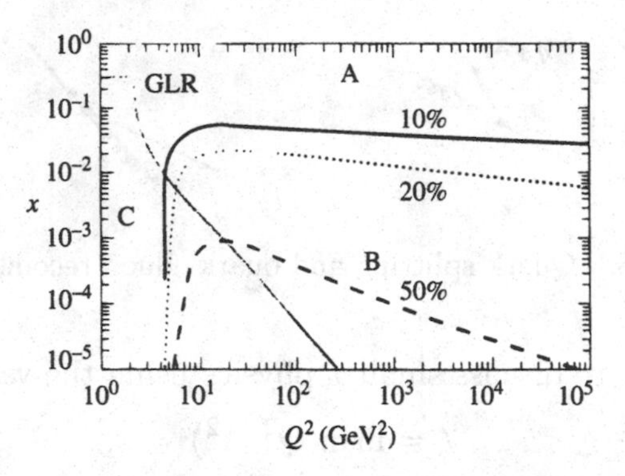

Fig. 23.19. Non-linear and non-perturbative regions in x–Q^2 plane. See text for explanation. (From Bartels, 1991.)

perturbative effects dominate, and there is a transition region B where perturbative QCD can still be used but the non-linear evolution equations are necessary. The latter contain a parameter C which is related to the distribution at the initial value $Q^2 = Q_0^2$ and which characterizes what is called the 'triple ladder vertex'. It is usually taken to be given by

$$C = \frac{3\pi Q_0^2}{16b\Lambda^2}. \tag{23.6.14}$$

Fig. 23.20 compares the evolution of $xG(x; Q^2)$ from $Q^2 = Q_0^2 = 4$ GeV^2 to $Q^2 = 10^4$ GeV^2 for two different input distributions at $Q^2 = Q_0^2$, using linear and non-linear evolution, the latter for two values of C. It is seen that increasing C leads to strong damping at very small x compared with the linear case.

According to Gribov, Levin and Ryskin there is a critical line in the y–ξ or x–Q^2 plane to the left of which the non-linear evolution equations fail and genuine non-perturbative effects set in. This is shown in Fig. 23.19 which also indicates lines along which the linear and non-linear evolutions differ from each other by 10, 20 and 50%. What is fascinating is the implication that some measurements at HERA will lie in both the non-linear and non-perturbative regions. Ultimately this will provide a wonderful springboard for the investigation of the transition from the perturbative to the non-perturbative regions of QCD. For new developments in this area see Bartels (1992).

23.7 Behaviour of distributions as $x \to 1$ in LLA

We shall illustrate the approach with the simplest case, the non-singlet evolution eqn (23.5.3). Since in LLA there is no flavour dependence we

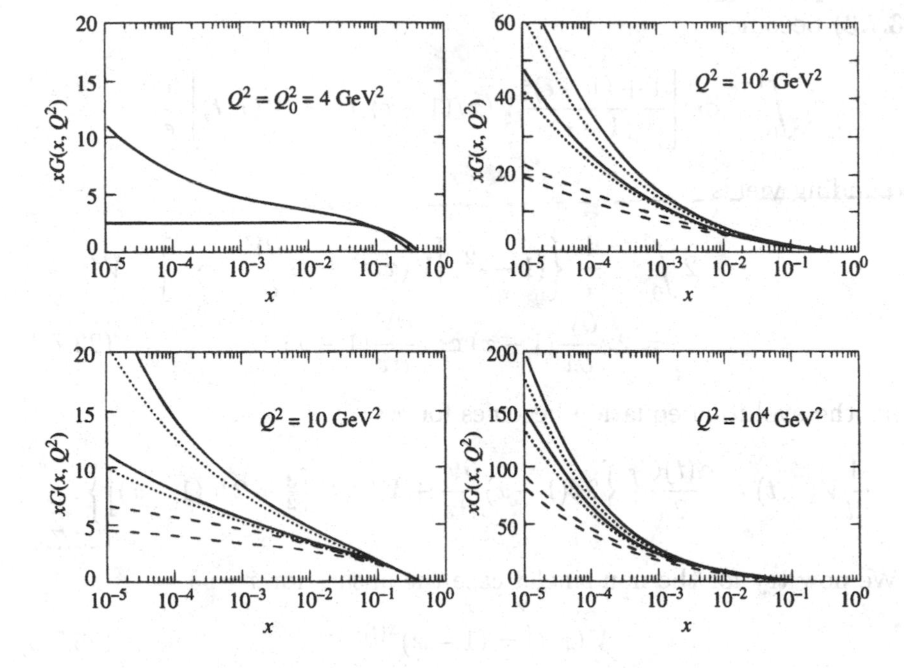

Fig. 23.20. Comparison of linear and non-linear evolution with two different inputs at $Q^2 = 4$ GeV^2. (—) linear; (- - -) non-linear with C given by (23.6.14); (···) non-linear with $C/10$. (From Bartels, 1991.)

drop the labels ij on V_{ij} and write

$$\frac{\mathrm{d}}{\mathrm{d}t}V(x,t) = \frac{\alpha(t)}{2\pi}\int_x^1 \frac{\mathrm{d}z}{z} P_{qq}(z) V(x/z,t) \tag{23.7.1}$$

where we have changed integration variables from y to $z = x/y$.

From (23.5.9 and 6) the RHS involves

$$\int_x^1 \mathrm{d}z \left(\frac{1+z^2}{z}\right) \frac{1}{(1-z)_+} V(x/z,t) + \tfrac{3}{2}V(x,t) \tag{23.7.2}$$

$$= \int_x^1 \mathrm{d}z \left[\frac{1+z^2}{z}V(x/z,t) - 2V(x,t)\right]\frac{1}{1-z}$$
$$- 2V(x,t)\int_0^x \frac{\mathrm{d}z}{1-z} + \tfrac{3}{2}V(x,t)$$
$$= \int_x^1 \mathrm{d}z \left[\frac{1+z^2}{z}V(x/z,t) - 2V(x,t)\right]\frac{1}{1-z} +$$
$$+ V(x,t)\left[\tfrac{3}{2} + 2\ln(1-x)\right]. \tag{23.7.3}$$

Since we are interested in $x \to 1$ we put $\epsilon = 1 - z$ so that the integral in

(23.7.3) becomes

$$\int_0^{1-x} \mathrm{d}\epsilon \left[\frac{1+(1-\epsilon)^2}{1-\epsilon} V(x/(1-\epsilon),t) - 2V(x,t)\right] \frac{1}{\epsilon}.$$

Expanding yields

$$\begin{aligned} &2\int_0^{1-x} \frac{\mathrm{d}\epsilon}{\epsilon} \left\{(1-\epsilon^2)\left[V(x,t) + x\epsilon\frac{\partial V}{\partial x} + \cdots\right] - V(x,t)\right\} \\ &\quad = 2x\frac{\partial V}{\partial x}(1-x) \simeq 2\frac{\partial V}{\partial x}(1-x). \end{aligned} \tag{23.7.4}$$

Thus the evolution equation becomes for $x \approx 1$

$$\frac{\mathrm{d}}{\mathrm{d}t}V(x,t) \approx \frac{\alpha(t)C_f}{2\pi}\left\{2x(1-x)\frac{\partial V}{\partial x} + V(x,t)\left[\tfrac{3}{2} + 2\ln(1-x)\right]\right\}. \tag{23.7.5}$$

We now try for the non-singlet case the ansatz for $x \to 1$

$$V(x,t) = (1-x)^{\lambda(t)_{\mathrm{NS}}} \tag{23.7.6}$$

and find the condition

$$\ln(1-x)\left[\frac{\mathrm{d}\lambda_{\mathrm{NS}}}{\mathrm{d}t} - \frac{C_F\alpha(t)}{\pi}\right] - \frac{C_F\alpha(t)}{\pi}\left[\tfrac{3}{2} + 2\lambda_{\mathrm{NS}}(t)\right] \approx 0 \tag{23.7.7}$$

which requires that

$$\frac{\mathrm{d}\lambda_{\mathrm{NS}}}{\mathrm{d}t} = \frac{C_F\alpha_s(t)}{\pi}, \tag{23.7.8}$$

Via (20.9.9) we get

$$\lambda(t)_{\mathrm{NS}} = \frac{C_F}{\pi b}\ln\left[\frac{\alpha_s(\mu^2)}{\alpha_s(t)}\right] + \text{const.} \tag{23.7.9}$$

Thus the power with which the distribution cuts off as $x \to 1$ increases like $\ln\ln Q^2$.

In similar fashion one can show that for gluons $\lambda_G(t) = \lambda_{\mathrm{NS}}(t) + 1$ and for sea quarks $\lambda_{\mathrm{sea}}(t) = \lambda_{\mathrm{NS}}(t) + 2$.

Of course the fact that the distributions cut off more and more strongly as $x \to 1$ in LLA does not guarantee that the above is an accurate description of the distributions in this region. Small correction beyond the LLA may become relatively more important. From (23.2.19) we see that the perturbative treatment cannot be justified if $(1-x)$ is too small, i.e. if $Q^2(1-x)/4$ is such that $\alpha_s[Q^2(1-x)/4]$ is not small.

23.8 Beyond the LLA

Let us return to eqn (23.4.15) and consider the terms, beyond the leading logarithm, but still of order α_s, which up to now have been neglected. In LLA we kept only terms of the form $\alpha_s \ln(Q^2/m^2)$ coming from the diagram of Fig. 23.12. In (23.4.15) there will be additional terms, arising from Figs. 23.12, 13 and 16, of the form $(\alpha_s/2\pi)f_i^q(u/x')$ and $(\alpha_s/2\pi)f_i^G(u/x')$ as well as terms involving powers of (m^2/Q^2) which can be neglected. Thus (23.4.15) will be altered to

$$\begin{aligned} F_i(x, Q^2) &= \sum_j \int_0^1 \frac{du}{u} f_i^{(j)}(x/u) \int_u^1 \frac{dx'}{x'} \bigg\{ \bigg[\delta(1-u/x') + \\ &\quad + \frac{t\alpha_s}{2\pi} P_{qq}(u/x') + \frac{\alpha_s}{2\pi} f_i^q(u/x') \bigg] q_j(x', \mu^2) + \\ &\quad + \bigg[\frac{t\alpha_s}{2\pi} P_{qG}(u/x') + \frac{\alpha_s}{2\pi} f_i^G(u/x') \bigg] G(x', \mu^2) \bigg\} . \end{aligned} \tag{23.8.1}$$

Now, contrary to the leading logarithmic terms, f_i^q and f_i^G are *not* universal. They depend upon the reaction in general and in the present case they depend upon which DIS scaling function F_i we are considering. Moreover they depend upon the renormalization scheme being used. Thus if we introduce the Q^2-dependent distribution functions exactly as before, i.e. via (23.4.16), then after some manipulation (23.4.15) becomes, to order α_s,

$$\begin{aligned} F_i(x, Q^2) &= \sum_j \int_0^1 \frac{du}{u} \hat{F}_i^{(j)}(x/u) \int_u^1 \frac{dx'}{x'} \bigg\{ \bigg[\delta(1-u/x') + \\ &\quad + \frac{\alpha_s}{2\pi} f_i^q(u/x') \bigg] q_j(x', Q^2) + \frac{\alpha_s}{2\pi} f_i^G(u/x') G(x', Q^2) \bigg\} \end{aligned} \tag{23.8.2}$$

where j runs over quarks and antiquarks.

If we utilize (23.4.13) this reads for $i = 1$

$$\begin{aligned} F_1(x, Q^2) &= \sum_j \frac{Q_j^2}{2} \int_x^1 \frac{dx'}{x'} \bigg\{ \bigg[\delta(1-x/x') + \\ &\quad + \frac{\alpha_s}{2\pi} f_1^q(x/x') \bigg] q_j(x', Q^2) + \frac{\alpha_s}{2\pi} f_1^G(x/x') G(x', Q^2) \bigg\} . \end{aligned} \tag{23.8.3}$$

Or, more simply,

$$
\begin{aligned}
F_1(x,Q^2) &= \sum_j \frac{Q_j^2}{2} q_j(x,Q^2) + \\
&+ \sum_j \frac{Q_j^2}{2} \int_x^1 \frac{\mathrm{d}x'}{x'} \left\{ \frac{\alpha_s}{2\pi} f_1^q(x/x') q_j(x',Q^2) + \right. \\
&\left. + \frac{\alpha_s}{2\pi} f_1^G(x/x') G(x',Q^2) \right\}.
\end{aligned}
\tag{23.8.4}
$$

which should be compared with the LLA result (23.4.17).

Now F_1 is a physical, measurable quantity, so cannot depend upon the renormalization scheme. Both f_1^q and f_1^G, as already mentioned, *do* depend upon the renormalization scheme. Thus in comparing (23.8.4) with experiment, the $q_j(x,Q^2)$ and $G(x,Q^2)$ that we extract from the data *will be renormalization scheme dependent.* Thus working beyond the LLA implies that the parton distributions depend upon the renormalization prescription, which then must be specified. Given this unpalatable fact, there are two very different strategies that have been adopted:

1. Specify carefully the renormalization scheme being used and extract the parton distributions from experimental data using (23.8.4) and analogous equations for other physical observables. The advantage of this approach is that the distribution functions are universal and can be used in all reactions provided only that the same renormalization scheme is used in each case. In this method the distributions should carry a label such as MS, $\overline{\text{MS}}$ etc.

2. For one particular physical observable, preferably one that can be experimentally measured to great accuracy—usually $F_2(x,Q^2)$—one absorbs *all* the QCD corrections into the distribution function. Thus one puts

$$
F_2(x,Q^2) \equiv x \sum_j Q_j^2 q_j(x,Q^2). \tag{23.8.5}
$$

This clearly gives a physical, and therefore renormalization scheme independent, definition of $q_j(x,Q^2)$. It should then strictly carry a label, say DIS, to indicate how it has been defined. This scheme has advantages and disadvantages. The distribution thus defined is clearly not universal and q_j^{DIS} cannot be used in other reactions without compensating for its non-universal part. Thus q_j^{DIS} no longer reflects just the characteristics of the quarks inside the hadron—it is process dependent.

There are, however, advantages in this prescription:

(*a*) The Adler sum rule [see (17.1.2) and discussion in Section 22.2]

$$\lim_{Q^2\to\infty}\int_0^1 \frac{\mathrm{d}x}{x}\left[F_2^{W^-\mathrm{p}}(x,Q^2)-F_2^{W^+\mathrm{p}}(x,Q^2)\right]=2 \qquad (23.8.6)$$

does not have any QCD corrections, and the LHS depends only on valence quark contributions, i.e. q_j–$\bar{q}_j$. Use of (23.8.5) will then guarantee that the valence quark sum rules (16.4.2–4) will continue to hold. But the momentum sum rule will no longer hold.

(*b*) If we calculate some other physical observable, say $F_1(x,Q^2)$, we will clearly find on the basis of (23.8.2) and (23.4.14)

$$\begin{aligned}F_1(x,Q^2) &= \sum_j \frac{Q_j^2}{2} q_j^{\mathrm{DIS}}(x',Q^2)+\\ &+\sum_j \frac{Q_j^2}{2}\int_x^1 \frac{\mathrm{d}x'}{x'}\left\{\frac{\alpha_s}{2\pi}\left[f_1^q(x/x')-f_2^q(x/x')\right]q_j^{\mathrm{DIS}}(x',Q^2)+\right.\\ &\left.+\frac{\alpha_s}{2\pi}\left[f_1^G(x/x')-f_2^G(x/x')\right]G^{\mathrm{DIS}}(x',Q^2)\right\}. \qquad (23.8.7)\end{aligned}$$

Now in this approach q_j^{DIS} and G^{DIS} are experimental quantities, as is $F_1(x,Q^2)$. Thus it must turn out, and it indeed does so, that the renormalization scheme dependence of the $f_{1,2}^q, f_{1,2}^G$ cancel out when the differences $f_1^q - f_2^q, f_1^G - f_2^G$ are taken. One finds [for references see Altarelli, 1982, Section 5] for the three scaling functions of DIS:

$$\begin{aligned}f_2^q(x)-f_1^q(x) &= 2C_F x = 8x/3\\ f_2^G(x)-f_1^G(x) &= (T/n_f)4x(1-x) = 2x(1-x) \qquad (23.8.8)\\ f_2^q(x)-f_3^q(x) &= C_F(1+x) = 4(1+x)/3\end{aligned}$$

(Recall that F_3 behaves as if it were *non-singlet*—it depends on the combinations q_j–$\bar{q}_j$—hence there is no gluon contribution.)

Despite these advantages we favour the first approach (in practice the $\overline{\mathrm{MS}}$ scheme), since the distribution functions at least do not depend upon the reaction.

The relationship between the two definitions of distributions follows trivially upon comparing the expression for $F_2(x,Q^2)$ in the two versions:

$$F_2(x,Q^2) = x\sum_j Q_j^2 q_j^{\mathrm{DIS}}(x,Q^2) \tag{23.8.9}$$

$$\begin{aligned} F_2(x,Q^2) &= x\sum_j Q_j^2 q_j^{\overline{\mathrm{MS}}}(x,Q^2) + \\ &\quad + x\sum_j Q_j^2 \int_x^1 \frac{\mathrm{d}x'}{x'}\frac{\alpha_s}{2\pi}\left\{ f_2^{q\overline{\mathrm{MS}}}(x/x')q_j^{\overline{\mathrm{MS}}}(x',Q^2) \right. \\ &\qquad \left. + f_2^{G\overline{\mathrm{MS}}}(x/x')G^{\overline{\mathrm{MS}}}(x',Q^2)\right\}. \end{aligned} \tag{23.8.10}$$

Thus

$$\begin{aligned} q_j^{\mathrm{DIS}}(x,Q^2) &= q_j^{\overline{\mathrm{MS}}}(x,Q^2) + \frac{\alpha_s}{2\pi}\int_x^1 \frac{\mathrm{d}x'}{x'}\left[f_2^{q\overline{\mathrm{MS}}}(x/x')q_j^{\overline{\mathrm{MS}}}(x',Q^2) \right. \\ &\qquad \left. + f_2^{G\overline{\mathrm{MS}}}(x/x')G^{\overline{\mathrm{MS}}}(x',Q^2)\right]. \end{aligned} \tag{23.8.11}$$

We have to face the unfortunate fact that one woman's 'quark distribution' may be a mixture of another man's 'quark' and 'gluon' distributions. Thus the beautiful simplicity of the original, naive parton model is to some extent lost beyond the leading logarithmic approximation.

> With the DIS or $\overline{\mathrm{MS}}$ definitions of the Q^2-dependent distribution functions, one can use (22.5.61) as a starting point to derive the evolution equations accurate to order α_s^2. The calculations are highly complicated. A general discussion can be found in Section 5 of Altarelli, 1982. For the evolution kernels or splitting functions beyond the LLA, see Furmanski and Petronzio, 1982; Curci, Furmanski and Petronzio, 1980; and Floratos, Lacaze and Kounnas, 1981. We shall not attempt to cover this very technical subject.

We shall end this section with some illustrations of the consequences of keeping non-leading terms. In the simple parton model and in the QCD improved version at LLA level we have $F_2(x,Q^2) = 2x\ F_1(x,Q^2)$. This ceases to hold beyond the LLA.

Let us define the *longitudinal* scaling function

$$F_{\mathrm{L}}(x,Q^2) \equiv F_2(x,Q^2) - 2x\ F_1(x,Q^2). \tag{23.8.12}$$

Then from (23.8.7 and 10), for electromagnetic reactions

$$F_{\rm L}^{\gamma p}(x,Q^2) = x\sum_j \frac{Q_j^2}{2}\int_x^1 \frac{dx'}{x'}\left\{\frac{\alpha_s}{2\pi}\left[f_1^q(x/x') - f_2^q(x/x')\right] q_j^{\rm DIS}(x',Q^2) + \right.$$
$$\left. + \frac{\alpha_s}{2\pi}\left[f_1^G(x/x') - f_2^G(x/x')\right] G^{\rm DIS}(x',Q^2)\right\}. \tag{23.8.13}$$

This can be rewritten using (23.8.8 and 9) as

$$F_{\rm L}^{\gamma p}(x,Q^2) = \frac{\alpha_s(Q^2)}{2\pi}x^2\int_x^1 \frac{dy}{y^3}\left\{\tfrac{8}{3}F_2^{\gamma p}(y,Q^2) + 2\Big(\sum_{\substack{\text{quarks}\\ \text{antiquarks}}} Q_j^2\Big)\times \right.$$
$$\left. \times\,(1-x/y)yG^{\rm DIS}(y,Q^2)\right\} \tag{23.8.14}$$

where, for 4 flavours, $\sum Q_j^2 = 20/9$.

Since F_2 is well measured, (23.8.14) offers the best possibility in principle to learn about the gluon distribution in DIS. This is extremely important since the gluon plays a key rôle in most hadronic large p_T reactions.

For the ratio $R = \sigma_{\rm L}/\sigma_T$ defined in (15.3.26), neglecting terms of order m^2/Q^2, one has

$$R = \frac{F_2}{2xF_1} - 1 = \frac{F_{\rm L}(x,Q^2)}{2x\,F_1(x,Q^2)} \tag{23.8.15}$$

so a measurement of R is essentially a measurement of $F_{\rm L}(x,Q^2)$. Unfortunately, as discussed in Sections 16.1 and 17.1, the data on R are not yet very good. The data in Fig. 17.4 are, however, consistent with (23.8.14).

A similar result holds for CC and NC reactions. One simply replaces, in (23.8.14),

$$\Big(\sum_{\substack{\text{quarks}\\ \text{antiquarks}}} Q_j^2\Big) \to \Big(\sum_{\substack{\text{quarks}\\ \text{antiquarks}}} \epsilon_j^2\Big) \tag{23.8.16}$$

where ϵ_j^2 is the coefficient of $q_j(x)$ in the simple parton model for F_2. Thus, for CC reactions the approximate expressions (16.4.34 and 35), in which quark mass and Kobayashi–Maskawa effects are neglected, indicate that for $F^{W^\pm p}$, $\epsilon_j^2 = 1$, so that for 4 flavours

$$\Big(\sum \epsilon_j^2\Big)_{\nu p \text{ or } \bar\nu p \text{ CC}} = 8. \tag{23.8.17}$$

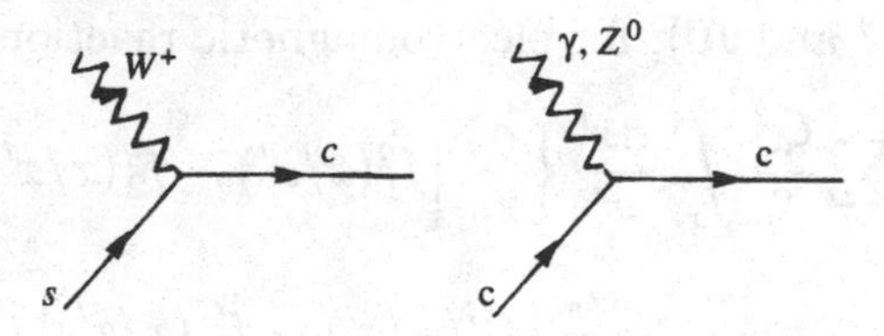

Fig. 23.21. Lowest order diagrams for charm production in CC and NC deep inelastic scattering.

Several studies have tried to extract the parton distributions from DIS data. The original analyses (Gluck, Hoffman and Reya, 1982; Duke and Owens, 1984; and Eichten *et al.*, 1984) utilized data available up to the early 1980s. More recently several further sets of distribution functions have been published (Diemoz *et al.*, 1988; Martin, Roberts and Stirling, 1989; Aurenche *et al.*, 1989; and Morfin and Tung, 1990) based upon data available up to 1989. In these Martin *et al.* and Aurenche *et al.* use the $\overline{\text{MS}}$ definition of the parton distributions, i.e. utilize (23.8.10), whereas Diemoz *et al.* use the DIS definition, i.e. utilize (23.8.9). Morfin and Tung present results in both definitions. For a general survey and assessment of the various distributions the reader is referred to Tung *et al.*, (1989).

We end this section with some comments upon the dangers of misusing parton distributions beyond the LLA.

Consider, for example, (23.8.11) written for one of the *small* sea-quark distributions, say the strange quark $s(x, Q^2)$. On the RHS of (23.8.11) we have the large gluon distribution $G(x, Q^2)$ multiplied by the small coupling $\alpha_s(Q^2)$, and the product may well be as big as or even larger than $s^{\text{DIS}}(x, Q^2)$ or $s^{\overline{\text{MS}}}(x, Q^2)$. Thus $s^{\text{DIS}}(x)$ or $s^{\overline{\text{MS}}}(x)$ can easily differ from each other by 100%. It is thus somewhat meaningless to measure $s(x, Q^2)$ using just the LLA formulae—the beyond-the-LLA corrections are not really small, i.e. are not really of order α_s compared to the LLA results. An analogous phenomenon occurs for the gluon distribution at large x, where it is supposed to be small. In the analogue of (23.8.11) $G^{\text{DIS}}(x, Q^2) - G^{\overline{\text{MS}}}(x, Q^2)$ receives contributions of order $\alpha_s \times$ valence quark distributions which are significant at large x. Thus giving $G(x, Q^2)$ without specifying the renormalization scheme can be quite misleading.

A further example of potential misuse is when the reaction under study singles out a small sea quark distribution in lowest order. For example in charm production in DIS the lowest order diagrams for CC and NC reactions are shown in Fig. 23.21. Because $s(x)$ and especially $c(x)$ are so small an equally or more important contribution will come from the diagrams in Fig. 23.22 even though they are nominally of order α_s.

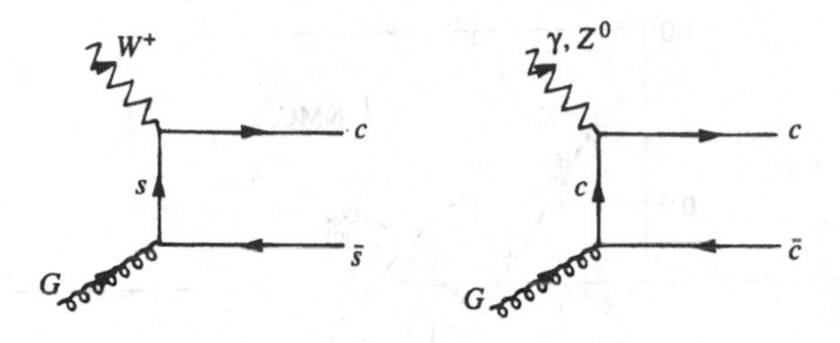

Fig. 23.22. Order α_s contributions to charm production in DIS. Crossed diagrams must also be included.

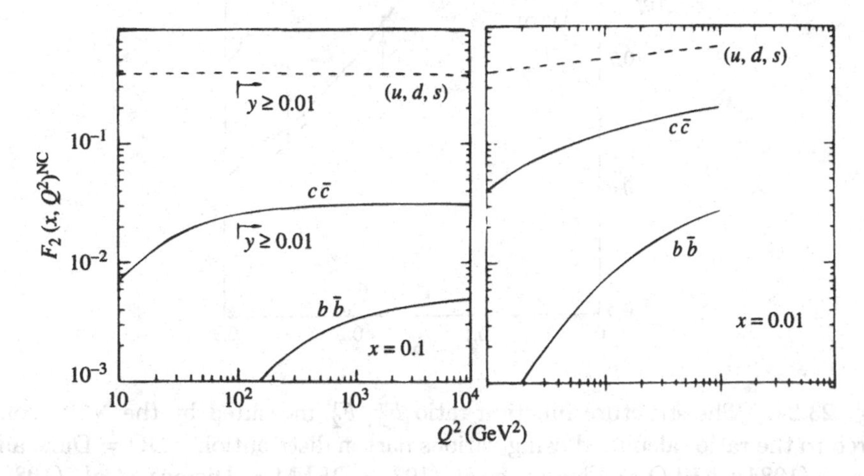

Fig. 23.23. Contributions of c and b quarks to F_2. (From Gluck, 1987.)

The issue is a subtle one. The LLA contribution from the diagrams in Fig. 23.22, i.e. terms like $\alpha_s \ln Q^2/m^2$, *are already included in the evolution equation for* $s(x, Q^2)$ and $c(x, Q^2)$. It is the beyond-the-LLA contributions which should also be included.

In our lengthy discussion of the detailed parton model in Section 16.4 and the comparison with experiment, Chapter 17, we assumed that any intrinsic charm and bottom quark distributions could be neglected. Clearly as Q^2 increases and as charm and bottom production thresholds are crossed, $c(x, Q^2)$ and $b(x, Q^2)$ will become less negligible. In calculating the contributions it is important to keep the masses of the heavy quarks non-zero if the results are to be used near charm and bottom thresholds. These threshold effects must be carefully taken into account when giving heavy quark distributions (Barone *et al.*, 1992). Some idea of the contributions to $F_2(x, Q^2)$ is given in Fig. 23.23. It is seen that even at $Q^2 \sim 10^3$ or 10^4 GeV^2 the contributions are very small.

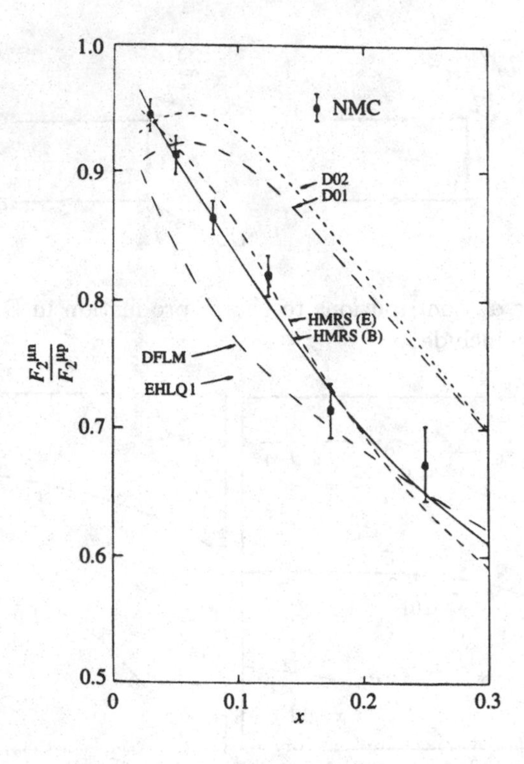

Fig. 23.24. The structure function ratio F_2^n/F_2^p measured by the NMC compared to the ratio calculated using various parton distributions: DO = Duke and Owens (1984); EHLQ = Eichten *et al.* (1984); DFLM = Diemoz *et al.* (1988); HMRS = Harriman *et al.* (1991).

23.9 Comparison with experiment in deep inelastic scattering

The nucleon structure functions, as measured in deep inelastic scattering, have been studied experimentally for more than 20 years now. There is a constant improvement in the accuracy of the measurements as well as a steady progression to larger values of Q^2 and smaller values of x. Throughout this period theorists have attempted to extract detailed information on the quark and gluon distributions at some initial Q_0^2 and then generated distributions at higher Q^2 via the Altarelli–Parisi evolution equations. These sets of distributions are, of course, very important and have been much used for calculating other hadronic reaction cross-sections, wherever perturbative QCD is applicable.

The impact of the newest generation of experiments by the NMC (1991) at CERN on our knowledge of the distributions is shown dramatically in Fig. 23.24, where it is seen that none of the distributions in use before the NMC experiment fits the data on F_2^n/F_2^p. The new distributions, labelled HMRS (B and E), were designed to accommodate the NMC data.

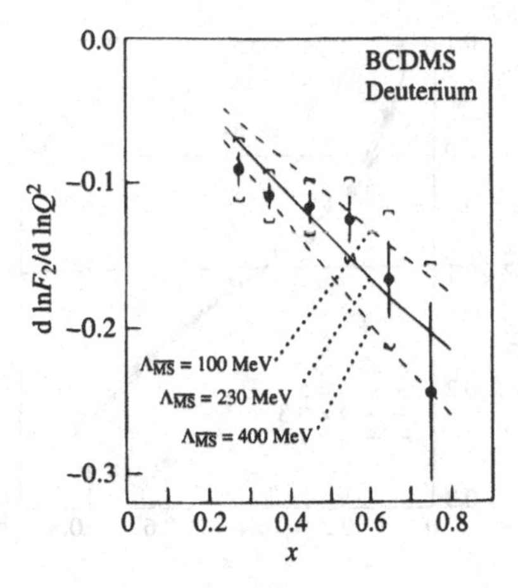

Fig. 23.25. BCDMS deuterium data for $x > 0.275$ compared with non-singlet QCD calculation in next-to-leading order in the $\overline{\text{MS}}$ scheme with 4 active flavours.

The recent BCDMS (1989) data on F_2 in μD scattering have been used to test the Q^2 dependence predicted by QCD. Only data with $Q^2 > 20$ GeV^2 were utilized so as to minimize any higher twist effects. The quantity that is compared with theory is $\mathrm{d}\ln F_2(x,Q^2)/\mathrm{d}\ln Q^2$ vs x, averaged over $Q^2 > 20$ GeV^2. In the first instance, because the gluon distribution is still relatively badly known the analysis is confined to the region $x > 0.275$ where gluon and sea effects should be negligible. This allows a more stringent test of the theory, i.e. of the Q^2 evolution, since it is then controlled by the non-singlet evolution equations which involve only well determined valence distributions. In Fig. 23.25 these data are compared with a calculation using four active flavours in next-to-leading order in the $\overline{\text{MS}}$ scheme. There is essentially only one free parameter, Λ_{QCD}, and it turns out that good agreement is obtained with

$$\Lambda^{(4)}_{\overline{\text{MS}}} = 205 \pm 22(\text{stat.}) \pm 60(\text{syst.})\ \text{MeV}. \tag{23.9.1}$$

This corresponds to a value for the strong coupling

$$\alpha_s^{\overline{\text{MS}}}(Q^2 = 100\ \text{GeV}^2) = 0.156 \pm 0.004(\text{stat.}) \pm 0.011(\text{syst.}). \tag{23.9.2}$$

In the second phase of the analysis a simple parametrization of the gluon distribution at $Q_0^2 = 5$ GeV^2 is assumed:

$$xG(x, Q_0^2) = A(1+\eta)(1-x)^\eta \tag{23.9.3}$$

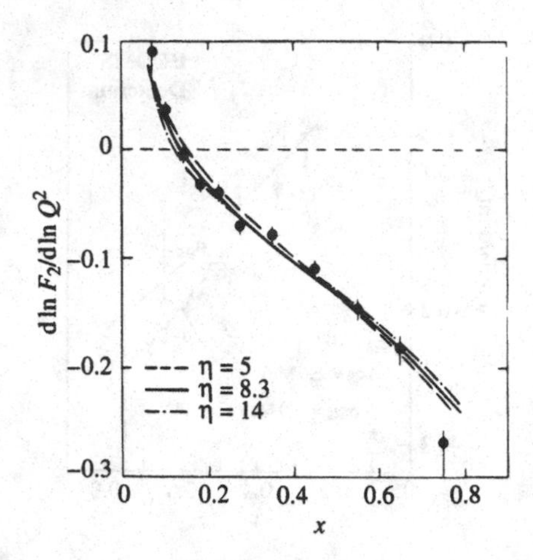

Fig. 23.26. BCDMS deuterium data for all x compared with singlet and non-singlet, next-to-leading order calculation in $\overline{\text{MS}}$ scheme with 4 active flavours. For meaning of η see eqn (23.9.3).

with η a free parameter and A is then fixed, as a function of η, from the momentum conservation sum rule. Both singlet and non-singlet evolution equations, in next-to-leading order, are then employed to calculate $d \ln F_2(x, Q^2)/d \ln Q^2$ for the full x-range of the experiment. The comparison with the data for three values of η is shown in Fig. 23.26.

The best fit corresponds to

$$\Lambda^{(4)}_{\overline{\text{MS}}} = 224 \pm 21 \text{ MeV} \qquad \text{and} \qquad \eta_{\overline{\text{MS}}}(Q_0^2) = 8.3 \pm 1.5. \qquad (23.9.4)$$

It is interesting to note that the same analysis, carried out in leading order only, i.e. in LLA, yields a significantly different value of the parameter η. One finds

$$\Lambda_{\text{LO}} = 215 \pm 27 \text{ MeV} \qquad \text{and} \qquad \eta_{\text{LO}}(Q_0^2) = 3.7 \pm 1.2. \qquad (23.9.5)$$

It is seen that the gluon distribution extracted is very sensitive to whether the calculation is done in LLA or beyond, but that the calculated curves are rather insensitive to the gluon distribution.

It is also interesting to note the correlation between Λ_{QCD} and η as shown in Fig. 23.27. Encouragingly, the correlation is weaker in the next-to-leading order calculation.

In Fig. 23.28 is shown the excellent agreement between QCD in next-to-leading order and the BCDMS hydrogen data on $F_2(x, Q^2)$.

A phenomenological higher twist term $C(x)/Q^2$ was included in the analysis and seems to be necessary at smaller values of Q^2, especially for the data at larger values of x.

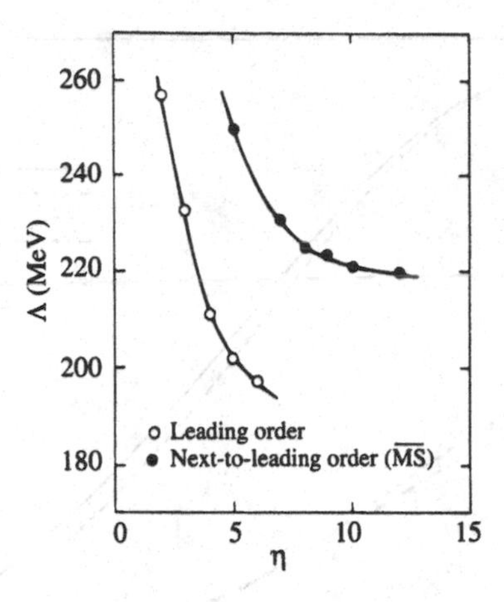

Fig. 23.27. Correlation between Λ_{QCD} and η [see eqn (23.9.3)] in fitting the BCDMS deuterium data.

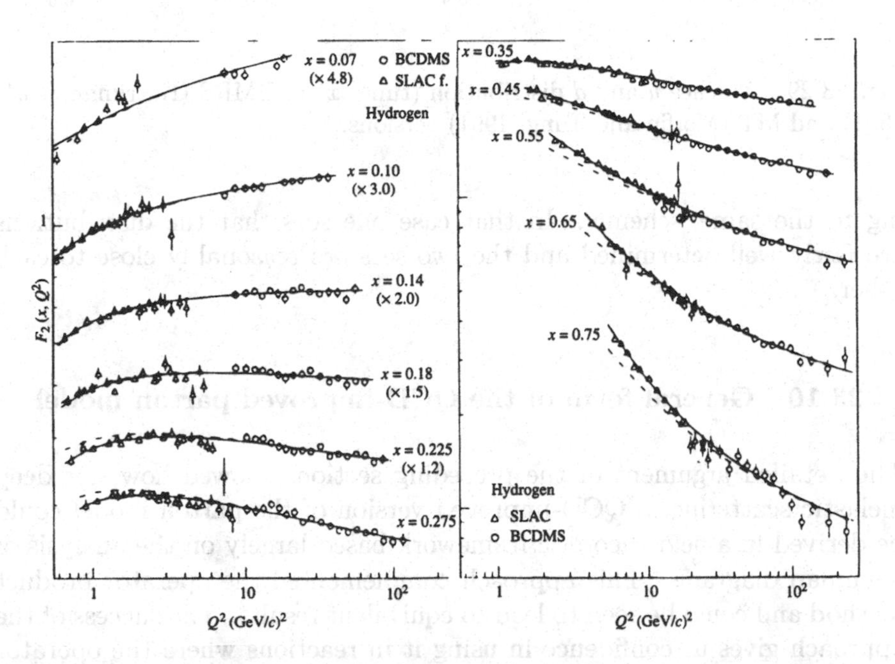

Fig. 23.28. QCD next-to-leading order fit to combined SLAC and BCDMS hydrogen data on $F_2(x, Q^2)$. Dashed line is pure QCD. Solid line includes higher twist term—see text.

Finally we show in Figs. 23.29–31 two recent sets of parton distributions (given in both $\overline{\text{MS}}$ and DIS schemes) which are compatible with all known deep inelastic scattering data (Owens and Tung, 1992). As stressed earlier, it is only meaningful to compare distributions belong-

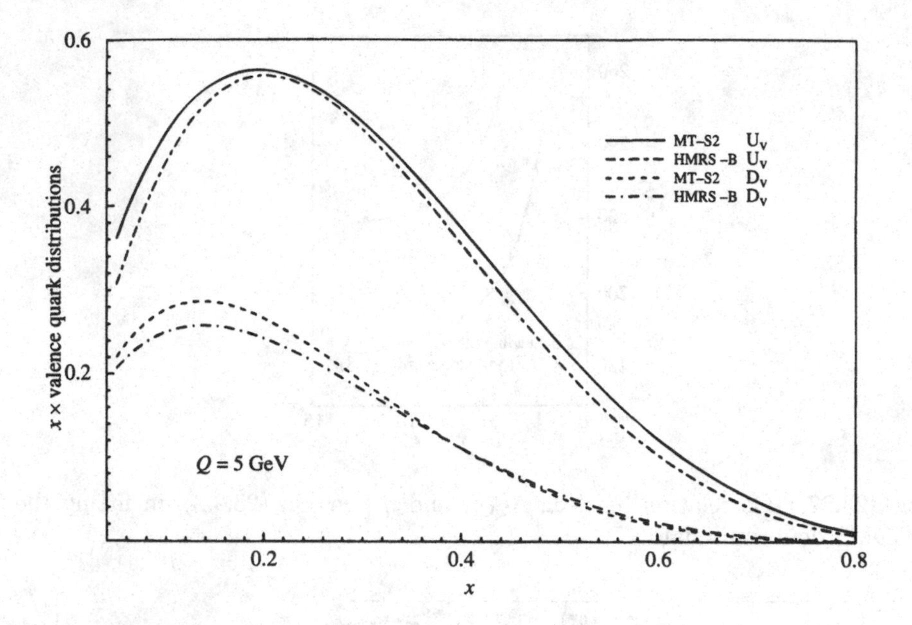

Fig. 23.29. Valence u and d distribution (times x) in HMRS (Harriman *et al.*, 1991) and MT (Morfin and Tung, 1991) versions.

ing to the same 'scheme'. In that case one sees that the distributions are fairly well determined and the two sets are reasonably close to each other.

23.10 General form of the QCD-improved parton model

The detailed argument of the preceding sections showed how, for deep inelastic scattering, a QCD-improved version of the parton model could be derived in a field theoretic framework based largely on the analysis of Feynman diagrams. This approach complemented the operator product method and could be seen to lead to equivalent results. The success of the approach gives us confidence in using it in reactions where the operator product expansion cannot be utilized.

Because of the similarity to the previous arguments we shall simply state the general results and comment upon them.

Consider a reaction

$$A + B \rightarrow C + X \tag{23.10.1}$$

in which hadron C is produced with large p_T. The momenta are labelled

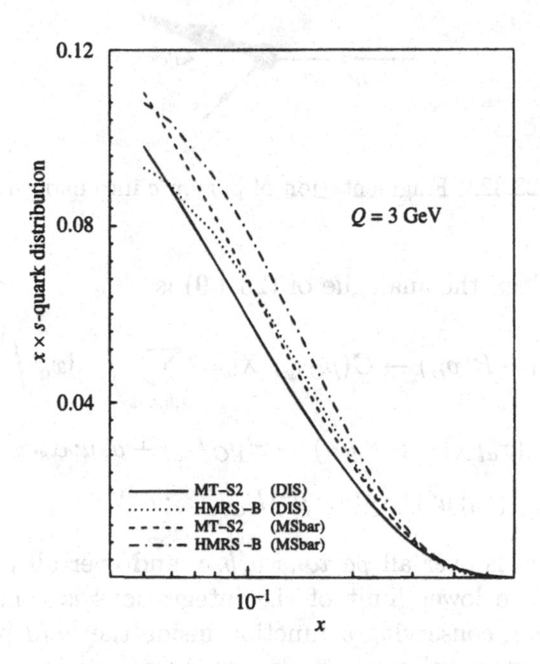

Fig. 23.30. As in Fig. 23.29 but for strange quark distribution (times x).

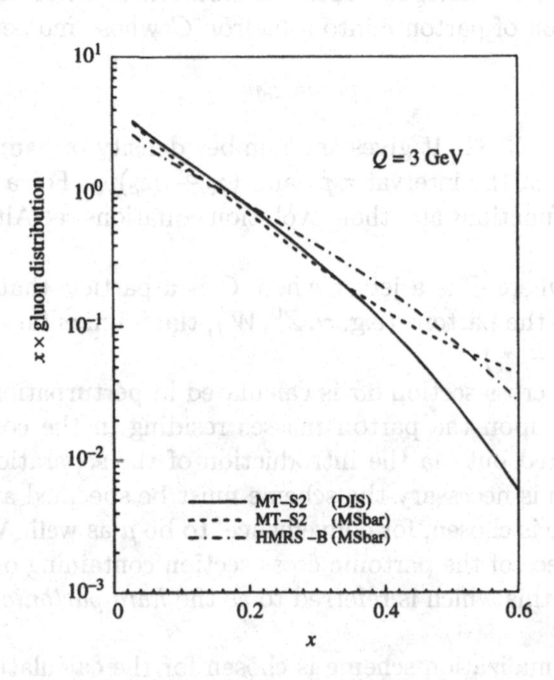

Fig. 23.31. As in Fig. 23.29 but for gluon distribution (times x).

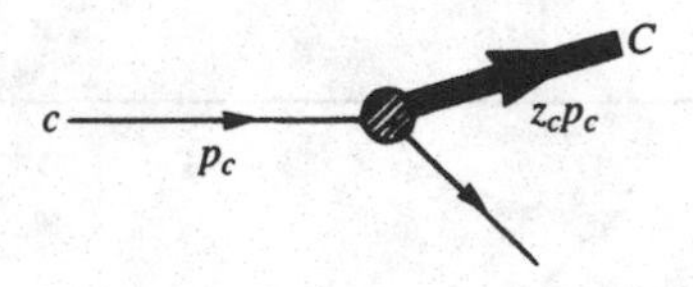

Fig. 23.32. Fragmentation of parton c into hadron C.

$p_A, p_B, p_C \ldots$. Then the analogue of (23.4.9) is

$$\begin{aligned} \mathrm{d}\sigma\left[A(p_A)+B(p_B)\to C(p_C)+\mathrm{X}\right] &= \sum_{a,b,c,d}\int^1 \mathrm{d}x_a \int^1 \mathrm{d}x_b \int^1 \mathrm{d}z_c \,\times \\ &\times \mathrm{d}\hat{\sigma}\left[a(x_a p_A)+b(x_b p_B)\to c(p_C/z_c)+d;\mu;\alpha_s(\mu^2)\right]\times \\ &\times f_{a/A}(x_a;\mu^2) f_{b/B}(x_b;\mu^2) D_{C/c}(z_c;\mu^2) \end{aligned} \tag{23.10.2}$$

in which the sum is over all *partons* a, b, c, and over all partons or sets of partons d. The lower limit of the integrations is controlled by the energy-momentum conserving δ function inside the *hard* partonic cross-section $\mathrm{d}\hat{\sigma}$. The renormalization scale and the separation or factorization scale are taken equal and are denoted by μ. The *fragmentation function* $D_{C/c}(z_c;\mu^2)$ is analogous to a distribution function and describes the fragmentation of parton c into a hadron C whose momentum is

$$p_C = z_c p_c \tag{23.10.3}$$

as shown in Fig. 23.32. It gives the number density of hadrons C whose momentum lies in the interval $z_c p_c$ and $(z_c + \mathrm{d}z_c)p_c$. For a discussion of fragmentation functions and their evolution equations see Altarelli (1982), Section 9.2.

In the case where C is a jet or where C is a particle that has a point-like coupling to the partons (e.g. γ, Z^0, W), the fragmentation function is replaced by $\delta(1-z_c)$.

The partonic cross-section $\mathrm{d}\hat{\sigma}$ is calculated in perturbation theory, and all dependence upon the parton masses residing in the collinear singularities is factored out via the introduction of the separation scale μ. If renormalization is necessary the scheme must be specified and the renormalization scale is chosen, for convenience, to be μ as well. What is left is then a finite piece of the partonic cross-section containing only hard subprocesses. It is this which is referred to as the *hard partonic* cross-section $\mathrm{d}\hat{\sigma}$.

Once a renormalization scheme is chosen for the calculation of $\mathrm{d}\hat{\sigma}$, the distribution and fragmentation functions used in (23.10.2) must be defined in the same scheme.

The perturbation series for $d\hat{\sigma}$ will in general be of the form:

$$d\hat{\sigma} \equiv a_0 \alpha_s^n \left(1 + \sum_{j=1} a_j \alpha_s^j\right) \qquad (23.10.4)$$

where the a_j are functions of the kinematic variables and where the exponent in the leading power α_s^n will depend upon the reaction. Thus for deep inelastic scattering $n = 0$ whereas for jet production in hadronic reactions $n = 2$.

The LLA corresponds to keeping just the lowest order term in $d\hat{\sigma}$ and using the lowest order evolution equations for the distributions. At this level $d\hat{\sigma}$ will not depend upon μ [except possibly through $\alpha_s(\mu^2)$] whereas the distribution functions do vary with μ. On the other hand the left-hand side of (23.10.2) is a physical quantity so cannot depend on μ. This is the spurious μ dependence discussed in Section 23.4. The situation is then not quite consistent and one chooses $\mu^2 \approx Q^2$, where Q^2 is the large scale in the reaction, so as to minimize the danger of large logarithms of Q^2/μ^2. In effect then the leading order (LO), or more correctly LLA approximation, boils down to using the Born approximation for $d\hat{\sigma}$ but with α_s replaced $\alpha_s(Q^2)$, where Q^2 is some measure of the large momentum transfer involved, and coupling this with distribution and fragmentation functions evaluated at the same large scale Q^2. When $d\hat{\sigma}$ is calculated to higher order the artificial dependence of the RHS of (23.10.2) upon μ diminishes.

Equation (23.10.2) is the basis of all calculations of large p_T reactions, jet production, etc. in the framework of perturbative QCD, which will be discussed in the next few chapters. The reader should be aware that many different notational conventions exist for the parton distribution and fragmentation functions. In this rather technical Chapter we have stuck to the logical, but cumbersome, notation $f_{a/A}(x)$ or $D_{C/c}(z)$. In more phenomenological sections these were often written $q_a^A(x)$ and $D_C^c(z)$ respectively, and in discussing the detailed parton model it is, of course, quite usual to simply replace $f_{u/p}(x)$ by $u(x)$ etc., where, by implication, one is referring specifically to distributions inside a proton.

We shall now turn to the Drell–Yan reaction to see the consequences of the above treatment.

23.11 QCD corrections to Drell–Yan and W production

The parton model result for the cross-section to produce lepton–antilepton pairs with 4-momentum q^μ and $q^2 = m^2$ in the reaction $h_A + h_B \to \ell^+\ell^- X$ was given in (17.4.33 and 34). The changes necessary for discussing the production of heavy vector mesons like Z^0 and W were explained in Section 17.4.3. We here consider the QCD corrections to the simple parton model results. Important new features appear and can be used to test the predictions of perturbative QCD. We utilize the diagrammatic approach discussed in much detail in the previous sections.

23.11.1 Drell–Yan production

To the lowest order terms

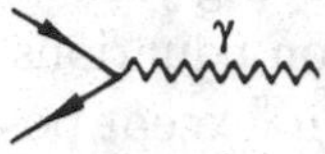

have to be added gluonic correction terms

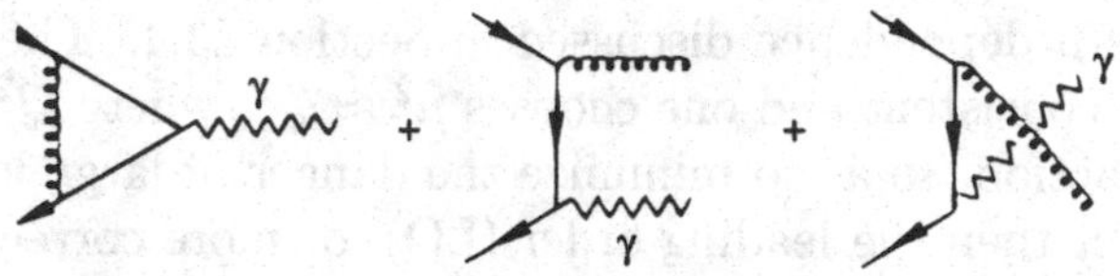

and terms coming from the gluon content of the hadrons

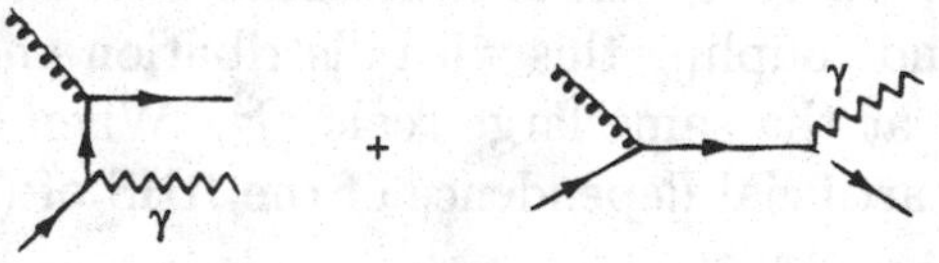

As expected from the general discussion in Section 23.10, in leading logarithmic approximation, one obtains again the formulae (17.4.33 and 34) but with $q_j^{\mathrm{A}}(x_{\mathrm{A}})\bar{q}_j^{\mathrm{B}}(x_{\mathrm{B}})$ replaced by $q_j^{\mathrm{A}}(x_{\mathrm{A}};m^2)\bar{q}_j^{\mathrm{B}}(x_{\mathrm{B}};m^2)$. Note that here in the Q^2-dependent parton distributions the rôle of the large scale Q^2 is played by m^2 so that the choice $Q^2 = m^2$ seems the most sensible. Given that the distribution functions are determined from studies of deep inelastic scattering and that their evolution is controlled by the Altarelli–Parisi equations (23.4.21 and 22), the comparison between theory and experiment for the Drell–Yan reaction provides, in principle, a significant test of both the parton picture and the QCD dynamics. One implication is that the scaling in τ [see eqns (17.4.31, 32)] will be somewhat spoilt by a logarithmic dependence on m^2, but given the smallness of the cross-section and its rapid decrease with m^2 this will be very difficult to detect.

As mentioned in Section 17.4 the experimental cross-sections are found to be about 1.5–2 times larger than the theoretical predictions, but *ratios* of cross-sections and the *scaling* of the cross-sections when considered as functions of $\tau = m^2/s$ are all in beautiful agreement with the results of the simple parton model.

Surprisingly, the next-to-leading order correction terms have a very big effect in $\mathrm{d}\sigma/\mathrm{d}m^2$. Aside from relatively unimportant contributions involving $G(x; m^2)$, the major effect, roughly speaking, is to multiply the quark–antiquark contribution by a factor (often referred to as the K factor) which, to lowest order, is

$$\left[1 + \frac{\alpha_s(m^2)}{2\pi} C_F \left(1 + \frac{4\pi^2}{3}\right)\right] \tag{23.11.1}$$

where, it should be recalled, $C_F = 4/3$. If we take $m^2 = 60\ (\mathrm{GeV}/c^2)^2$ we have $\alpha_s(m^2) \approx 0.2$ so the correction factor (23.11.1) is 'huge', ≈ 1.5, i.e. a 50% change. Now given the disagreement between theory and experiment mentioned above, it is, of course, very encouraging to find that the improved theory is so much closer to experiment. At the same time it is disconcerting to discover that an $O(\alpha_s)$ correction can produce a change not that far from 100%! The source of the large term in (23.11.1) can be traced to the vertex correction in the diagrams shown above. The expressions (17.4.33, 34), with inclusion of next-to-leading order effects, become

$$\begin{aligned}
\frac{\mathrm{d}\sigma}{\mathrm{d}m^2} = \frac{4\pi\alpha^2}{9m^2 s} \int_0^1 \frac{\mathrm{d}x'_{\mathrm{A}}}{x'_{\mathrm{A}}} \int_0^1 \frac{\mathrm{d}x'_{\mathrm{B}}}{x'_{\mathrm{B}}} \sum_{\text{flavours}} Q_j^2 \Big[q_j^{\mathrm{A}}(x'_{\mathrm{A}}; m^2) \bar{q}_j^{\mathrm{B}}(x'_{\mathrm{B}}; m^2) + \\
+ \bar{q}_j^{\mathrm{A}}(x'_{\mathrm{A}}; m^2) q_j^{\mathrm{B}}(x'_{\mathrm{B}}; m^2) \Big] \left[\delta(1-z) + \theta(1-z) \frac{\alpha_s(m^2)}{2\pi} f^q_{\mathrm{DY}}(z) \right] + \\
+ \sum_{\text{flavours}} Q_j^2 \Big\{ \Big[q_j^{\mathrm{A}}(x'_{\mathrm{A}}; m^2) + \bar{q}_j^{\mathrm{A}}(x'_{\mathrm{A}}; m^2) \Big] G^{\mathrm{B}}(x'_{\mathrm{B}}; m^2) + \\
+ G^{\mathrm{A}}(x'_{\mathrm{A}}; m^2) \Big[q_j^{\mathrm{B}}(x'_{\mathrm{B}}; m^2) + \bar{q}_j^{\mathrm{B}}(x'_{\mathrm{B}}; m^2) \Big] \Big\} \theta(1-z) \frac{\alpha_s(m^2)}{2\pi} f^q_{\mathrm{DY}}(z)
\end{aligned} \tag{23.11.2}$$

where

$$z \equiv \tau / x'_{\mathrm{A}} x'_{\mathrm{B}} \tag{23.11.3}$$

and the functions $f^q_{\mathrm{DY}}(z), f^G_{\mathrm{DY}}(z)$, analogous to the f^q_i, f^G_i introduced in Section 23.8, depend upon the renormalization scheme being used. As stressed earlier, they must be evaluated in the same scheme used to specify the distribution functions.

If the distribution functions are obtained from deep inelastic scattering using the DIS convention [see eqn (23.8.5)] then f^q_{DY} and f^G_{DY} in (23.11.2) must be replaced by

$$\begin{aligned} f_{\mathrm{DY}}^{q}(z) - 2f_2^q(z) &= C_F\left\{\left(1+\frac{4\pi^2}{3}\right)\delta(1-z) + \right. \\ &\left. + 2(1+z^2)\left[\frac{\ln(1-z)}{1-z}\right]_+ + \frac{3}{(1-z)_+} - 6 - 4z\right\} \\ f_{\mathrm{DY}}^{G}(z) - f_2^G(z) &= \frac{T}{n_f}\left\{[z^2+(1-z)^2]\ln(1-z) + \right. \\ &\left. + 9z^2/2 - 5z + 3/2\right\}. \end{aligned} \tag{23.11.4}$$

The factor 2 in front of $f_2^q(z)$ appears because in a Drell–Yan reaction each of the two partons participating in the reaction can radiate a gluon.

The effect of the vertex correction, since it involves the same kinematics as the simple parton model diagram, is visible in the $\delta(1-z)$ term in (23.11.4). The worrying aspect of this very large perturbative correction is to some extent alleviated by the knowledge that a major part of this term, namely $(\alpha_s/2\pi)C_F\pi^2\delta(1-z)$, arises from continuing the vertex contribution from the spacelike region $q^2 = -Q^2$ of deep inelastic scattering to the timelike region $q^2 = m^2$ of the Drell–Yan reaction, and that this contribution can be summed to all orders. The result is

$$\begin{aligned} \exp[(\alpha_s/2\pi)C_F\pi^2] &= 1 + (\alpha_s/2\pi)C_F\pi^2 + \cdots \\ &= 1.52 \quad \text{for} \quad \alpha_s = 0.2. \end{aligned} \tag{23.11.5}$$

For the data discussed in Section 17.4, i.e. for πp and pp initiated reactions, the relatively low values of $\sqrt{s}$ available imply that the minimum values of τ that can be explored are not very small [see Table 17.2]. In this region the next-to-leading order corrections are dominated by the vertex term and, as mentioned, effectively multiply the parton model result by a constant K factor. At $\bar{\mathrm{p}}$p colliders it will be possible to go down to much smaller values of τ where the gluon contribution in (23.11.2) becomes more important. The overall correction factor is found to decrease as τ decreases. Tests of this feature in the near future will be of great interest.

In Fig. 23.33 we show a comparison between recent Drell–Yan data for 800 GeV protons incident upon a Cu target (E605 experiment at Fermilab: Brown *et al.*, 1989) and a next-to-leading order calculation using the HMRS distributions (Harriman *et al.*, 1991). The agreement is seen to be excellent.

The cross-section for the Drell–Yan reaction at large m^2 and $\sqrt{s}$ is exceedingly small, but eventually it should be possible to look at lepton–antilepton pairs whose mass is comparable with M_Z. In that case also

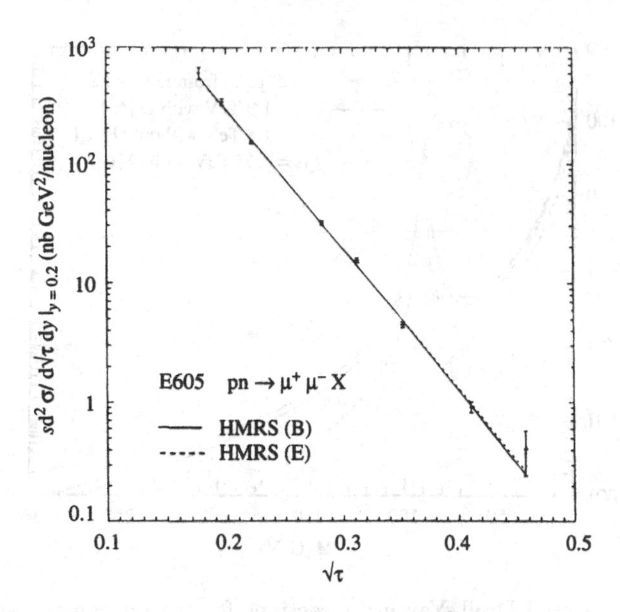

Fig. 23.33. Drell–Yan data from E605 collaboration compared with next-to-leading order theory. (From Harriman *et al.*, 1991.)

the Z propagator must be taken into account (see Section 17.4.3) and beautiful effects should be visible, as shown in Fig. 23.34.

23.11.2 Transverse momentum distribution of Drell–Yan pairs

A very characteristic signal for the QCD corrections should be the production of lepton pairs with large q_T (see Section 17.4.2) arising from the radiated gluon or quark in the extra diagrams shown above, being emitted with a large transverse momentum, this being compensated by an equal and opposite q_T for the virtual photon. For events of this type, the vertex correction diagram does not contribute.

The cross-section to produce the Drell–Yan pair or virtual γ with transverse momentum q_T is calculated via the general formula (23.10.2) wherein C is now a 'γ' and the fragmentation function is replaced by $\delta(1-z_c)$. The partonic cross-sections are controlled by the following squared and spin-summed and averaged Feynman matrix elements:

$$\overline{|\mathcal{M}(q\bar{q}\to \text{`}\gamma\text{'}G)|^2} = \pi\alpha_s e^2 Q_j^2\,\frac{8}{9}\cdot\frac{\hat{t}^2+\hat{u}^2+2m^2\hat{s}}{\hat{t}\hat{u}}$$

$$\overline{|\mathcal{M}(G\bar{q}\to \text{`}\gamma\text{'}q)|^2} = \pi\alpha_s e^2 Q_j^2\,\frac{1}{3}\cdot\frac{\hat{s}^2+\hat{u}^2+2\hat{t}m^2}{-\hat{s}\hat{u}} \qquad (23.11.6)$$

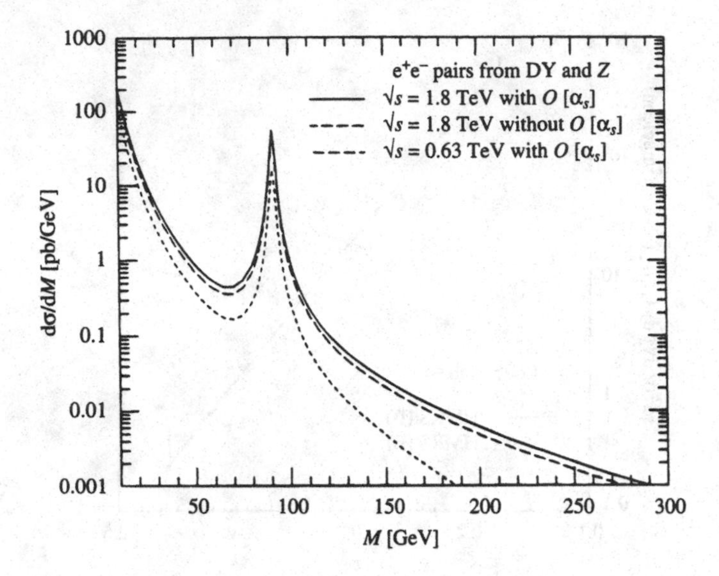

Fig. 23.34. Expected Drell–Yan cross-sections for lepton pair masses comparable with M_Z. (From Ellis and Stirling, 1990.)

where for the partonic reaction

$$a(p_a) + b(p_b) \to `\gamma\text{'}(p_\gamma) + d(p_d)$$

$$\hat{s} = (p_a + p_b)^2, \qquad \hat{t} = (p_a - p_c)^2, \qquad \hat{u} = (p_a - p_d)^2 \qquad (23.11.7)$$

and Q_j is the charge of the quark.

As expected the partonic cross-section diverges at small angles, i.e. for $q_T \to 0$. So the above can only be trusted in the perturbative region $q_T^2 \gg \mu^2$. For the mean q_T one finds (Altarelli, Ellis and Martinelli, 1979)

$$\langle q_T^2 \rangle = \alpha_s(m^2)\ s\, f[\tau, \alpha(m^2)] + \text{ constant}, \qquad (23.11.8)$$

where the unknown constant reflects the intrinsic (non-perturbative) parton transverse momentum and f is calculable. At fixed τ we thus expect $\langle q_T \rangle$ to rise roughly like $\sqrt{s}$. The data, averaged over τ, displayed in Fig. 23.35 definitely show the $\sqrt{s}$ growth, and this seems to be one of the most convincing pieces of evidence in favour of QCD. It is interesting to note that at large q_T it is the quark–gluon collisions ('Compton scattering') that dominate, as shown in Fig. 23.36 taken from Halzen and Scott (1978).

The above required a convolution of the partonic $d\hat{\sigma}$ with the parton distributions. Strictly speaking it only applies to the case where $q_T^2 \approx m^2$,

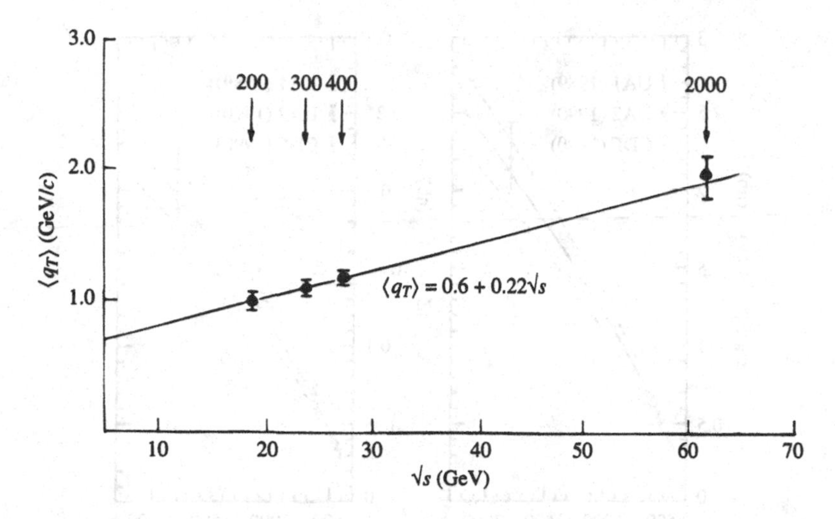

Fig. 23.35. Variation of mean transverse momentum of lepton pairs with energy. Arrows indicate equivalent laboratory energy for a fixed-target experiment. (From Altarelli, 1979.)

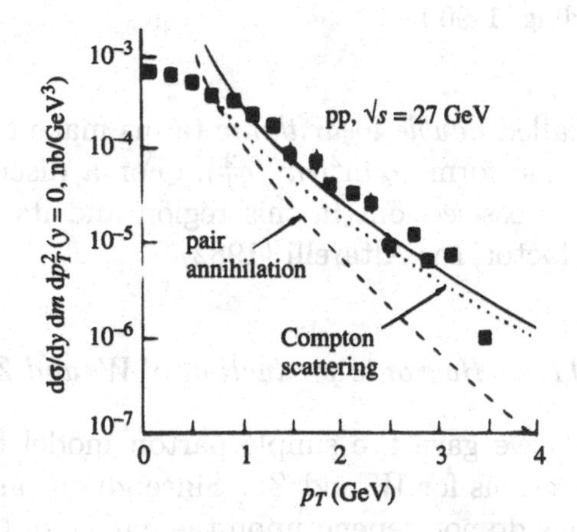

Fig. 23.36. The QCD calculation for lepton pairs with mass 7.5 GeV/c^2 is shown separated into its two components: quark–antiquark annihilation and quark–gluon Compton scattering. (From Halzen and Scott, 1978.)

so that there is only one large scale in the problem. But when m^2 is sufficiently large there is an important and interesting kinematic region in which

$$\mu^2 \ll q_T^2 \ll m^2 \tag{23.11.9}$$

where perturbative QCD should be applicable. The analysis in this region is complicated by the existence of the two large but disparate scales q_T^2 and m^2. Terms of the type $\ln(m^2/q_T^2)$ cannot be neglected compared with

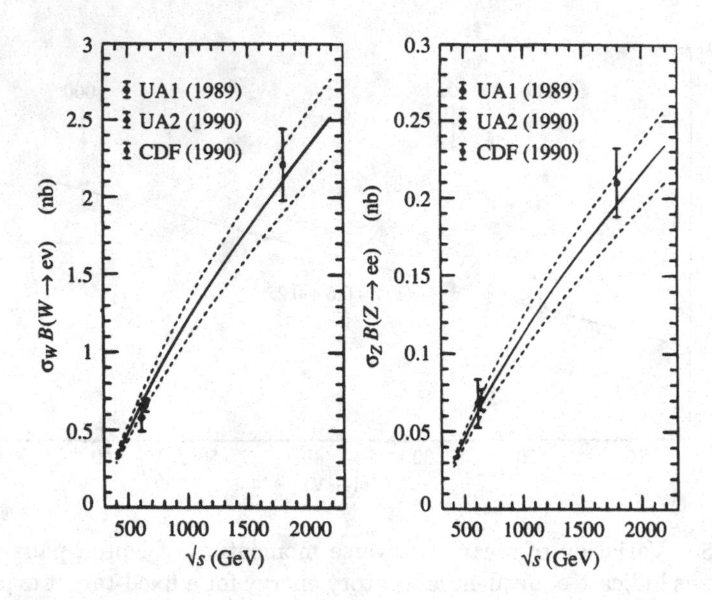

Fig. 23.37. Comparison of W and Z cross-sections with theory (see text). (From Ellis and Stirling, 1990.)

$\ln(q_T^2/\mu^2)$ and so-called *double logarithmic* terms make their appearance, i.e. expressions of the form $\alpha_s \ln^2(m^2/q_T^2)$. For a discussion of the q_T^2 dependence of the cross-sections in this region and its connection with the Sudakov form factor, see Altarelli (1982).

23.11.3 Hadronic production of W and Z^0

In (17.4.43 and 44) we gave the simple parton model formulae for the production cross-sections for W and Z^0. Since gluons are 'flavour-blind' the QCD corrections do not depend upon the nature of the vector meson being produced and are the same as for Drell–Yan pair production.

Thus in LLA one simply uses the expressions (17.4.43 and 44), but the luminosity is now computed via (17.4.34) using Q^2-dependent parton distributions $q_i^{\mathrm{A}}(x_{\mathrm{A}}; M_V^2)$ etc. where $M_V = M_W$ or M_Z. Thus e.g. in W production $\mathcal{L}_{ij}^{\mathrm{AB}}$ is not just a function of τ_W but also depends weakly on M_W^2. Since the cross-section now depends upon parton distributions evolved up to the huge scale $Q^2 = M_W^2$, the comparison between experiment and theory constitutes a serious check of the QCD evolution equations. In Fig. 23.37 the data on $\sigma_W \times BR(W \to \mathrm{e}\nu)$ and $\sigma_Z \times BR(Z \to \mathrm{e}^+\mathrm{e}^-)$ are compared with a calculation using the HMRS (B) parton distributions. A 10% error band for the theoretical calculation is indicated. It is seen that the agreement is excellent.

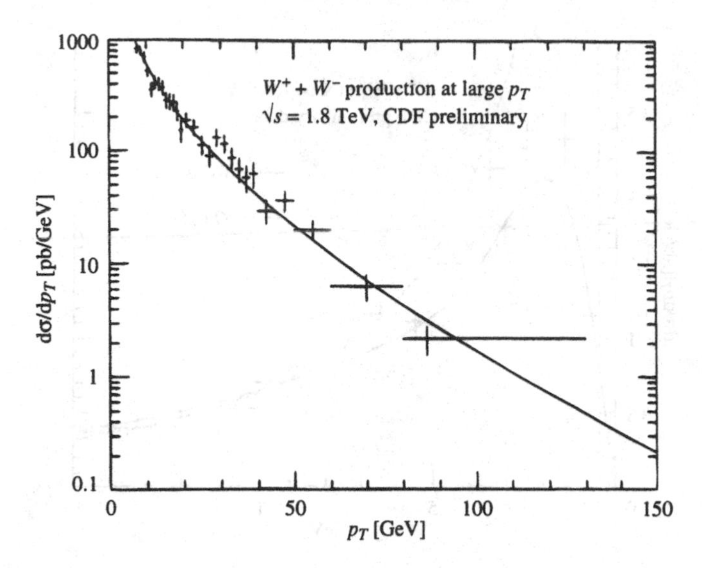

Fig. 23.38. Large transverse momentum distribution of *W*s produced in $\bar{p}p$ collisions at $\sqrt{s} = 1.8$ TeV. For curve see text. (From Ellis and Stirling, 1990.)

23.11.4 *Transverse momentum distribution of W and Z^0*

In order to discuss the p_T distribution of the produced W or Z^0, at large p_T, one simply replaces the electromagnetic coupling by the relevant weak boson coupling in (23.11.6), and replaces the lepton pair mass m^2 by M_V^2. Thus

$$W: \quad e^2Q_j^2 \quad \rightarrow \quad \sqrt{2}GM_W^2; \quad m^2 \rightarrow M_W^2 \tag{23.11.10}$$

$$Z^0: \quad e^2Q_j^2 \quad \rightarrow \quad \sqrt{2}GM_Z^2[(g_V^j)^2 + (g_A^j)^2]; \quad m^2 \rightarrow M_Z^2. \tag{23.11.11}$$

As seen in Fig. 23.38 the large p_T data are not yet very accurate. Nonetheless they are nicely compatible with the next-to-leading order calculation of Arnold and Reno (1989), using the HMRS (B) parton distributions.

Finally, as for Drell–Yan pairs, the situation for $\mu^2 \ll p_T^2 \ll M_W^2$ is more complicated, involving the summation of the 'double logarithmic' terms. Details can be found in Collins and Soper (1987). An interesting comparison between the rather meagre data and theoretical calculations (i) to $O(\alpha_s^2)$ and (ii) with summation of double logarithms to all orders, is shown in Fig. 23.39.

It is seen that the data show signs of 'turning over' at smaller p_T in agreement with the expectations of the 'summed' theory. The comparison with better data in the future will be of great interest.

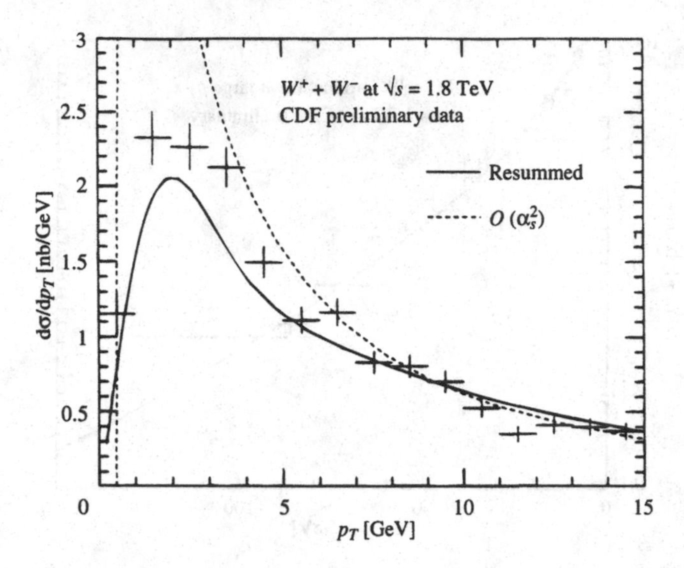

Fig. 23.39. Small transverse momentum distribution of Ws produced in $\bar{p}p$ collisions at $\sqrt{s} = 1.8$ TeV. For curves see text. (From Ellis and Stirling, 1990.)

23.12 Summary

We have indicated how the Feynman diagram approach leads to a QCD-improved parton model for high energy large momentum transfer reactions even in situations where one does not have recourse to the operator product expansion. When there is just one large scale Q^2 involved, probably the most important element is the replacement of the parton distribution functions $f_{j/\mathrm{A}}(x)$ by the scale-dependent distributions $f_{j/\mathrm{A}}(x; Q^2)$. The evolution in Q^2 of these functions is intimately controlled by the dynamics of QCD and offers, ultimately, a very direct test of the theory. Because the evolution is logarithmic only relatively small effects have been seen up to the time of writing. But HERA has just begun functioning and a new gigantic range of Q^2 will become accessible to experiment. It is with great interest that we await nature's verdict on QCD.

The generic results obtained in this chapter form the basis for the discussion of large p_T reactions and jet physics to be discussed in the next few chapters.

24
Large p_T phenomena and jets in hadronic reactions

Jets have by now become a major component of high energy particle physics. It is widely (if not universally) accepted that jets are implied by perturbative QCD and that they are the simplest and perhaps best evidence supporting it. This means that jets are to be found in all large p_T phenomena irrespective of whether they are initiated by hadronic or by leptonic processes. In spite of this, for pedagogical and historical reasons we shall present the discussion of jets in hadronic and in e^+e^- reactions separately.

A drastic selection has had to be made to reduce the material on jets to an acceptable size. For additional material see Cavasinni (1990), Altarelli (1989) and Jacob (1990) and references therein. [See also the various contributions in the CERN 89-08 (vol 1–3) edited by Altarelli, Kleiss and Verzegnassi, 1989].

Note that a complete set of formulae for the cross-sections of all the $2 \to 2$ partonic reactions discussed herein can be found in Appendix 7.

24.1 Introduction

We here discuss the evidence that hadronic reactions involving large transfers of transverse momentum (p_T) are controlled by the direct collision of constituents within the colliding hadrons, with subsequent fragmentation of these constituents into showers of hadrons. Naturally, it is supposed that these constituents are the quarks and gluons discussed in previous chapters.

It is well known that high energy hadronic interactions are dominated by the production of a large number of particles, mostly confined to the nearly forward direction, i.e. dominated by events in which the produced secondaries have small p_T leading to the conclusion that strong interactions at high energy are generally rather 'soft'.

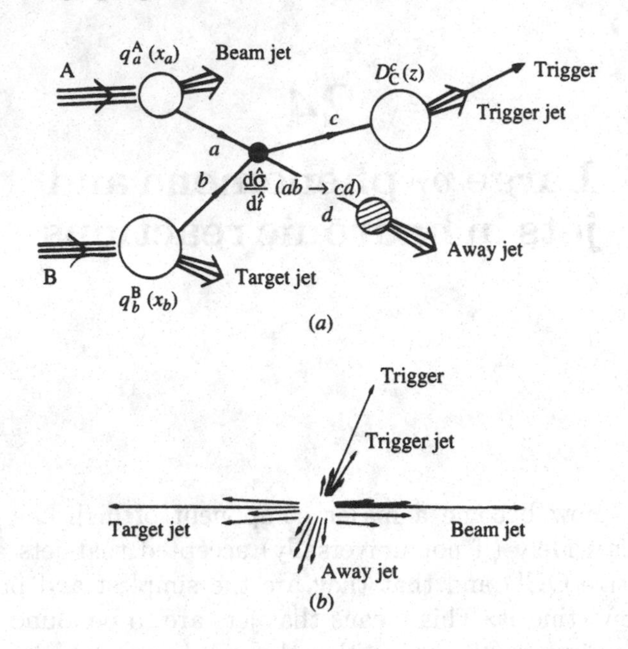

Fig. 24.1. Idealization of a large p_T hadronic interaction producing four jets.

Whereas it is easy to understand qualitatively the origin of these events at high energies (the two fast particles being excited on passing each other and subsequently de-exciting by bremsstrahlung), a quantitative treatment is prohibitively difficult being, in essence, a collective effect. This, in QCD language, amounts to dealing with non-perturbative effects.

Where, however, one may have a better chance of detecting the basic dynamic properties of the interaction is in those events that occur at large p_T values. It is hoped that these events will allow an unravelling of the inner structure of the hadron and will teach us how the constituents interact.

The evidence that we will discuss in this chapter suggests that large p_T events proceed via 'hard scattering' involving the collision of just one constituent from each initial particle. This is in agreement with the whole philosophy of the parton model discussed previously (Chapters 15–17). The partons involved are scattered through large p_T and are supposed to materialize as a set of fairly well collimated hadrons called a 'jet'. This is the natural outcome of essentially any quark or parton model.

An idealized description of such a situation is given in Fig. 24.1, which shows that the simplest configuration of a large p_T hadronic process implies the existence of at least *four jets*; two of these, the beam and target jets, have long been familiar in multiparticle production physics and will not be discussed here.

In Fig. 24.1 the terminology is self-explanatory.

The inclusive hadronic reaction

$$A + B \rightarrow C + X \tag{24.1.1}$$

is expressed (see Section 21.3) in terms of: (i) the elementary elastic quark–quark cross-section $\mathrm{d}\hat{\sigma}/\mathrm{d}\hat{t}$ for the reaction

$$a + b \rightarrow c + d \tag{24.1.2}$$

(ii) the density function $q_a^A(x_a)$ for finding quark of *flavour* a in the hadron A, and (iii) the fragmentation function $D_C^c(z)$ for quark c to produce hadron C.

A 'trigger' particle with high p_T signals a possible jet and defines it as the 'trigger' jet (or 'towards') as opposed to the 'away' jet.

The case of $\mathrm{e}^+\mathrm{e}^-$ annihilation into hadrons is expected to proceed in essentially the same way but without the presence of beam and target jets.

At higher energies, 3, 4 and even more high p_T jets are expected to be produced. They correspond to the radiation of hard gluons from the quarks.

After a brief historical survey, in this chapter we discuss the evidence that large p_T events in hadronic collisions are indeed due to hard collisions and the properties of jets in these reactions.

24.2 Historical survey. Hard *qq* scattering

It was discovered long ago in cosmic ray experiments that there is a roughly exponential cut-off in the p_T distribution of secondary particles for $p_T < 1$ GeV/c.

In the early 1970s it was widely believed that this exponential cut-off would continue to hold at higher p_T values. However, this expectation was challenged (Berman, Bjorken and Kogut, 1971) after the success of the parton model, on the grounds that hard parton–parton collisions should produce a tail in the large p_T distribution, showing up as a transition from an exponential to an inverse-power-like behaviour.

Generally speaking, this should be expected whenever the basic process is the direct collision of two point-like *elementary* objects (partons in our case).

The suggested change of behaviour was dramatically confirmed by the ISR data shown in Fig. 24.2 where the straight line is the extrapolation of the exponential form $\exp(-6p_T)$ which fits the small p_T data.

It seems reasonable that masses should be irrelevant in hard collisions implying that cross-sections should be scale free. This led (Berman, Bjorken and Kogut, 1971) to the prediction of a power behaviour p_T^{-4}

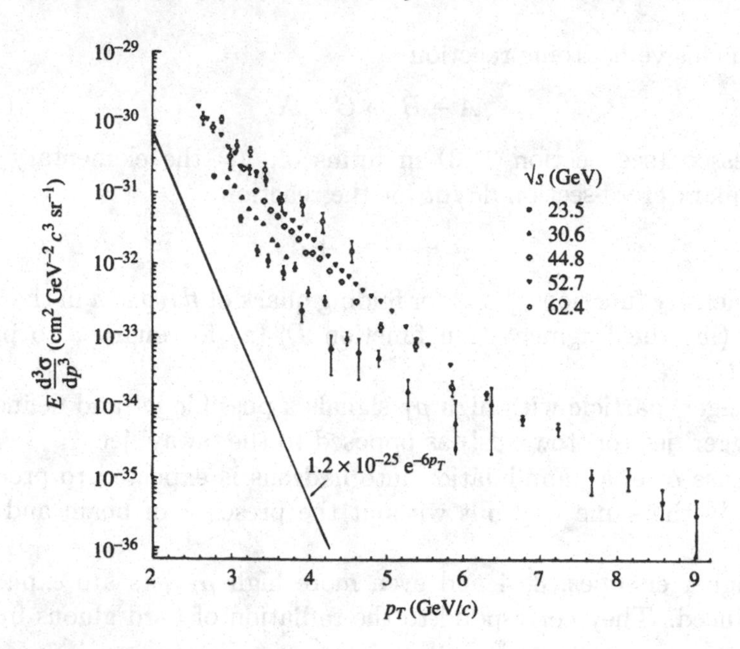

Fig. 24.2. Invariant cross-section for pp $\rightarrow \pi^0 X$ at large p_T compared with the extrapolation (solid curve) of the exponential behaviour at lower p_T. (Data from CERN-Columbia-Rockefeller-Saclay collaboration.)

for $\mathrm{d}\hat{\sigma}/\mathrm{d}\hat{t}$ at large p_T. In fact, in QCD, assuming single-gluon exchange to be the dominant process, one gets (Sivers and Cutler, 1978; Combridge, Kripfganz and Ranft, 1977) from Appendix 7* a generalization of the QED result for the scattering of spin $\frac{1}{2}$ objects used in Section 8.3

$$\frac{\mathrm{d}\hat{\sigma}}{\mathrm{d}\hat{t}}(q_a q_b \rightarrow q_a q_b) = \frac{4\pi\alpha_s^2}{9\,\hat{s}^2}\left[\frac{\hat{s}^2+\hat{u}^2}{\hat{t}^2} + \delta_{ab}\left(\frac{\hat{s}^2+\hat{t}^2}{\hat{u}^2} - \frac{2\hat{s}^2}{3\hat{u}\hat{t}}\right)\right] \tag{24.2.1}$$

for the colour-averaged quark–quark cross-section, where $\hat{s}$, $\hat{t}$, $\hat{u}$ are the usual Mandelstam variables but defined for the quark process (24.1.2) [see (24.3.3)].

If we simply assume that the hadronic inclusive cross-section for $A + B \rightarrow C + X$ is proportional to the basic quark inclusive cross-section for $a + b \rightarrow c + X$ (which may be a rather drastic assumption, see Section 24.3.1) one obtains, at 90° in the CM, the prediction (reminiscent of the Rutherford formula)

$$E\frac{\mathrm{d}^3\sigma}{\mathrm{d}^3\boldsymbol{p}} \simeq \frac{3\alpha_s^2}{p_T^4}f(x_T), \tag{24.2.2}$$

* Note that in Appendix 7 i, j, k, l, are used as flavour labels, rather than a, b, c, d. The latter notation is more suitable in the present chapter.

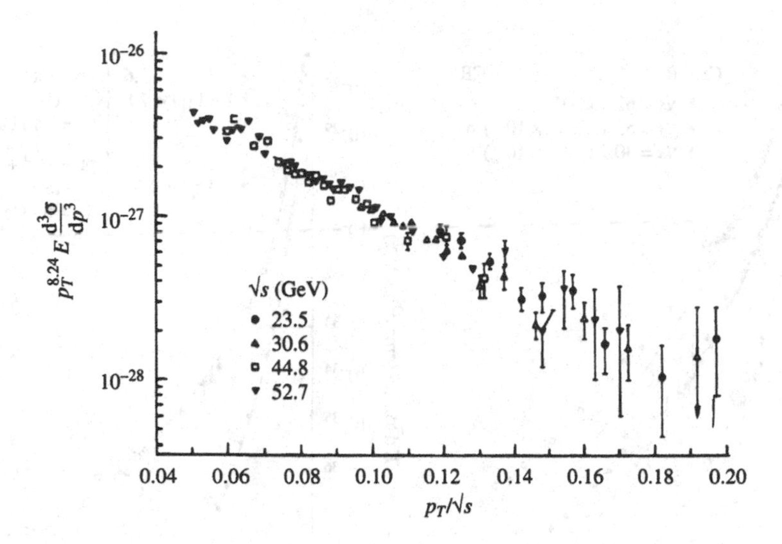

Fig. 24.3. Large p_T data for pp $\to \pi^0 X$ plotted so as to check the scaling form (24.2.4). (Data from CERN-Columbia-Rockefeller-Saclay collaboration.)

where

$$x_T \equiv p_T/(p_T)_{\max} \simeq 2p_T/\sqrt{s}. \tag{24.2.3}$$

In Fig. 24.3, the ISR data from CERN-Columbia-Rockefeller-Saclay collaboration (up to $\sqrt{s} \simeq 52.7$ GeV) are plotted versus x_T assuming the general scaling form

$$E\frac{d^3\sigma}{d^3p}\bigg|_{p_T \gg 1\mathrm{GeV}} \simeq p_T^{-n}(1-x_T)^m , \tag{24.2.4}$$

where m and n are left as free parameters.

The data appear to lie, more or less, on a single curve, but the best fit gives $m \simeq 10$ and $n \simeq 8.24$, i.e. much larger than the value $n = 4$ required in (24.2.2).

These results triggered an intense theoretical search for alternative mechanisms where the basic processes could be $qM \to qM$ (M = meson) or $q\bar{q} \to MM$ rather than $qq \to qq$.

However, data at larger p_T values ($p_T \geq 6$ GeV/c) indicate that: (i) the simple form (24.2.4) is probably inadequate; and (ii) the very large p_T points fall much less rapidly than p_T^{-8}.

This is seen in Fig. 24.4(a) where the dashed curve represents the fit (24.2.4) to the lower p_T data. The data from the CERN-Columbia-Rockefeller-Saclay (CCRS) and the CERN-Columbia-Oxford-Rockefeller (CCOR) collaborations are superimposed. The data of the latter collaboration at $\sqrt{s} = 62.4$ GeV are in reasonable agreement with those of the

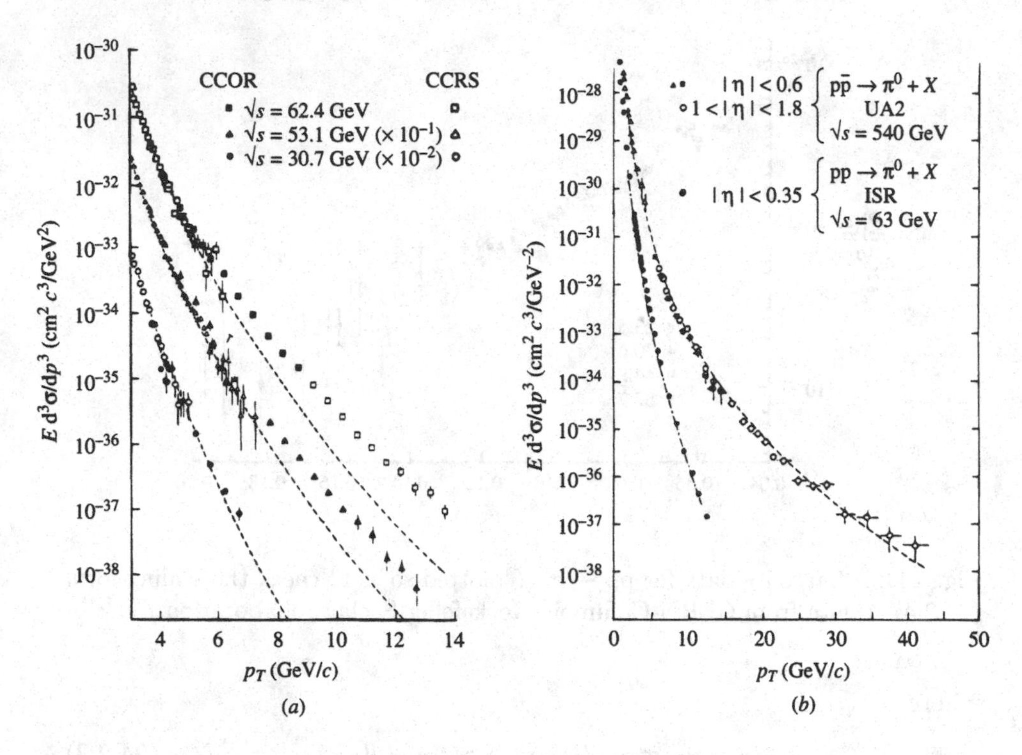

Fig. 24.4. (*a*) Very large p_T data for pp $\to \pi^0 X$ compared with the scaling form (dashed curve) which fitted the data of Fig. 24.3. (Data from CERN-Saclay-Zurich collaboration.) (*b*) Comparison between ISR and UA2 data at various rapidities. The dashed curve of the UA2 data corresponds to $n \simeq 6$.

Athens-Brookhaven-CERN-Syracuse (ABCS) and of the CERN-Saclay-Zurich (CSZ) collaborations extending up to $p_T \simeq 15$–16 GeV/c.

Using the data at $\sqrt{s} = 52.7$ GeV and $\sqrt{s} = 62.4$ GeV to separate the p_T and the x_T structure it turns out that for $x_T < 0.25$ the average value for n obtained by the various groups (CCOR, ABCS) is compatible with $n \simeq 8$ whereas for $x_T > 0.25$ the values for n range from 4–5 (CCOR, ABCS) to $n \simeq 6.6$ (CSZ). It seems to be established that the n relevant for very large p_T is moving closer to the prediction of $n = 4$ from 'elementary' qq scattering.

This trend is confirmed by the UA2 data shown in Fig. 24.4(*b*) where the dashed curve corresponds to $n \simeq 6$. Notice the spectacular increase with energy of the large p_T data as one goes from the ISR to the CERN Sp$\bar{\text{p}}$$S$ collider (Fig. 24.4(*b*)).

It is interesting to note that already in 1976 it was claimed by Gaisser (1976) that cosmic ray data (Matano *et al.*, 1968, 1975) are in rough agreement with a p_T^{-4} form (Fig. 24.5).

In summary, although it cannot be said that the data on single particle

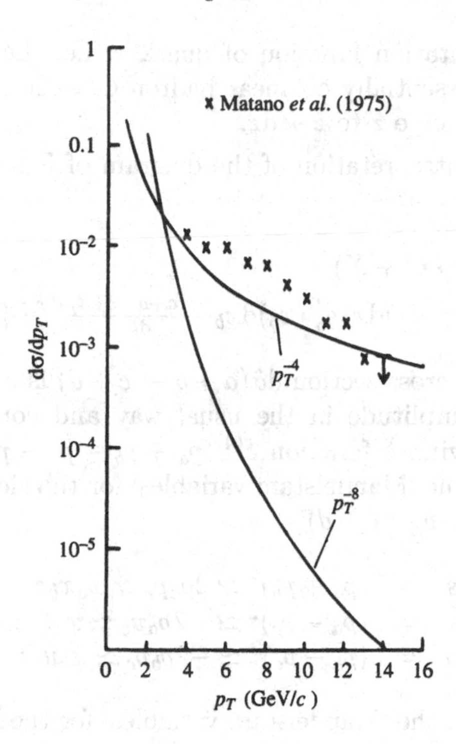

Fig. 24.5. Cosmic ray data (arbitrary normalization) at large p_T. (Data from Matano *et al.*, 1975.)

inclusive reactions follow the QCD prediction, there is at least the hint that as p_T increases the data are moving towards the expected behaviour.

24.3 From quarks to hadrons

We now proceed to a more quantitative interpretation of the picture of a high p_T process given in Fig. 24.1.

24.3.1 Inclusive reactions

In the previous section we assumed that the inclusive cross-section $E_C \mathrm{d}^3\sigma^{(\mathrm{in})}/\mathrm{d}^3 p_C$ for the hadronic process

$$A + B \to C + X \tag{24.3.1}$$

where C has a large p_T in the frame in which A and B are collinear, was proportional to the quark cross-section $a + b \to c + X$. We ignore momentarily the quark transverse momentum and denote by $q_a^A(x_a)$ the number density of constituents of flavour a within the hadron A with longitudinal momentum fraction in the range $x_a \to x_a + \mathrm{d}x_a$ and by

$D_C^c(z)$ the fragmentation function of quark c, i.e. the probability that quark c emits an essentially collinear hadron C with an energy fraction $z \equiv E_C/E_c$ in the range z to $z + \mathrm{d}z$.

A probabilistic interpretation of the diagram of Fig. 24.1 leads to the factorized form

$$\mathrm{d}\sigma(A+B \to C+X) = \\ \sum_{\text{partons}} q_a^A(x_a)\mathrm{d}x_a q_b^B(x_b)\mathrm{d}x_b \frac{\mathrm{d}\hat{\sigma}(a+b\to c+d)}{\mathrm{d}t}\frac{\mathrm{d}\varphi}{2\pi} D_C^c(z)\mathrm{d}z, \quad (24.3.2)$$

where the partonic cross-section $\mathrm{d}\hat{\sigma}(a+b \to c+d)$ is expressed in terms of the Feynman amplitude in the usual way and contains an energy-momentum conserving δ function $\delta^{(4)}(p_a + p_b - p_c - p_d)$. Also, we denote with a caret ^ the 'Mandelstam variables' for the elementary reaction among partons $(a + b \to c + d)$

$$\begin{cases} \hat{s} &= (p_a+p_b)^2 \simeq 2p_a p_b \simeq x_a x_b s \\ \hat{t} &= (p_a-p_c)^2 \simeq -2p_a p_c \simeq x_a t/z \\ \hat{u} &= (p_b-p_c)^2 \simeq -2p_b p_c \simeq x_b u/z \, , \end{cases} \quad (24.3.3)$$

where s, t and u are the 'Mandelstam variables' for the hadronic reaction

$$\begin{cases} s &= (p_A+p_B)^2 \simeq 2p_A \cdot p_B \\ t &= (p_A-p_C)^2 \simeq -2p_A \cdot p_C \\ u &= (p_B-p_C)^2 \simeq -2p_B \cdot p_C \, . \end{cases} \quad (24.3.4)$$

The above assumes that masses are negligible (which may not be the case for heavy flavour production) in which case

$$\hat{s} + \hat{t} + \hat{u} = 0. \quad (24.3.5)$$

Equation (24.3.2) is rather self-evident; its detailed derivation is given in Blankenbecler *et al.* (1976). Note that for given x_a, x_b, the energy E_c is fixed by energy conservation in the partonic process, and thus, for given E_C, this implies that z is determined. One finds

$$z = \frac{p_C}{x_a x_b \sqrt{s}} \{x_a(1-\cos\theta_C) + x_b(1+\cos\theta_C)\} \, . \quad (24.3.6)$$

Equation (24.3.2) may be recast to exhibit the invariant phase space element d^3p_C/E_C. Assuming collinearity between the quark c and the hadron C, we have, using (24.3.3) and (24.3.4)

$$\frac{z}{2}\,\mathrm{d}z\,\mathrm{d}\hat{t}\,\mathrm{d}\phi \simeq z\mathrm{d}z\, E_c^2 \mathrm{d}\Omega = E_C \mathrm{d}E_C \mathrm{d}\Omega = \frac{\mathrm{d}^3 p_C}{E_C} \quad (24.3.7)$$

so that eqn (24.3.2) becomes

$$E_C \frac{d^3\sigma(AB \to CX)}{d^3 p_C} = \frac{1}{\pi} \sum \int_0^1 dx_a \int_0^1 dx_b q_a^A(x_a) q_b^B(x_b) \times$$
$$\times \frac{1}{z} D_C^c(z) \frac{d\hat{\sigma}(ab \to cd)}{d\hat{t}}. \qquad (24.3.8)$$

Hadronization of the parton d and of the remainder of the incoming hadrons is implicit in the summation over all partons a, b, c and d in (24.3.2) or (24.3.8). As long as p_T is sufficiently large, the possibility that hadron C is produced as one of the remnants of the incoming hadrons is negligible.

The above formulae can, in principle, be generalized to take account of the transverse momentum components p_{Ta} and p_{Tb} of partons a and b inside hadrons A and B as well as a possible transverse momentum of C relative to c, by introducing factors like

$$\tilde{q}_a^A(p_{Ta})\, \tilde{q}_b^B(p_{Tb})\, w_C^c(p_{Tc})\, d^2p_{Ta}\, d^2p_{Tb}\, d^2p_{Tc}$$

under the integral sign.

It should be stressed that if the above complications can be ignored, if $d\hat{\sigma}/d\hat{t}$ is indeed perturbatively calculable and if eqn (24.3.8) is trustworthy, then the use of distribution functions taken from deep inelastic scattering leaves only the fragmentation function unknown in our calculation of inclusive hadronic cross-sections. In practice, this assumes a reliable knowledge of *all* distribution functions which, as already discussed (see Chapter 17), has not yet been achieved.

The fragmentation function $D_C^c(z)$ can be studied in any reaction in which one hadron of type C is produced by one parton of type c. For instance, within the parton treatment of lepton induced inelastic scattering, the determination of $D_C^c(z)$ would necessitate measuring semi-inclusive processes such as

$$lh \to l'CX, \qquad (24.3.9)$$

where h is any hadron. Compared with the standard deep inelastic reaction studied previously,

$$lh \to l'X \qquad (24.3.10)$$

reaction (24.3.9) requires three additional variables to describe the process. Semi-inclusive data of this type should come from HERA.

Similarly, $D_C^c(z)$ could be determined from inclusive neutrino processes $\nu h \to CX$ or from $e^+e^- \to CX$. The latter corresponds to taking $q_a^A(x) \to q_{e^-}^{e^-}(x) = \delta(1-x)$, $q_b^B = q_{e^+}^{e^+}(y) = \delta(1-y)$ and $z = 2p_C/\sqrt{s}$

in (24.3.8) so that one obtains

$$\frac{\mathrm{d}\sigma(\mathrm{e}^+\mathrm{e}^- \to CX)}{\mathrm{d}z} = \sum_c \sigma(\mathrm{e}^+\mathrm{e}^- \to c\bar{c}) D_C^c(z) \qquad (24.3.11)$$

which provides a rather direct determination of $D_C^c(z)$.

The naive parton formulae (24.3.8), (24.3.11) retain their formal structure when QCD corrections are taken into account provided that the distribution functions $q(x)$ are replaced by their Q^2-dependent counterparts and α_s in the parton cross-sections is replaced by the effective coupling constant $\alpha_s(Q^2)$ with $Q^2 \simeq p_T^2$. This highly non-trivial result was explained in Chapter 23 where it was emphasized that care must be taken to utilize the same renormalization scheme in calculating the partonic cross-sections and in defining the Q^2-dependent parton distributions.

Equation (24.3.8) thus becomes

$$E_C \frac{\mathrm{d}^3\sigma(AB \to CX)}{\mathrm{d}^3 p_C} =$$
$$\frac{1}{\pi} \sum \int_0^1 \mathrm{d}x_a \mathrm{d}x_b q_a^A(x_a; Q^2) q_b^B(x_b; Q^2) \frac{1}{z} D_C^c(z; Q^2) \frac{\mathrm{d}\hat{\sigma}(ab \to cd)}{\mathrm{d}\hat{t}}, \qquad (24.3.12)$$

where the sum runs over all partons (quarks and gluons).

The case of jet production will be further discussed in Sections 24.5 and 24.7.

24.3.2 Exclusive reactions

The generalization of the previous scheme to exclusive reactions like

$$A + B \to C + D \qquad (24.3.13)$$

depends upon whether the use of perturbative QCD is justified, i.e. if Q^2 is large enough or $\alpha_s(Q^2)$ is small enough. It is believed that this is the case if we restrict our analysis to large angle scattering at high energies. The main scheme that has been proposed (Brodsky and Farrar, 1975; Brodsky and Lepage, 1979) assumes again factorization of the basic hadronization process. However, there is a crucial difference.

While for inclusive processes it was assumed that one could add elementary *cross-sections*, for exclusive reactions one must sum scattering amplitudes and therefore requires wave-functions rather than just number densities. Thus, for reaction (24.3.13) we define

$$\begin{aligned} M^{AB\to CD}(s,\theta) &= \sum_{i,j,k,l} \int [\mathrm{d}x_a][\mathrm{d}x_b][\mathrm{d}x_c][\mathrm{d}x_d] \times \\ &\times \Phi_C^*(x_{c_1}, x_{c_2}, \ldots, x_{c_k}, \ldots; Q^2) \Phi_D^*(x_{d_1}, x_{d_2}, \ldots, x_{d_l}, \ldots; Q^2) \times \\ &\times \hat{M}(x_{a_i}, x_{b_j}, x_{c_k}, x_{d_l}; s; \theta) \times \end{aligned}$$

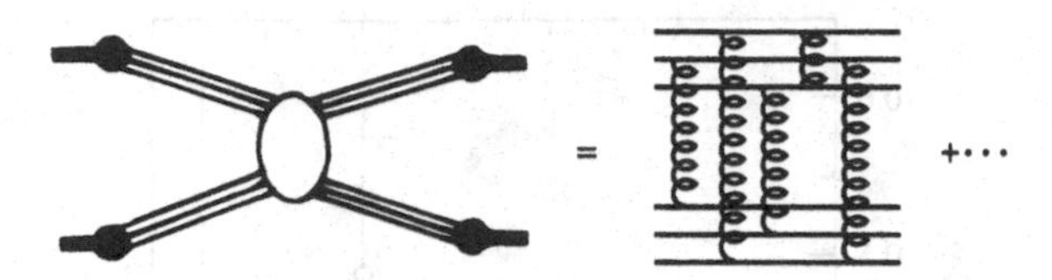

Fig. 24.6.

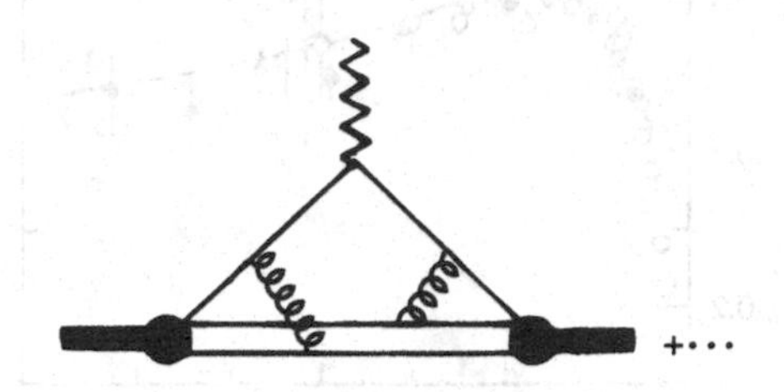

Fig. 24.7.

$$\times \Phi_A(x_{a_1}, x_{a_2}, \ldots, x_{a_i}, \ldots; Q^2) \times$$
$$\times \Phi_B(x_{b_1}, x_{b_2}, \ldots, x_{b_j}, \ldots; Q^2), \qquad (24.3.14)$$

where, for example,

$$[\mathrm{d}x_a] = \prod_{i=1}^{n_a} \mathrm{d}x_{a_i}\, \delta\left(1 - \sum_{i=1}^{n_a} x_{a_i}\right) \qquad (24.3.15)$$

and n_a is the number of valence quarks a in hadron A. In (24.3.14), $\Phi_A(x_{a_i}, Q^2)$ is an effective wave-function which contains all non-perturbative and hadronization effects. The squared modulus of this function, integrated over all but one of the x_a is, essentially, the number density of partons $q_a^A(x_a, Q^2)$ introduced previously.

In general, there is a large number of Feynman diagrams for each reaction. For example, in pp $\to$ pp we require the scattering of three quarks on three quarks (see Fig. 24.6). So, very little is known quantitatively. The scheme has met with some success when the situation is reasonably simple, e.g. in the behaviour of the electromagnetic form factor of the nucleon at large q^2. One draws the lowest order diagram (Fig. 24.7) which leads for the proton form factor to

$$G_M^p(Q^2) \propto \left[\alpha_s(Q^2)/Q^2\right]^2. \qquad (24.3.16)$$

The Q^2 dependence (see Fig. 24.8) seems to be in agreement with the data (Arnold and Reno, 1989). In fact, there is some question as to whether the result emerging from a detailed calculation has the right magnitude (Isgur and Llewellyn-Smith, 1984).

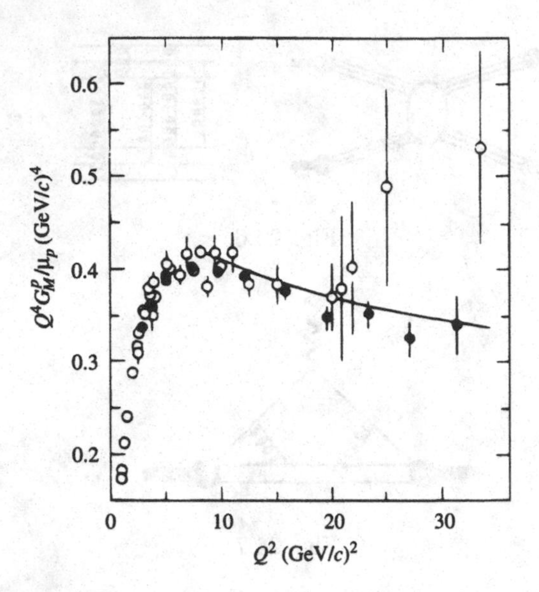

Fig. 24.8. Extracted values of $Q^4 G_M^p/\mu_p$ vs Q^2. Open circles show older data. The curve shows a perturbative QCD prediction for $\Lambda_{QCD} = 100$ MeV.

As an example of the complications in scattering reactions, for pp $\to$ pp in lowest order QCD, one must exchange a minimum of five gluons amongst the six valence quarks in all possible ways (Fig. 24.6). There results about 500,000 different diagrams! Also, singularities arise when integrating over $[\mathrm{d}x_i]$.

Qualitatively, however, one can resort to 'counting rules', i.e. to simply counting the minimal number of exchanges in a perturbative approach. For reaction $A\,B \to C\,D$ one finds on this basis at fixed θ and high energy

$$\frac{\mathrm{d}\sigma(AB \to CD)}{\mathrm{d}t} \sim f(\theta)\left(\frac{\alpha(s)}{s}\right)^{n_A+n_B+n_C+n_D-2}, \qquad (24.3.17)$$

where n_h is the number of constituents of particle h ($n = 1$ for leptons, $n = 2$ for mesons and $n = 3$ for baryons). As an example, for the elastic reaction pp $\to$ pp, at large angles one gets

$$\frac{\mathrm{d}\sigma(\mathrm{pp} \to \mathrm{pp})}{\mathrm{d}t} \propto (\alpha(s)/s)^{10}. \qquad (24.3.18)$$

Fig. 24.9 shows the good agreement between eqn (24.3.18) and the very large s data (dashed lines). At intermediate energies, however, eqn (24.3.18) fails to give good agreement with the data.

The so-called helicity conservation rule is another result which follows from the scheme discussed above as a consequence of the vector nature of

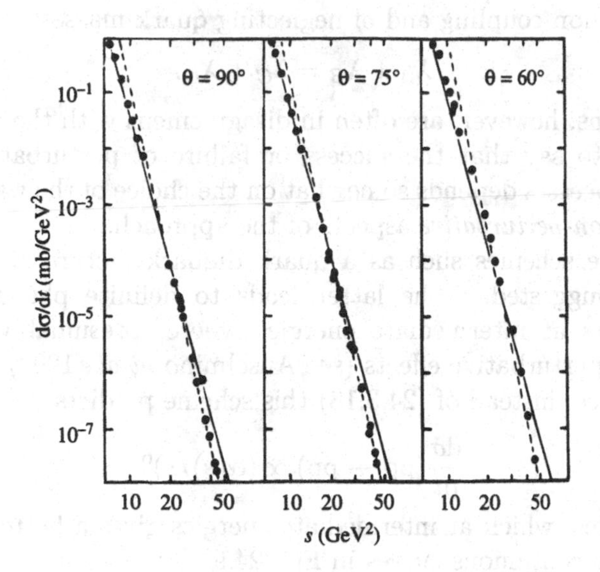

Fig. 24.9. pp → pp data at fixed θ as a function of s. For the curves, see text.

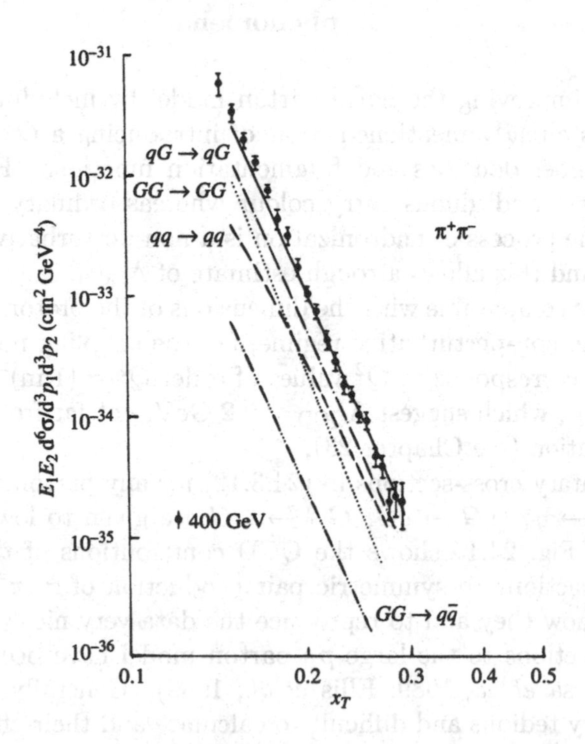

Fig. 24.10. Comparison of QCD calculation with data at 400 GeV on symmetric pair production. (From Jacob, 1979.)

the quark–gluon coupling and of neglecting quark masses:

$$\lambda_A + \lambda_B = \lambda_C + \lambda_D. \tag{24.3.19}$$

Its predictions, however, are often in disagreement with the data.

It is fair to say that the success or failure of perturbative QCD in exclusive processes depends somewhat on the choice of the wave function, i.e. on the *non-perturbative* aspects of the approach.

Alternative schemes such as a quark–diquark picture of the nucleon have been suggested. The latter leads to definite phenomenological improvements at intermediate energies where, presumably, a diquark mimics non-perturbative effects (see Anselmino *et al.*, 1992).

For instance, instead of (24.3.18) this scheme predicts

$$\frac{\mathrm{d}\sigma}{\mathrm{d}t}(\mathrm{pp} \to \mathrm{pp}) \propto (\alpha(s)/s)^6 \tag{24.3.20}$$

at fixed angles, which at intermediate energies gives a better account of the data (see continuous curves in Fig. 24.9).

24.4 Comments on the QCD interpretation of large p_T phenomena

One effect of improving the naive parton model by including QCD corrections is, as already mentioned, that of introducing a Q^2 dependence in parton number densities and fragmentation functions. Furthermore, in QCD, quarks and gluons carry colour whereas ordinary hadrons are colourless. The process of hadronization is a non-perturbative 'large distance' effect and this allows a rough estimate of Λ_{QCD}.

At distances comparable with the dimensions of the proton ($\sim$ 1fm), we must be in the non-perturbative regime, i.e. the coupling must be large. This distance corresponds to Q^2 values of order $Q^2 \approx (1\mathrm{fm})^{-2}$ $\simeq (0.2\,\mathrm{GeV}/c)^2$, which suggest $\Lambda_{\mathrm{QCD}} \approx 0.2$ GeV, not far from the experimental indication (see Chapter 23).

The elementary cross-sections in (24.3.12) for any partonic process $qq \to qq$, $q\bar{q} \to q\bar{q}$, $qG \to qG$, $GG \to GG$ are given to lowest order in Appendix 7. Fig. 24.10 shows the QCD contributions of these various elementary reactions to symmetric pair production of $\pi^+\pi^-$ in $\mathrm{p}Be \to \pi^+\pi^- X$ and how they add to reproduce the data very nicely.

$O(\alpha_s^3)$ corrections to the large p_T parton model have been calculated recently (Aversa *et al.*, 1989; Ellis *et al.*, 1989). Generally, non-leading orders are very tedious and difficult to calculate and their effect is sometimes taken into account by introducing correction factors, the so-called K factors already encountered (Section 17.4, see also Furman, 1982). Attempts have been made to calculate K to all orders.

24.4.1 Evidence for jets

The first direct measurements of hadronic jets were performed at the CERN ISR by triggering on the transverse energy deposition in electromagnetic and hadronic calorimeters with coverage of a sufficiently large solid angle compared with the size of the expected jet so as to provide clean evidence of the latter.

It was realized at a very early stage of large p_T physics that events with one large p_T particle (the 'trigger') can, roughly speaking, be interpreted as consisting of three general components:

(i) a component similar to typical small momentum transfer processes, but at a scaled-down energy

$$\sqrt{s_{\text{eff}}} \simeq \sqrt{s} - 2p_T \qquad (24.4.1)$$

(ii) a component including the large p_T trigger particle and a few other particles with small transverse momentum relative to it,

(iii) a component made up of particles produced over a fairly large cone with a mean azimuth opposite to that of the trigger particle.

These components are interpreted as the beam and the target jet, the towards jet and the away jet, respectively, which were anticipated as a consequence of the parton model (see Fig. 24.1).

The first component is expected on the ground that triggering on just one large p_T particle can hardly spoil the leading particle effect typical of multiple production at high energy. In the CM one will then have a narrow cone of particles in the forward (beam jet) and backward (target jet) directions. Nothing new is to be learned from these jets once the energy has been appropriately scaled down (24.4.1). One usually ignores these jets when specifying the number of jets in an event.

To the extent that the dominant process is either $qq \to qq$ or $q\bar{q} \to q\bar{q}$, one expects jet physics to be dominated by two jets. This will indeed turn out to be so up to certain energies. We postpone to Section 25.2.1 the discussion of the expected angular distribution of these two jets. Higher order contributions, such as gluon radiation (e.g. $q\bar{q} \to q\bar{q}G$) will become increasingly important with increasing energy.

It should be realized that in talking about a jet we are sidestepping some severe problems; for example, by what mechanism does the coloured quark neutralize its colour quantum number in order to produce a colour singlet jet of real physical hadrons? There are also technical problems in defining both theoretically and experimentally what constitutes a jet and in avoiding biases in jet identification. Even the total momentum of a jet cannot be easily measured because of the difficulty in detecting neutral particles.

A natural assumption is that those fragments which follow the trigger particle define the towards jet while those on the opposite side define the away jet. We shall see in the following how one attempts to measure the 'jettiness' of the distribution of particles.

The trigger side jet axis is defined by the vector sum of the momenta of all the particles (trigger included) seen in the trigger region:

$$\boldsymbol{p}_{\text{jet}} = \sum_i \boldsymbol{p}_i \tag{24.4.2}$$

With this definition, and given that one is triggering on a particle with large p_T, the jet axis is essentially along the trigger direction since it is experimentally observed that the trigger gives the dominant contribution to the vector sum (24.4.2), i.e. $\sum_i \boldsymbol{p}_i \simeq \boldsymbol{p}_{\text{trigger}}$. The fact that one is triggering on a particle with large p_T could be responsible for a bias in identifying jet formation. This is often referred to as 'trigger bias'. The first evidence for jet production using a technique that was free from instrument bias was obtained by UA2 at the CERN $\bar{\text{p}}$p collider (UA2, 1982, 1984).

The distribution of the observed total energy deposited in a calorimeter transverse to the beam (the total transverse energy $\sum E_T$) measured by the UA2 calorimeter, Fig. 24.11, shows a clear departure from an exponential when $\sum E_T$ exceeds 60 GeV.

An investigation of the pattern of energy distribution in the events was also carried out by UA2 by tagging clusters of energy deposition in each event, and by ordering them in decreasing transverse energies ($E_T^1 > E_T^2 > E_T^3$). Fig. 24.12 shows the mean values of the fractions $h_1 = E_T^1/\sum E_T$ and $h_2 = (E_T^1 + E_T^2)/\sum E_T$ as a function of $\sum E_T$. As $\sum E_T$ increases (i.e. for $\sum E_T > 60$ GeV), a very substantial fraction of $\sum E_T$ is shared, on the average, by two clusters with roughly equal transverse energies. These energy clusters appear to be associated with collimated multiparticle systems, as was found by reconstructing the charged-particle tracks in these events.

At the ISR, the direct demonstration of jets in large $\sum E_T$ events was more difficult because of the lower energy. The $\sum E_T$ region where jet–jet events have been found to become dominant at the $\bar{\text{p}}$p collider (~ 60 GeV) was just at the end of the ISR phase space.

Fig. 24.12 shows that the large $\sum E_T$ data are almost saturated by a two-cluster contribution giving clear evidence that two-jet production dominates in this energy interval.

24.4.2 Inclusive jet production

In the spirit of Fig. 24.1 the cross-section for inclusive jet production

$$A + B \to \text{jet} + X \tag{24.4.3}$$

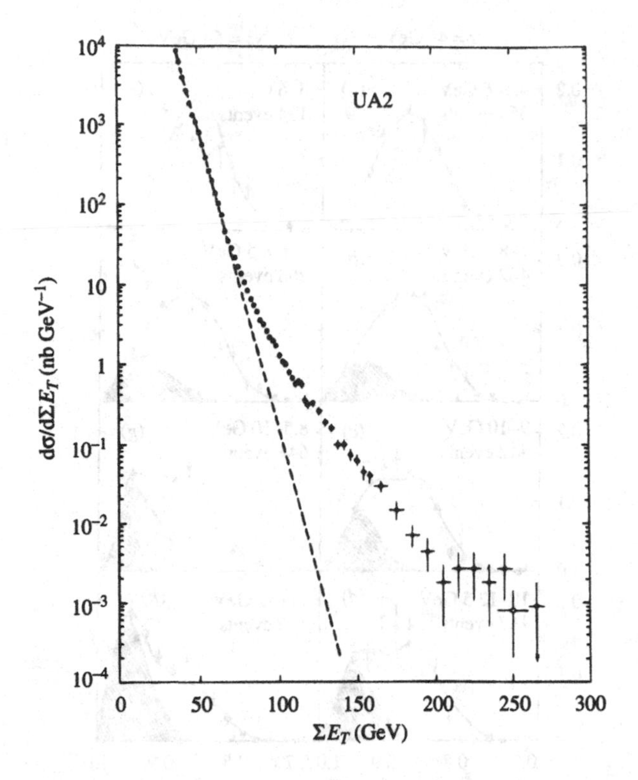

Fig. 24.11. Total transverse energy distribution ($\sum E_T$) measured in UA2 at $\sqrt{s} = 540$ GeV. A clear departure from an exponential behaviour is visible for $\sum E_T > 60$ GeV.

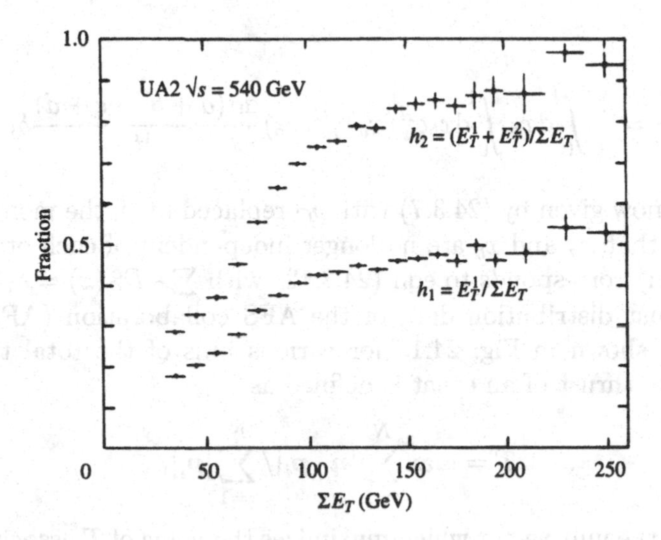

Fig. 24.12. Fraction of total transverse energy carried by the two leading jets $[(E_T^1 + E_T^2)/\sum E_T]$ and by the leading jet.

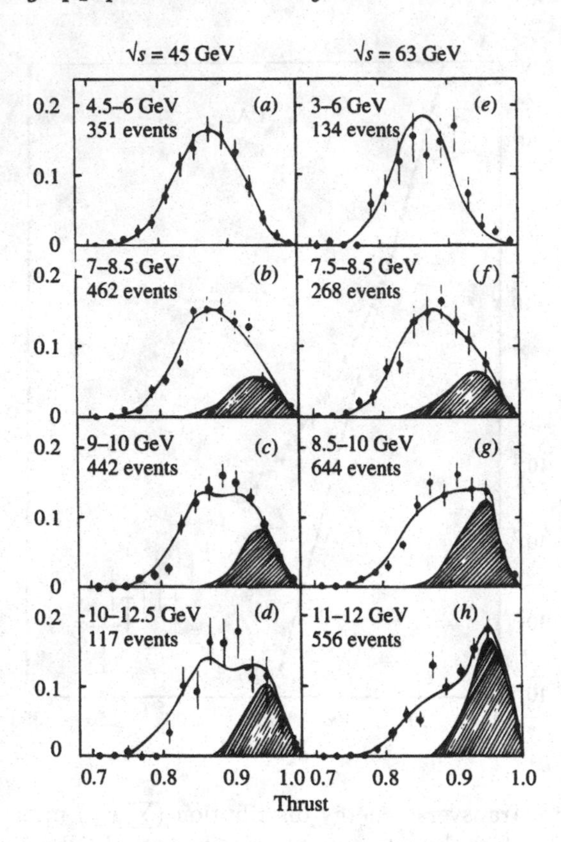

Fig. 24.13. Thrust distribution at two ISR energies for various bins of $\sum E_T$.

is given by

$$E\frac{\mathrm{d}^3\sigma^{\mathrm{Jet}}}{\mathrm{d}^3\boldsymbol{p}} = \frac{1}{\pi}\int_0^1 \mathrm{d}x_a \int_0^1 \mathrm{d}x_b q_a^A(x_a) q_b^B(x_b) \frac{\mathrm{d}\hat{\sigma}(a+b\to c+d)}{\mathrm{d}t}\delta(1-z), \tag{24.4.4}$$

where z is now given by (24.3.7) with p_C replaced by p, the momentum of the jet, so that x_a and x_b are no longer independent of each other. Note that (24.4.4) corresponds to eqn (24.3.12) with $\sum_C D_C^c(z) = \delta(1-z)$.

The thrust distribution data of the AFS collaboration (AFS, 1982a, 1983a) are shown in Fig. 24.13 for various bins of the total transverse energy. The thrust of an event is defined as

$$T = \max \sum_{i=1}^{N} |\boldsymbol{n}\cdot\boldsymbol{p}_i| / \sum_{i=1}^{N} |\boldsymbol{p}_i|, \tag{24.4.5}$$

where $\boldsymbol{n}$ is the unit vector which maximizes the value of T associated with the group of particles in a jet. Thrust is discussed in detail in Chapter 25. The closer T is to 1, the more jet-like the event.

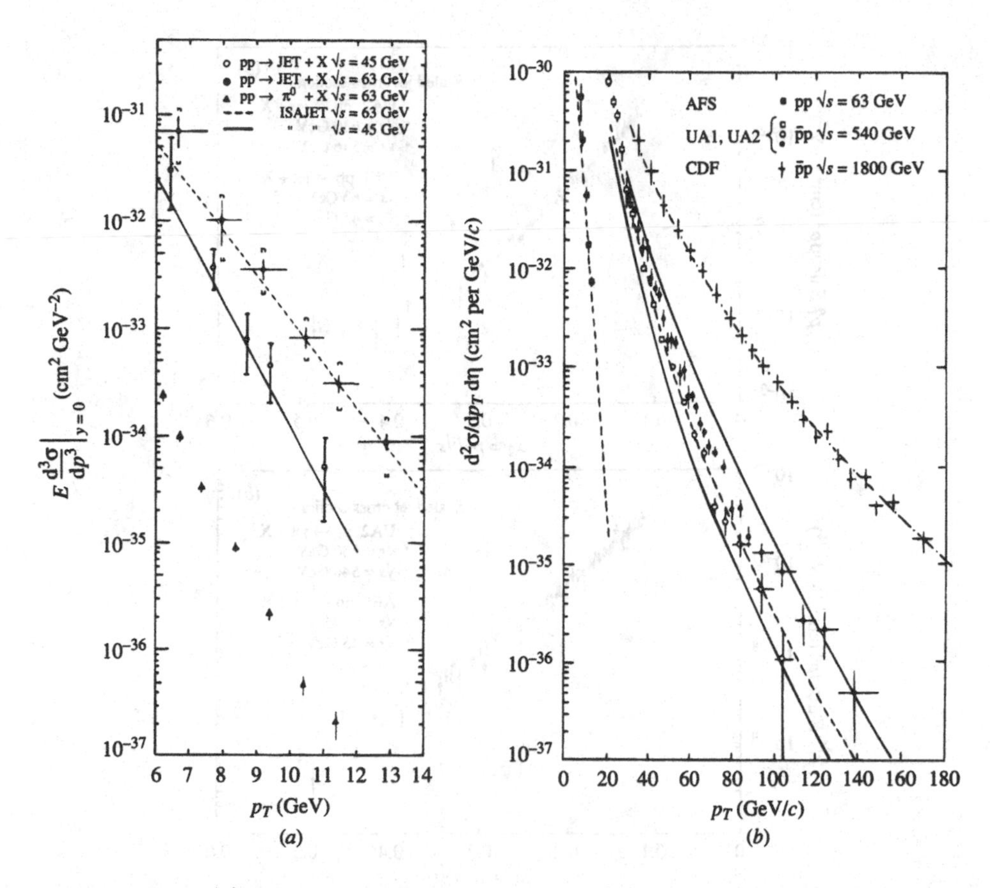

Fig. 24.14. (a) Inclusive jet cross-section at ISR energies vs. p_T. For curves see text. (b) Comparison of inclusive jet cross sections with QCD calculations at various energies. For curves see text.

As can be seen in the data, the jettiness seems to increase with $\sum E_T$ at least at $\sqrt{s} = 63$ GeV. The curves of Fig. 24.13 are QCD-based Monte Carlo calculations and the shaded area shows how the contribution from events with a leading jet increases with $\sum E_T$.

From the ISR data on correlations, it was predicted (Jacob and Landshoff, 1976) that for a given p_T, the true event rate for jet production would be at least two orders of magnitude higher than single particle yields. This is borne out by the data of Fig. 24.14.

Fig. 24.14(a) shows the jet cross-section extracted from the data of Fig. 24.13. These data are compared with the inclusive π^0 yields at $\sqrt{s} = 63$ GeV. The jet/π^0 ratio increases from ~200 to ~1500 as p_T increases from 6 to 14 GeV/c. The agreement with the QCD calculation supplemented by the so-called 'Isajet' hadronization Monte Carlo program is excellent.

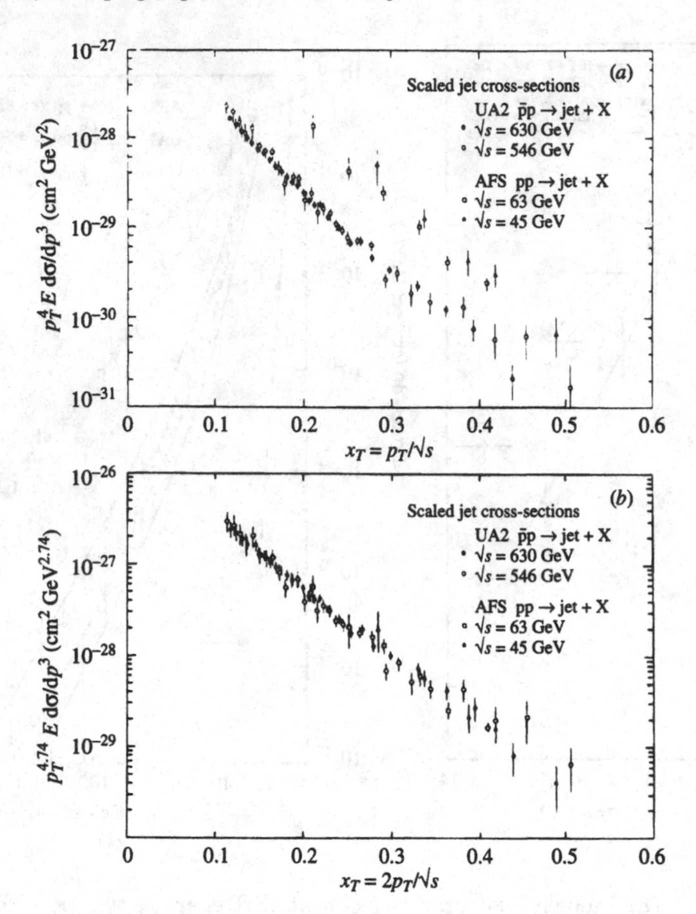

Fig. 24.15. Scaled jet cross-sections (see text).

Fig. 24.14(b) shows the *predicted* (Horgan and Jacob, 1981) dramatic increase of the jet cross-section as $\sqrt{s}$ increases from 63 to 540 to 1800 GeV. The QCD predictions at 540 and 1800 GeV are shown as dashed and dash-dotted lines. The solid curves at 540 GeV indicate the uncertainty in the calculations. The principal uncertainty lies in the choice of the value of Q^2 used in $\alpha_s(Q^2)$ and in $q(x, Q^2)$ and also in the imprecision in our knowledge of the distribution functions.

Note that the data follow the predictions through seven orders of magnitude.

The comparison of these data with the scaled form (24.2.2) is shown in Fig. 24.15. In Fig. 24.15(a) the naive choice $n = 4$ is made. Naive scaling is not so good but $n = 4.74$ gives an excellent scaling and the x_T dependence is well described by

$$f(x_T) = A(1 - x_T)^m / x_T^2 \qquad (24.4.6)$$

We conclude that jet cross-sections conform much better to the asymptotic parton model predictions than the single particle yields discussed in Section 24.2.

24.4.3 Transverse momentum distribution with respect to the jet axis

From our intuitive picture of a jet, one would expect each constituent of a jet (mostly pions) to be produced with limited transverse momentum *relative to the jet axis.* There is some evidence that the particles in the away side jet are produced symmetrically around the jet axis (Angelis *et al.*, 1979).

In Fig. 24.16 the hadrons' mean transverse momentum relative to the jet axis is plotted as a function of the particle momentum for different values of the trigger momentum. This mean value shows little increase (Angelis *et al.*, 1979) as the trigger momentum increases (Fig. 24.16) and is generally in agreement with the value

$$\langle q_T \rangle \simeq 0.55 \pm 0.05 \text{ GeV}/c \tag{24.4.7}$$

quoted by the CERN-Saclay collaboration, where q_T denotes the particle's momentum relative to the jet axis. The distribution of events in $q_{T\theta}$ and $q_{T\varphi}$ has also been analysed. If the latter data are compared with Monte Carlo curves obtained by assuming no correlation of the associated hadrons (neither in θ nor in φ) with the trigger particle, the data are found to lie well below these curves thus showing the existence of rather strong correlations, presumably indicative of a jet structure.

It is worth noting in Fig. 24.16 that there is little dependence on the trigger momentum.

24.5 Two-jet production at large p_T

Two jet cross-sections follow from the various elementary reactions $qq \to qq$, $q\bar{q} \to q\bar{q}$, $qG \to qG$, $GG \to GG$ etc. The squared matrix elements (summed and averaged over spins) for these various reactions are compared in Fig. 24.17 as function of the parton CM scattering angle $\hat{\theta}$.

By generalizing to two jets the treatment of the one-jet case, we write for

$$A + B \to \text{jet}(p_a) + \text{jet}(p_b) + X \tag{24.5.1}$$

$$\frac{\mathrm{d}^3\sigma^{\text{2jets}}}{\mathrm{d}x_a \mathrm{d}x_b \mathrm{d}\cos\theta^*} = \sum_{\text{partons}} q_a^A(x_a) q_b^B(x_b) \frac{\mathrm{d}\hat{\sigma}(ab \to cd)}{\mathrm{d}\cos\hat{\theta}}. \tag{24.5.2}$$

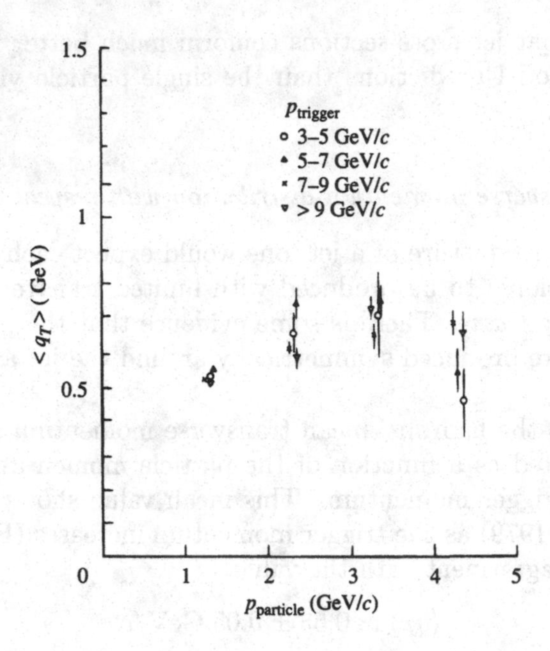

Fig. 24.16. Mean transverse momentum relative to the jet axis vs particle momentum for various values of p_{trigger}. (From Jacob, 1979.)

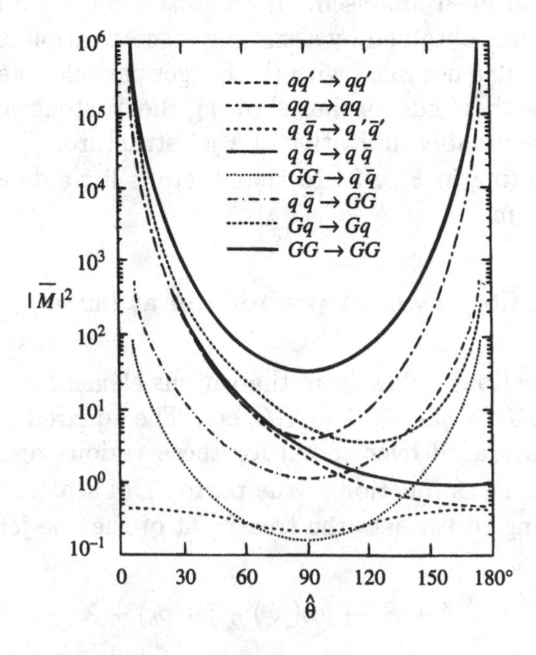

Fig. 24.17. The dependence of the matrix elements squared $\overline{|M|}^2$ on the parton scattering angle for interactions of quarks (q and q'), antiquarks ($\bar{q}$ and $\bar{q}'$) and vector gluons (G) in lowest order QCD.

Here θ^* is the polar angle of the jets relative to the $A-B$ collision axis, as measured in the CM of the two jets. To the extent that we ignore the parton transverse momenta, we should have $\theta^* = \hat{\theta}$. From the above we have several interesting observations.

24.5.1 Jet angular distribution

For each of the elementary reactions of Fig. 24.17, the $\hat{\theta}$ distribution has the familiar Rutherford scattering behaviour at small angles which is characteristic of the t-channel exchange of a vector particle

$$\frac{\mathrm{d}\hat{\sigma}}{\mathrm{d}\cos\hat{\theta}} \propto \left(\frac{1}{\sin^4\hat{\theta}/2}\right) \tag{24.5.3}$$

Since we do not identify the final partons, a given jet seen at angle θ^* could have come equally from parton c at angle $\hat{\theta}$ (see Fig. 24.1) or c at angle $\pi-\hat{\theta}$ which implies d at angle $\hat{\theta}$ ($d \neq c$).

Thus, we must use the sum

$$\frac{\mathrm{d}\hat{\sigma}(\hat{\theta})}{\mathrm{d}\cos\hat{\theta}} + \frac{\mathrm{d}\hat{\sigma}(\pi-\hat{\theta})}{\mathrm{d}\cos\hat{\theta}}. \tag{24.5.4}$$

This corresponds to adding terms with $\hat{u} \leftrightarrow \hat{t}$ in the formula for $\mathrm{d}\hat{\sigma}$ when the final state contains *distinct* partons.

Now it turns out that for these symmetrized cross-sections, the reactions $qq' \to qq'$, $q\bar{q} \to q\bar{q}$, $qG \to qG$, $GG \to GG$ (the latter is automatically symmetric) totally dominate and, moreover, to a very good approximation (Fig. 24.18)

$$\begin{aligned} &qq' \to qq'|_{\mathrm{sym}} = q\bar{q} \to q\bar{q}|_{\mathrm{sym}} = \\ &\quad \tfrac{4}{9}(qG \to qG)_{\mathrm{sym}} = \tfrac{4}{9}(\bar{q}G \to \bar{q}G)_{\mathrm{sym}} = \tfrac{16}{81}(GG \to GG). \end{aligned} \tag{24.5.5}$$

The net result is that the x_a, x_b and $\hat{\theta} = \theta^*$ behaviours factorize, and for a $\bar{\mathrm{p}}\mathrm{p}$ initiated reaction one has

$$\frac{\mathrm{d}^3\sigma^{\mathrm{2jets}}}{\mathrm{d}x_a\mathrm{d}x_b\mathrm{d}\cos\theta^*} \simeq F(x_a)F(x_b)\left(\frac{\mathrm{d}\hat{\sigma}}{\mathrm{d}\cos\hat{\theta}}\right)_{GG\to GG}, \tag{24.5.6}$$

where

$$F(x) = G(x) + \tfrac{4}{9}\left[Q(x) + \bar{Q}(x)\right] \tag{24.5.7}$$

with

$$Q(x) = \sum_{\text{flavour } j} q_j(x). \tag{24.5.8}$$

The jet angular distribution has been measured by the UA1 collaboration (UA1, 1986) in a sample of events with large jet–jet invariant mass. Tests

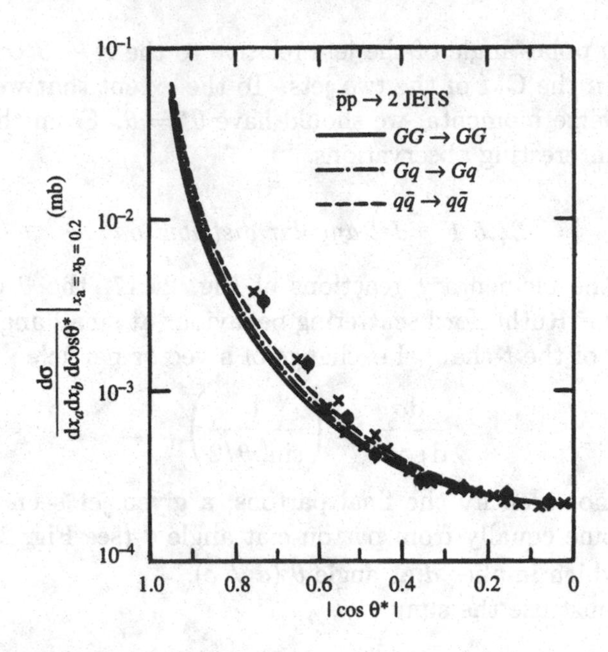

Fig. 24.18. Almost universal distribution of $2 \to 2$ parton cross-sections is compared with two-jet data.

of the angular dependence are shown in Fig. 24.19 at fixed mass bins for the jet–jet invariant mass, i.e. fixed $\sqrt{\hat{s}}$.

Given that the θ^* behaviour seems good, one can integrate over a range of θ^* to improve statistics and then measure $F(x)$. The result of this is discussed in the following.

24.5.2 Tests of the Q^2 evolution

Given the x dependence at some value Q_0^2 one numerically integrates the Altarelli–Parisi eqns (22.2.53, 54) to find the distributions at some higher Q^2.

A fit to the deep inelastic ν and $\bar{\nu}$ data (CDHS) for $F_2(x, Q^2)$ and $xF_3(x, Q^2)$ is shown in Fig. 24.20 (Diemoz *et al.*, 1986). The leading order QCD fit gives $\Lambda_{\rm QCD} \cong 300 \pm 150$ MeV. This 'scale-breaking' was discussed initially in Chapter 17 and in more detail in Section 23.9.

Fig. 24.21 shows the extracted distribution functions at various Q^2 from the CDHS and CHARM collaborations. Finally in Fig. 24.22 we show the combination (24.5.7) extracted from the $\bar{\rm p}{\rm p} \to 2\,$jet analysis of UA1. This corresponds to a huge value of Q^2: $Q^2 \simeq 2000\,({\rm GeV}/c)^2$. It is compared with $F(x, Q^2)$ calculated using distribution functions compatible with those shown in Fig. 24.21 with a choice of two values of Q^2. It seems

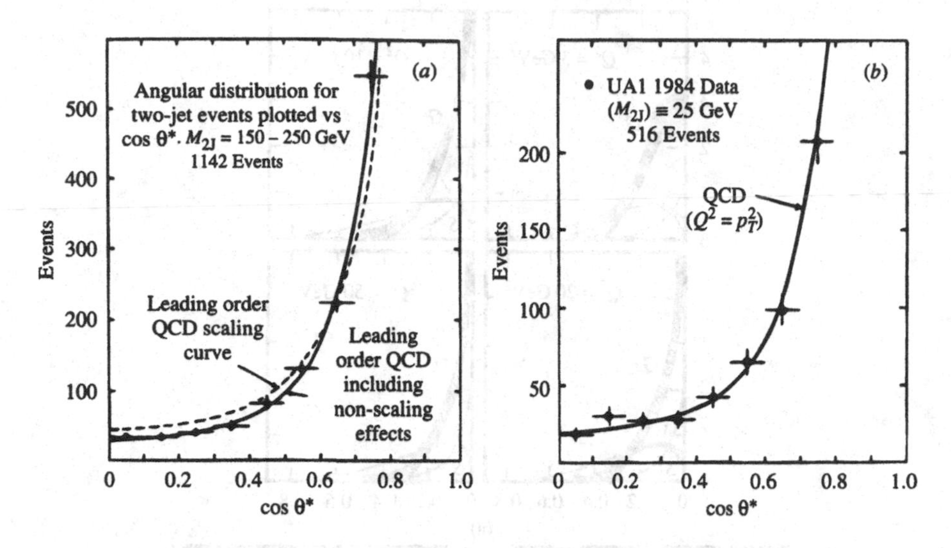

Fig. 24.19. Two-jet angular distribution as a function of $\cos\theta^*$ for (a) m_{2J} = 150–250 GeV and (b) $m_{2J} \approx 25$ GeV.

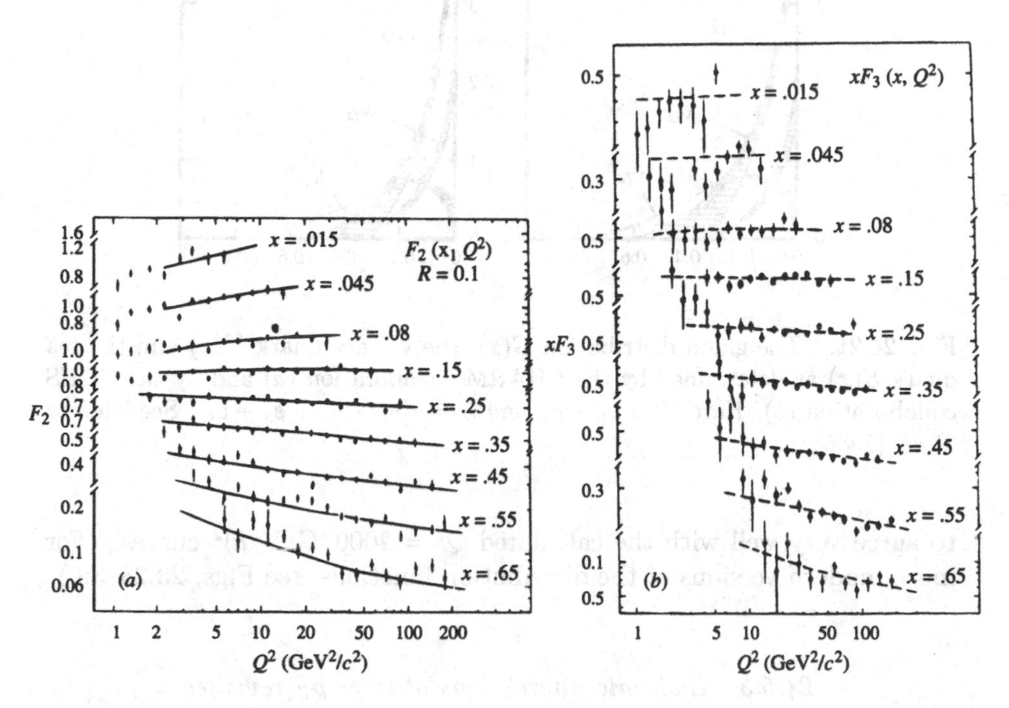

Fig. 24.20. Scaling function versus Q^2 for different bins of x, as measured by the CDHS collaboration. (a) F_2; (b) xF_3. The solid lines are the result of a leading-order QCD fit to the data.

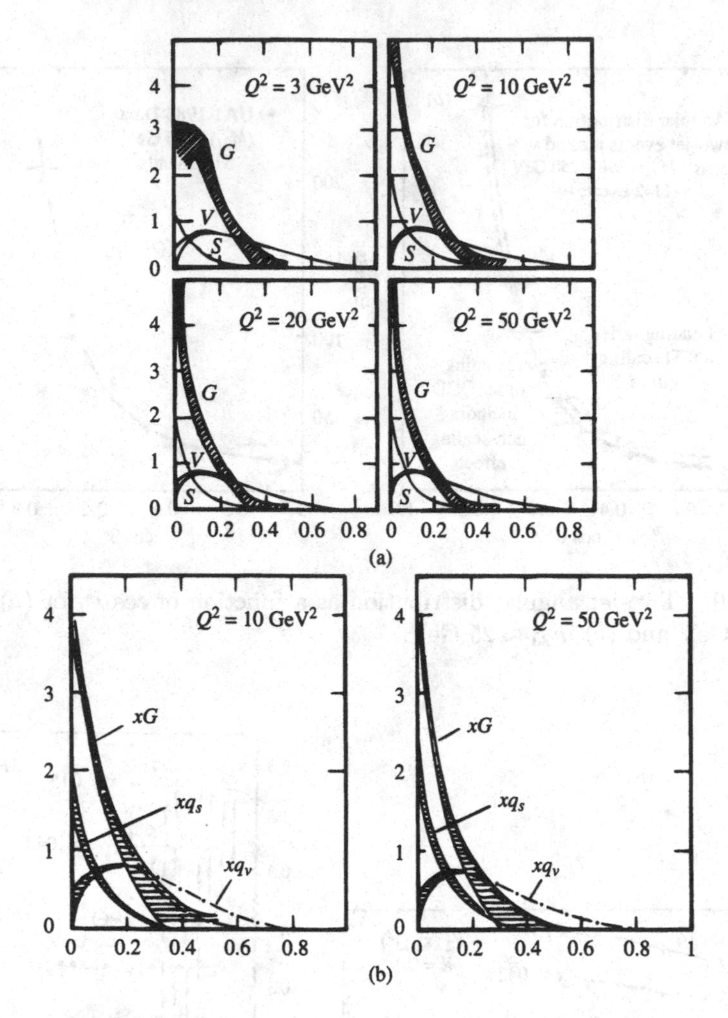

Fig. 24.21. The gluon distribution $G(x)$, the valence quark $V(x)$ and the sea quark $S(x)$ as determined by the CHARM collaboration (*a*) and by the CDHS collaboration (*b*). Here $V \equiv u_v + d_v$ and $S = 2[u_s + d_s + s_s + c_s]$. See Diemoz *et al.* (1986).

to agree very well with the calculated $Q^2 = 2000\,(\text{GeV}/c)^2$ curves. (For more modern versions of the distribution functions, see Figs. 23.28–30.)

24.5.3 Hadronic interactions at large p_T revisited

As was explained in Section 23.10, the correct prescription to include the lowest order, or LLA, corrections to the parton model calculations is to take:

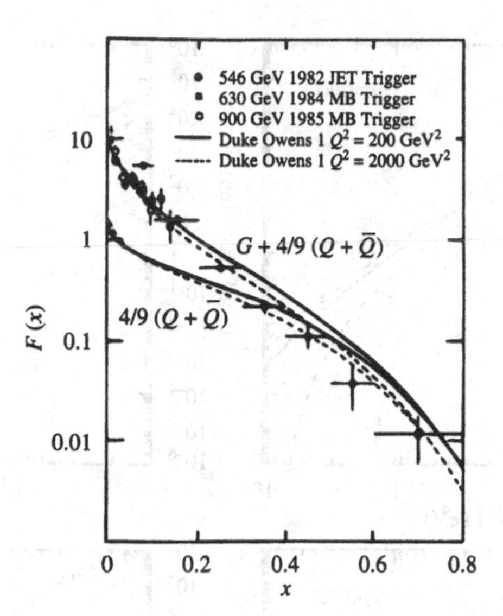

Fig. 24.22. The quantity $F(x)$ (eqn 24.5.7) as obtained from UA1 data in $\bar{\text{p}}\text{p} \to 2\text{jets}$ at $Q^2 \simeq 2000\,(\text{GeV}/c)^2$. Curves are distribution functions calculated for various values of Q^2.

(a) distribution and fragmentation functions appropriate to the size of p_T^2, i.e. $q(x, Q^2)$ with $Q^2 = p_T^2$,

(b) lowest order parton–parton cross-sections but with $\alpha_s \to \alpha_s(Q^2)$.

The latter does not have a major effect on the angular distribution. The consequences of (a) are important. Fig. 24.22 already shows better agreement with the calculated distribution functions at the relevant scale $Q^2 \simeq 2000\,(\text{GeV}/c)^2$. Consider what might happen at a Super Collider: say 20 TeV + 20 TeV. In principle, one might expect events like $\text{p} + \text{p} \to \text{jet}(p_1) + \text{jet}(p_2)$ with, for CM $y \simeq 0$, $p_1 \sim 20$ TeV/c. But since the distributions are shifting to the left, more and more partons have smaller and smaller x. So most partonic collisions will be at much lower energies, i.e. $\hat{s} \ll s$.

This is seen in Fig. 24.23 (Eichten *et al.*, 1984); τ is defined as $\tau = \hat{s}/s$ and $(\tau/\hat{s})\mathrm{d}\mathcal{L}/\mathrm{d}\tau$ is the number of parton–parton collisions of a given $\hat{s}$ per unit range of $\hat{s}$ per hadronic collision. The curves correspond to various CM energy values of the collider (in TeV).

24.6 Prompt photons

If in a QCD diagram one of the strong couplings is replaced by an electromagnetic one a direct photon is produced.

These hard photon reactions provide one of the very few precision tests

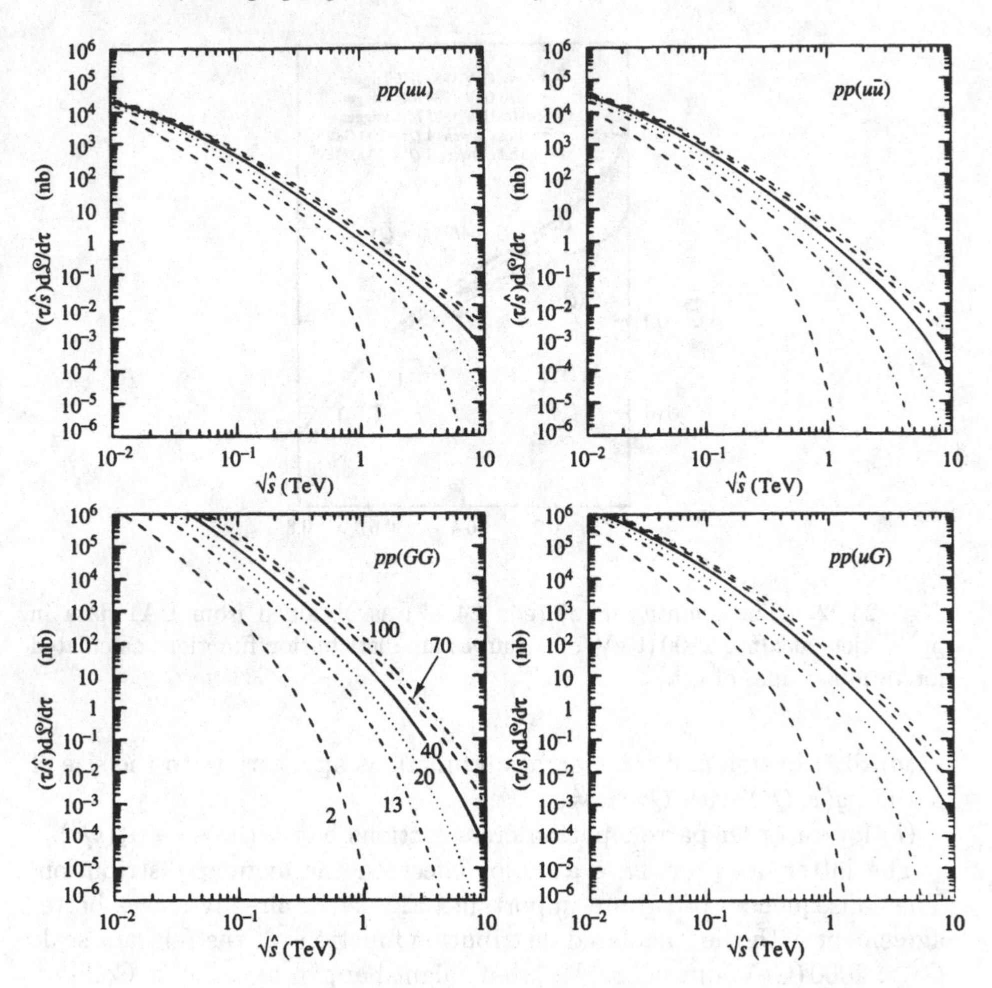

Fig. 24.23. The number of parton–parton collisions $(\tau/\hat{s})d\mathcal{L}/d\tau$ as function of $\hat{s}^{1/2}$. See text (from Eichten *et al.*, 1984).

of QCD in hadronic collisions. The lowest order diagrams are shown in Fig. 24.24. Either of these two diagrams may dominate depending on the nature and on the energy of the colliding hadrons. A competing process at order α_s would be the em bremsstrahlung from a scattered quark. In this case, however, a jet would accompany the photon. This contribution is rather small (AFS, 1982b, 1983b; CCOR, 1981).

The major experimental problem is to distinguish the direct photon from γs originating from the decay of π^0s and ηs. Use is made of the fact that the direct γ has a lower conversion probability than the γs from π^0 decay. Fig. 24.25 shows the γ/π^0 ratio compared to QCD estimates (solid curve for pp and dashed one for $\bar{\text{p}}$p). Fig. 24.26 shows the UA2 p_T distribution of prompt photons (Lubrano, 1990), compared with the

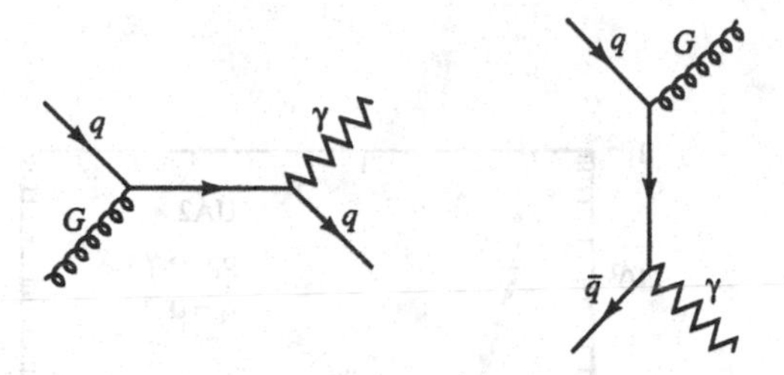

Fig. 24.24. Dominant diagrams for prompt photon production.

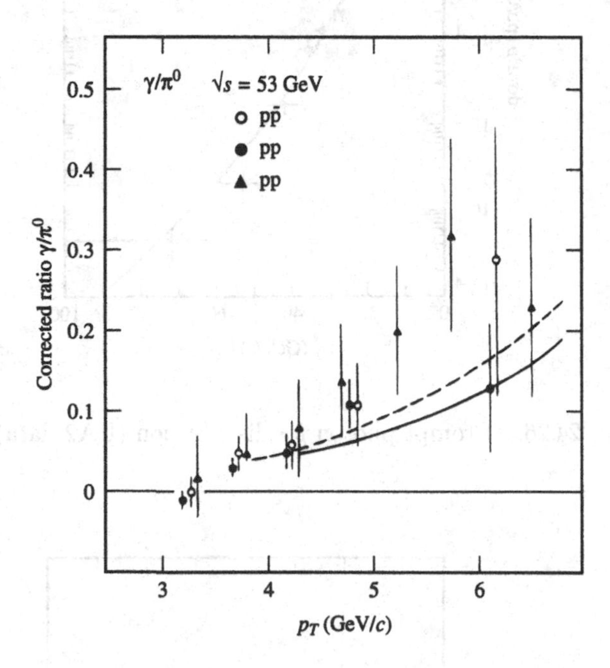

Fig. 24.25. γ/π^0 ratio as a function of p_T at $s^{1/2} = 53$ GeV for pp and p̄p compared with QCD predictions (solid curve pp, dashed curve p̄p).

next to leading order QCD predictions from Aurenche *et al.* (1984). The agreement is remarkable.

Since both leading order processes (Fig. 24.24) involve fermion exchange (in the s and t channels, respectively), the angular distribution is less singular at small angles than the corresponding jet distribution of equation (24.5.3). Indeed, one finds

$$\frac{\mathrm{d}\sigma(\gamma+\mathrm{jet})}{\mathrm{d}\cos\theta^*} \Big/ \frac{\mathrm{d}\sigma(\pi^0+\mathrm{jet})}{\mathrm{d}\cos\theta^*} \propto 1 - \cos\theta^* \qquad (24.6.1)$$

which is in agreement with the UA2 data (Fig. 24.27).

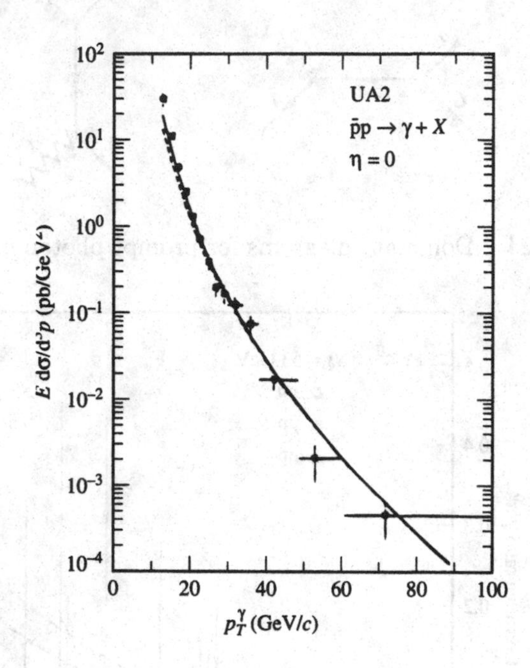

Fig. 24.26. Prompt photon p_T distribution (UA2 data).

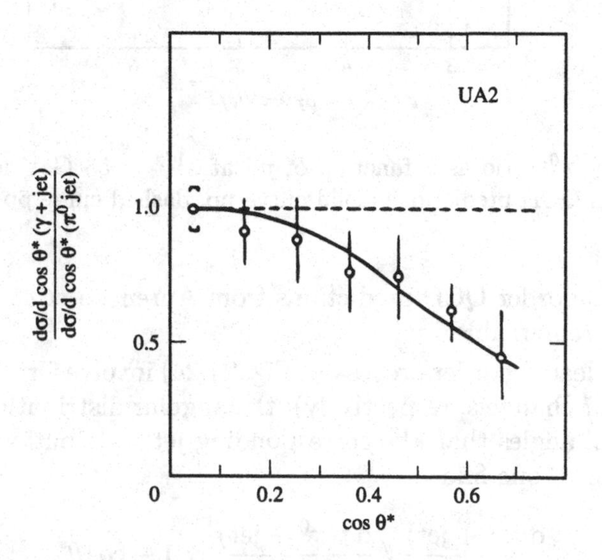

Fig. 24.27. Angular distribution to fit the UA2 data on prompt photons.

24.7 Two and more jets in the final state

Two jet events in hadronic reactions have long been detected. Three and more jets are expected, within QCD, from gluon radiation. The lowest order QCD cross-section for n-jet production is proportional to $[\alpha_s(Q^2)]^n$.

Thus, in principle, clean detection of three jet events may allow a measurement of $\alpha(s)$ via, e.g.

$$\frac{N(3\,\text{jet events})}{N(2\,\text{jet events})} = \left(\frac{\sigma^{3\text{jet}}}{\sigma^{2\text{jet}}}\right)_{\text{QCD}} \propto \alpha_s(Q^2). \tag{24.7.1}$$

Even assuming that the various uncertainties (such as luminosity uncertainty, energy scale uncertainty etc.) cancel in the above ratio, what is determined in (24.7.1) is really $\alpha_s K_3/K_2$ where K_i are the usual K factors due to the higher order corrections to $\sigma^{2\text{jet}}$ and $\sigma^{3\text{jet}}$ (see Section 23.11).

In addition, it is far from unambiguous what precise scale value of Q^2 should be put into $\alpha_s(Q^2)$. For example, the 'obvious' choice $Q^2 = s$ leads to difficulties as we shall now show.

From the experimental results (UA1, 1985; UA2, 1985)

$$\alpha_s K_3/K_2 = 0.24 \pm 0.01 \pm 0.04\ (\text{UA2}) \tag{24.7.2}$$

$$= 0.23 \pm 0.02 \pm 0.04\ (\text{UA1}) \tag{24.7.3}$$

If now we take $K_3/K_2 = 1 + O(\alpha_s)$, the value of $\alpha_s(Q^2)$ emerging from (24.7.2) is much higher than the expected value of $\alpha_s(Q^2 = s = (546\ \text{GeV})^2)$. The analogous question in e^+e^- collisions is discussed in Chapter 25.

Turning now to events with more than three jets, it is only at Tevatron energies that unequivocal evidence for four-jet events has been gathered. In Fig. 24.28 the two-jet and four-jet lego plot events as seen by the CDF detector appear impressively clean (CDF, 1989b).

24.8 Jet fragmentation

The study of jet fragmentation gives, in principle, a means to test several QCD assumptions, such as whether the jets are independent of the initial state (qq or e^+e^-), Q^2 scale-breaking effects etc. As an example, Fig. 24.29 shows the longitudinal jet fragmentation function

$$D(z) = \frac{1}{N_{\text{jet}}}\frac{dn}{dz} \tag{24.8.1}$$

as measured by the AFS collaboration at the ISR (Cavasinni, 1990) where n is the number of particles at $z = 2p_L/M$ (in this case $N_{\text{jet}} = 2$; p_L is the momentum component of the particle along the jet direction and M

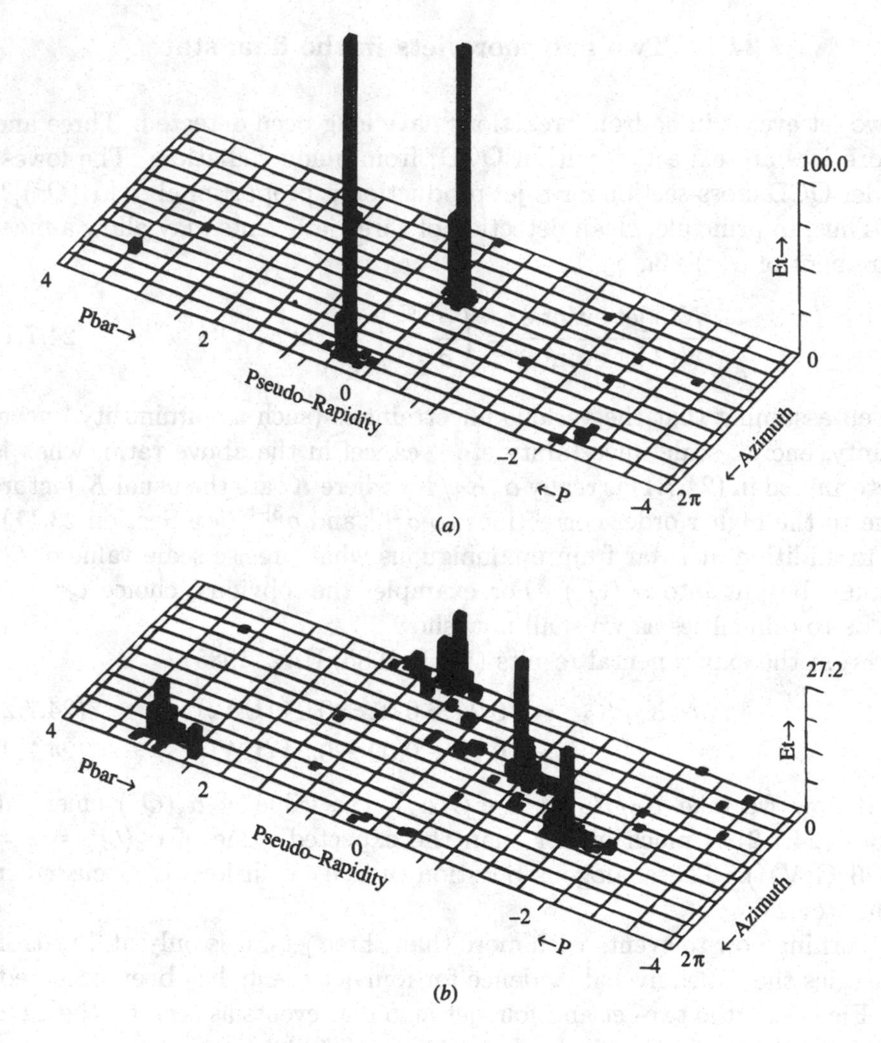

Fig. 24.28. (*a*) One of the highest E_T jet events seen at the CDF detector, with an invariant mass of approximately 630 GeV. (*b*) A four-jet event as seen in the CDF detector. All four jets have E_Ts in excess of 40 GeV.

the di-jet invariant mass; $M/2$ is a measure of the jet energy). This curve shows good agreement between ISR and e^+e^- data at comparable CM energies. The disagreement with the much higher energy UA1 data suggests that here a large contribution from gluon jets is occurring.

24.9 Comments on $O(\alpha_s^3)$ corrections and conclusions

Calculations of the hadronic yields to next-to-leading order α_s^3 have recently become available (Ellis *et al.*, 1989, 1990; Aversa *et al.*, 1989, 1990).

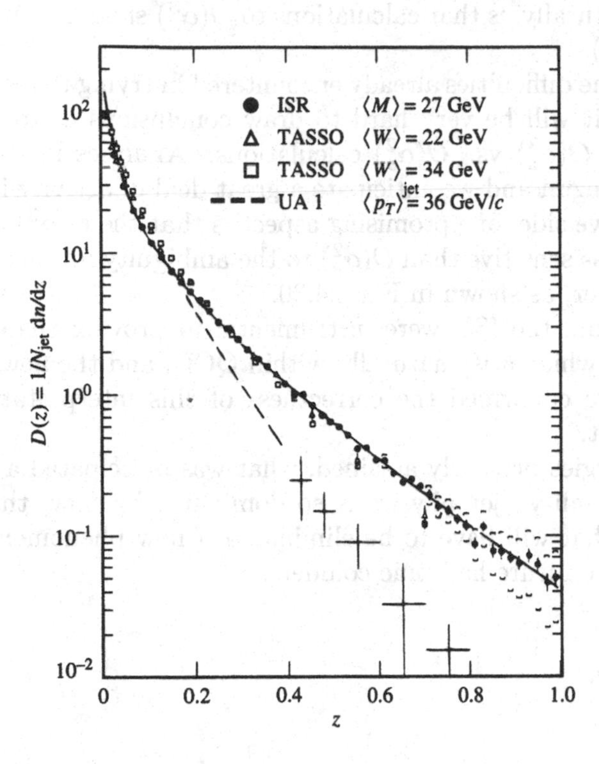

Fig. 24.29. Jet fragmentation as function of z; data from the ISR and UA1 compared with e^+e^- data.

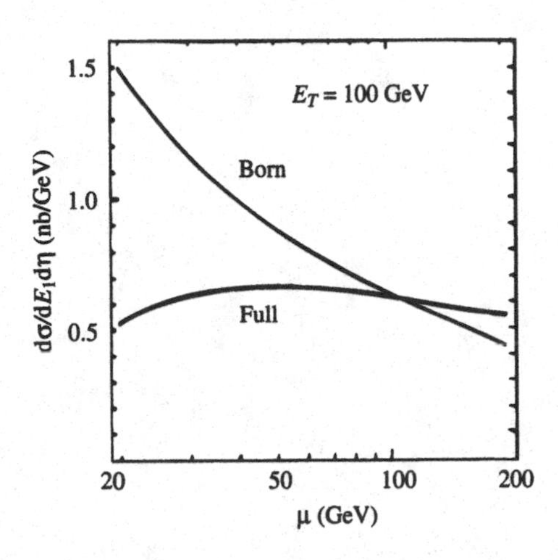

Fig. 24.30. The scale dependence of the inclusive jet yield at the Born-term level, i.e. $O(\alpha_s^2)$, and at the next level, $O(\alpha_s^3)$; $E_T = 100$ GeV.

The hope, naturally, is that calculations to $O(\alpha_s^3)$ should be more reliable than to $O(\alpha_s^2)$.

However, the difficulties already encountered in trying to test the theory suggest that it will be very hard to draw conclusions as to the relative merits of the $O(\alpha_s^3)$ vs. $O(\alpha_s^2)$ calculations. Analyses in this direction have barely begun and we anticipate a great deal of activity in this area. On the positive side, one promising aspect is that the $O(\alpha_s^3)$ calculations seem much less sensitive than $O(\alpha_s^2)$ to the ambiguity in the choice of the scale parameter, as shown in Fig. 24.30.

In conclusion, the ISR were instrumental in proving the existence of hadronic jets which arise naturally within QCD; and the newer hadronic machines have confirmed the correctness of this interpretation beyond possible doubt.

At the energies presently attained, what was anticipated a decade ago has become reality: jet physics is so dominant, by now, that it is the background that will have to be eliminated if new phenomena are to be unveiled at the future hadronic colliders.

25
Jets and hadrons in e^+e^- physics

25.1 Introduction

Although the earliest evidence for jets came from hadronic physics, it is in e^+e^- collisions that jets can be studied most cleanly. Thus, the first hint of planar (i.e. three-jet) configurations implying the existence of gluon jets came from PETRA in 1979. The field is, by now, dominated by LEP data at which energies several jets can be produced.

In this chapter we briefly discuss the historical developments and then turn to the most recent results. Some mention is made of models designed to bridge the gap between the QCD perturbative hard reactions and the non-perturbative process of hadronization. All these models use Monte Carlo simulations based on probabilistic descriptions; thus they cannot describe the subtle quantum mechanical effects that might occur. One should be able to attack this problem analytically but it may be a long time before a viable approach is found.

25.2 General outline of e^+e^- jets

Evidence for two-jet formation in the multihadronic final states in e^+e^- reactions above 5 GeV/c first emerged from the SLAC-LBL magnetic detector (Hanson *et al.*, 1975; many excellent reviews of this group's data have appeared, see, for example, Hanson, 1976) at SPEAR and was based on the following pieces of data: (i) analysis of the mean sphericity variable (see below) to reconstruct the jet axis and comparison between a pure (isotropic) phase space model and a jet model; (ii) determination of the angular distribution of the jet axis as expected from the decay of an e^+e^- initial state into final hadrons through a vector meson intermediate state, and consistent with partons being spin $\frac{1}{2}$ particles; (iii) analysis of the mean transverse momentum $\langle p_T \rangle$ of the final hadrons with respect to the jet axis; (iv) analysis of the single inclusive x_F ($x_F = 2p/E_{\rm CM}$) distribution at energies around $\sqrt{s} \equiv E_{\rm CM} \geq 6$ GeV; (v) analysis of the inclusive distribution in variables *relative to jet axis* ($x_\parallel, x_T$ and the rapidity y) showing an approximate scaling in $x_\parallel = 2p_\parallel/E_{\rm CM}$ (where $p_\parallel$ is the momentum parallel to the jet axis), an approximate constancy of $\langle p_T \rangle$ as $E_{\rm CM}$ increases, and the development of a rapidity plateau. Some of these data will be discussed in what follows.

The general picture emerging from SPEAR data, namely that the hadronic final state in e^+e^- collisions (aside from the absence of beam and target jets) is similar to that found in hadronic reactions, has been corroborated by the data from PETRA and PEP and, more recently, from LEP.

Several variables have been introduced to specify the jet-like nature of an event. For example:

$$\text{Sphericity} \equiv S' = \frac{3}{2}\min_{\boldsymbol{n}} \left(\frac{\sum_i \boldsymbol{p}_{Ti}^2}{\sum_i \boldsymbol{p}_i^2} \right), \tag{25.2.1}$$

where $\boldsymbol{n}$ is an arbitrary unit vector relative to which $\boldsymbol{p}_{Ti}$ is measured;

$$\text{Thrust} \equiv T = \max_{\boldsymbol{n}} \left(\frac{\sum_i |\boldsymbol{p}_i \cdot \boldsymbol{n}|}{\sum_i |\boldsymbol{p}_i|} \right) \tag{25.2.2}$$

$$\text{Spherocity} \equiv S = \left(\frac{4}{\pi}\right) \min_{\boldsymbol{n}} \left(\frac{\sum_i |\boldsymbol{p}_{Ti}|}{\sum_i |\boldsymbol{p}_i|} \right)^2 \tag{25.2.3}$$

$$\text{Acoplanarity} \equiv A = 4\min_{\boldsymbol{n}} \left(\frac{\sum_i |\boldsymbol{p}_{\text{out}i}|}{\sum_i |\boldsymbol{p}_i|} \right)^2, \tag{25.2.4}$$

where $\boldsymbol{p}_{\text{out}i}$ is measured transverse to a plane with normal $\boldsymbol{n}$. In these the sum is over all detected particles, and $\boldsymbol{n}$ is varied until the desired maximum or minimum is found.

For an ideal two-jet event one would have $S' = 0$, $T = 1$, $S = 0$ and $A = 0$, whereas an isotropic distribution has $S' = 1$, $T = \frac{1}{2}$, $S = 1$ and $A = 1$.

Although sphericity is a rather simple variable to measure experimentally, it can be argued on theoretical grounds that it ought to be infinite because of infrared divergences which arise in perturbative QCD as a result of emission of parallel (i.e. collinear) massless gluons and quarks (see Section 23.3.2). Thus a comparison between theory and experiment using S' is, strictly speaking, meaningless. But in practice, spherocity, which does not suffer from these difficulties, does not generally differ much numerically from sphericity. Indeed, for practical purposes, the use of all these variables is nearly equally good, and we shall use them in a fairly arbitrary way at our convenience. We will come back later on to the use of these and other similar variables (global event shape variables). See Section 25.7.

Generally speaking, the picture which emerges from the data is the following:

(*a*) At moderately high energies ($\sqrt{s} \sim 5$–15 GeV), hadron production is dominated by two-jet events where the jet cone seems to shrink with energy.

This can be understood qualitatively from the fact that, as in usual hadronic reactions, $\langle p_T \rangle$ is about constant and $\langle p_\parallel \rangle$ is expected to grow, roughly speaking, as $\langle p_\parallel \rangle \simeq \sqrt{s}/\langle n(s) \rangle$ (where $\langle n \rangle$, the average multiplicity, grows like $\ln^2 s$). This suggests that the cone opening should decrease as

$$\langle \lambda \rangle = \frac{\langle p_T \rangle}{\langle p_\parallel \rangle} \simeq \frac{\langle p_T \rangle \langle n \rangle}{\sqrt{s}} \simeq \frac{1}{\sqrt{s}}. \tag{25.2.5}$$

As is clear from the definition of sphericity (24.2.1), one would then expect roughly $S' \simeq \langle \lambda \rangle^2 \simeq O(1/s)$. The data from the PLUTO and TASSO collaborations at PETRA, show that S' does decrease (the jet cone opening drops from ~31° at 4 GeV to ~18° at 27.4 GeV) but with a slower energy dependence. This latter, it will turn out, is linked to jet broadening due to hard gluon emission.

(*b*) With increasing energy $\sqrt{s}$, three and more jet events become increasingly important. At LEP energies, three-jet production is a very important effect, with a sizeable contribution from four and more jets (see Section 25.11),

25.2.1 *Angular distribution of hadrons produced in e⁺e⁻ collisions*

We first consider hadrons produced at energies where the simplest mechanism

$$e^+e^- \to \gamma \to q\bar{q} \tag{25.2.6}$$

dominates. [At higher energies interference with Z^0 exchange will become important (see Section 8.5.1).] The comparison of the angular distributions with data is at the origin of the claim that partons are spin $\frac{1}{2}$ fermions, and also plays an important rôle in a number of other issues.

Consider a reaction such as e⁺e⁻ → hadrons, which is assumed to proceed via one photon to yield a final state hX:

$$e^+e^- \to \gamma \to hX. \tag{25.2.7}$$

The treatment is a slight generalization of that of Section 8.3, where the reaction $e^+e^- \to \mu^+\mu^-$ was considered.

In (25.2.7) h and X can both be genuine particles so that the final state is a *bona fide* two-body state, or h may represent a particle that is detected while X stands for everything else (inclusive h production), or they can be two jets of collimated particles.

Just as we did in Section 8.5.1, we assume the beam energy to be much larger than the electron mass. As a consequence of the γ_μ coupling to e$^+$e$^-$ the first step of reaction (25.2.7) is to produce a 'photon' with helicity $\lambda = \pm 1$.

Let us choose an axis system with OZ along the e$^+$ beam and OY along the magnetic guide field, i.e. perpendicular to the plane of the circulating beam. Let h be produced with polar angles (θ, φ), with helicity μ_h. Then, with obvious labels for the helicities of the various particles, we have

$$A^{hX}(\theta,\varphi) = M^{hX}_{\mu_h\mu_X} M^{\mathrm{e^+e^-}}_{\lambda_+\lambda_-} \mathrm{e}^{-\mathrm{i}\varphi(\lambda-\mu)} d^1_{\lambda\mu}(\theta), \tag{25.2.8}$$

where $\lambda = \lambda_+ - \lambda_-$ and $\mu = \mu_h - \mu_X$.

The helicity of the intermediate photon does not appear in the amplitudes since for a one-particle state $\lambda = J_z$ and by rotational invariance the amplitude cannot depend on J_z.

Thus the unpolarized cross-section for $\mathrm{e^+e^-} \to \gamma \to hX$ will be given by

$$\frac{\mathrm{d}\sigma}{\mathrm{d}\Omega}(\mathrm{e^+e^-} \to hX) = N \sum_{\mu=0,\pm1} \sum_{\lambda=\pm1} \left(d^1_{\lambda\mu}(\theta)\right)^2 \times \sigma^\gamma_\mu(\mathrm{e^+e^-} \to hX), \tag{25.2.9}$$

where we have used, as a consequence of parity invariance, $|M^{\mathrm{e^+e^-}}_{\lambda_+\lambda_-}| = |M^{\mathrm{e^+e^-}}_{-\lambda_+-\lambda_-}|$. $\sigma^\gamma_\mu(\mathrm{e^+e^-} \to hX)$ is the total e$^+$e$^-$ cross-section to produce an hX state with helicity μ via one photon and N is a normalization factor to guarantee that $\sigma_{\mathrm{tot}} = \sigma_1 + \sigma_{-1} + \sigma_0$. It turns out that $N = 3/8\pi$. Parity conservation actually makes $\sigma_1 = \sigma_{-1}$. Using

$$d^1_{1\pm1}(\theta) = \frac{1}{2}(1 \pm \cos\theta), \quad d^1_{10}(\theta) = \frac{1}{\sqrt{2}}\sin\theta \tag{25.2.10}$$

one gets

$$\begin{aligned}\frac{\mathrm{d}\sigma}{\mathrm{d}\Omega}(\mathrm{e^+e^-} \to hX) &= \frac{3}{8\pi}[(1+\cos^2\theta)\sigma^\gamma_T(\mathrm{e^+e^-} \to hX) \\ &\quad + (1-\cos^2\theta)\sigma^\gamma_L(\mathrm{e^+e^-} \to hX)],\end{aligned} \tag{25.2.11}$$

where we have defined the 'longitudinal' (sometimes also called 'scalar') cross-section

$$\sigma^\gamma_L(\mathrm{e^+e^-} \to hX) = \sigma^\gamma_0(\mathrm{e^+e^-} \to hX) \tag{25.2.12}$$

and the 'transverse' cross-section

$$\sigma^\gamma_T(\mathrm{e^+e^-} \to hX) = \tfrac{1}{2}[\sigma^\gamma_1(\mathrm{e^+e^-} \to hX) + \sigma^\gamma_{-1}(\mathrm{e^+e^-} \to hX)]. \tag{25.2.13}$$

It is often convenient to parametrize the angular distribution (25.2.11) as

$$\frac{d\sigma}{d\Omega}(e^+e^- \to hX) = \frac{3}{8\pi}(1+\alpha\cos^2\theta)[\sigma_T^\gamma(e^+e^- \to hX) + \sigma_L^\gamma(e^+e^- \to hX)], \tag{25.2.14}$$

where

$$-1 \le \alpha \equiv \frac{\sigma_T^\gamma(e^+e^- \to hX) - \sigma_L^\gamma(e^+e^- \to hX)}{\sigma_T^\gamma(e^+e^- \to hX) + \sigma_L^\gamma(e^+e^- \to hX)} \le 1. \tag{25.2.15}$$

Note that Z^0 exchange would yield an additional parity-violating term linear in $\cos\theta$ in (25.2.14) (see Section 8.5.1).

Powers of $\cos\theta$ higher than two could arise only for $J > 1$ (they are, for instance, produced by 2γ intermediate states).

The case $e^+e^- \to \bar{f}f$ where f is any *elementary* spin $\frac{1}{2}$ particle leads to

$$\frac{\sigma_L^\gamma(e^+e^- \to \bar{f}f)}{\sigma_T^\gamma(e^+e^- \to \bar{f}f)} = \frac{4m_f^2}{q^2}, \tag{25.2.16}$$

where q^2 is the total CM squared of the produced particles.

The reader should not forget that if the final state is e^+e^- then the reaction $e^+e^- \to e^+e^-$ can also go via exchange of a space-like photon (see Fig. 8.5 and eqn (8.3.2)).

Equations (25.2.14, 16) coincide with (8.5.2) for the case $e^+e^- \to \mu^+\mu^-$ and have been discussed in detail in Chapter 15 when studying the crossed reaction $e\mu \to e\mu$, in which case q^2 is the squared CM momentum transfer.

Note that when f is a spin $\frac{1}{2}$ elementary particle it follows from (25.2.16) that

$$\lim_{q\to\infty} \frac{\sigma_L^\gamma(e^+e^- \to \bar{f}f)}{\sigma_T^\gamma(e^+e^- \to \bar{f}f)} = 0. \tag{25.2.17}$$

If, on the other hand, two spinless hadrons (say pions) are produced, only $\mu = 0$ contributes to (25.2.9) so that

$$\sigma_L^\gamma(e^+e^- \to \pi^+\pi^-) = 0 \tag{25.2.18}$$

and

$$\frac{d\sigma}{d\Omega}(e^+e^- \to \gamma \to \pi^+\pi^-) \propto \sin^2\theta. \tag{25.2.19}$$

From (25.2.14) we conclude that the angular distribution of the final hadrons allows the determination of $\sigma_T^\gamma(e^+e^- \to hX)/\sigma_L^\gamma(e^+e^- \to hX)$. In principle, the limited θ acceptance of the spectrometer often demands a somewhat more careful investigation. The additional information depends upon the fact that in e^+e^- devices the beams are naturally polarized transversely (see Section 8.1).

The best way to handle the calculation of the differential cross-section in such a situation is by density matrix methods, for which the reader is referred to Bourrely, Leader and Soffer (1980). Here we shall sketch how the result arises by a less sophisticated argument.

If the degree of polarization of the e^+ beam is P then the number of e^+s to be found with spin 'up' or 'down', i.e. along or opposite to OY, satisfy the relation $P = (N_\uparrow - N_\downarrow)/(N_\uparrow + N_\downarrow)$ so that $N_\uparrow^{e^+}/N_\downarrow^{e^+} = (1+P)/(1-P)$. For the electrons, being polarized to the same degree but in the opposite direction, $N_\uparrow^{e^-}/N_\downarrow^{e^-} = (1-P)/(1+P)$.

The relative number of collisions taking place in the four possible e^+e^- spin configurations is then $N_{\uparrow\uparrow} : N_{\downarrow\downarrow} : N_{\uparrow\downarrow} : N_{\downarrow\uparrow} = (1-P^2) : (1-P^2) : (1+P)^2 : (1-P)^2$. The transitions from these states, with spin quantized along OY, can be related to the helicity transitions on noting that

$$|\uparrow \text{ or } \downarrow\rangle_{e^+} = \frac{1 \text{ or } i}{\sqrt{2}}\{|+\rangle_{e^+} \pm i|-\rangle_{e^+}\}$$

$$|\uparrow \text{ or } \downarrow\rangle_{e^-} = \frac{i \text{ or } (-1)}{\sqrt{2}}\{|+\rangle_{e^-} \mp i|-\rangle_{e^-}\}$$

Then, using (25.2.8) one finds for the polarized cross-section

$$\begin{aligned}\frac{d\sigma}{d\Omega}(e^+e^- \to hX)_{\text{Pol}} \propto\ & (1+P^2)[\sigma_T^\gamma(1+\cos^2\theta + \sin^2\theta\cos 2\varphi) + \\ & + \sigma_L^\gamma \sin^2\theta(1-\cos 2\varphi)] + (1-P^2)[\sigma_T^\gamma(1+\cos^2\theta - \\ & - \sin^2\theta\cos 2\varphi) + \sigma_L^\gamma\sin^2\theta(1+\cos 2\varphi)] \qquad (25.2.20)\end{aligned}$$

and adjusting the normalization so as to regain (25.2.14) when $P = 0$ we get

$$\begin{aligned}\frac{d\sigma}{d\Omega}(e^+e^- \to \gamma \to hX)_{\text{Pol}} =\ & \frac{3}{8\pi}(\sigma_T^\gamma + \sigma_L^\gamma)[1 + \alpha(\cos^2\theta + \\ & + P^2\sin^2\theta\cos 2\varphi)], \qquad (25.2.21)\end{aligned}$$

where α is defined in (25.2.15). Equation (25.2.21) is the most general form of angular distribution from single inclusive hadron production in e^+e^- collisions via one photon.

25.3 SPEAR two-jet events

At SPEAR energies ($\sqrt{s} \simeq 7.5$ GeV), jet events are expected to proceed via the virtual photon mechanism $e^+e^- \to \gamma \to q\bar{q}$ where the quarks fragment into hadrons.

As repeatedly emphasized, we have no understanding as to how the colour of each quark or antiquark is neutralized in producing the colour

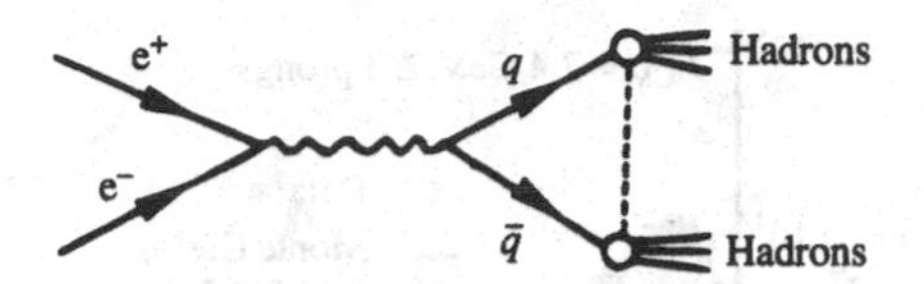

Fig. 25.1. Dominant mechanism for $e^+e^- \to 2$ jets.

singlet hadrons. But it is reasonable to assume that this is a *soft* process (symbolized in Fig. 25.1) so that the kinematics of the jets reflects the kinematics of the $q\bar{q}$ from which they originate.

25.3.1 Sphericity

The, by now historical, method used by the SLAC–LBL group to detect jet formation was to minimize the sum of squares of transverse momenta by diagonalizing the three-dimensional tensor

$$T^{ij} = \sum_f (\delta^{ij} \boldsymbol{p}_f^2 - p_f^i p_f^j), \tag{25.3.1}$$

where f are the final (charged) particles and i, j denote the space components of their momenta. The problem is, in essence, to find the eigenvalues λ_1, λ_2, λ_3 which are the sums of squares of transverse momenta with respect to the three eigenvector directions. The smallest such eigenvalue λ_3 is the minimum of the sums of squares of transverse momenta and the corresponding eigenvector is defined to be the jet axis.

According to the previous discussion, a jet will, in practice, occur when the sphericity

$$S' = \frac{3}{2}\left(\frac{\sum_i \boldsymbol{p}_{Ti}^2}{\sum_i \boldsymbol{p}_i^2}\right)_{\text{min}} = \frac{3\lambda_3}{\lambda_1 + \lambda_2 + \lambda_3} \tag{25.3.2}$$

is close to zero while a purely isotropic distribution will correspond to $S' = 1$.

A Monte Carlo simulation was used to discriminate between a Lorentz invariant phase space model and a jet model. The comparison of the p_T distribution of these models with the data at $\sqrt{s} \equiv E_{\text{CM}} = 7.4$ GeV is shown in Fig. 25.2.

The mean $\langle p_T \rangle$ in this analysis turns out to be of the order $\sim$ 325–360 MeV/c, similar to the values found in typical hadronic reactions.

Evidence that the data favour a jet structure is shown in Fig. 25.3, which gives the sphericity distribution at several energies. The mean sphericity decreases (as expected for a jet distribution), whereas phase

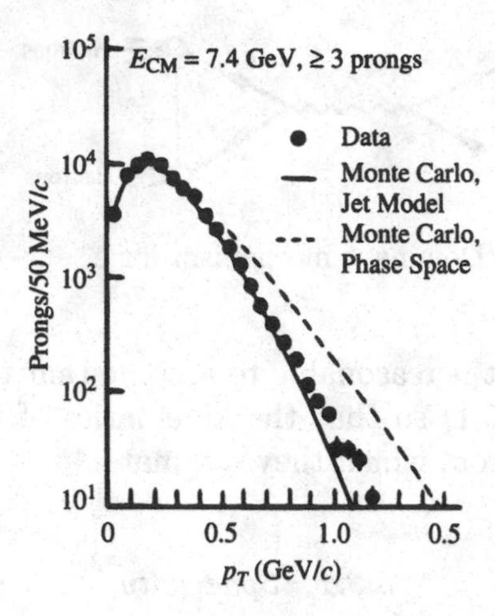

Fig. 25.2. Transverse momentum distribution of charged particles in e^+e^- collisions at 7.4 GeV. (For curves see text.) (From Hanson, 1976.)

space predicts a slow increase of $\langle S' \rangle$. The theoretical curves in Fig. 25.3 have been corrected for acceptance and detection efficiency.

The fall with energy of $\langle S' \rangle$ is evidence that the cone opening of the jet decreases.

25.3.2 Jet axis

The angular distribution for e^+e^- to go to a final state of two elementary particles of spin $\frac{1}{2}$ or 0 was given in equation (25.2.21).

Recall that $\alpha = 1$ implies spin $\frac{1}{2}$ for the final particles and a $1 + \cos^2\theta$ distribution, while $\alpha = -1$ implies spin 0 and a $\sin^2\theta$ distribution.

The use of the azimuthal distribution (25.2.21) is, in practice, very helpful for determining α in those cases when the θ acceptance is too small.

Comparing with the data, at $\sqrt{s} = 6.2$ GeV the beams are unpolarized and the φ distribution is flat in agreement with (25.2.21). At $\sqrt{s} =$ 7.4 GeV, $P^2 = 0.5$, and the distribution (Fig. 25.4) agrees with the $\cos 2\varphi$ form predicted from (25.2.21); one finds

$$\alpha = 0.97 \pm 0.1 \tag{25.3.3}$$

when correction for the incomplete acceptance of the detector and loss of neutral particles is duly accounted for. This value (corresponding to $\sigma_L/\sigma_T \simeq 0.02 \pm 0.07$) is one of the cleanest pieces of evidence for spin $\frac{1}{2}$ constituents.

Fig. 25.5 shows the dependence of $\sigma_L^\gamma/\sigma_T^\gamma$ on $x_F \equiv 2p/\sqrt{s}$ and α for the experiment of Fig. 25.4.

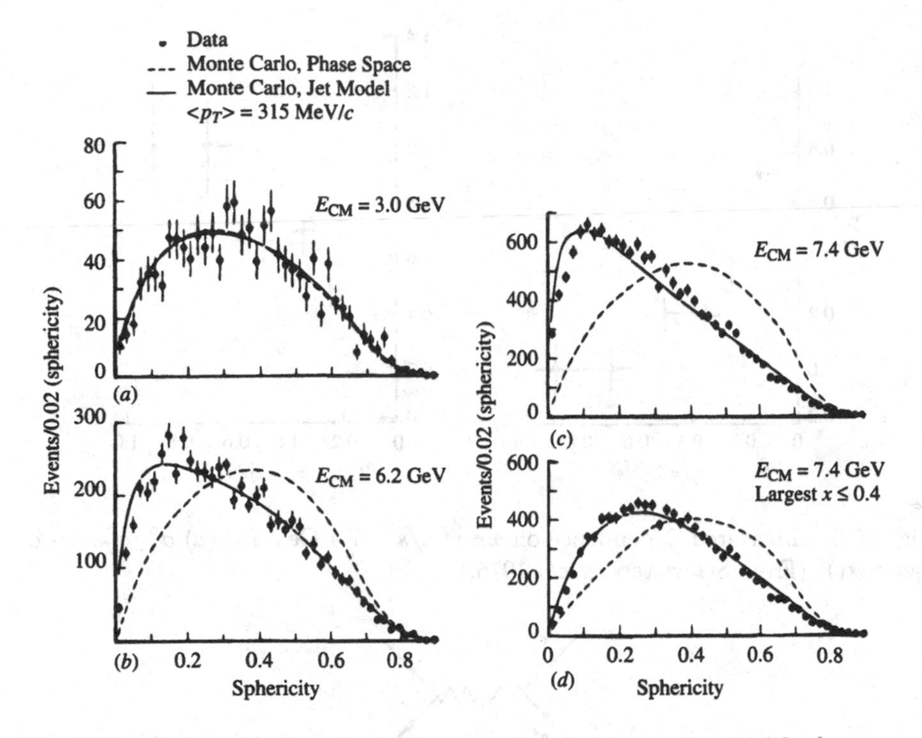

Fig. 25.3. Sphericity distribution at several energies compared with phase space and uncorrelated jet models. (From Hanson, 1976.)

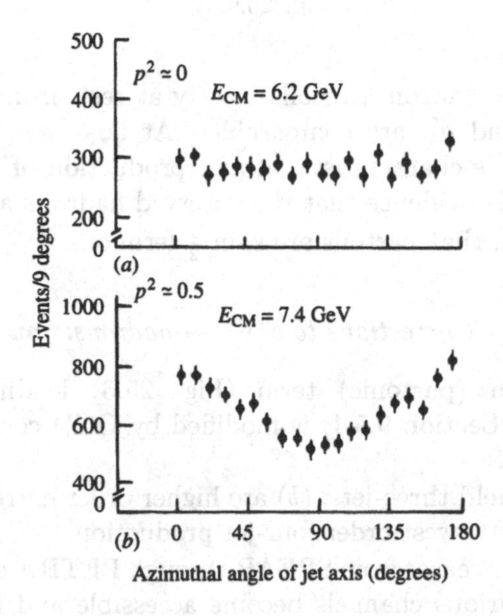

Fig. 25.4. Angular distribution of the jet axis in e^+e^- collisions at E_{CM} = 6.2 GeV, where the polarization is zero, and at E_{CM} = 7.4 GeV where $P^2 = 0.5$. (From Hanson, 1976.)

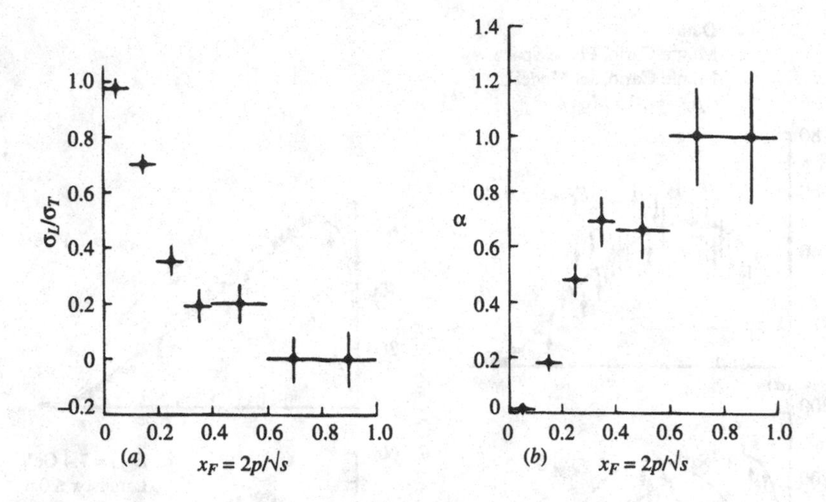

Fig. 25.5. Measured dependence on x_F at $\sqrt{s} = 7.4$ GeV for (a) $\sigma_L^\gamma/\sigma_T^\gamma$, (b) α (see text). (From Schwitters *et al.*, 1975.)

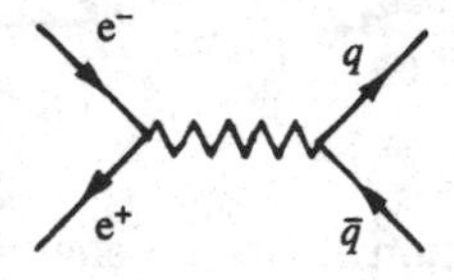

Fig. 25.6.

At small x_F the hadron h recoils, nearly at rest, from a very massive system and σ_L^γ and σ_T^γ are comparable. At large x_F ($x_F > 0.3$), σ_T^γ dominates, which is characteristic of pair production of elementary spin $\frac{1}{2}$ particles. This is evidence that the observed hadrons are emitted from spin $\frac{1}{2}$ objects, i.e. that partons are spin $\frac{1}{2}$ fermions.

25.3.3 Corrections to $e^+e^- \to$ *hadrons: multijets*

The simple Born (partonic) term (Fig. 25.6) leading to $R(s) = 3\sum_{\text{flavour}} Q_f^2$ (see Section 9.5.4) is modified by QCD corrections such as given in Fig. 25.7.

Diagrams (a) yield three-jets; (b) are higher order corrections to three-jet production; (c) lowest order four-jet production.

As the energy increases from SPEAR through PETRA and PEP to LEP energies, those various channels become accessible and lead to multiple jets. The question of how to separate two- from three- or four-jet events will be dealt with later.

The sum of all diagrams up to order α_s^3 contributing to the total $e^+e^- \to$

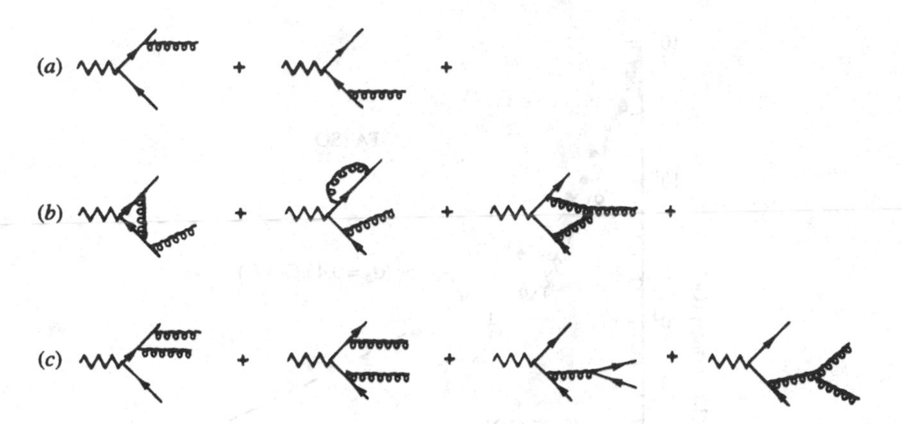

Fig. 25.7. Three- and four-jet contributions.

$\gamma \to$ hadrons cross-section has recently been recalculated and the result is (Gorishny *et al.*, 1991; Surguladze and Samuel, 1991)

$$R(s) = 3\sum_f Q_f^2[1 + \alpha_s/\pi + r_1(\alpha_s/\pi)^2 + r_2(\alpha_s/\pi)^3] + O(\alpha_s^4), \quad (25.3.4)$$

where $\alpha_s = \alpha_s(Q^2 = s)$ is defined in the $\overline{\text{MS}}$ renormalization scheme and

$$\begin{aligned} r_1 &= 1.9857 - 0.1153n_f \\ r_2 &= -6.6368 - 1.2001n_f - 0.0052n_f^2 - 1.2395\frac{(\sum Q_f)^2}{3\sum Q_f^2}. \end{aligned} \quad (25.3.5)$$

For five active flavours, this gives

$$R(s) = \frac{11}{3}\left[1 + \frac{\alpha_s}{\pi} + 1.4092\left(\frac{\alpha_s}{\pi}\right)^2 - 12.8046\left(\frac{\alpha_s}{\pi}\right)^3\right] \quad (25.3.6)$$

(note that the above does not include contributions from $e^+e^- \to Z^0 \to$ hadrons and, therefore, should not be compared directly with LEP data).

When considering gluon radiation there are new features that emerge and that show in the data either as three-jet production (see Section 25.4) or as jet-broadening.

To understand the latter phenomenon, consider that, with increasing energy, more and more gluons can be radiated. Each emission results in a momentum kick with a component perpendicular to the motion of the emitting quark.

Thus, if we measure the $\langle p_T^2 \rangle$ of particles in a jet relative to the jet axis, we expect it to increase with energy. TASSO data (Figs. 25.8, 9) support this jet-broadening.

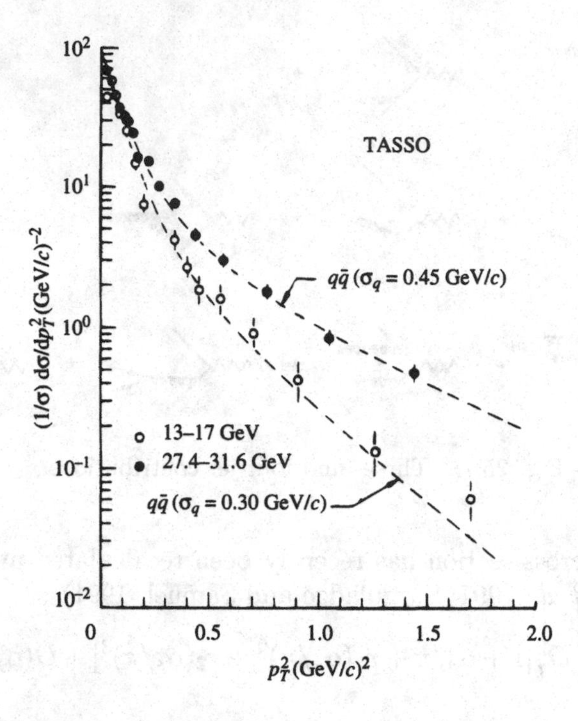

Fig. 25.8. Transverse momentum distributions in e^+e^- collisions for two energy ranges. (For curves see text.) (From Wolf, 1980.)

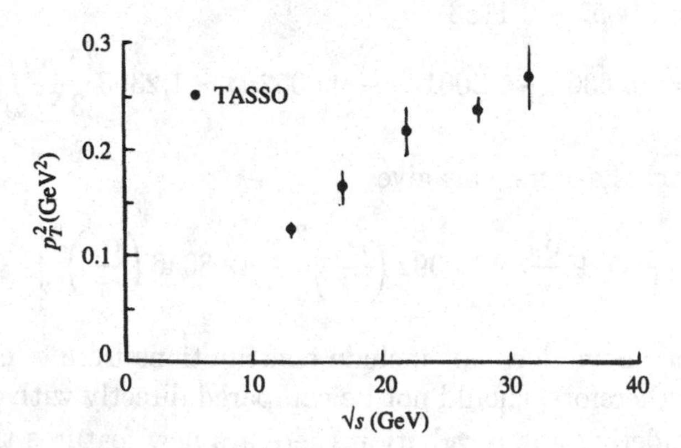

Fig. 25.9. Mean p_T^2 in e^+e^- collisions as function of CM energy. (From Wolf, 1980.)

When fitting the data to a Gaussian form

$$\frac{d\sigma}{dp_T^2} \propto \exp(-p_T^2/2\sigma_q^2) \qquad (25.3.7)$$

jet broadening amounts to σ_q (i.e. $\langle p_T \rangle$) growing with CM energy.

The same effect is also seen in Fig. 25.9, where $\langle p_T^2 \rangle$ grows as we move from $\sqrt{s} \sim 10$ GeV to ~ 30 GeV.

To check that these results are in keeping with theoretical predictions, we indicate that one expects $\langle p_T \rangle \sim \alpha_s \sqrt{s}$ from one gluon radiation.

This behaviour can be understood rather simply since, analogously to the emission of photons from an electron, the gluon distribution radiated in a quark process is approximately given by

$$\frac{\mathrm{d}\sigma(qq \to qqG)}{\mathrm{d}p \, \mathrm{d}\cos\theta} \propto \frac{\alpha_s \sin^2\theta}{p(1-\cos\theta)^2} \sigma_{qq \to qq}, \tag{25.3.8}$$

where p and θ are the gluon momentum and its emission angle relative to the quark direction. The cross-section integrated over θ diverges. This is a typical infrared divergence which would be cancelled by divergent contributions from vertex and self-energy corrections in $qq \to qq$. The expression for average transverse momentum of the hard gluon jet does not diverge, so we ignore this problem. We have then

$$\begin{aligned} \langle p_T \rangle &= \frac{\alpha_s \sigma_{qq \to qq} \int \frac{p \sin^4\theta}{p(1-\cos\theta)^2} \mathrm{d}p \, \mathrm{d}\theta}{\sigma_{qq \to qq}} \\ &\sim \alpha_s \sqrt{s}. \end{aligned} \tag{25.3.9}$$

Higher order effects modify this somewhat.

25.4 Planar events: evidence for three jets

We have seen that there is clear evidence that the dominant process in large p_T hadronic e^+e^- collisions is the production of multihadron states in the form of two jets. The principal mechanism for this is supposed to be the production of a $\bar{q}q$ final state which then materializes into hadrons. With increase of energy, other processes such as $\bar{q}qG$ and, eventually, $\bar{q}qGG$, $\bar{q}q\bar{q}q$, ... should play a rôle.

We shall now examine in detail the evidence for three-jet events at the highest PETRA energies from several groups using the detectors TASSO, PLUTO, JADE and Mark J.

The TASSO, PLUTO and JADE groups analyze their data by constructing the momentum tensor

$$M^{ij} = \sum_f p_f^i p_f^j, \tag{25.4.1}$$

the sum being performed over all charged final particles, and introducing the eigenvalues

$$\Lambda_i = \sum_f (\boldsymbol{p}_f \cdot \boldsymbol{n}_i)^2 \quad (i = 1, 2, 3) \tag{25.4.2}$$

where $\boldsymbol{n}_i$ are the unit eigenvectors of M ordered in such a way that $\Lambda_1 < \Lambda_2 < \Lambda_3$. $\boldsymbol{n}_1$ is thus the direction in which the sum of the momentum components is minimized so that $\boldsymbol{n}_2, \boldsymbol{n}_3$ define the plane of the event with $\boldsymbol{n}_3$ giving the jet axis.

In terms of the normalized eigenvalues

$$Q_i = \frac{\Lambda_i}{\sum_f p_f^2}, \quad \sum_{i=1}^{3} Q_i = 1, \tag{25.4.3}$$

the sphericity (25.2.1)and acoplanarity (25.2.4) are given by

$$S' = \tfrac{3}{2}(Q_1 + Q_2) = \tfrac{3}{2}(1 - Q_3) \tag{25.4.4}$$

$$A = \tfrac{3}{2}Q_1. \tag{25.4.5}$$

Collinear events, as already mentioned, correspond to $S' = 0$; non-collinear coplanar events to $S' \neq 0, A \simeq 0$; non-coplanar events have S' and A both non-zero.

It is interesting to compare the distribution in

$$\langle p_T^2 \rangle_{\text{out}} = \frac{1}{N} \sum_{f=1}^{N} (\boldsymbol{p}_f \cdot \boldsymbol{n}_1)^2 \tag{25.4.6}$$

(i.e. the mean of the square of the momentum components normal to the event plane), with those in

$$\langle p_T^2 \rangle_{\text{in}} = \frac{1}{N} \sum_{f=1}^{N} (\boldsymbol{p}_f \cdot \boldsymbol{n}_2)^2 \tag{25.4.7}$$

(i.e. the mean of the square of the momentum components in the event plane and perpendicular to the jet axis).

The data, at various energies, from TASSO, are shown in Fig. 25.10.

The data in $\langle p_T^2 \rangle_{\text{out}}$ show little variation in going from 13–17 GeV to 27–31 GeV, whereas a sizable flattening takes place in the distributions in $\langle p_T^2 \rangle_{\text{in}}$. The curves in Fig. 25.10 are based on a $q\bar{q}$ model.

At lower energies the $q\bar{q}$ predictions agree with the data for the $\langle p_T^2 \rangle_{\text{in}}$ distribution with $\sigma_q = 300$ MeV/c, but even the choice $\sigma_q \simeq 450$ MeV/c gives poor agreement at the higher energies. On the other hand, good agreement with the $q\bar{q}$ prediction ($\sigma_q = 300$ MeV/c) is found at all energies for the $\langle p_T^2 \rangle_{\text{out}}$ distributions.

Similar conclusions were reached by the groups PLUTO and JADE.

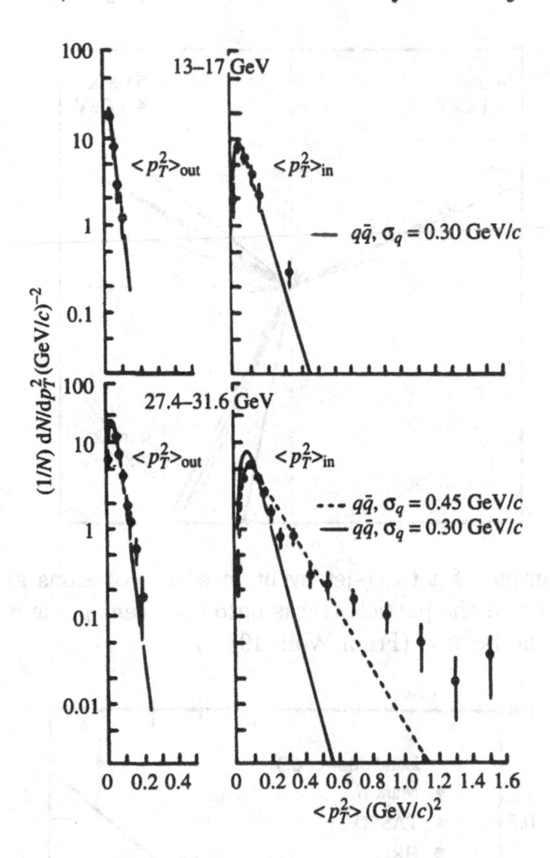

Fig. 25.10. Distributions in $\langle p_T^2 \rangle_{in}$ and $\langle p_T^2 \rangle_{out}$, as defined in the text, for e^+e^- collisions in two energy ranges. (From Wolf, 1980.)

A somewhat different approach was used by the Mark J group (Wolf, 1980) based on a new variable, 'oblateness' (O), defined (25.10.1) as the difference of the major-minor axis in the study of the energy flow in the events. At high energies (27–31 GeV) an excess of oblateness (i.e. planar events) is obtained.

All the above data show the existence of planar events, and a detailed analysis suggests that up to 13–17 GeV the events can be described as a two-jet structure, whereas at the highest energies three-jet events (exhibiting nearly the same p_T^2 distribution) appear.

A typical three-jet event from TASSO is shown in Fig. 25.11.

The previous conclusions are well corroborated by the data at higher CM energies (see Fig. 25.12 where the curves come from a model to be discussed in Section 25.8.2).

More detailed considerations on the rate of three-jet events at LEP will be discussed in Section 25.11.

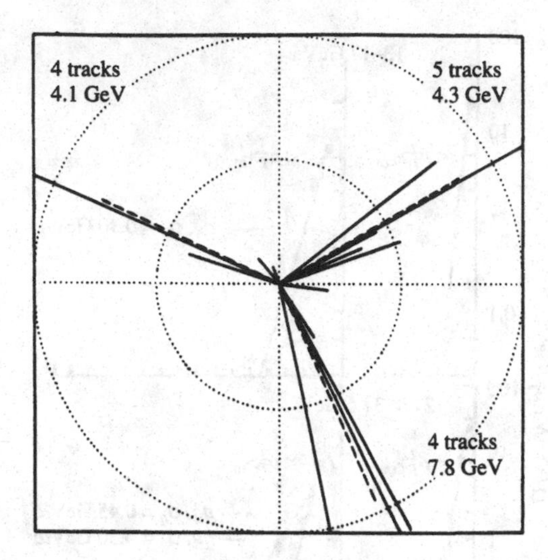

Fig. 25.11. Example of a three-jet event in e$^+$e$^-$ collisions at PETRA. The lines are projections of the particle tracks onto the event plane which is roughly perpendicular to the beams. (From Wolf, 1980.)

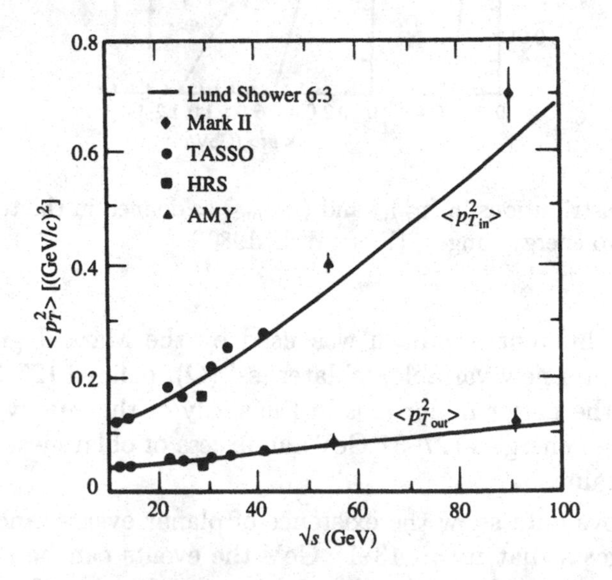

Fig. 25.12. Fit to the $\langle p_T^2 \rangle$ data with the Lund model (from de Boer, 1991).

25.5 Tests of QCD up to LEP energies

As we have seen, LEP has brought spectacular confirmation of the electroweak sector of the standard model (SM). By contrast, the QCD aspects of the SM are much less well tested since it is generally believed that most theoretical predictions are uncertain by more than ~20%. Schematically, the greatest causes of uncertainty are: (i) the calculations are difficult to perform beyond lowest order (this is partly connected with the non-Abelian nature of QCD which is one of the key points that one would like to verify), (ii) the process of fragmentation is not well understood (for this, Monte Carlo simulations are used), and (iii) the connection between partons and jets is not perfectly unambiguous.

Going to LEP energies, QCD makes predictions for the following types of observables:

(*a*) the total hadronic width;

(*b*) shapes and correlations in two-jet events;

(*c*) three-jet quantities, such as the global event variables, thrust, sphericity, oblateness, energy–energy correlations etc.;

(*d*) four- and five-jet quantities;

(*e*) multiplicities;

(*f*) fragmentation functions.

We shall concentrate mainly on point (*c*), with a brief discussion of (*a*), (*d*) and (*e*). Points (*a*) and (*f*) were also dealt with in Chapters 8, 17 and 23 respectively.

For more details, the reader is referred to Kunszt and Nason (1989) and references therein.

Among the top priorities in the discussion is, of course, the determination of the effective coupling constant $\alpha_s(Q^2)$ for which the next-to-leading order form recommended by the Particle Data Group (1990), and which follows upon expanding (21.7.11), is

$$\alpha_s(Q^2) = \frac{12\pi}{(33-2n_f)\ln(Q^2/\Lambda^2)}\left[1 - 6\frac{153-19n_f}{(33-2n_f)^2}\frac{\ln(\ln Q^2/\Lambda^2)}{\ln(Q^2/\Lambda^2)}\right]. \tag{25.5.1}$$

This amounts, of course, to the determination of Λ.

We begin with a discussion of the total hadronic width. Thereafter, we digress for a brief survey of basic Monte Carlo techniques which are an essential element in implementing the other QCD predictions listed above.

25.6 The total hadronic width at the Z^0

This is the simplest quantity to analyse. Denoting by Γ_h^0 the Born approximation to $Z^0 \to$ hadrons obtained from

$$\Gamma_h^0 = \sum_f \Gamma_{Z\to f\bar{f}}, \tag{25.6.1}$$

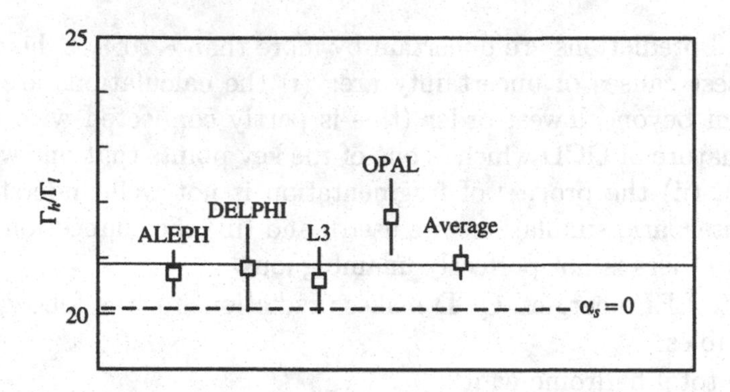

Fig. 25.13. The quantity Γ_h/Γ_1 as measured at LEP.

where $\Gamma_{Z\to f\bar{f}}$ was given in (8.6.9) to order α_s^3 one finds in the $\overline{\text{MS}}$ scheme

$$\Gamma_h = \Gamma_h^0 \left[1 + \alpha_s/\pi + \bar{r}_1(\alpha_s/\pi)^2 + \bar{r}_2(\alpha_s/\pi)^3\right], \tag{25.6.2}$$

where α_s is short for $\alpha_s(M_Z^2)$ and where $\bar{r}_1$ and $\bar{r}_2$ differ slightly from the values r_1 and r_2 given in (25.3.6) on account of the vector and axial-vector coupling of the Z^0. These differences depend on m_t/M_Z and m_H/M_Z; formulae can be found in Kunzst and Nason (1989).

For $m_t \simeq 127$ GeV/c^2, $\bar{r}_1 \simeq 1.97$. Fitting to the hadronic width yields $\alpha_s^{\overline{\text{MS}}}(M_Z^2) \simeq 0.12$.

Figure 25.13 compares this with the fit using $\alpha_s = 0$.

Figure 25.14 shows the dependence of Γ_h upon m_t and m_H (Altarelli, 1990).

25.7 Basic Monte Carlo formulations

With the growing complexity of the final states in reactions as the energy increases, and with the ever-increasing complexity of detectors, it has become virtually impossible to go analytically from a theoretical formula to what this implies should be seen in an experiment. Thus even if the theory yields a precise formula for some multiparticle differential cross-section, the integration of this formula over the available phase space, taking into account the experimental cuts, is inevitably too complicated to do analytically. In this situation Monte Carlo methods can be used to estimate the actual cross-sections corresponding to the experimental conditions.

Another use, closely related in fact, is as an 'event generator'. A Monte Carlo program can provide simulated data, i.e. can simulate the events expected experimentally on the grounds of some given theoretical input.

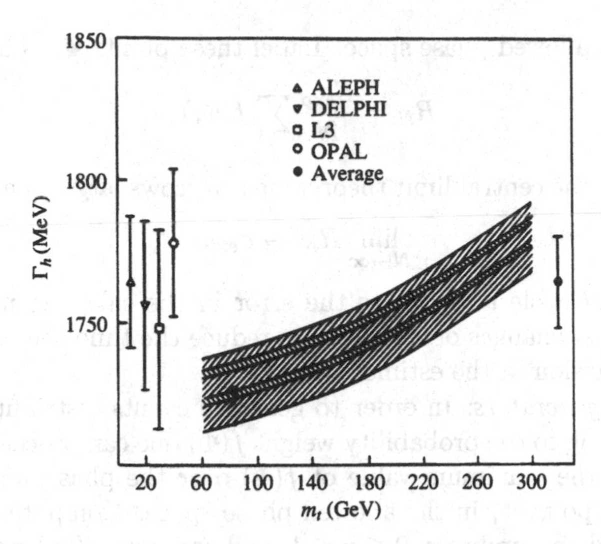

Fig. 25.14. The prediction of the standard model for the hadronic width Γ_h (obtained for $m_Z = 91.177 \pm 0.031$ GeV, $m_H = 40 - 1000$ GeV, $\alpha_s = 0.12^{+0.01}_{-0.02}$) as a function of m_t is compared with the LEP results. The central band is from δm_H. The two narrow intermediate bands are from δm_Z. The external bands are from $\delta\alpha_s$. All uncertainties are added linearly.

This aspect is especially useful in designing an experiment, for example, in deciding which regions of phase space should have good detection efficiencies.

We sketch very briefly the underlying ideas.

(*a*) *Cross-section estimates*: Suppose we have a formula for some fully differential cross-section

$$\frac{\mathrm{d}\sigma}{\mathrm{d}^3\boldsymbol{p}_1\mathrm{d}^3\boldsymbol{p}_2\cdots\mathrm{d}^3\boldsymbol{p}_n} = f(\boldsymbol{p}_1, \boldsymbol{p}_2, \ldots, \boldsymbol{p}_n) \tag{25.7.1}$$

and we wish to integrate over some region of phase-space Φ_{exp} specified by the experimental cuts, i.e. to evaluate

$$\sigma_{\mathrm{exp}} = \int_{\Phi_{\mathrm{exp}}} f(\boldsymbol{p}_1 \cdots \boldsymbol{p}_n)\mathrm{d}\Phi, \tag{25.7.2}$$

where $\mathrm{d}\Phi$ is short for $\mathrm{d}^3\boldsymbol{p}_1/(2\pi)^3 2E_1 \ldots \mathrm{d}^3\boldsymbol{p}_n/(2\pi)^3 2E_n$. Let the volume of this phase space be

$$S_{\mathrm{exp}} = \int_{\Phi_{\mathrm{exp}}} \mathrm{d}\Phi \tag{25.7.3}$$

which we assume known.

Then the integral (25.7.2) is estimated as follows. Choose, *at random*, *N* points in the allowed phase space. Label these points Φ_i. Then calculate

$$R_N = \frac{S_{\text{exp}}}{N} \sum_i f(\Phi_i). \tag{25.7.4}$$

According to the central limit theorem, as N grows large, one should have

$$\lim_{N\to\infty} R_N \to \sigma_{\text{exp}}. \tag{25.7.5}$$

It is also possible to estimate the error in the calculation of σ_{exp}. In practice, clever changes of variable can reduce the time needed to achieve a chosen precision in the estimate.

(*b*) *Event generators*: In order to generate events distributed in phase space according to the probability weight $f(\Phi)$ one can proceed as follows. Let f_{max} be the maximum value of $f(\Phi)$ over the phase space. Choose at random a point Φ_i in the allowed phase space. Compute $f(\Phi_i)$. Now choose a random number r, $0 \leq r \leq 1$, and compare $f(\Phi_i)$ with $r f_{\text{max}}$. If $f(\Phi_i) > r f_{\text{max}}$ accept, i.e. keep, this 'event'. If $f(\Phi_i) < r\, f_{\text{max}}$ reject the 'event'. Proceeding in this way one ends up with the collection of accepted 'events'. For N large enough they can be shown to be distributed through phase space according to $f(\Phi)$.

For a more precise treatment and for access to the literature, see Kleiss (1989).

25.8 QCD Monte Carlo programs

We give here a brief description of the ambitious programs which attempt to translate the information from the basic partonic process governing a reaction into details about final state hadrons, i.e. 'fragmentation' programs. The schematic structure of the program to generate hadronic events in an e$^+$e$^-$ collision is shown in Fig. 25.15. The physical interpretation is:

(i) an e$^+$e$^-$ pair annihilates into a virtual γ or a Z^0 and a $q\bar{q}$ pair is produced;

(ii) the $q\bar{q}$ pair radiates gluons which in turn may radiate partons (i.e. quarks and gluons);

(iii) the partons convert into colourless hadrons;

(iv) for jet studies, the same jet reconstruction procedure as is used on the experimental data (see Section 25.11) should be applied to the Monte Carlo events.

In region (ii) perturbative QCD is, in principle, applicable, but an exact treatment for more than four partons in the final partonic state is hopeless. In region (iii) the hadronization is non-perturbative and has to be modelled phenomenologically, thereby introducing several parameters

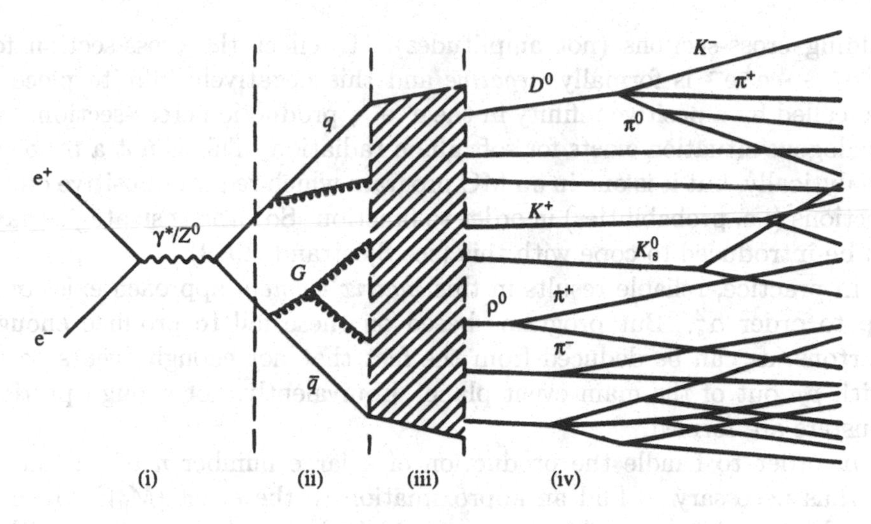

Fig. 25.15. Schematic illustration of $e^+e^- \to$ hadrons.

into the program. The models are designed to produce as many of what are believed to be the basic features of QCD as possible. The hope, clearly, is that this does not too much alter the results that one would obtain with *bona fide* QCD calculations.

As a consequence of the difficulties outlined above, all 'tests of QCD' which rely on Monte Carlo are somewhat biased by the subjective inclination to believe or not to believe the result obtained — '*intelligo ut credam*' ('I understand provided I believe' (St Augustine)).

Generally speaking, comparison of MC results with data for global parameters (such as those in Section 25.2) is, hopefully, testing the underlying hard, perturbative phase, whereas comparison with particle content and multiplicities should be very sensitive to the hadronization model.

In the following we outline just the principal features and differences of the QCD Monte Carlo programs. For details and access to the literature see Sjöstrand (1989) and Paige (1989).

25.8.1 *The perturbative phase*

Ideally one should compute the Feynman diagrams M_n to produce 2, 3, 4, ..., n partons, shown in Fig. 25.7. The events generated by a MC program using cross-sections based upon these would then be fed into the hadronization phase. But even given the exact M_n, the procedure is not straightforward on account of infrared singularities. It is well known from QED that infinities coming from internal loop corrections to, say, $e^+e^- \to e^+e^-$ are cancelled by infinities coming from the real emission of infrared photons in $e^+e^- \to e^+e^-\gamma$. These cancellations occur when adding cross-sections (not amplitudes). In effect the cross-section for $e^+e^- \to e^+e^-$ is formally *negative* and this negatively infinite piece is cancelled by a positive infinity in the $e^+e^-\gamma$ production cross-section. An analogous situation exists for soft gluon radiation. This is not a problem analytically, but it is one in an MC program which requires positive cross-sections (i.e. probabilities) in order to function. So various strategies have to be introduced to cope with this (see Sjöstrand, 1989).

In practice, reliable results in this *matrix element* approach exist only up to order α_s^2. But programs based on these fail to produce enough partons, as can be deduced from the fact that not enough events occur with p_T out of the main event plane. Equivalently, not enough particle clusters are formed.

In order to handle the production of a large number n of partons it is thus necessary to find an approximation to the exact $|M_n|^2$ which is simple enough to extend to large n and which, hopefully, is a reasonably good approximation to the true n parton cross-section.

The so-called *parton shower* is a generalization of the parton splitting approach used originally in deep inelastic scattering (Section 22.2.5). It amounts, in the present context, to the statement that the *dominant* term in the partonic cross-section σ_{n+1}, for $i+l \to j_1+j_2\dots j_n+k$ to produce $n+1$ partons, corresponding to the Feynman diagram Fig. 25.16, is given by a classical convolution of the probability for the parton i to split into partons j and k as shown, with the cross-section σ_n for $j+l \to j_1+j_2\dots j_n$. With $p_i = (p,0,0,p)$, $p_j = (zp+\frac{p^2}{2zp}, p_\perp, zp)$ one finds for the leading term

$$\sigma_{n+1} \approx \int \mathrm{d}p_\perp^2\,\mathrm{d}z \left[\frac{\alpha_s(p_\perp^2)}{2\pi p_\perp^2} P_{ji}(z)\right] \sigma_n. \tag{25.8.1}$$

In the deep inelastic case, the above is derived for essentially massless partons and the P_{ji} are the usual splitting functions given in (23.5.6, 7, 11, 12). In the Feynman diagrams appearing in e^+e^- collisions the internal parton lines can have large values of p^2. For example, in Fig. 25.17, t_1 could be of order s. It can be argued that a highly virtual parton (p_i) will mainly decay into partons (p_j and p_k) with $p_j^2, p_k^2 \ll p_i^2$, and that one reasonable estimate of the probability for this splitting is a modified form of (25.8.1) namely

$$\frac{\alpha_s(p_i^2)}{2\pi p_i^2} P_{ji}(z_i), \tag{25.8.2}$$

where now

$$z_i = \frac{E_j + |\boldsymbol{p}_j|}{E_i + |\boldsymbol{p}_i|}. \tag{25.8.3}$$

(Slightly different formulae are used in different Monte Carlo programs.)

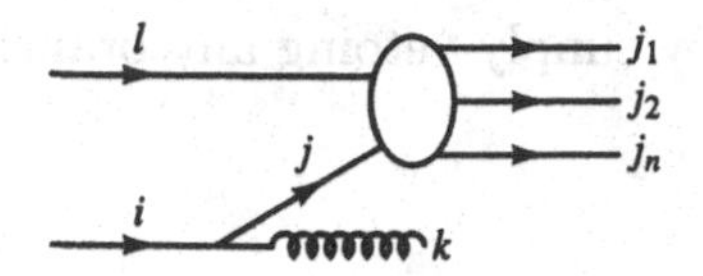

Fig. 25.16. Example of an n-jet diagram.

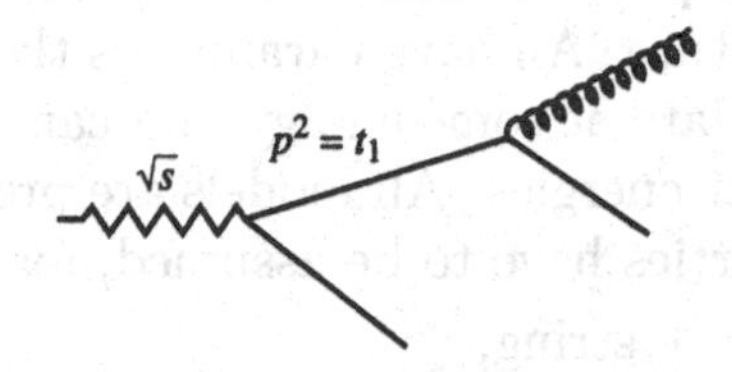

Fig. 25.17.

Note that several different recipes for the denominator in (25.8.2), for the argument of α_s and for the definition of z_i are to be found in the literature.

The above describes branching into hard partons. A complicating issue is that one should take into account the possible emission of soft radiation between the hard branchings; see Paige (1989). One can find a function (related to the so-called Sudakov form factor) which gives the probability $\Xi(t)$ that the first hard branching in a parton which initially had $p^2 = t_0$ will take place at some $p^2 = t < t_0$. To account for this, the events are generated by choosing the initial $p_1^2 = t_1^0$ within the kinematic limits, then choosing $t_1 < t_1^0$ according to the distribution $\Xi(t)$, allowing parton 1 to split with calculated probability into partons 2 and 3, then randomly choosing initial values $p_2^2 = t_2^0$ and $p_3^2 = t_3^0$ within the kinematically allowed domain, allowing them to split and continuing on until a value $p_n^2 = m^2$ is reached. At this point the perturbative branching must stop.

Two of the most commonly quoted Monte Carlos are HERWIG (Marchesini and Webber, 1984, 1988) and JETSET (Bengtson and Sjöstrand, 1987).

In HERWIG a more sophisticated form of parton shower is used which takes into account a property of the QCD Feynman amplitudes which tends to order the angles θ_n between the decay partons from successive branchings in such a way that $\theta_{n+1} < \theta_n$. This is implemented by replacing p_i^2 in the denominator of (25.8.2) by $E_i^2\left(\frac{p_j \cdot p_k}{E_j E_k}\right)^2$ for the branching $i \to j + k$.

In JETSET the first branching of the parton shower is constrained to agree with the explicit three-parton matrix elements. Angular ordering of some kind is achieved by simply vetoing any branching for which $\theta_{n+1} > \theta_n$.

25.8.2 The hadronization phase

Various models of how partons hadronize or fragment into hadrons have been used in the literature. All have parameters that have to be adjusted to reproduce existing data at some energy and can then be used for making predictions at other energies. All models are probabilistic and certain basic branching properties have to be assumed, for example:

(*a*) string $\to$ hadron + string;

(*b*) cluster$\to$ hadron + cluster;

with all variations on the theme (e.g. cluster $\to$ cluster + cluster...).

At each branching, all quantum numbers are to be conserved.

Here, we briefly introduce the basic ideas of string fragmentation (SF) and cluster fragmentation (CF) models.

(*a*) String fragmentation

The most popular of these models is the so-called Lund model and SF is now used in JETSET.

Under the assumption that QCD confines and that the confinement potential grows linearly, the energy in a $q\bar{q}$ colour dipole increases as the colour charges separate. As the partons move away from each other, the resulting colour flux tube is stretched while its transverse dimensions remain, more or less, the size of a hadron. The simplest Lorentz covariant and causal description of this physical picture is a relativistic string with no transverse degrees of freedom (Artru, 1983). As, for example, a q and $\bar{q}$ separate, the potential energy stored in the string increases until the string breaks with the creation of a $q'\bar{q}'$ pair. Effectively the original $q\bar{q}$ system has split into two colourless objects $q\bar{q}'$ and $q'\bar{q}$. This process continues until the invariant masses of the pairs are of the order of a hadron mass (or, in other versions, the mass of some cluster, typically a few GeV/c^2).

In the Lund model, the break-up of the string is attributed to quantum tunnelling with a gaussian probability $\exp(-\pi m_T^2/\kappa)$ where κ is the string tension and $m_T^2 = m^2 + p_T^2$ is the $q'\bar{q}'$ transverse mass. Note that this depresses the formation of heavy flavours as experimentally observed.

In this kind of scheme, baryons can be produced via the intermediate creation of $(q'q'')(\bar{q}'\bar{q}'')$ or of diquark pairs.

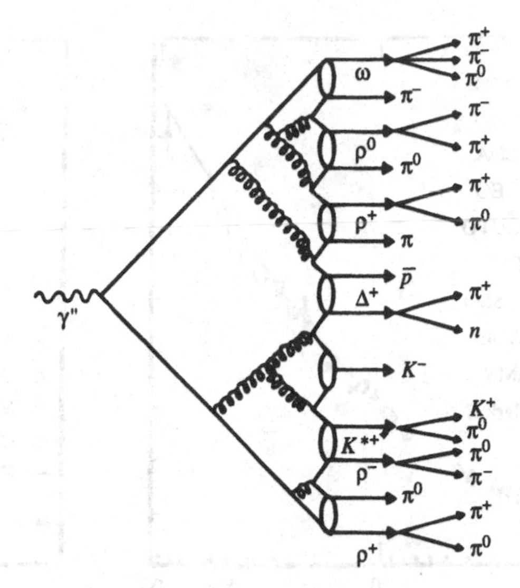

Fig. 25.18. HERWIG's cluster fragmentation scenario: showing shower evolution, $G \to q\bar{q}$ branchings, cluster formation, and cluster decay.

(b) Cluster fragmentation

Cluster fragmentation ideas are used in many models the most popular being HERWIG. In HERWIG, in the final stage of the parton shower the final gluons are forced to split into $q\bar{q}$ pairs which recombine with other nearby quarks to give colourless clusters which finally decay into hadrons (Fig. 25.18).

25.9 Multiplicity

The mean multiplicity measured by the LEP groups shows the expected increase with energy, suggested by data at lower energies. In fact, the LEP value is in good agreement with the prediction from the analytic QCD form (Kunszt and Nason, 1989)

$$\ln\langle n_{ch}(s)\rangle \approx \frac{A}{\sqrt{\alpha_s(s)}} + B \ln \alpha_s(s), \tag{25.9.1}$$

where

$$A = \frac{\sqrt{96\pi}}{\beta}; \; B = \frac{1}{4} + \frac{10n_f}{27\beta}; \; \beta = 11 - \frac{2}{3}n_f \tag{25.9.2}$$

which gives $\langle n_{ch}\rangle \sim 20.4$ at $s = M_Z^2$ to be compared with $\langle n_{ch}\rangle_{\rm expt} \simeq 20$–21. The data are shown in Fig. 25.19 together with predictions (*a*) from the Lund model and (*b*) from HERWIG at 91 GeV.

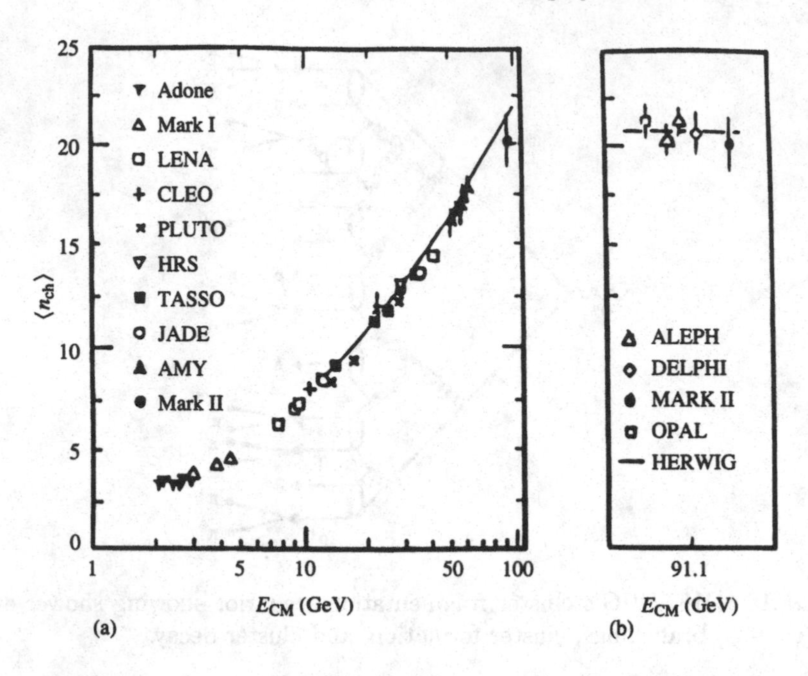

Fig. 25.19. Mean charged particle multiplicity in e^+e^- annihilation as function of the centre of mass energy; (a) the prediction of the Lund parton shower model and (b) the prediction of HERWIG at 91 GeV.

25.10 Global event-shape analysis

Given a set of experimental events one can evaluate the various global parameters which can be used to characterize the event. As stressed earlier one should, strictly speaking, only utilize parameters which are infrared and collinear-safe. For example, sphericity (25.2.1) is infrared-safe but not collinear-safe whereas spherocity (25.2.3) is both.

Tables of theoretical values of the global parameters are available based upon Monte Carlo calculations.

As an example, thrust (Section 25.2) equals $\frac{1}{2}$ for an isotropic configuration and 1 for a two-jet event. For a three-jet configuration, $2/3 \leq T \leq 1$ whereas $1/\sqrt{3} \leq T \leq 1$ for a four particle final state.

The data on thrust distribution is shown in Fig. 25.20 (for the curves, see Section 25.9.2). The parameters of the model, adjusted to fit LEP data, yield a successful description also at PETRA and PEP energies.

Fig. 25.20 clearly indicates a mixture of at least 2, 3 and 4 jet events, and good agreement between experimental data and the Monte Carlo calculation curves.

Several other event parameters have been introduced besides the ones

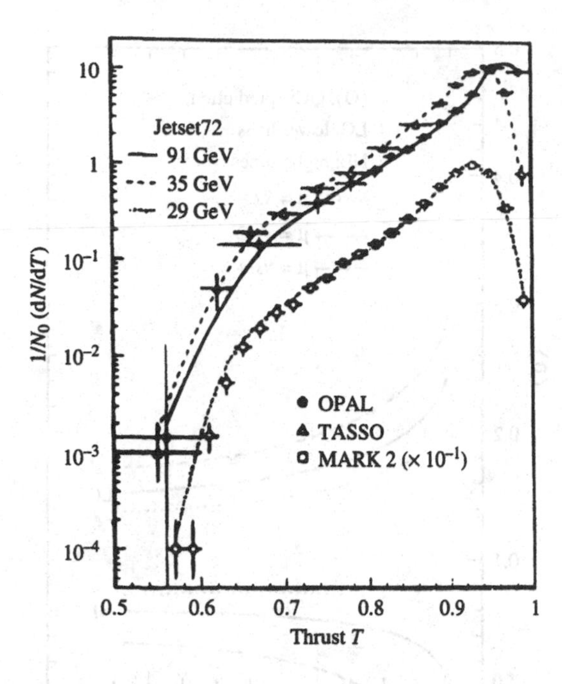

Fig. 25.20. A global fit of the LEP, PETRA, and PEP results on the thrust distribution (JETSET 7.2).

already defined (25.2.1–4). The most frequently used of these are:

$$(a)\quad \text{Oblateness} \equiv O \equiv T_{\text{major}} - T_{\text{minor}}, \tag{25.10.1}$$

where T_{major} is defined by maximizing the expression

$$T_{\text{major}} = \frac{\sum_i |\boldsymbol{E}_i \cdot \hat{\boldsymbol{n}}_{\text{major}}|}{\sum_i E_i} \tag{25.10.2}$$

by varying the unit vector $\hat{\boldsymbol{n}}_{\text{major}}$, subject to the restrictions that $\hat{\boldsymbol{n}}_{\text{major}} \cdot \hat{\boldsymbol{n}}_T = 0$ where $\hat{\boldsymbol{n}}_T$ is the unit vector which maximizes the thrust in equation (25.2.2), and T_{minor} by maximizing an analogous expression by varying $\hat{\boldsymbol{n}}_{\text{minor}}$ subject now to the restriction $\hat{\boldsymbol{n}}_{\text{minor}} \cdot \hat{\boldsymbol{n}}_T = 0$ and $\hat{\boldsymbol{n}}_{\text{minor}} \cdot \hat{\boldsymbol{n}}_{\text{major}} = 0$.

Note that because there is no detailed particle identification, one uses the vector $\boldsymbol{E}_i$ (instead of the particle momentum) whose magnitude is the energy flow into the ith calorimeter segment and whose direction is from the interaction point to the ith segment. Oblateness is zero for collinear and spherical events and it is comparatively large for planar events. For 3-jet events its value cannot exceed $1/\sqrt{3}$ and for a 4-jet planar event its

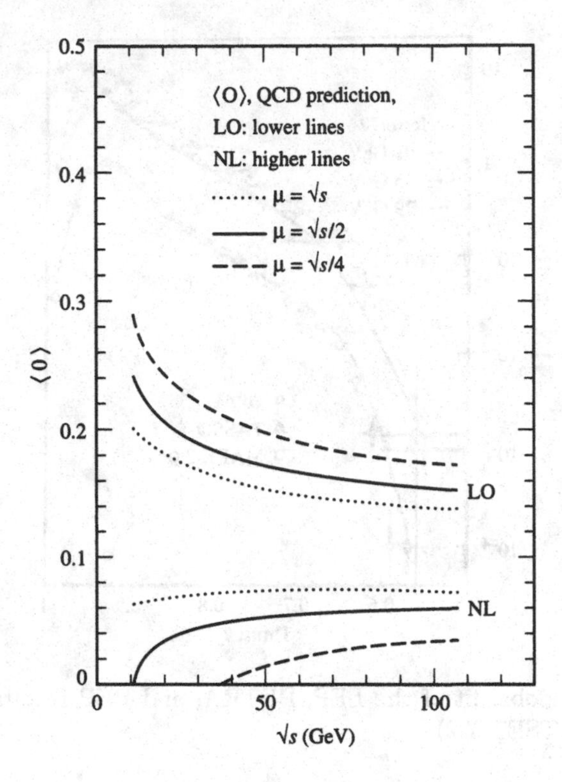

Fig. 25.21. Average value of oblateness versus the energy LO and NL results.

maximum value is $1/\sqrt{2}$. The MC calculated average value of oblateness vs. $\sqrt{s}$ is shown in Fig. 25.21 using leading order (LO) and next-to-leading order (NL) matrix element with various choices of the energy scale used in the effective coupling constant $\alpha_s(\mu^2)$.

(*b*) The energy–energy correlation (EEC) (Basham *et al.*, 1979) is defined as

$$\frac{\mathrm{d}\Sigma}{\mathrm{d}\cos\chi} = \int \mathrm{d}\Omega_1 \mathrm{d}\Omega_2 \frac{\mathrm{d}\Sigma}{\mathrm{d}\Omega_1 \mathrm{d}\Omega_2} \delta(\mathbf{\Omega_1} \cdot \mathbf{\Omega_2} - \cos\chi), \tag{25.10.3}$$

where $\mathrm{d}\Sigma/\mathrm{d}\Omega_1\mathrm{d}\Omega_2$ is obtained as follows.

For the jth event amongst a total of N events, let $\mathrm{d}E^j/\mathrm{d}\Omega_1$ and $\mathrm{d}E^j/\mathrm{d}\Omega_2$ be the energy flows per unit solid angle in the directions $\mathbf{d\Omega_1}$, $\mathbf{d\Omega_2}$ respectively. Then

$$\frac{1}{\sigma}\frac{\mathrm{d}\Sigma}{\mathrm{d}\Omega_1\mathrm{d}\Omega_2} \equiv \frac{1}{N}\sum_{j=1}^{N}\left(\frac{1}{s}\frac{\mathrm{d}E^j}{\mathrm{d}\Omega_1}\right)\left(\frac{1}{s}\frac{\mathrm{d}E^j}{\mathrm{d}\Omega_2}\right). \tag{25.10.4}$$

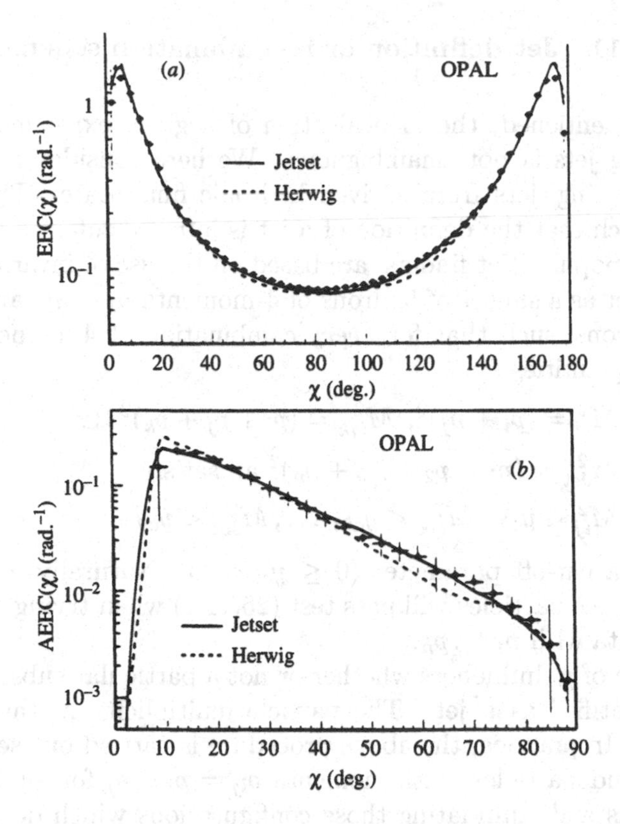

Fig. 25.22. EEC and AEEC distributions at LEP by OPAL, compared with JETSET and HERWIG fits (Dydak, 1991).

The definition (25.10.3) gives an average EEC where one integrates over all $\boldsymbol{\Omega}_1$ and $\boldsymbol{\Omega}_2$ directions keeping their relative angle fixed. MC calculated values for $\mathrm{d}\Sigma/\mathrm{d}\cos\chi$ have been given (see e.g. Kunszt and Nason, 1989).

(*c*) The asymmetry of energy–energy correlation (AEEC) defined as

$$\frac{\mathrm{d}\Sigma_A}{\mathrm{d}\cos\chi} \equiv \frac{\mathrm{d}\Sigma(180^\circ - \chi)}{\mathrm{d}\cos\chi} - \frac{\mathrm{d}\Sigma(\chi)}{\mathrm{d}\cos\chi} \tag{25.10.5}$$

has also been introduced on the grounds that it is less affected by non-perturbative hadronization effects and radiative corrections.

A comparison between measured and Monte Carlo calculated EEC and AEEC distributions is shown in Fig. 25.22. Generally, the agreement is good.

Several other 'jet-hunting' variables have been defined (the C and D parameters, triple energy correlation, heavy jet mass etc.). For all these, see Kunszt and Nason (1989).

25.11 Jet definition or recombination schemes

As already mentioned, the identification of a given experimental event as containing jets is not unambiguous. We here consider the problem of 'reconstructing' jets from a given hadronic final state. The schemes should be such that the definition of a jet is infrared safe.

The most popular 'jet finders' are based on the use of invariant masses to define a jet as a subset of hadrons of 4-momenta $p_1 \cdots p_n$ amongst the N final hadrons, such that for *every* combination of 4-momenta in the subset, the quantities

$$\begin{aligned}
&M_{ij}^2 = (p_i + p_j)^2, \; M_{ijk}^2 = (p_i + p_j + p_k)^2 \dots \\
&M_{\text{Jet}}^2 = (p_1 + p_2 + \dots + p_n)^2 \text{ all satisfy} \qquad (25.11.1) \\
&M_{ij}^2 < y_c s, \; M_{ijk}^2 < y_c s, \; \dots, M_{\text{Jet}}^2 < y_c s
\end{aligned}$$

where y_c is a cut-off parameter ($0 \leq y_c < 1$). Naturally, none of the remaining $N - n$ particles will pass test (25.11.1) when trying to combine their momenta with $p_1 \dots p_n$.

The choice of y_c influences whether or not a particular subset of events is to be identified as a jet. The particle multiplicity in the above jet would be n. In practice, the above procedure is carried out sequentially, forming 'pseudoparticles' with momenta $p_{ij} = p_i + p_j$ for all the N final state hadrons and eliminating those configurations which do not satisfy (25.11.1).

The choice of y_c is dictated by the desire to compare data with MC calculations of the 2-jet, 3-jet, 4-jet, ..., rates. In these calculations, too small a choice for y_c leads to partonic configurations where potentially infrared-singular cross-sections can be negative. For MC simulation at LEP energies it is recommended that $y_c \geq 0.01$ is safe. On the other hand, experience at PETRA energies suggests the need for a low value of y_c in order that manifestly 3-jet events (see e.g. Fig. 25.11) be identified as such.

The reader is warned that the handling of the kinematics is different in the various MC jet-algorithm programs, again because of the need to avoid potentially infrared-singular negative cross-sections. For example, in JADE, $(p_i + p_j)^2$ is taken to mean $2E_iE_j(1 - \cos\theta_{ij})$, i.e. parton masses are neglected.

The N-jet fractions (as fitted by L3) are shown in Fig. 25.23 and compared with a MC simulation. Good agreement is achieved over a wide range of y_c but the value of α_s involved, if interpreted as $\alpha_s(M_Z^2)$, is too large to be in agreement with the lower energy measurements of α_s and its expected energy dependence (25.5.1). Possible reasons for this are the following.

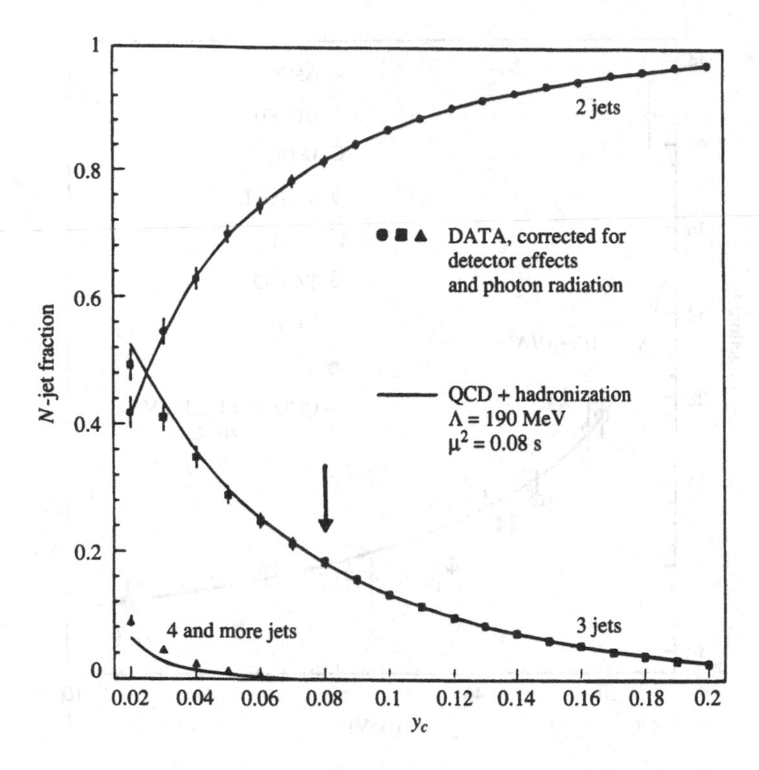

Fig. 25.23. The N-jet rates determined by L3 as function of y_c.

In lowest order for each type of event, the 2-jet, 3-jet and 4-jet cross-sections are proportional to α_s^0, α_s and α_s^2 respectively (see Fig. 25.7). However, it is found that the order α_s^2 corrections to the 3-jet rate are important.

On the other hand, the 4-jet rate is only known in lowest order (α_s^2) so that, if higher order corrections are important, the fit to the data could be distorting the value of α_s.

Also, since not all radiated gluons are hard, some softening of the scale μ^2 at which $\alpha_s(\mu^2)$ is operative seems sensible. With $\Lambda = 190$ MeV it turns out that the fit to the data requires $\mu^2 \approx 0.08$ s.

With the same choice of scale, the energy dependence of the 3-jet rate (defined with $y_c = 0.08$) is reasonably well reproduced, as shown in Fig. 25.24.

Recently a new clustering algorithm claimed to be immune from most of the diseases common to standard jet algorithms (such as JADE) has been proposed. This new algorithm is called $k_\perp$ (see Catani, 1992, and references therein) and the basic innovations are described very briefly in what follows.

The new algorithm tries to simulate some of QCD dynamical properties.

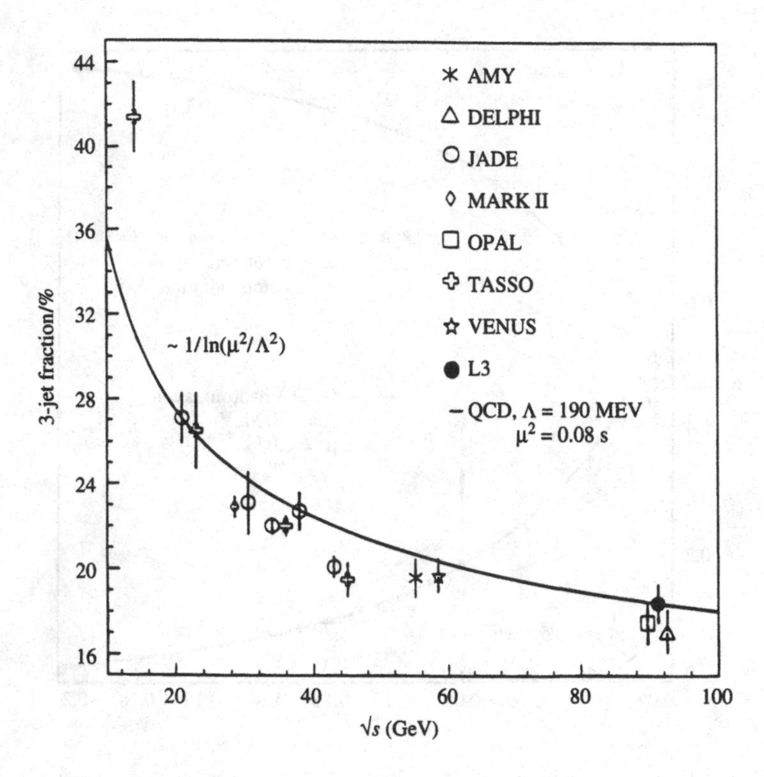

Fig. 25.24. The energy dependence of 3-jet production (Dydak, 1991).

The main observation is that jets, in the naive parton model, have a cylindrical shape in the sense that, with increasing total CM energy, they are produced with a limited fixed mean transverse momentum $\langle k_\perp \rangle$ with respect to the jet axis. Within QCD, however, as we have seen, the mean transverse momentum increases with energy so that the shape, as it should be intuitively, is *conical* rather than cylindrical. The idea is then to replace the invariant mass of the original JADE algorithm by the transverse momentum as the jet resolution variable, hence the name.

After this choice is made, the procedure parallels very much that of the JADE algorithm, namely: (i) a resolution parameter y_c is chosen; (ii) for every pair of hadrons a transverse momentum resolution variable y_{kl} computed; (iii) if y_{ij} is the smallest of the y_{kl} computed according to (ii) above and if $y_{ij} < y_c$, combine (p_i, p_j) into a single jet (a *pseudoparticle*) p_{ij} according to the chosen recombination prescription; and, finally, (iv) repeat the above steps until all pairs of objects (particles and/or pseudo particles) have $y_c > y_{kl}$. Whatever objects remain at this stage are called *jets*.

Without entering into the details of the procedure, for which we refer the reader to the literature, the preliminary indications seem to be that the scale at which the running coupling constant must be evaluated in order to reproduce the data does not have to be rescaled as enormously as in the jet algorithms discussed earlier in this section, the origin for this rescaling being an attractive soft gluon correlation artificially induced by the JADE algorithm itself (other jet schemes may also induce repulsive gluon correlations).

The hope is that these results will soon be confirmed and that the whole jet calculus will become a standard component of the perturbative QCD machinery enabling one to reliably compute high energy hadronic reactions.

25.12 Particle flow patterns in 3-jet events

It is found that the flow of particles into angular regions measured relative to the jet directions, shows interesting features. These are presumably linked to the complicated phenomena of angular ordering and coherent emission of soft gluons in QCD (Marchesini and Webber, 1988).

Figure 25.25 shows a projection onto the event plane of the distribution in angle of particles produced in the event plane, as measured from the most energetic jet 1 which is at $\theta = 0°$. The second (2) and the least energetic jet (3) (assumed to be the gluon) are indicated. Clearly the number of particles between 1 and 2 and between 1 and 3 are much smaller than between 2 and 3.

Monte Carlo simulations which do not include angular ordering (labelled IF for independent fragmentation) fail to reproduce these data. The curves in Fig. 25.25 show that both HERWIG and JETSET are reasonably successful. A different way of demonstrating the same effect is shown in Fig. 25.26 where the jets are ordered in energy $E_1 > E_2 > E_3$. Assuming the least energetic jet to be the gluon, the soft radiation in the $q\bar{q}G$ event lies predominantly between the qG and the $\bar{q}G$ jets. It is interesting that in the $q\bar{q}\gamma$ case the soft radiation occurs mostly between the $q\bar{q}$ jets.

Attempts have been made to explain this in terms of what is called 'colour drag', i.e. the coloured quarks are dragged towards the coloured gluon. Because the photon is uncharged, this effect does not occur in QED (Dokshitzer *et al.*, 1985).

25.13 To what extent is QCD being tested?

It is widely accepted that e^+e^- physics provides the best evidence for jets. It is less clear to what extent there is hard evidence in favour of perturbative QCD. Although it is encouraging that no explicit contradiction

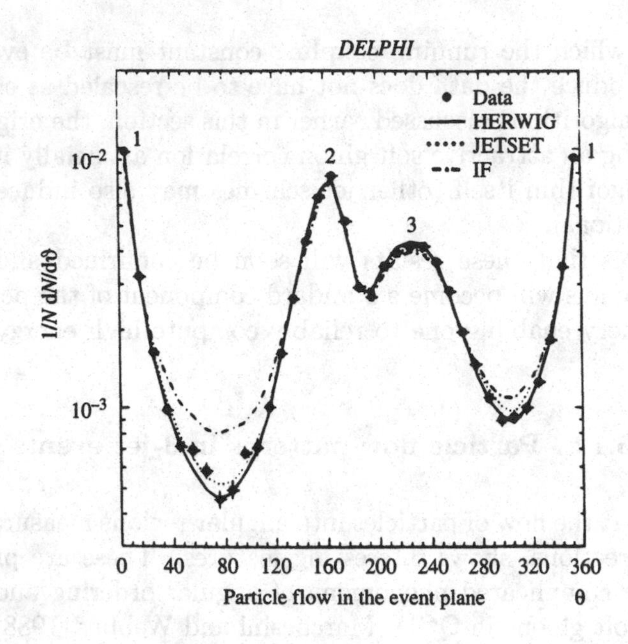

Fig. 25.25. The particle flow in 3-jet events. Data from DELPHI.

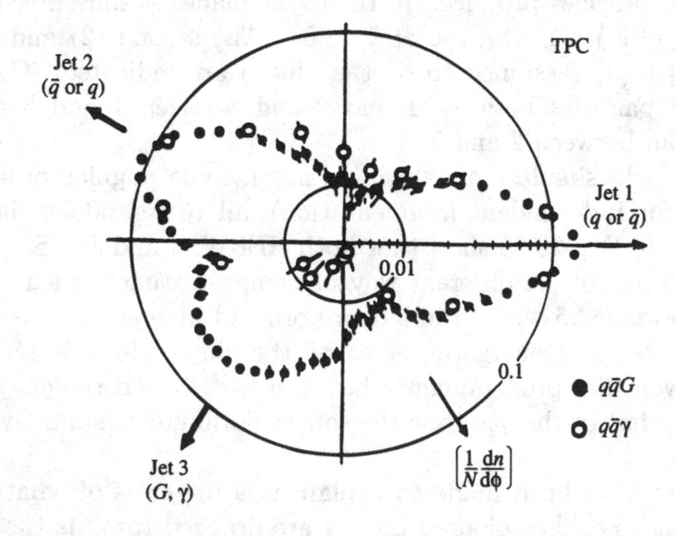

Fig. 25.26. Particle flow as a function of angle in the plane of the event.

with the predictions of perturbative QCD has so far emerged, care must be taken in assessing how successful the theory is given the many foreign ingredients that have to be used before a comparison can be made with the data.

The determination of α_s and of the scale Λ is tricky and care is required in defining clearly what is being determined. Let us just recall the main

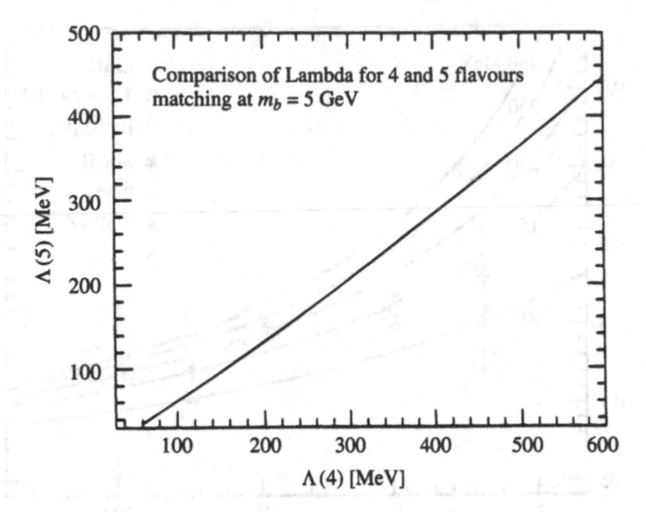

Fig. 25.27. Comparison for Λ for 4 and 5 light quark flavours, with matching at $m_b = 5$ GeV.

points: (i) Each determination of Λ is based upon a calculation to some order (e.g. LO or NLO) with data and this leads to different values of Λ. (ii) The value of Λ depends on the number of active flavours. Assuming continuity of $\alpha_s(Q^2)$, Λ must change discontinuously when crossing a quark threshold. Matching $\alpha^{(4)}(s) = \alpha^{(5)}(s)$ at $s = m_b^2 \simeq 25(\text{GeV}/c^2)^2$ yields the comparison between $\Lambda(5)$ and $\Lambda(4)$ shown in Fig. 25.27. (iii) Λ depends on the renormalization scheme.

From a compilation of various data (e^+e^- total cross-sections, $\Upsilon \to \gamma X$ and deep inelastic scattering), the picture that emerges for $\alpha_s(Q^2)$ and for $\Lambda(5)$ is summarized in Fig. 25.28. Given the large errors, this leads to

$$100\,\text{MeV} \leq \Lambda_{\overline{\text{MS}}}(5) \leq 250\text{ MeV} \tag{25.13.1}$$

but, as one can see, from Fig. 25.28 one is hard put to claim that α_s decreases with Q^2. More convincing evidence of the running of α_s is needed and better data with reduced errors are necessary for this.

Since the gluon is a key object in QCD, pinning down its properties from experiment is of great importance. The three-jet data from PETRA (Brandelik, 1980) were convincingly used to show that the gluon has spin 1 and not 0.

In lowest order the cross-section for $\text{e}^+\text{e}^- \to q\bar{q}G$ is given as a function of the scaled energies $x_1 = 2E_q/\sqrt{s}$, $x_2 = 2E_{\bar{q}}/\sqrt{s}$ and $x_3 = 2E_G/\sqrt{s}$ with $0 < x_i < 1$ and $\sum_i x_i = 2$ as

$$\frac{1}{\sigma_0}\frac{\text{d}^2\sigma}{\text{d}x_1\text{d}x_2} = \frac{\alpha_s}{2\pi}C_F\frac{x_1^2 + x_2^2}{(1-x_1)(1-x_2)} \tag{25.13.2}$$

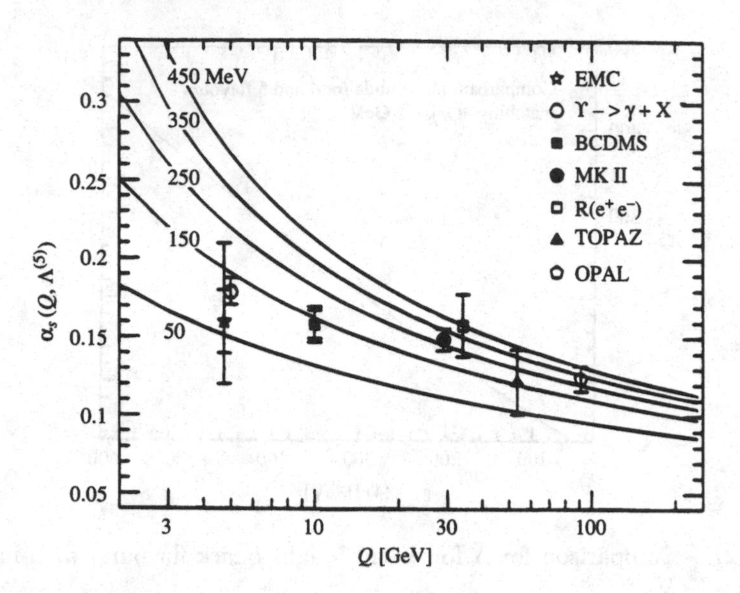

Fig. 25.28. Measurements of α_s compared with predictions for various values of $\Lambda(5)$.

where σ_0 is the lowest order cross-section for $e^+e^- \to q\bar{q}$ and $C_F = \frac{4}{3}$ is the colour factor.

The dependence on x_i is very different for a spin-zero gluon; one has

$$\frac{1}{\sigma_0}\frac{d^2\sigma}{dx_1 dx_2} \propto \frac{(2-x_1-x_2)^2}{(1-x_1)(1-x_2)} \tag{25.13.3}$$

and this is excluded by the data (Brandelik, 1980).

Equally important is to check the non-Abelian nature of the gluon coupling, i.e. the three-gluon vertex. The latter contributes an additional 4-jet final state (through the last diagram of Fig. 25.7) as compared with Abelian theories in which this coupling is missing.

Various analyses of LEP data compare fits using QCD (i.e. the triple gluon vertex) with Abelian toy models in which the gluon self-coupling is switched off. Calculations of the 4-jet rate are claimed to be *closer* to data for the non-Abelian case (L3, 1990; DELPHI, 1990).

26

Low p_T or 'soft' hadronic physics

The general subject of purely hadronic reactions, being somewhat removed from the main thread of this book, has not been covered. In any case it would deserve an extensive discussion in itself. In what follows we shall have to make occasional references to ideas stemming from Regge theory. For a complete treatment of the general principles the reader is referred to the classical texts (Eden, 1967; Collins, 1977). For a more *modern* presentation see Collins and Martin (1984).

Here, we shall confine ourselves to an outline of the recent developments that have occurred. Note that it is not possible to apply perturbative QCD arguments to these reactions; they are inherently non-perturbative. Consequently our discussion is largely phenomenological.

26.1 The total and elastic cross-sections

While the pp total cross-section $\sigma_{\text{tot}}(\text{pp})$ has been measured only up to ISR energies ($\sqrt{s} \sim 63$ GeV), the p$\bar{\text{p}}$ one has been measured up to CERN Sp$\bar{\text{p}}$S and FNAL Tevatron energies ($\sqrt{s} = 546$ and 900 GeV and $\sqrt{s} = 1.8$ TeV respectively).

The standard method of measuring σ_{tot} is to use the optical theorem (see Chapter 21 of Gasiorowicz, 1976) and extrapolate the measured elastic differential cross-section to $t = 0$ via

$$\begin{aligned} \sigma^2_{\text{tot}} &= \frac{16\pi}{1+\varrho^2}\frac{\text{d}\sigma}{\text{d}t}\bigg|_{t=0} \\ &= \frac{16\pi}{1+\varrho^2}\frac{\text{d}N_{\text{el}}(t)}{\text{d}t}\bigg|_{t=0}\frac{1}{\mathcal{L}} \end{aligned} \tag{26.1.1}$$

where $\mathcal{L}$ is the luminosity and ϱ is the ratio of the real to the imaginary forward scattering amplitude (strictly, the spin-averaged amplitude). Because beam luminosity measurements are not very accurate, some ex-

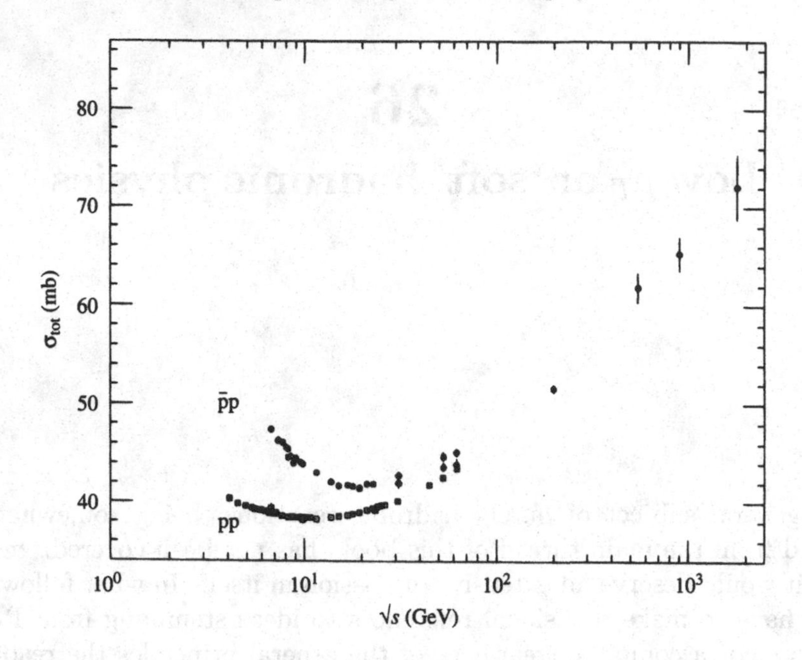

Fig. 26.1. pp and p̄p total cross-sections up to the highest energies presently available.

perimental groups also use a luminosity independent method. This is obtained by writing

$$\sigma_{\text{tot}} = \frac{(N_{\text{el}} + N_{\text{inel}})}{\mathcal{L}} \tag{26.1.2}$$

and using it for one of the factors of σ_{tot} in (26.1.1) to get

$$\sigma_{\text{tot}} = \frac{16\pi}{1+\varrho^2} \frac{\mathrm{d}N_{\text{el}}/\mathrm{d}t|_{t=0}}{N_{\text{el}} + N_{\text{inel}}} \tag{26.1.3}$$

In both (26.1.1) and (26.1.3) one needs either to measure or guess the value of ϱ. Measurements of ϱ require detection of Coulomb-nuclear interference. Published values of $\sigma_{\text{tot}}(\text{pp})$ and $\sigma_{\text{tot}}(\bar{\text{p}}\text{p})$ up to the highest energies are shown in Fig. 26.1.

Until recently almost everybody would have agreed that the high energy growth of the data is best reproduced by a $\ln^2 s$ behaviour. In part, this choice is motivated by its giving a good account of the data and, in part, by theoretical prejudices. A $\ln^2 s$ functional form, in fact, is the fastest growth permitted by the celebrated Froissart bound.

Also, most physicists believed that the pp and p̄p total cross-sections would become equal asymptotically as suggested by the Pomeranchuk theorem and as is compatible with the data in Fig. 26.1. This has been put into question by the UA4 measurement of ϱ to be discussed in Section 26.3.

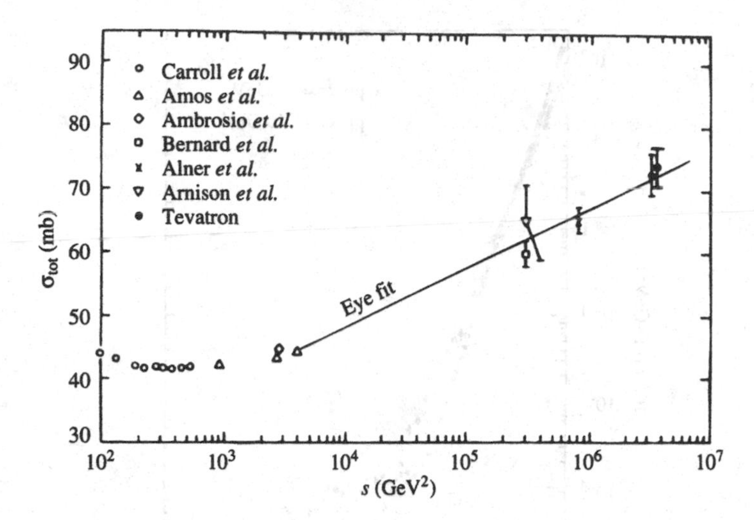

Fig. 26.2. 'Linear' $\ln s$ behaviour of $\sigma_{\text{tot}}(\bar{p}p)$ in the high energy domain (from Leader, 1992).

Note that it can be shown that the ratio $\sigma_{\text{tot}}(\text{pp})/\sigma_{\text{tot}}(\bar{\text{p}}\text{p})$ must tend to one asymptotically but that this does not imply equal cross-sections if they are both growing with energy.

The Fermilab total cross-section at $\sqrt{s} = 1.8$ TeV has varied significantly with time [a value ~85 mb was announced at Moriond in 1989; the printed version had been 75.4 mb; it was 78.3 mb at Blois III in 1989 and 72.1±2 mb was published more recently (Huth, 1991)]. A consensus now seems to have been reached between the luminosity dependent measurement by CDF [measures $\sigma_t^2(1+\varrho^2)$ and uses $\varrho = 0.145$] and the luminosity independent one by E710 [measures $\sigma_t(1+\varrho^2)$ and also $\varrho = 0.140 \pm 0.069$]. The values announced by CDF (by S. White) and by E710 (by S. Shukla) at Blois IV [see, e.g., Leader (1992); see also Huth (1991)] were

$$\left.\begin{array}{lrcl} \text{E710}: & \sigma_{\text{tot}}(\bar{\text{p}}\text{p}) & = & 72.8 \pm 3.1 \text{ mb} \\ \text{CDF}: & \sigma_{\text{tot}}(\bar{\text{p}}\text{p}) & = & 72.0 \pm 3.6 \text{ mb} \end{array}\right\} \tag{26.1.4}$$

The data from the Tevatron at FNAL could be construed as casting doubt on the $\ln^2 s$ growth, but the significance of this is not yet clear. That the *high energy* data look linear in $\ln s$ is shown in Fig. 26.2.

Another point which deserves attention is the ratio $\sigma_{\text{el}}/\sigma_{\text{tot}}$ of the elastic to the total cross-sections. This ratio was thought to be constant (~0.175) up to ISR energies but has risen to 0.215±0.005 at the Sp$\bar{\text{p}}$S CERN collider and to 0.228±0.012 at the FNAL Tevatron. This rise has strong implications for theoretical models (Giffon *et al.*, 1984).

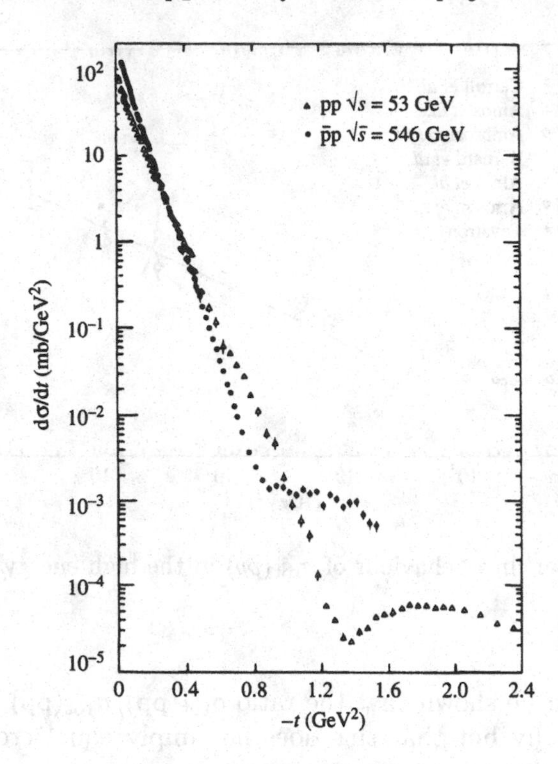

Fig. 26.3. pp and p̄p differential cross-sections at high energies.

26.2 The differential cross-section

At small values of momentum transfer the CERN collider data (Fig. 26.3) show no surprises, but at large $|t|$ the p̄p angular distribution shows interesting deviations from a simple extrapolation of the pp and p̄p data at lower energies (Bozzo *et al.*, 1984a, b).

The diffraction minimum has moved to a smaller value of $|t|$ and a high *shoulder* has appeared at $t \simeq -1.2\ \text{GeV}^2$. Taken in conjunction with ISR data on p̄p there seems to be growing evidence that pp and p̄p elastic scattering are still unequal even at these huge energies, implying that the crossing-odd amplitude (F_-) is not negligible, a conclusion very much at variance with standard Regge-type arguments which would suggest

$$\left|\frac{F_-}{F_+}\right| \sim (\sqrt{s}\ln^2 s)^{-1} \tag{26.2.1}$$

If F_- were to remain comparable to F_+ asymptotically it would imply that $\sigma_{\text{tot}}(\text{pp}) \neq \sigma_{\text{tot}}(\bar{\text{p}}\text{p})$ as $s \to \infty$. With both cross-sections growing, however, their ratio does nonetheless approach one asymptotically.

For a survey of the situation and a discussion of the intriguing possibility that also $|F_-|$ grows like $s \ln^2 s$ (the so-called *Odderon* scenario) see Gauron, Leader and Nicolescu (1985, 1990). Other interesting possibilities are discussed in Donnachie and Landshoff (1983, 1984).

At the time of writing no data in the dip region had yet appeared from the Tevatron.

26.3 The real to imaginary ratio

In the 'standard' picture of high energy elastic diffraction the crossing-odd amplitude (F_-) becomes negligible compared with the crossing-even amplitude (F_+) as the energy increases. Analyticity then implies that in the forward direction at $t = 0$,

$$\varrho \equiv \frac{\mathrm{Re} F_+(s,0)}{\mathrm{Im} F_+(s,0)} \to \frac{\mathrm{const}}{\ln s} \to 0 \tag{26.3.1}$$

On the basis of measurements at lower energies, and a careful quantitative analysis, using dispersion relations (in which it is assumed that $\sigma_{\mathrm{tot}}(\mathrm{pp}) = \sigma_{\mathrm{tot}}(\bar{\mathrm{p}}\mathrm{p})$ asymptotically), it was expected that $\varrho \simeq 0.12$ at $\sqrt{s} = 546$ GeV.

A few years ago, however, an unusually large value for ρ was reported by the UA4 group at CERN (UA4, 1987) using Coulomb interference techniques, namely

$$\varrho = 0.24 \pm 0.04 \tag{26.3.2}$$

Taken at face value, (26.3.2) would have remarkable implications, such as either the opening up of a new threshold beyond ISR energies or the validity of the odderon scenario. These possibilities are exemplified in Fig. 26.4.

The value (26.3.2) has come under renewed scrutiny since E710 and CDF released their preliminary measurements of ϱ at the Tevatron. The value found by E710 (1992), for instance,

$$\varrho_{\bar{\mathrm{p}}\mathrm{p}}(1.8 \text{ TeV}) = 0.126 \pm 0.067 \tag{26.3.3}$$

is barely compatible with (26.3.2). Barring more or less extraordinary explanations which would make (26.3.2) and (26.3.3) compatible, the danger of too simple an extrapolation in t towards $t = 0$, for example with an exponential, may be responsible for such different results [for a complete discussion see Leader (1992)]. A selection of data is shown in Fig. 26.5.

More precise results are needed to settle the issue; at the time of writing, the UA4 group is in the process of remeasuring ϱ at $\sqrt{s} = 546$ GeV.

Given the difficulty of this kind of Coulomb interference measurements and their interpretation, we cannot sufficiently stress the importance of a

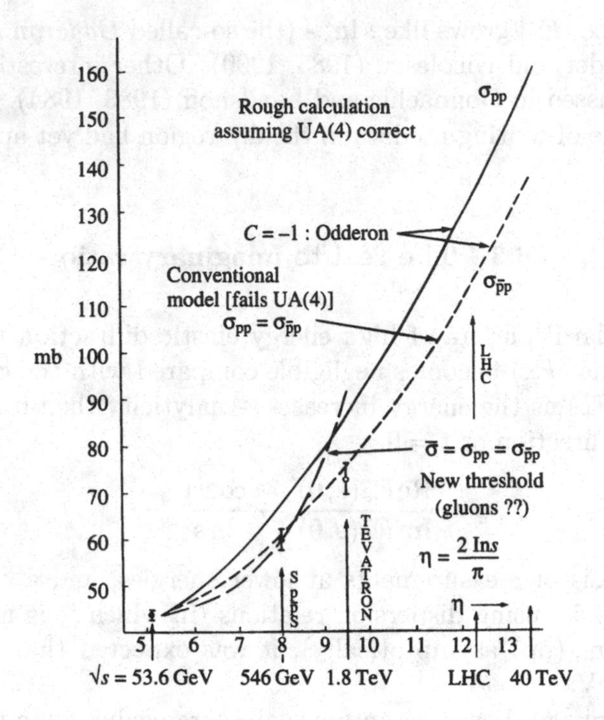

Fig. 26.4. *Conventional* vs. *odderon* and *new threshold* scenarios. The cross-sections expected are plotted as function of $\eta = 2\ln s/\pi$. This shows the importance of measuring $\sigma_{\text{tot}}(\text{pp})$ at high energies (from Leader, 1992).

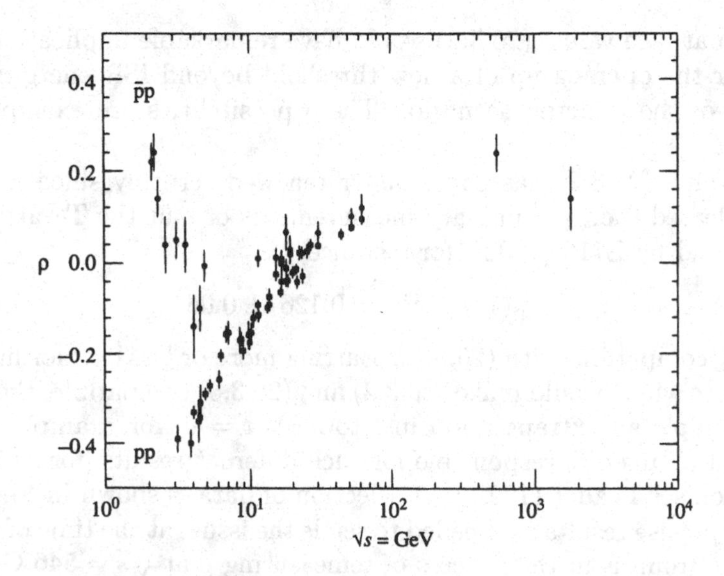

Fig. 26.5. The ϱ parameter (26.3.1) for pp and p̄p (compilation by the authors).

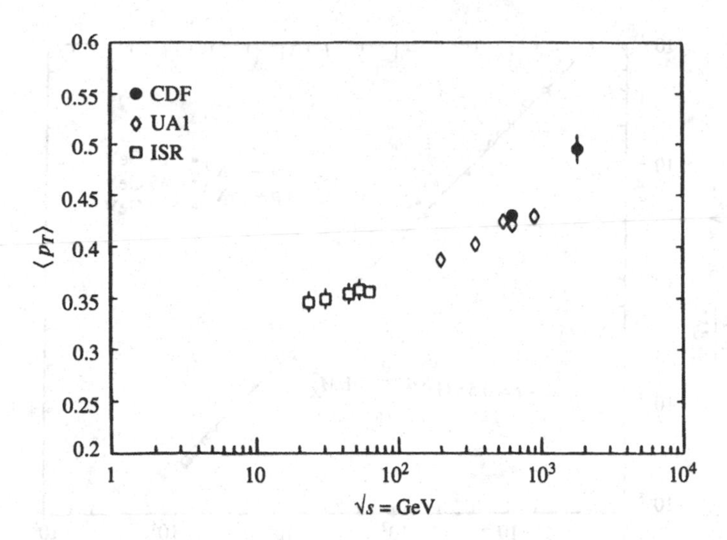

Fig. 26.6. Energy dependence of $\langle p_T \rangle$. UA1 results are averaged over jet and non-jet samples; ISR values are averaged over particle types. (From CDF, 1988.)

direct comparison of pp and $\bar{\text{p}}$p total cross-sections at very high energies (see Fig. 26.4).

26.4 The inclusive p_T distribution

At the ISR the mean p_T for inclusive reactions was $\langle p_T \rangle \sim 0.3-0.4$ GeV/c, and seemed to be increasing like $\ln s$. At CERN collider energies $\langle p_T \rangle \sim 0.5$ GeV/c and the value seems to be larger the heavier the inclusively produced particle is. The trend with energy has now been confirmed by the CDF Tevatron data (CDF, 1988), see Fig. 26.6. This trend is compatible also with cosmic ray data which had suggested a significant growth in $\langle p_T \rangle$ at a time when accelerator results seemed to be compatible with a constant behaviour.

26.5 Diffractive dissociation

Although diffractive dissociation is, in itself, a large subject, we are only able to make a few brief comments here.

A typical example of a diffractive reaction is

$$\bar{\text{p}}\text{p} \rightarrow \bar{\text{p}}\text{X} \tag{26.5.1}$$

which has been studied in the kinematic region where $M_X^2 \ll s$ by both the UA4 and UA5 groups at CERN (Palladino, 1985). In this region the

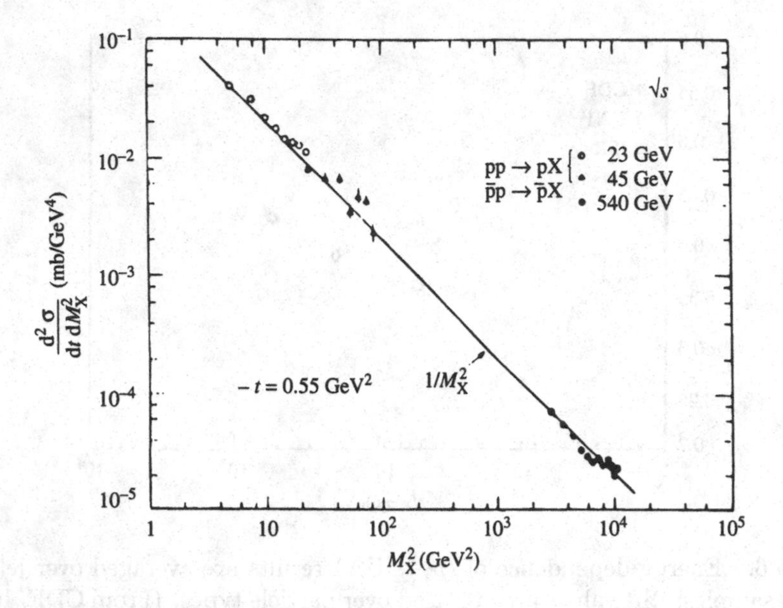

Fig. 26.7. Single diffractive dissociation (from Palladino, 1985).

reaction is viewed as a pseudo-elastic diffractive process (referred to as *single diffractive*) and is expected to have certain features in common with elastic scattering, in particular that $d\sigma/dt$ at fixed M_X^2 and small t does not decrease with s. This is borne out by the measurements (Fig. 26.7). These also show that at fixed t the cross-section drops off like $1/M_X^2$ a behaviour which had been discovered at lower energies and which is expected theoretically on the basis of the so-called *triple-Pomeron* mechanism. See Collins, (1977).

Diffractive events are also possible in deep inelastic lepton–hadron scattering and studies of these may yield information on the partonic structure of the Pomeron [see Ingelman and Schlein (1985), Donnachie and Landshoff (1984)], since, like elastic scattering, it can be interpreted as arising from Pomeron exchange.

26.6 The average multiplicity

The average multiplicity of charged particles produced was originally believed to be growing like $\ln s$. Experiments at the ISR suggested a $\ln^2 s$ growth and this appears to be confirmed by the CERN collider measurement. An example of a fit to the data in Fig. 26.8 is:

$$\begin{aligned}\langle n(s)_{ch}\rangle &= (0.80 \pm 0.12) + (0.47 \pm 0.05)\ln s + \\ &+ (0.114 \pm 0.005)\ln^2 s\end{aligned} \tag{26.6.1}$$

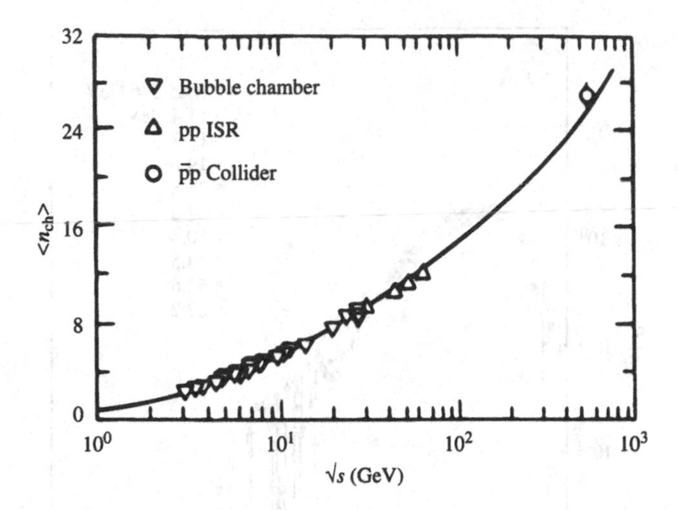

Fig. 26.8. Average multiplicity growth from *low* up to Sp̄pS collider energies (from UA5, 1986).

but there is nothing fundamental in this kind of fit. [In (26.6.1) s is measured in units of $(\mathrm{GeV})^2$.]

It also appears that the same multiplicity formula $n(s)_{ch}$ fits the charged particle production in the single diffractive type events discussed above, provided one takes $s = M_X^2$ in equation (26.6.1) (UA5, 1986).

26.7 The multiplicity distribution of charged particles

It was thought for a long time that the distribution of multiplicity n followed a very simple law known as 'KNO scaling' (Koba, Nielsen and Olesen, 1972) according to which the probability of finding n particles in an event was given by

$$P_n(s) = \frac{1}{\langle n\rangle}\Psi(n/\langle n\rangle) \tag{26.7.1}$$

where Ψ is a universal function. The CERN collider data, however, gave the first clear evidence that the KNO behaviour no longer holds; Fig. 26.9 (UA5, 1983, 1984 and 1986).

A very good fit to the data is achieved with a *negative binomial distribution* (Giovannini, 1973; Giovannini and Van Hove, 1986)

$$P_n(s) = \begin{pmatrix} n+k-1 \\ k-1 \end{pmatrix} \left[\frac{\langle n\rangle/k}{1+\langle n\rangle/k}\right]^n (1+\langle n\rangle/k)^{-k} \tag{26.7.2}$$

where the parameter k is found to have the value (UA5, 1985)

$$k = 0.028\ \ln(s/32) \tag{26.7.3}$$

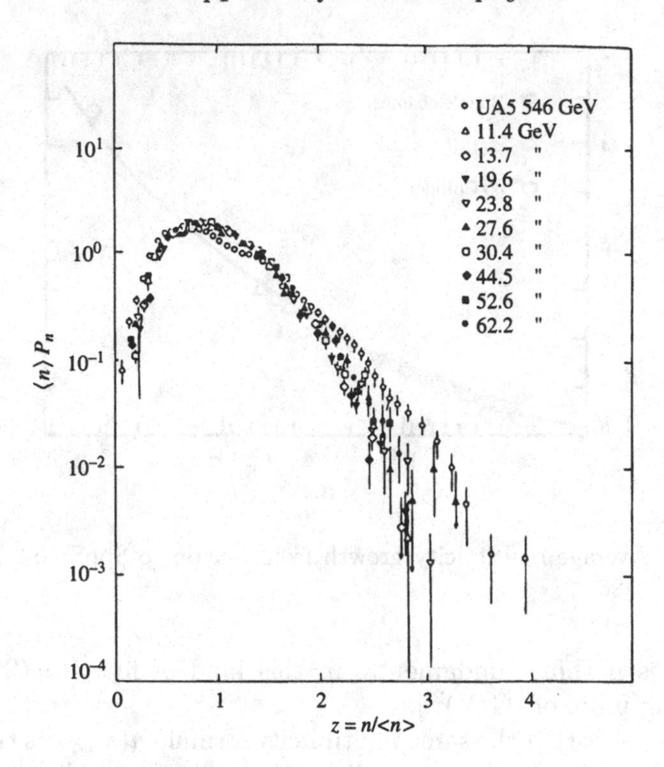

Fig. 26.9. Breaking of KNO scaling in going from 'low' energies up to the S$\bar{p}$pS CERN collider data (data from UA5, 1983, 1984 and 1986).

A typical example of this kind of fit is shown in Fig. 26.10.

It is too early to assess in a definitive way whether this distribution has any fundamental significance. Tests at higher energies will be interesting. Equations (26.7.2, 3) predict $k \sim 3.3$, $\langle n \rangle \sim 35$ at $\sqrt{s} = 900$ GeV and $k \sim 2$, $\langle n \rangle \sim 95$ at $\sqrt{s} = 40$ TeV.

26.8 Conclusions

The $\bar{p}p$ colliders have proved tremendous machines in advancing our physics knowledge, in providing lots of confirmations of *expected* high energy phenomena and very valuable information on hadronic physics both in the large p_T (Chapter 24) and in the low p_T domain. As already mentioned, this chapter is only intended to provide an extremely succinct introduction to the subject of low p_T physics and the reader is urged to go to the specialized literature quoted earlier for a more complete presentation.

We draw the reader's attention to the recent great revival of interest in low p_T physics as a consequence of its links to the behaviour of the parton

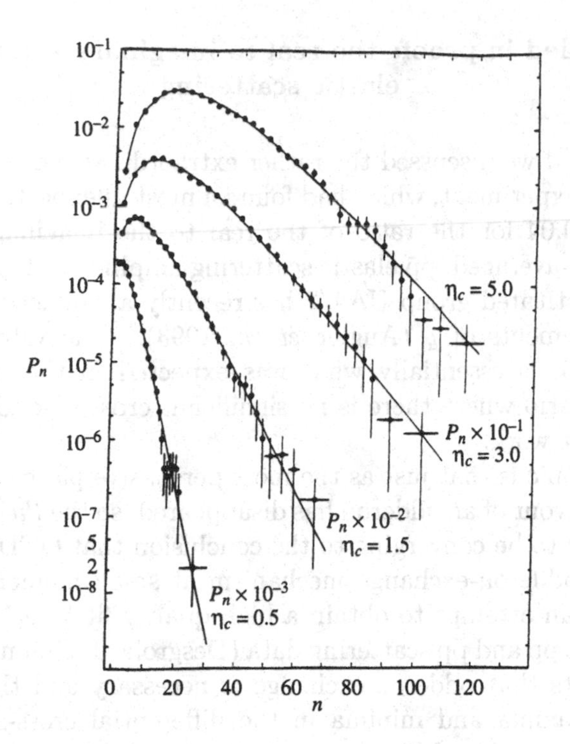

Fig. 26.10. Charged multiplicity distributions P_n at the S$\bar{p}$pS collider ($\sqrt{s}$ =560 GeV) for various values of pseudorapidity η_c. Curves are from equation (26.7.2) (from UA5, 1985).

number densities $q(x)$, $G(x)$ as $x \to 0$ (see Levin and Ryskin, 1990) (this connection was briefly discussed in Section 16.6; see also Section 23.6). Also, there is new and promising work on the QCD interpretation of the Pomeron, which uses integral equation techniques to go beyond a perturbative treatment (Lipatov, 1974, 1992). Progress in this field looks very encouraging.

Note added in proof: the real to imaginary ratio, ϱ, in $p\bar{p}$ elastic scattering

In Section 26.3 we discussed the rather extraordinary implications of the CERN UA4 experiment, which had found a most unexpectedly large value $\varrho = 0.24 \pm 0.04$ for the ratio of the real to the imaginary part of the forward, spin-averaged, $p\bar{p}$ elastic scattering amplitude at $\sqrt{s} = 546$ GeV.

The reconstituted group UA4/2 has recently announced the results of its remeasurements of ϱ (Augier *et al.*, 1993). The value found, $\varrho = 0.135 \pm 0.015$, is essentially what was expected in the traditional dynamical scenario where there is no significant crossing-odd (or odderon) mechanism at work.

What is ironic is that just as the most persuasive piece of *experimental* evidence in favour of an odderon has disappeared, so the *theoretical* studies of QCD seem to be converging to the conclusion that QCD does produce an effective odderon-exchange mechanism at small momentum transfer. In addition, an attempt to obtain a high quality fit to *all* available high energy elastic pp and $p\bar{p}$ scattering data (Desgrolard, Giffon and Predazzi, 1994) suggests that odderon-exchange *is* necessary and that interesting secondary maxima and minima in the differential cross-section should appear as the region of higher momentum transfers is explored. Some evidence ought to show up already at RHIC energies in the $\sqrt{s} = 500$ GeV pp angular distribution.

All of this emphasizes most strongly how important it is to be able to compare the total cross-sections for pp and $p\bar{p}$ at very high energies, something which will be become possible when RHIC comes into operation.

27
Some non-perturbative aspects of gauge theories

The vast majority of phenomena that can be studied in a field theory rely upon perturbation theory. For strong interactions, governed by a large coupling constant, this obviously leaves one in a somewhat hopeless situation.

In QCD the miracle of asymptotic freedom, that is, the fact that the *effective* coupling becomes weak at short distances, allows us to use perturbative methods to study many interesting and important physical processes, provided only that we restrict ourselves to kinematic regions where the perturbative approach can be justified. As is clear from the subject matter of this book there is a huge industry devoted to perturbative QCD (PQCD). Nonetheless the *vast majority* of events in a typical hadronic reaction fall into a kinematical regime where we cannot use PQCD. For example, in our long discussion of electroweak interactions we were forced to treat the hadronic matrix elements of the electroweak currents as incalculable, parameterizing them in terms of constants or functions to be determined from experiment. We invoked symmetry arguments to limit the number of unknown parameters but we did not produce any dynamical scheme for calculating them. Thus matrix elements of the type $\langle 0|h|\pi\rangle$ etc. or KM matrix elements like $\langle \mathrm{n}|h^{\mu}_{+}|\mathrm{p}\rangle$ etc. ought, in principle, all to be calculable; yet they cannot be attacked perturbatively. Even the fundamental issue of confinement is beyond the realm of PQCD.

It is thus of the utmost importance to try to find approaches to QCD which do not utilize a perturbative expansion in the coupling constant. Much has been achieved in the past decade, including a major breakthrough via a semi-classical approach based upon the concept of *instantons.* We shall only have space to present a brief introduction to some of the most active areas of research: QCD sum rules, the lattice approach, instantons and baryon non-conservation in electroweak theory. The subject of high energy scattering at *small momentum transfer* was dealt with briefly in Chapter 26.

27.1 QCD sum rules

Our aim is to give the reader only a general idea of the method. We do not present any detailed discussion of techniques or calculations. For these, and for a very lucid explanation of the whole approach, the reader should consult the original paper of Shifman, Vainshtein and Zakharov (1978). More modern developments are covered in Reinders, Rubinstein and Azachi, (1985).

One attempts to calculate various low-energy hadronic properties (masses, decay-widths, form factors, etc.) with sum rules that utilize asymptotic freedom and some information (non-perturbative) about the structure of the true QCD vacuum. The sum rules relate properties of physical states at small time-like values of k^2 to the behaviour in the perturbative region of large, negative space-like k^2.

As an example consider the correlation function $\Pi_{\mu\nu}$ of two electromagnetic currents introduced in Section 17.4.1.

It can be written as

$$\Pi_{\mu\nu}(k^2) = \mathrm{i}\int \mathrm{d}^4 z \langle 0|T[J_\mu^{\mathrm{em}}(z)J_\nu^{\mathrm{em}}(0)]|0\rangle \mathrm{e}^{\mathrm{i}kz}. \tag{27.1.1}$$

The perturbative contributions to $\Pi_{\mu\nu}$

would only be reliable if the coupling constant α_s were small. In Chapter 22 we showed how the renormalization group gives corrections to the simplest perturbative diagrams, resulting in α_s being replaced by the effective coupling constant $\alpha_s(k^2)$ which goes to zero as $k^2 \to -\infty$ (asymptotic freedom). The improved calculation of $\Pi(k^2)$ is thus reliable as $k^2 \to -\infty$.

However, the above diagrams and their renormalization group improvement are not the whole story. In Section 22.2.1 we discussed the Wilson operator product expansion for a product of two currents and we used it to study deep inelastic scattering. There it was the light-cone behaviour $z^2 \to 0$ that was important and that behaviour was controlled by operators of lowest twist: $\tau =$ (dimension—spin). In the present analysis, if we make a Wilson expansion of the product of currents in eqn (27.1.1) and we then take matrix elements between *vacuum* states, only the spin-zero operators will survive, and it is thus the operators of lowest *dimension* that will control the small z limit relevant to large k^2. The operators of lowest dimension that occur for $\Pi_{\mu\nu}(k^2)$ are: the unit operator $I(d=0)$, $m_q\bar{\Psi}\Psi$ $(d=4)$, $G^a_{\mu\nu}G^a_{\mu\nu}$ $(d=4)$ etc. (see table in Section 22.2.1).

All operators, other than the unit operator, on dimensional grounds, will give contributions of order $(M^2/k^2)^{d/2}$, where M is some typical hadronic mass-scale, and these become negligible as $k^2 \to -\infty$. The perturbative result valid for $k^2 \to -\infty$ follows from just keeping the operator I. But the other operators are important for moderate values of k^2. However, there is a subtle point. If we use the standard perturbative form for the fields and vacuum state then $\langle 0|\bar{\Psi}\Psi|0\rangle = 0$ and $\langle 0|G^a_{\mu\nu}G^a_{\mu\nu}|0\rangle = 0$ etc. So we might think these terms do not even exist.

However, we know from chiral symmetry breaking, i.e. from the fact that $m_\varrho \neq m_{a_1}$ and that the pion exists, that expressions like $\langle\bar{\Psi}\Psi\rangle \equiv \langle \text{true vac.}|\bar{\Psi}\Psi|\text{true vac.}\rangle \neq 0$. We cannot calculate these matrix elements. They reflect the deepest non-perturbative aspects of the theory. But we can calculate the k^2 dependence of the coefficients of these operators in the operator product expansion. Thus we can produce a QCD calculation of $\Pi(k^2)$ [call it $\Pi^{\text{QCD}}(k^2)$] which includes the perturbative term plus corrections to it that involve powers of $(k^2)^{-d/2}$ multiplied by unknown numbers, the unknown values of the vacuum matrix elements. (This is similar in spirit to Section 21.2.1.)

Schematically then

$$\Pi^{\text{QCD}}(k^2) = \Pi^{\text{pert}}(k^2) + a\frac{\langle GG\rangle}{k^4} + b\frac{\langle m_q\bar{\Psi}\Psi\rangle}{k^4} + \cdots \tag{27.1.2}$$

where a and b are known.

What is the point of this if the coefficients are not calculable? First one may study *many* different physically interesting current correlation functions and they all involve the same basic set of unknown vacuum matrix elements. Secondly, some of the matrix elements are already estimated phenomenologically from current algebra studies using e.g. PCAC.

Finally then we have an expression for $\Pi^{\text{QCD}}(k^2)$ valid not just for $k^2 \to -\infty$ but for a wide range of k^2, and involving a small number of unknown coefficients.

We now use the fact that $\Pi(k^2)$ satisfies a dispersion relation

$$\Pi(k^2) = \frac{1}{\pi}\int_0^\infty \frac{\text{Im}\Pi(s')}{s'-k^2}\text{d}s' \tag{27.1.3}$$

together with the fact that $\text{Im}\Pi(s')$ for $s' > 0$ is directly proportional to the cross-section for $\text{e}^+\text{e}^- \to$ hadrons. (Analogous statements will hold for other kinds of correlation functions.) If now there exist resonances R in the channel $\text{e}^+\text{e}^- \to$ hadrons, e.g. the ϱ-meson, they will show up as sharp bumps in $\text{Im}\Pi(s')$ near $s' = m_R^2$. Thus via e.g. (27.1.3.), $\Pi(k^2)$ can be expressed in terms of the properties of resonances and of the behaviour of cross-sections. Let us call this calculation of $\Pi(k^2)$, $\Pi^{\text{DISP}}(k^2)$.

The demand that

$$\Pi^{\text{DISP}}(k^2) \simeq \Pi^{\text{QCD}}(k^2) \tag{27.1.4}$$

for reasonable values of k^2 then provides all sorts of relations between the properties of different hadrons.

The technique of QCD sum rules has proved most productive. Much of classical meson spectroscopy seems to be well accounted for and predictions have been made for the properties of glueballs. It is not yet clear whether nature is in agreement with the latter. For baryon spectroscopy the situation is not so clean. All in all though, this is perhaps the most coherent *analytical* approach to the fundamental complications of non-perturbative QCD.

27.2 Lattice approach to QCD

Since ultraviolet divergences in field theory come from the singular behaviour of products of field operators at the same space-time point [see Section 22.2.1] an interesting way to regularize such theories is to imagine space-time as broken up into a lattice of discrete separated points. An enormous effort has gone into trying to derive quantitative results from field theory in the non-perturbative regime by using finite size lattices with the hope that the lattices are big enough to approximate real spacetime. Once on the lattice the theory involves a finite number of dynamical variables, i.e. the field at each lattice point, so there is some hope of solving the dynamics using very powerful computers and sophisticated numerical techniques.

The origin of these developments goes back to the pioneering works of Wegner (1971) and Wilson (1974, 1975). There is by now a large literature and the interested reader may consult several books (Creutz, 1983; Rebbi, 1983) and review articles (Kogut, 1983; Creutz, Jacobs and Rebbi, 1983). More modern developments are covered in the excellent proceedings of a recent symposium 'From Actions to Answers' [see Boulder, 1990]. See also the proceedings of the annual Symposia on Lattice Field Theory (Lattice, 1989, 1990) and Ukawa (1991).

The infinities in a quantum field theory arise because products of field operators* like $\hat{\phi}(x)\hat{\phi}(0)$ diverge as $x \to 0$. The usual methods of regularization depend upon the possibility of expanding in a small parameter, e.g. $\alpha \equiv 1/137$ in QED, and as such are not suitable for a theory of strong interactions. In the lattice formulation space-time is taken to consist of discrete points and the ultraviolet divergences that would appear if space-time points were continuous are absent. Moreover, similarities exist between the lattice version of a field theory and a thermodynamical system. For example, for an $SU(N)$ gauge theory, the analogue of

* We use ^ to emphasize the operator nature of ϕ.

$\beta = 1/kT$ which appears in the thermodynamic partition function is the combination

$$\hat{\beta} \equiv 2N_C/g_B^2 \tag{27.2.1}$$

where N_C is the number of colours and g_B is the bare coupling constant. (In the literature $2N_C/g_B^2$ is denoted by β. We shall use $\hat{\beta}$ to avoid confusion with the β function which occurs in the renormalization analysis.) Strong coupling thus corresponds to high T and one can utilize the high temperature expansion methods of thermodynamics to derive results in the non-perturbative domain of field theory.

The entire approach, however, is based upon the Feynman path integral formulation of quantum field theory, a subject which we have not had space to discuss in this volume. In the briefest of terms one deals there with the action $S(\phi)$ of the theory expressed in terms of the various fields $\phi(x)$ ... treated as ordinary classical (non-operator) functions.

Let $f(\hat{\phi})$ be any function of the field *operators* $\hat{\phi}(x)$. Then the vacuum expectation value of $f(\hat{\phi})$ is given by

$$\langle 0|f(\hat{\phi})|0\rangle = \frac{1}{Z}\int \mathrm{D}\phi f(\phi)\mathrm{e}^{\frac{\mathrm{i}}{\hbar}S(\phi)} \tag{27.2.2}$$

where $Z = \int \mathrm{D}\phi\ \mathrm{e}^{\frac{\mathrm{i}}{\hbar}S(\phi)}$ and the integral $\int \mathrm{D}\phi(x)$ means a functional integral over all functions $\phi(x)$ satisfying certain boundary conditions at infinity.

When space-time is discretized, the functional integral becomes a product of ordinary integrals $\int \mathrm{d}\phi_j$ over the values ϕ_j of ϕ at the jth lattice point. By rotating to imaginary time the i disappears in (27.2.2) and the integrals over the ϕ_j then look just like a sum over configurations used in statistical thermodynamics, with Z analogous to the partition function.

In writing down a suitable discrete version of S there are some subtleties. There are often several different action functions which all reduce to the original S, as the lattice spacing $\boldsymbol{a} \to 0$. For bosons this seems to be unimportant — the simplest procedure is just to replace derivatives by finite differences. For fermions there are two kinds of problem. Firstly a straightforward discretization using a chiral invariant action always leads to an action which when $\boldsymbol{a} \to 0$ produces a spectrum with twice as many fermions as possessed by the original theory. Various lattice actions which avoid this problem have been suggested, of which the most popular are the Wilson and the Kogut–Susskind models. These give up explicit chiral invariance for $\boldsymbol{a} \neq 0$; a somewhat worrying matter given that chiral invariance is supposed to be an important approximate symmetry of nature. [For a general discussion, see Karsten and Smit, 1981.] Secondly, in the path integral formulation of field theory, which underlies the whole lattice approach, the 'classical' fermion fields are not true numbers. They are

Grassmann variables—non-commuting 'numbers'—so cannot be directly simulated on a computer. However, it is possible to formally integrate over the fermion fields and thereby transmute the problem into one of inverting what is called the *Dirac operator.* The latter, in practice, means inverting a huge matrix, so that computer time becomes a serious issue. [For a review, see Weingarten, 1989.] For this reason much of the earlier work replaced the Dirac operator by the unit operator, which physically corresponds to eliminating all fermion–antifermion loop diagrams. This is referred to as the *quenched* approximation. In recent years much progress has been made in dealing with the fermionic degrees of freedom, though it remains a difficult matter.

For the gauge fields it turns out that to get a gauge invariant discrete theory the gauge fields $A_\mu^a(x)$ must not be localized at a lattice point j, but rather associated with the links between lattice points. For example for $SU(2)$, if $\mathrm{d}x_{ij}^\mu$ is the vector along the link from point i to one of its neighbours j, then the variables that play a rôle in the discrete theory are the matrices

$$U_{ij} = \exp\{\mathrm{i}g_B A_\mu^b(x)\tau_b \mathrm{d}x_{ij}^\mu\}, \tag{27.2.3}$$

where the τ_b are the Pauli matrices.

To actually calculate vacuum expectation values on a given lattice one utilizes Monte Carlo numerical methods (the so-called Metropolis algorithm), since a true sum over all configurations would take a hopelessly long time even with a super-computer.

Of particular importance are products of the U_{ij} corresponding to a path through a sequence of neighbouring sites and returning to the initial site, i.e. corresponding to a closed loop. If U_l is the product for the closed loop l then it can be shown that

$$W_l \equiv \mathrm{Tr}[U_l] \tag{27.2.4}$$

is gauge-invariant. It is called the *Wilson loop factor.*

There remain two crucial questions: (i) how to get the results of the continuum theory from the results of the lattice calculation, (ii) how to check whether the numerical simulation is reliable.

Suppose that some physical quantity, say the mass of a meson, m, has been calculated on a lattice with lattice spacing $\boldsymbol{a}$ and coupling constant g_B. To start with we do not know what actual physical length $\boldsymbol{a}$ corresponds to. The lattice spacing is playing the rôle of the cut-off λ in Section 20.3 ($\boldsymbol{a} \propto 1/\lambda$), and for given $\boldsymbol{a}$ the maximum momentum that appears is $\pi/\boldsymbol{a}$. The coupling constant g_B is the coupling in the original Lagrangian and is thus the bare coupling constant; this might suggest that the theory is not being renormalized. But that is not true. We do not treat g_B as a *constant.* As was stressed in Section 20.3, to obtain finite results g_B must vary as the cut-off λ changes. On the lattice we will have, analogously, that $g_B = g_B(\boldsymbol{a})$, i.e. that the value of g_B depends upon the value of $\boldsymbol{a}$.

Clearly the value of m will depend upon g_B and $\boldsymbol{a}$. To approach the continuum theory we must let $\boldsymbol{a} \to 0$ and the dependence of g_B upon $\boldsymbol{a}$ is thus crucial.

One way to find this dependence would be to fix $\boldsymbol{a}$, calculate the mass of some meson A, say m_A

$$m_A = f_A(\boldsymbol{a}, g_B) \tag{27.2.5}$$

and adjust g_B so that it equals the correct experimental value (on the assumption that that is possible). One could then demand that as $\boldsymbol{a} \to 0$ the result

$$m_A^{\text{expt}} = f_A[\boldsymbol{a}, g_B(\boldsymbol{a})] \tag{27.2.6}$$

continues to hold. This would fix the dependence of g_B on $\boldsymbol{a}$. Thereafter, other physical quantities could be calculated and the limit $\boldsymbol{a} \to 0$ taken. It should be noted that this is not necessarily a simple matter. Since $\boldsymbol{a}$ is the only parameter with dimensions (we use $\hbar = c = 1$), (27.2.6) must have the form

$$m_A = \frac{1}{\boldsymbol{a}} F_A(g_B) \tag{27.2.7}$$

so that to get a finite answer as $\boldsymbol{a} \to 0$ we must have $g_B \to g_{CR}$, a critical value for which $F_A(g_{CR}) = 0$.

For asymptotically free theories like QCD the critical value is $g_B = 0$. To see this consider first a regularization scheme with ultraviolet cut-off λ and write eqn (21.5.2) in the form

$$g_B(\lambda) = Z(\lambda/\mu) g_\mu. \tag{27.2.8}$$

Then it is easy to check that analogously to (21.5.3)

$$\frac{\lambda}{2} \cdot \frac{\mathrm{d}}{\mathrm{d}\lambda} \alpha_B(\lambda^2) = \beta[\alpha_B(\lambda^2)] \tag{27.2.9}$$

with β the same function as in (21.5.3). The analogue of (21.7.2) is then, for some λ_0,

$$\alpha_B(\lambda^2) = \frac{\alpha_B(\lambda_0^2)}{1 + b\alpha_B(\lambda_0^2)\ln(\lambda^2/\lambda_0^2)} \tag{27.2.10}$$

so that indeed $\alpha_B(\lambda^2) \to 0$ at $\lambda \to \infty$. With a lattice regularization scheme $\boldsymbol{a}$ plays the rôle of $1/\lambda$ and one has that $\alpha_B(\boldsymbol{a}) \to 0$ or $\hat{\beta}(\boldsymbol{a}) \to \infty$ as $\boldsymbol{a} \to 0$.

In analogy to what was done in Section 21.7, we can define a lattice version of Λ_{QCD} via

$$\Lambda^2_{\text{LAT}} = \frac{1}{a^2} \exp\left\{-\frac{1}{b}\left[\frac{1}{\alpha_B(a)} + b' \ln[b\alpha_B(a)]\right]\right\} \equiv \frac{1}{a^2} E^2(\alpha_B) \quad (27.2.11)$$

which, to next-to-leading order, is independent of a.

Returning to (27.2.7), we can now write for sufficiently small α_B

$$F_A(g_B) = am_A = \left(\frac{m_A}{\Lambda_{\text{LAT}}}\right) a\Lambda_{\text{LAT}} \quad (27.2.12)$$

$$= \left(\frac{m_A}{\Lambda_{\text{LAT}}}\right) E(\alpha_B) \quad (27.2.13)$$

which tells us how the numerically determined function $F_A(g_B)$ *ought* to behave as g_B is varied. This can be used as a test that the lattice calculation is producing results consistent with the continuum theory. Once this is so, (27.2.13) can be used to measure the value of Λ_{LAT}, and all other dimensional quantities can then be expressed in terms of Λ_{LAT} via (27.2.11).

By considering a renormalization scheme based on lattice regularization, it is possible to relate Λ_{LAT} to the Λs used in the continuum theory. One finds, for example (see Kawai, Nakayama and Seo, 1981),

$$\Lambda^{\overline{\text{MS}}} \approx 50\Lambda_{\text{LAT}}. \quad (27.2.14)$$

Once the physical value of Λ_{LAT} is known, (27.2.11) can be used to compute the physical length corresponding to a for any (small) value of α_B.

Now we have already stressed that a decreases as α_B decreases and that the continuum limit corresponds to $a \to 0$. However, the size of the lattice, i.e. the number of sites on it, is severely limited by computer time. If the number of sites on the hypercubic lattice is n^4 then at present only $n \lesssim 24$ is possible. That means that the linear dimensions of the lattice are $\approx 24a$. If typically we are studying some static property of a hadron, since its size is about 1 fm, we would like the lattice to be considerably larger so that our results are not influenced by edge effects. If we demand a lattice of size say 2 fm, this implies that a is of order 0.1 fm, not at all really small, and corresponds to a momentum cut-off of only 2 GeV. Thus the choice of

$$\hat{\beta} = \frac{2N_C}{g_B^2} = \frac{3}{2\pi\alpha_B} = \frac{0.48}{\alpha_B} \quad (27.2.15)$$

is constrained by conflicting demands. We want $\hat{\beta}$ large since this will correspond to a small lattice spacing, but then the total lattice size will be small and vice versa. Typically values of $\hat{\beta} \simeq 6$ are used and turn out to correspond to $\boldsymbol{a} \approx 0.12$ fm.

Despite these tricky points, and despite the difficulty with fermions, many interesting results have been obtained, mainly in the quenched approximation, i.e. ignoring the fermionic degrees of freedom. It is not yet clear to what extent the inclusion of these will alter the results.

It is important to realize that what can be measured on the lattice is the vacuum expectation value of any function of the field operators as given by (27.2.2). The *art* of doing successful lattice simulations lies in the choice of the operator so as best to expose the physical quantity one is trying to evaluate. We shall look at one example just to give a taste of the approach. Suppose we wish to find the mass m_A of the lowest mass quantum A belonging to some field ϕ_A. Consider

$$C(t) \equiv \langle 0|\phi_A(\boldsymbol{x}=0,t)\phi_A^\dagger(0)|0\rangle. \tag{27.2.16}$$

Inserting a complete set of states $|n\rangle$ and using (17.4.3) gives

$$\begin{aligned} C(t) &= \sum_n \langle 0|\phi_A(\boldsymbol{x}=0,t)|n\rangle\langle n|\phi_A^\dagger(0)|0\rangle \\ &= \sum_n |\langle 0|\phi_A(0)|n\rangle|^2 \mathrm{e}^{-\mathrm{i}E_n t}. \end{aligned} \tag{27.2.17}$$

Going now to Euclidean time τ, i.e. putting $t = -\mathrm{i}\tau$, we have for the Euclidean lattice

$$\langle 0|\phi_A(\boldsymbol{x}=0,\tau)\phi_A^\dagger(0)|0\rangle = \sum_n |\langle 0|\phi_A(0)|n\rangle|^2 \mathrm{e}^{-E_n\tau} \tag{27.2.18}$$

In the limit $\tau \to \infty$ the term with the minimum value of E_n, i.e. $E_n = m_A$ survives, so that

$$\langle 0|\phi_A(\boldsymbol{x}=0,\tau)\phi_A^\dagger(0)|0\rangle \overset{\tau\to\infty}{\longrightarrow} |\langle 0|\phi_A(0)|A\rangle|^2 \mathrm{e}^{-m_A\tau} \tag{27.2.19}$$

from which m_A can be measured.

All lattice calculations use variants of this procedure with products of 2, 3 or even 4 fields or currents.

A selection of the most interesting results, some of which may be regarded as controversial, follows:

(i) Abelian systems [$U(1)$ group] show a confining phase in the strong coupling regime and a non-confining continuum limit representing electrodynamics (as one would expect);

(ii) non-Abelian systems [$SU(N)$ groups] seem to confine at large separations while behaving as asymptotically free at small distances. Although this conclusion rests on numerical considerations only, its importance is quite obvious, as it is, so far, the only 'realistic' evidence of confinement for non-Abelian gauge theories. [For doubts about these conclusions, however, see Stuller (1987).]

J^{PC}	Value of Mass (MeV/c^2)
0^{++}	~ 1000
2^{++}	1620 ± 100
0^{-+}	1420^{+240}_{-170}
1^{-+}	1730 ± 220
0^{--}	2880 ± 300
1^{+-}	2980 ± 300

Table 27.1. Glueball masses within lattice calculations.

(iii) There is a broad series of predictions on glueball masses (see Table 27.1) and on the QCD parameter Λ introduced in Section 21.7. All these predictions are not unreasonable in the sense that they compare quite well with other kinds of investigations (bags, potential models, QCD sum rules) and Λ is found where one expects it to be (a few hundred MeV).

(iv) Hadron masses have been calculated within lattice QCD for a variety of particles. Usually the ϱ and π masses are reputed to be reliably calculated in this approach and they are used to fit the values of the scale parameter and of the quark mass. The proton mass turns out to be too large in quenched approximation. The inclusion of quark loops seems to help somewhat but the picture is not yet clear. The $\Delta - p$ mass difference is too small in quenched approximation, but this seems to be less of a problem than the proton mass.

(v) The behaviour of the potential between two static sources is an important element in understanding the dynamics of a theory.

If QCD is a confining theory then one expects the potential between say a quark and an antiquark to have the behaviour

$$V(r) \xrightarrow{r \to \infty} \sigma r. \tag{27.2.20}$$

This corresponds to a constant attractive force at large separation of magnitude σ, usually referred to as the *string tension.* For small r one expects to recover the perturbative Coulombic form

$$V(r) \xrightarrow{r \to 0} c/r. \tag{27.2.21}$$

The shape of the potential can be determined from the behaviour of the expectation value $\langle W_l \rangle$ of a Wilson loop factor taken for a rectangular loop in a space-time plane of dimensions $n_s \boldsymbol{a}$ in the spatial direction and $n_t \boldsymbol{a}$ in the time direction, with $n_t \gg n_s$. It can be shown (Wilson, 1974)

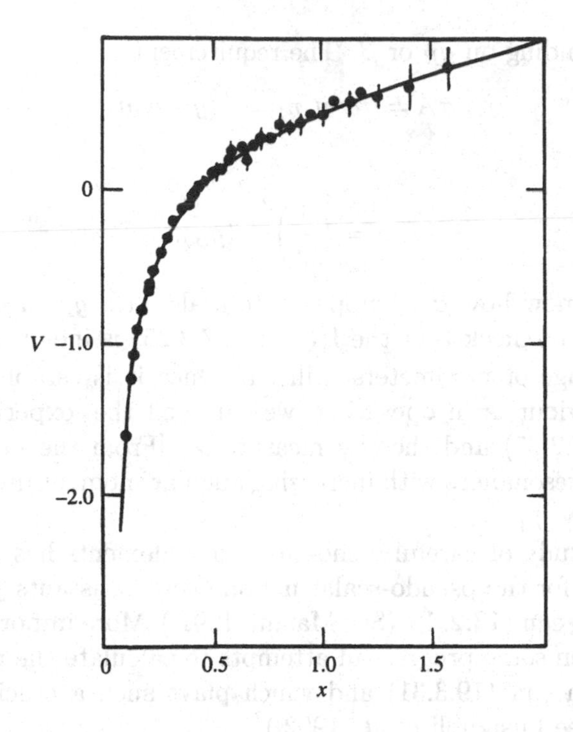

Fig. 27.1. The potential between static colour charges versus distance. The points represent the Monte Carlo data of Stack (1984). The bold line corresponds to the linear+Coulomb fit to the data.

that

$$\langle W_l \rangle = \exp\{-n_t \boldsymbol{a} V(r)\}, \tag{27.2.22}$$

where $r = n_s \boldsymbol{a}$.

The behaviour $V(r) \to \sigma r$ for large r would thus correspond to finding

$$\langle W_l \rangle \sim \exp\{-\sigma A\} \tag{27.2.23}$$

for large $n_s \boldsymbol{a}$, where $A = (n_t \boldsymbol{a}) \times (n_s \boldsymbol{a})$ is the area enclosed by the loop. Clearly the practical requirement of using a lattice satisfying $n_t \gg n_s$ yet $n_s \boldsymbol{a}$ large (say ≈ 1 fm) is non-trivial.

The behaviour of the potential between two static sources has been estimated and found to be compatible with a Coulombic form at small distances which turns into a linear form at large separations. See Fig. 27.1 taken from Stack (1984).

The string tension can also be used to determine the physical length corresponding to the lattice spacing $\boldsymbol{a}$. In the lattice calculation the exponent in (27.2.23) will be of the form $T(g_B) n_s n_t$, where $T(g_B)$ is some number depending on g_B or $\hat{\beta}$. The requirement

$$\sigma A = \sigma \boldsymbol{a}^2 n_s n_t = T(g_B) n_s n_t \tag{27.2.24}$$

then yields

$$\sigma = \left(\frac{1}{\boldsymbol{a}}\right)^2 T(g_B). \tag{27.2.25}$$

Since we know how $\boldsymbol{a}$ is supposed to scale with g_B in QCD, one can use (27.2.11) to check that the RHS of (27.2.25) is independent of g_B or $\hat{\beta}$ for the range of parameters utilized. Once it is established that the scaling behaviour is in operation, we can feed the 'experimental' value of σ into (27.2.25) and thereby measure $\boldsymbol{a}$. [From the mass spacing of sequences of resonances with increasing angular momentum one estimates $\sigma \simeq (420\ \mathrm{MeV}^2)$.]

(vi) The study of carefully chosen matrix elements has led to encouraging results for the pseudo-scalar meson decay constants f_π, f_K and f_D introduced in eqn (13.2.5). (See Maiani, 1991.) More important, perhaps, there has been some progress in attempts to calculate the parameter B_K introduced in eqn (19.3.31) and which plays such a crucial rôle in CP violation. (See Lusignoli *et al.*, 1992.)

(vii) Results are starting to appear on the matrix elements of the electroweak current involved in semi-leptonic decay reactions, e.g. in $D^0 \to K^- \mathrm{e}^+ \nu_\mathrm{e}$ or $D^0 \to \pi^- \mathrm{e}^+ \nu_\mathrm{e}$. Once these are known with reasonable accuracy, it will be possible to use the above reactions in a direct determination of the KM matrix elements V_{cs} and V_{cd} (Martinelli, 1991).

In summary, many interesting studies have been made using the lattice approach and information has been obtained on parameters and matrix elements which are essentially non-perturbative. A major effort is under way to increase computing power often by the design and building of dedicated processors. It seems certain that within the next few years it will be possible to improve the accuracy of the present calculations to a point where any disagreement between theory and experiment will have to be taken quite seriously.

27.3 The vacuum in quantum mechanics and instantons

In almost all texts on quantum field theory one writes expansions for fields in terms of destruction and creation operators as in Appendix 1. The vacuum state $|0\rangle$, which is by definition the state of lowest energy, contains no quanta and is thus annihilated by all destruction operators

$$a_j |0\rangle = 0.$$

As a consequence the vacuum expectation value of any field (i.e. the mean value of any field when the system is in the vacuum state) is zero:

$$\langle 0|\phi(\boldsymbol{x},t)|0\rangle = 0. \tag{27.3.1}$$

By the correspondence principle this then reflects a classical situation in which all fields are zero. Since a perturbative expansion must, by definition, lead to small changes, we are limited in this approach to describing quantum fluctuation about this classical situation.

In the above the vacuum state is assumed to be unique. However, there are many cases in physics where the lowest energy state is degenerate, and one such case was analysed in detail in Chapter 3 when dealing with the Higgs mechanism. Clearly it was of the utmost importance there to know what the actual ground state of the system was. This is true in general—to have any chance of doing meaningful calculations in a field theory one must have some idea, at least, of the structure of the vacuum state.

Our aim will be to give a qualitative discussion of the general problem of degenerate vacuum states and its relevance to QCD and to electroweak theory. To illustrate the mechanism involved we shall first consider an ordinary quantum mechanical particle moving in one dimension in a periodic potential.

27.3.1 *An example in one-dimensional motion*

Consider the one-dimensional periodic potential shown in Fig. 27.2. Classically there are an infinite number of ground states, with the particle sitting at rest at the positions

$$x_n = na, \qquad n = 0, \pm 1, \pm 2, \cdots \tag{27.3.2}$$

Quantum mechanically, if the barrier between each well is 'large', the classical ground states will correspond roughly to Gaussian wave packets centred at the points x_n, i.e. wave functions

$$\Psi_n(x) = \Psi(x - x_n) \tag{27.3.3}$$

with Ψ a Gaussian, each having energy $\approx \frac{1}{2}\hbar\omega$, where ω would be related to the shape of the well near its minimum. Any linear combination

$$\boldsymbol{\Psi}(x) = \sum_n c_n \Psi_n(x) = \sum_n c_n \Psi(x - x_n) \tag{27.3.4}$$

is again an approximate solution with energy $\approx \frac{1}{2}\hbar\omega$.

However, the periodic symmetry of the problem demands that

$$\boldsymbol{\Psi}(x + a) = \text{phase} \times \boldsymbol{\Psi}(x) = \mathrm{e}^{i\theta}\boldsymbol{\Psi}(x) \tag{27.3.5}$$

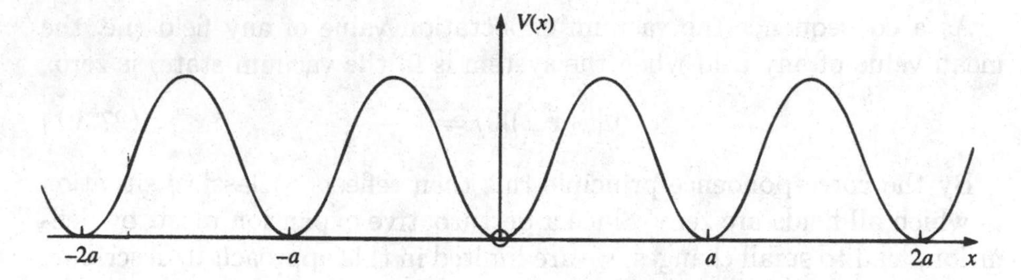

Fig. 27.2. A periodic potential in one dimension.

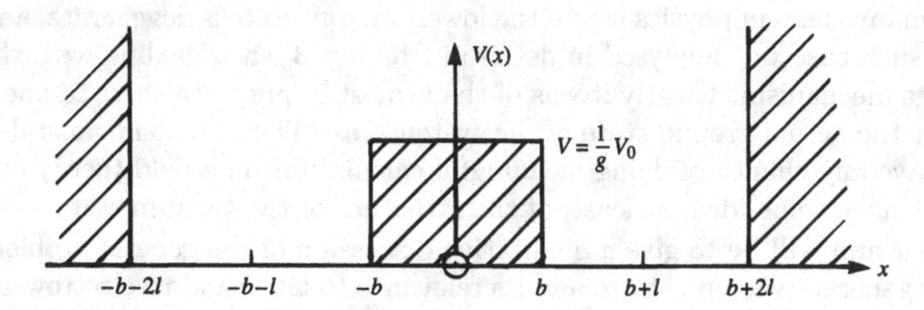

Fig. 27.3. Symmetric well potentials separated by a rectangular barrier.

which implies taking superpositions (27.3.4) of the form

$$\Psi_\theta(x) = \sum_n \mathrm{e}^{in\theta}\Psi(x - x_n) \tag{27.3.6}$$

labelled by the parameter θ.

Now comes the crucial point. The solutions (27.3.6) for different θ are not exactly degenerate because no matter how large the barriers are there is always some tunneling between adjacent wells, and the energy E_θ depends upon θ. The ground state will thus correspond to a particular value of θ. The $\Psi_\theta(x)$ are called Bloch states.

To the extent that the Gaussians $\Psi(x - x_n)$ are reasonably sharply localized in the region $x \approx x_n$ one might hope to approximate them by position eigenstates. This is very often done and (27.3.6) usually appears in the literature in the form

$$|\theta\rangle = \sum_n \mathrm{e}^{in\theta}|x_n\rangle. \tag{27.3.7}$$

The state $|\theta\rangle$ of least energy as θ is varied will give an approximation to the true ground state. It is a semi-classical approximation to the vacuum.

The simplest example of the above phenomenon is a pair of rectangular wells symmetric about $x = 0$ as shown in Fig. 27.3.

If we call the height of the central barrier $V = \frac{V_0}{g}$, where V_0 is some fixed energy, then $g \to 0$ corresponds to an infinite barrier between the wells. The ground state is then exactly doubly degenerate with energy, $E_0 = \pi^2\hbar^2/8ml^2$.

For $g \neq 0$ the barrier is finite. There is tunneling and the degeneracy of the ground state is broken. The relevant symmetry in this case is parity and the approximate wave functions are

$$\Psi_\pm(x) = \Psi(x - b - l) \pm \Psi(x + b + l), \tag{27.3.8}$$

where

$$\begin{aligned} \Psi(x) &= \cos\left(\tfrac{\pi x}{2l}\right) \quad \text{for} \quad -l \leq x \leq l \\ &= 0 \quad \text{for} \quad |x| > l. \end{aligned} \tag{27.3.9}$$

It is a simple matter to solve for the energies $E_\pm$ and for a large barrier (i.e. small g) one gets

$$\Delta E \equiv E_- - E_+ \simeq \frac{4E_0\sqrt{g}}{l\kappa_0} \exp\left[-\frac{2\kappa_0 b}{\sqrt{g}}\right], \tag{27.3.10}$$

where $\hbar\kappa_0 \equiv \sqrt{2mV_0}$.

Note that the formula for ΔE is not analytic at $g = 0$. Thus we could not obtain ΔE via a perturbative expansion in powers of g.

There is a particularly interesting way of looking at the above based upon the Feynman path integral formalism for quantum mechanics. In that approach operators do not appear explicitly and transition amplitudes are calculated by integrating an expression of the form $\exp\left\{\frac{i}{\hbar}S[x]\right\}$, where S is the *classical* action function, over all possible 'paths' $x(t)$ relevant to the transition under study. For example, if the transition is from a position eigenstate $|x_1\rangle$ at $t = t_1$, to $|x_2\rangle$ at $t = t_2$ then the path functions $x(t)$ must satisfy $x(t_1) = x_1$ and $x(t_2) = x_2$ and S is the classical action from t_1 to t_2 along the path $x(t)$.

Since this summation over paths is generally impossible to carry out exactly, one approximates the result by looking for particular types of path which, hopefully, dominate the summation. If, for example, there exists a classically possible trajectory $x_{cl}(t)$ connecting x_1 at $t = t_1$ with x_2 at $t = t_2$, then by Hamilton's principle the action will (usually) be a minimum for $x(t) = x_{cl}(t)$ and this path and paths close to it will dominate the sum over paths. One can then obtain an approximate result by keeping just this contribution and small fluctuations about it. However, for the study of problems like degenerate ground states we need to calculate transition amplitudes through barriers, i.e. such that no classical path connects x_1 at t_1 to x_2 at t_2. There is a clever trick which enables one to find paths which dominate. One formally continues the times t_1, t_2 to imaginary values $t_1 \to -i\tau_1$, $t_2 \to -i\tau_2$ ($\tau_{1,2}$ real) so that the exponent in the path

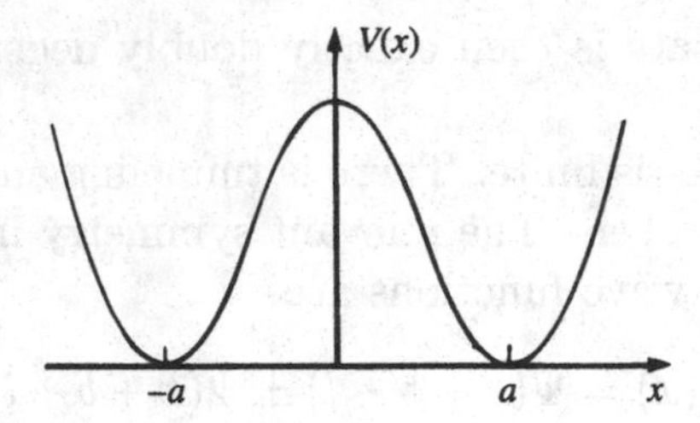

Fig. 27.4. A symmetric pair of wells in one dimension.

integral becomes

$$\begin{aligned}\frac{\mathrm{i}}{\hbar}S[x]_{t=-\mathrm{i}\tau_1}^{t=-\mathrm{i}\tau_2} &= \frac{\mathrm{i}}{\hbar}\int_{-\mathrm{i}\tau_1}^{-\mathrm{i}\tau_2}\mathrm{d}t\; L(x,\dot{x},t)\\ &= \frac{\mathrm{i}}{\hbar}\int_{-\mathrm{i}\tau_1}^{-\mathrm{i}\tau_2}\left[\frac{1}{2m}\left(\frac{\mathrm{d}x}{\mathrm{d}t}\right)^2 - V(x)\right]\mathrm{d}t.\end{aligned}$$

Changing the integration variable by putting $t = -\mathrm{i}\tau$ we have

$$\begin{aligned}\frac{\mathrm{i}}{\hbar}S[x]_{t=-\mathrm{i}\tau_1}^{t=-\mathrm{i}\tau_2} &= \frac{1}{\hbar}\int_{\tau_1}^{\tau_2}\left[\frac{1}{2m}\left(\frac{\mathrm{d}x}{-\mathrm{i}\mathrm{d}\tau}\right)^2 - V(x)\right]\mathrm{d}\tau\\ &= -\frac{1}{\hbar}S_E[x]_{\tau_1}^{\tau_2},\end{aligned} \tag{27.3.11}$$

where S_E, called the Euclidean action, is

$$S_E[x]_{\tau_1}^{\tau_2} = \int_{\tau_1}^{\tau_2}\left[\frac{1}{2m}\left(\frac{\mathrm{d}x}{\mathrm{d}\tau}\right)^2 + V(x)\right]\mathrm{d}\tau, \tag{27.3.12}$$

in which τ plays rôle of the 'Euclidean time'.

It is just the action for the 'mirror image' of the problem under study, i.e. it corresponds to a potential $= -V(x)$! For *this* dynamical problem, regarding τ as the true time variable, what was a barrier has become a well, so that classical paths will now exist connecting x_1 at $\tau = \tau_1$ to x_2 at $\tau = \tau_2$. One can now use these classical paths to obtain an approximation to the sum over paths and then finally analytically continue to $\tau_1 \to \mathrm{i}t_1$, $\tau_2 \to \mathrm{i}t_2$ in the approximate answer. In some cases, for example in the calculation of ΔE, it turns out to be unnecessary to continue back to real times.

Consider the pair of wells shown in Fig. 27.4.

It can be shown that ΔE relevant to finding the lowest energy state can be obtained from the transition amplitude $\langle x = a, t = T | x = -a, t = 0\rangle$ in the limit $T \to \infty$. By the above this will be dominated by classical paths connecting $x = -a$ at $\tau = 0$ to $x = a$ at $\tau = T$ in the mirror problem whose potential is shown in Fig. 27.5.

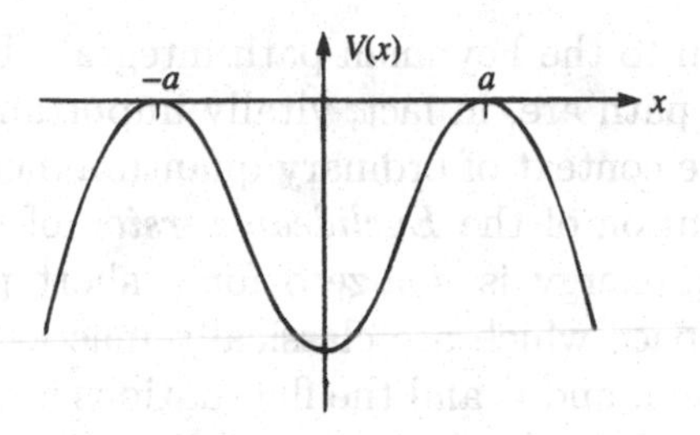

Fig. 27.5. Potential of the mirror or Euclidean problem corresponding to the potential in Fig. 27.4.

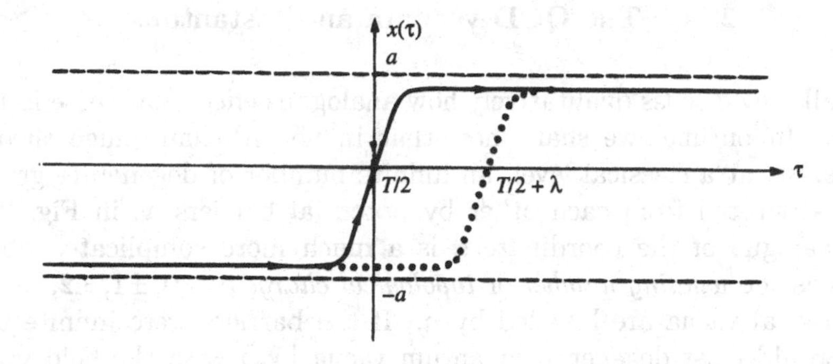

Fig. 27.6. Trajectory $x(\tau)$ (solid line) for an instanton moving from $-a$ to a. For dotted curve see text.

The classical motions are rather peculiar. In the simplest case, the particle starts at $-a$, accelerates, decelerates and reaches a, but must take an infinitely long time to do this because the potential is flat at a and $-a$. Hence its velocity at $-a$ and at a must be infinitesimal. Its total energy is therefore essentially zero. Thus for an infinite time the particle is effectively at rest, and for a short finite time it has a significant speed. The form of $x(\tau)$ might be as shown in Fig. 27.6.

A particle moving in this way is called an *instanton* since its kinetic energy is non-zero only for a very short time on the scale $T \to \infty$. The instanton may either be thought of as the path in *real* time from $-a$ to a in the *mirror* or *Euclidean* problem or the path from $-a$ to a at *imaginary* times in the *original* problem. More complicated paths are also possible, e.g. $-a \to a$ at $t = T/3$ then back to $-a$ at $t = 2T/3$ then to a at $t = T$ etc.

Now the Euclidean action is a minimum for the instanton path shown in Fig. 27.6. But we can easily imagine an infinite set of paths connecting $-a$ to a for which the Euclidean action will differ very little from its minimum value. An example is shown by the dashed curve in Fig. 27.6. This path differs from the instanton path only over a finite region of time. Since there are infinitely many such paths we expect that they will give an essential contribution to the Feynman path integral. These 'fluctuations' about the instanton path are, in fact, vitally important.

In summary, in the context of ordinary quantum mechanics, an instanton is a classical solution of the *Euclidean version* of the problem under study, whose kinetic energy is non-zero for a short period of time. It connects points in space which are classically inaccessible to each other in the *original* problem, and it and the fluctuations around it ought to be important for finding the true quantum vacuum when several degenerate classical ground states exist, all classically inaccessible to each other.

27.4 The QCD vacuum and instantons

We shall now discuss qualitatively how analogous effects can arise in field theory. In outline, we shall show that in non-Abelian gauge theories there exist, at a classical level, an infinite number of degenerate ground states separated from each other by potential barriers as in Fig. 27.2. The analogue of the coordinate x is a much more complicated object known as the *winding number* or *topological charge* $n = 0, \pm 1, \pm 2, \ldots$ and the classical vacua are labelled by n. If the barriers were infinite then there would exist degenerate quantum vacua $|\Psi_n\rangle$ with the field values distributed in a Gaussian fashion around the classical field configurations corresponding to the classical vacuum n.

Analogously to (27.3.6) the symmetry of the problem demands that we take a superposition

$$|\theta\rangle = \sum_n e^{in\theta} |\Psi_n\rangle \tag{27.4.1}$$

and, as before, tunneling will break the degeneracy so that the true vacuum corresponds to a particular value of θ.

The semi-classical approximation to the vacuum state corresponds to replacing the states $|\Psi_n\rangle$ by field-eigenstates $|A_n\rangle$ in which the fields have exactly the values of the classical vacuum fields of winding number n.

Our discussion will, necessarily, be rather compressed. For a more detailed and complete, yet extremely lucid treatment of this subject, the reader should consult the review article of Jackiw (1980).

27.4.1 Degenerate vacua in classical field theory

A very important element in the path integral treatment of field theory is the transition amplitude to go from the ground state (i.e. vacuum state) of the system at $t = t_1$ to the ground state again at $t = t_2$. In addition, as discussed at the beginning of this chapter, in perturbation theory one is studying fluctuations about the vacuum state. It is thus essential to have some idea of the structure of the vacuum state. We shall see that in QCD one has degenerate ground states classically and this leads to an analogue of the Bloch states (27.3.6 or 7) for the QCD ground state. This, it turns out, has a rather surprising and indeed worrisome consequence, leading to the unwanted possibility of CP non-invariance of the *strong* interactions!

In a field theory the fields are generally regarded as the analogue of the position variables as explained in Section 20.2. But for a gauge theory, it is the gauge potentials A^a_μ rather than $G^a_{\mu\nu}$ that are given this rôle. The ground state or states of the *classical* gauge theory will be given by certain configurations of the matter field and of the $A^a_\mu(x)$ which minimize the total energy. They will be time independent so as to reduce the kinetic energy to zero. And they will be such as to minimize the (colour) electric and (colour) magnetic field energy, i.e. they will correspond to having zero electric and magnetic fields. Of course, the potentials are not unique because of gauge invariance, and it turns out that that is the source of a major distinction between Abelian theories like QED and non-Abelian theories like QCD.

Since the essential aspects of the problem lie in the description of the gauge fields, we shall, in the following, ignore the other matter fields. Thus we shall discuss a pure gauge field theory and, unless specifically stated, we shall be working at a purely classical level.

• *The Abelian case*

To get some feeling for the matter let us consider the classical electromagnetic field. The electric and magnetic fields are related to the field tensor $F^{\mu\nu}$ via

$$F^{\mu\nu} = \begin{pmatrix} 0 & -E_x & -E_y & -E_z \\ E_x & 0 & -B_z & B_y \\ E_y & B_z & 0 & -B_x \\ E_z & -B_y & B_x & 0 \end{pmatrix}. \tag{27.4.2}$$

The *dual field tensor* $\tilde{F}^{\mu\nu}$ is defined as

$$\begin{aligned} \tilde{F}^{\mu\nu} &= \tfrac{1}{2}\epsilon^{\mu\nu\varrho\sigma}F_{\varrho\sigma} \\ &= \begin{pmatrix} 0 & -B_x & -B_y & -B_z \\ B_x & 0 & E_z & -E_y \\ B_y & -E_z & 0 & E_x \\ B_z & E_y & -E_x & 0 \end{pmatrix}. \end{aligned} \tag{27.4.3}$$

Maxwell's inhomogeneous pair of equations can be written as

$$\partial_\mu F^{\mu\nu} = eJ^\nu_{\text{em}} \tag{27.4.4}$$

and the homogeneous pair as

$$\partial_\mu \tilde{F}^{\mu\nu} = 0. \tag{27.4.5}$$

When there are no charges or currents present we get, on taking $\nu = 0$,

$$\nabla \cdot \boldsymbol{E} = 0 \quad \text{and} \quad \nabla \cdot \boldsymbol{B} = 0 \tag{27.4.6}$$

i.e. that the $\boldsymbol{B}$-field is divergenceless and that $\boldsymbol{E}$ is divergenceless when there are no charges present.

The field energy is given by

$$E = \tfrac{1}{2} \int \mathrm{d}^3\boldsymbol{r}\{\boldsymbol{E}^2 + \boldsymbol{B}^2\} \tag{27.4.7}$$

so that the ground state has $\boldsymbol{E} = \boldsymbol{B} = 0$. What does the latter imply for $A^\mu(x)$? To answer this recall [see Section 2.3] that the whole theory, including the interaction with charged matter fields, is invariant under the $U(1)$ group of local gauge transformation

$$\Phi_j(x) \to \mathrm{e}^{-\mathrm{i}q_j\theta(x)}\Phi_j(x), \tag{27.4.8}$$

where q_j is the electric charge (in units of e) of the quanta of the field Φ_j. The electromagnetic potentials transform inhomogeneously

$$A_\mu \to A_\mu(x) + \frac{1}{e}\partial_\mu\theta(x) \tag{27.4.9}$$

whereas $\boldsymbol{E}$ and $\boldsymbol{B}$ remain unchanged. Thus when $A_\mu(x)$ is a *pure gauge* i.e.

$$A_\mu(x) = \frac{1}{e}\partial_\mu\theta(x) \tag{27.4.10}$$

we will have $\boldsymbol{E} = \boldsymbol{B} = 0$.

Let us now exercise our freedom to choose a gauge by demanding that

$$A_0(x) = 0. \tag{27.4.11}$$

Clearly (27.4.10) will be compatible with this only if we restrict θ to be time-independent, i.e. we take as our pure gauge potentials

$$A_0 = 0; \qquad \boldsymbol{A}(\boldsymbol{r}) = \frac{1}{e}\nabla\theta(\boldsymbol{r}). \tag{27.4.12}$$

Consider now the spatial behaviour of such a pure gauge $\boldsymbol{A}(\boldsymbol{r})$. It is physically sensible here (and essential for the non-Abelian case) to regard all points at spatial infinity, i.e. at $r = \infty$, as physically equivalent. If we lived in a two-dimensional world this would imply that the circle at infinity is effectively one point so that the two-dimensional plane becomes topologically equivalent to the two-dimensional surface of an ordinary sphere. (Think of wrapping a rubber plane around a sphere.) The two-dimensional surface of a sphere in three dimensions is referred to as S^2. A circle is denoted by S^1. S^n is the n-dimensional 'surface' of a sphere in $(n+1)$-dimensional space. Somewhat confusingly these surfaces are called 'n-spheres'. Thus S^2 is not a 'sphere' in two dimensions, but the two-dimensional surface of a three-dimensional sphere.

With this equivalence of all points at infinity our three-dimensional physical space becomes topologically equivalent to S^3.

On the other hand, the elements of the gauge group $U(1)$ are unimodular complex numbers, so can be regarded as points on a circle of radius one, i.e. on S^1.

Each choice of a pure gauge $\boldsymbol{A}(\boldsymbol{r})$ thus corresponds to a choice of

$$U(\boldsymbol{r}) = \mathrm{e}^{-\mathrm{i}q_j\theta(\boldsymbol{r})}, \tag{27.4.13}$$

which maps points of S^3 onto points of S^1.

Now it can be shown that all such mappings can be smoothly transformed into each other and, in particular, into the unit mapping $U(\boldsymbol{r}) = I$ which corresponds to having

$$\boldsymbol{A}(\boldsymbol{r}) = 0. \tag{27.4.14}$$

Thus the different pure gauges (27.4.12) are all equivalent to (27.4.14). There is effectively one ground state and we can choose to work in the convenient gauge where it is specified by (27.4.14).

• *The non-Abelian case*

Let us turn now to the non-Abelian case. We can introduce colour-electric $\boldsymbol{E}^a$ and colour-magnetic $\boldsymbol{B}^a$ fields in complete analogy to (27.4.2). For each colour a we put

$$G_a^{\mu\nu} = \begin{pmatrix} 0 & -E_x^a & -E_y^a & -E_z^a \\ E_x^a & 0 & -B_z^a & B_y^a \\ E_y^a & B_z^a & 0 & -B_x^a \\ E_z^a & -B_y^a & B_x^a & 0 \end{pmatrix} \tag{27.4.15}$$

and, for the dual field tensor, have

$$\tilde{G}_a^{\mu\nu} = \tfrac{1}{2}\epsilon^{\mu\nu\varrho\sigma}G^a_{\varrho\sigma} = \begin{pmatrix} 0 & -B_x^a & -B_y^a & -B_z^a \\ B_x^a & 0 & E_z^a & -E_y^a \\ B_y^a & -E_z^a & 0 & E_x^a \\ B_z^a & E_y^a & -E_x^a & 0 \end{pmatrix}. \tag{27.4.16}$$

In the absence of colour-charged matter fields the classical equations of motion (21.3.8) become

$$\mathrm{D}_\mu^{ab} G_b^{\mu\nu} = 0. \tag{27.4.17}$$

It can also be checked that the analogue of (27.4.5) holds:

$$\mathrm{D}_\mu^{ab} \tilde{G}_b^{\mu\nu} = 0. \tag{27.4.18}$$

Taking $\nu = 0$ and using (21.3.5), these yield

$$\nabla \cdot \boldsymbol{E}^a = g f_{abc} \boldsymbol{A}^b \cdot \boldsymbol{E}^c \tag{27.4.19}$$

and

$$\nabla \cdot \boldsymbol{B}^a = g f_{abc} \boldsymbol{A}^b \cdot \boldsymbol{B}^c \qquad (27.4.20)$$

in stark contrast to the electromagnetic result (27.4.6).

We note from these that the equations of motion for the non-Abelian case cannot be written purely in terms of $\boldsymbol{E}^a$ and $\boldsymbol{B}^a$. Here the potentials A_μ^a play a more fundamental rôle than in electromagnetism.

We shall concentrate only on (27.4.19). The reason it differs from the electromagnetic case is that the non-Abelian gauge fields have non-zero colour charge.

Recalling that in the presence of an electric charge density ϱ, the Maxwell equation for $\nabla \cdot \boldsymbol{E}$ reads

$$\nabla \cdot \boldsymbol{E} = \varrho, \qquad (27.4.21)$$

we see from (27.4.19) that the colour charge density of the pure Yang–Mills gauge field is given by

$$\varrho_a = g f_{abc} \boldsymbol{A}^b \cdot \boldsymbol{E}^c \qquad (27.4.22)$$

so that the total colour charge is

$$Q_a = g f_{abc} \int \mathrm{d}^3\boldsymbol{r}\, \boldsymbol{A}^b \cdot \boldsymbol{E}^c. \qquad (27.4.23)$$

Using (27.4.19) we recast (27.4.23) into a form which will be more useful to us:

$$\begin{aligned} Q_a &= \int \mathrm{d}^3\boldsymbol{r}\, \nabla \cdot \boldsymbol{E}^a \\ &= \int \mathrm{d}\boldsymbol{S} \cdot \boldsymbol{E}^a, \end{aligned} \qquad (27.4.24)$$

where the last integral is over a surface at spatial infinity.

Let us now try to identify the gauge potentials $A_\mu^a(x)$ corresponding to the ground state of the classical theory. As before we choose to work in gauges in which $A_0^a(x) = 0$.

The field energy can be shown to be given by the analogue of (27.4.7):

$$E = \tfrac{1}{2} \int \mathrm{d}^3\boldsymbol{r} \{(\boldsymbol{E}^a)^2 + (\boldsymbol{B}^a)^2\}, \qquad (27.4.25)$$

where a sum over colour is implied.

Now recall that the transformation of the gauge potentials under finite gauge transformations was most simply expressed in terms of quantities like $\boldsymbol{L} \cdot \boldsymbol{W}_\mu$ in (2.3.34) where the matrices L_a ($a = 1, \cdots, N$) represented the generators of the gauge group $SU(N)$. We therefore define matrix gauge potentials

$$\underline{A}_\mu(x) \equiv L_a A_\mu^a(x), \qquad (27.4.26)$$

where for $SU(3)$ of colour the L_a would be expressed in terms of the eight Gell-Mann matrices: $L_a = \lambda^a/2$. [The reader is warned that often the matrix potentials are defined using L_a/i instead of (27.4.26), i.e. $\sigma^a/2\mathrm{i}$ for SU(2) and $\lambda^a/2i$ for SU(3).] Similarly we define $\underline{G}^{\mu\nu}$, $\underline{\boldsymbol{E}}$, $\underline{\boldsymbol{B}}$ etc.

With $A_0^a(x) = 0$ the theory is still invariant under time-independent gauge transformations

$$U(\boldsymbol{r}) = \mathrm{e}^{-\mathrm{i}\boldsymbol{L}\cdot\boldsymbol{\theta}(\boldsymbol{r})} \tag{27.4.27}$$

and the spatial components of the matrix gauge potential transform as

$$\underline{\boldsymbol{A}} \to U\underline{\boldsymbol{A}}U^{-1} + \frac{\mathrm{i}}{g}(\nabla U)U^{-1}. \tag{27.4.28}$$

Now the field energy is clearly zero when $\underline{A}_\mu = 0$. A short calculation shows that it is also zero when $\underline{\boldsymbol{A}}$ is a pure gauge, i.e. when

$$\underline{\boldsymbol{A}} = \frac{\mathrm{i}}{g}(\nabla U)U^{-1}. \tag{27.4.29}$$

(This is fairly obvious since (27.4.29) is the gauge transform of $\underline{A}_\mu = 0$.)

The ground state is thus specified by the gauge transformation matrix $U(\boldsymbol{r})$. In the non-Abelian case, in contrast to the Abelian case, we shall see that there exist physically distinct families or classes of $U(\boldsymbol{r})$ and hence the classical ground state is degenerate.

We first explain why it is necessary to demand that the $U(\boldsymbol{r})$ have a unique, angle-independent limit as $r \to \infty$. (This is tantamount to saying that all points at spatial infinity are equivalent.) In a non-Abelian theory we know that $G^a_{\mu\nu}$ is *not* gauge invariant [see discussion after (2.3.37)]. As a consequence the colour-electric and magnetic fields are not invariant:

$$\underline{\boldsymbol{E}} \to U\underline{\boldsymbol{E}}U^{-1}, \qquad \underline{\boldsymbol{B}} \to U\underline{\boldsymbol{B}}U^{-1}. \tag{27.4.30}$$

It follows that under a gauge transformation the charge matrix in (27.4.24) becomes:

$$\underline{Q} \to \int \mathrm{d}\boldsymbol{S} \cdot U(\boldsymbol{r})\underline{\boldsymbol{E}}U^{-1}(\boldsymbol{r})|_{r\to\infty}. \tag{27.4.31}$$

If $U(\boldsymbol{r})$ depended arbitrarily on angles as $r \to \infty$ the result for (27.4.31) would be arbitrary, so the charge would be ill-defined. But if we insist that

$$\mathcal{L}_{r\to\infty} U(\mathbf{r}) \equiv U_\infty = \text{multiple of } I \tag{27.4.32}$$

we find that

$$\underline{Q} \to U_\infty \underline{Q} U_\infty^{-1} = \underline{Q} \tag{27.4.33}$$

i.e. $\underline{Q}$ is well defined and invariant.

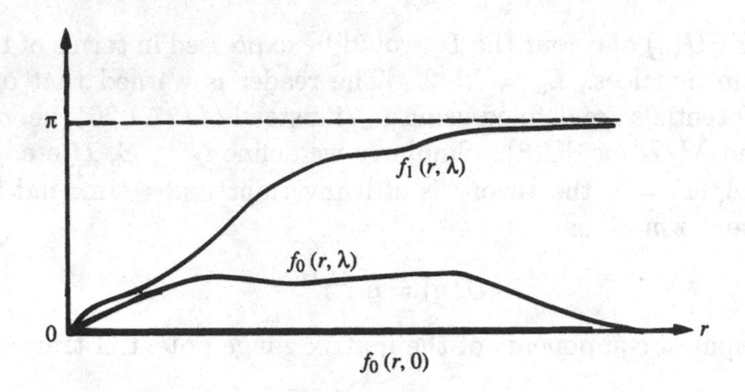

Fig. 27.7. Examples of the functions $f_n(r,\lambda)$.

To illustrate the source of degeneracy of the ground state we shall discuss a colour $SU(2)$ theory for simplicity. As a consequence of a remarkable theorem of Bott, which implies that if the gauge group G possesses an $SU(2)$ subgroup then all results can be obtained merely by considering this $SU(2)$ subgroup, everything derived for $SU(2)$ will go over directly to $SU(3)$.

For $SU(2)$ the representatives of the generators L_j are just proportional to the Pauli matrices, so that the gauge transformations (27.4.27) become

$$U(\boldsymbol{r}) = \mathrm{e}^{-\mathrm{i}\boldsymbol{\sigma}\cdot\boldsymbol{\theta}(\boldsymbol{r})/2}, \tag{27.4.34}$$

These will not have an angle-independent limit as $r \to \infty$ in general.

Consider therefore the class of $U(\boldsymbol{r})$ of the more restricted form

$$U(\boldsymbol{r}) = \mathrm{e}^{\mathrm{i}\boldsymbol{\sigma}\cdot\hat{\boldsymbol{r}}f(r)} \tag{27.4.35}$$

where $\hat{\mathbf{r}}$ is a unit vector. We can write this as

$$U(\boldsymbol{r}) = \cos f(r) + \mathrm{i}\boldsymbol{\sigma}\cdot\hat{\boldsymbol{r}}\sin f(r) \tag{27.4.36}$$

so that $U(\mathbf{r})$ will be independent of angles for $r \to \infty$ provided $\sin f(r) = 0$ as $r \to \infty$, i.e. provided

$$\mathcal{L}_{r\to\infty} f(r) = 0, \pm\pi, \pm 2\pi \cdots \tag{27.4.37}$$

which also implies $U(\boldsymbol{r}) = \pm I$ as $r \to \infty$.

Moreover, to be single valued at $\boldsymbol{r} = 0$, we shall demand that $f(0) = 0$.

Consider now the subset of these restricted gauge transformations specified by the set of well-behaved functions labelled by some parameter λ: $f_0(r,\lambda)$ with $f_0(0,\lambda) = f_0(\infty,\lambda) = 0$, and such that $f_0(r,\lambda = 0) = 0$ for all r. $f_0(r,\lambda)$ will have some form as indicated in Fig. 27.7.

Clearly $f_0(r,\lambda)$ can be continuously deformed into $f_0(r,0) = 0$. In other words the set of gauge transformations

$$U_0(\boldsymbol{r}) = \mathrm{e}^{\mathrm{i}\boldsymbol{\sigma}\cdot\hat{\boldsymbol{r}}f_0(r,\lambda)} \tag{27.4.38}$$

can be continuously deformed into the trivial transformation

$$U(\boldsymbol{r}) = I. \tag{27.4.39}$$

On the contrary, for the set of functions $f_1(r,\lambda)$ with $f_1(0,\lambda) = 0$ but $f_1(\infty,\lambda) = \pi$, it is clear from Fig. 27.7 that $f_1(r,\lambda)$ cannot be continuously deformed into $f_0(r,\lambda)$.

The generalization is obvious. The gauge transformations (27.4.36) fall into classes, labelled by an integer n ($n = 0, \pm 1, \pm 2, \cdots$)

$$U_n(\boldsymbol{r}) = \mathrm{e}^{\mathrm{i}\boldsymbol{\sigma}\cdot\hat{\boldsymbol{r}} f_n(r)}. \tag{27.4.40}$$

such that $f_n(\infty) = n\pi$.

Note that we are very much in the habit of building up a finite symmetry transformation by iterating infinitesimal ones, the latter being expressed in terms of the *generator* of the transformation. It can be shown that this is not possible for a gauge transformation with $n \neq 0$. For this reason transformations with $n = 0$ are sometimes called *small* and those with $n \neq 0$ *large* gauge transformations.

Now the elements of the group $SU(2)$ can be thought of as the points lying on a three-dimensional sphere since (27.4.36) can be written as

$$U = a + \boldsymbol{b}\cdot\boldsymbol{\sigma} \tag{27.4.41}$$

with

$$a^2 + \boldsymbol{b}^2 = 1. \tag{27.4.42}$$

Thus U is specified by the coordinates $(a, \boldsymbol{b})$ of points in a four-dimensional Euclidean space which lie on the three-dimensional sphere (27.4.42).

Hence (27.4.40) provides classes of mappings from physical space (S^3) to the group manifold (S^3). These classes are called *homotopy* classes and the integer n is called the *winding number.*

Miraculous as it may seem, it turns out that the winding number of these gauge transformations for $SU(2)$ is given by *volume* integral

$$n = \frac{1}{24\pi^2}\int \mathrm{d}^3\boldsymbol{r}\,\epsilon_{ijk}\mathrm{Tr}\{(U\partial_i U^{-1})(U\partial_j U^{-1})(U\partial_k U^{-1})\}. \tag{27.4.43}$$

To end this section, we return to (27.4.29) and conclude that there exist distinct pure gauge configurations of the potentials

$$\boldsymbol{A}_n = \frac{\mathrm{i}}{g}(\nabla U_n)U_n^{-1} \tag{27.4.44}$$

all of which correspond to zero field energy. In other words, the classical theory has degenerate vacua labelled by the winding number n (also called the topological charge).

27.4.2 The θ-vacuum in QCD

We shall finally justify the claim made at the beginning of Section 27.4 that in the semi-classical approximation to the quantized field theory the QCD vacuum should be taken to be

$$|\theta\rangle = \sum_n \mathrm{e}^{in\theta}|\underline{\boldsymbol{A}}_n\rangle, \tag{27.4.45}$$

where the state $|\underline{\boldsymbol{A}}_n\rangle$ is an eigenstate of the matrix gauge potential operators $\hat{\underline{\boldsymbol{A}}}_n(x)$ in which the matrix gauge potentials have precisely the values $\underline{\boldsymbol{A}}_n(x)$ corresponding to the classical ground state of winding number n.

The crucial point is that, just like in the one-dimensional periodic potential example, there is tunneling between the states $|\underline{\boldsymbol{A}}_n\rangle$. In the path integral formalism the tunneling is generated through the existence of instanton configurations of the potentials $\underline{A}_\nu(\tau, \boldsymbol{r})$ (ν = integer) which have the following properties:

(i) they correspond to finite Euclidean action (recall that τ is the Euclidean time),

(ii) they are solutions of the Euclidean space classical field equations of motion with essentially zero Euclidean field energy,

(iii) they evolve from configurations $\underline{\boldsymbol{A}}_n(\boldsymbol{r})$ at $\tau = -\infty$ to configurations $\underline{\boldsymbol{A}}_{n+\nu}(\boldsymbol{r})$ at $\tau = \infty$.

The Euclidean action is given by (see Appendix 8 for Euclidean conventions)

$$S_E = \tfrac{1}{2}\int \mathrm{d}\tau \mathrm{d}^3\boldsymbol{r}\, \mathrm{Tr}(\underline{G}^E_{\alpha\beta}\, \underline{G}^E_{\alpha\beta}), \tag{27.4.46}$$

where α, β are Euclidean indices: $\alpha, \beta = x, y, z, \tau$. S_E can be finite only if $\underline{G}^E_{\alpha\beta} \to 0$ as $R \equiv \sqrt{\tau^2 + r^2} \to \infty$. This requires that the $\underline{A}_\mu$ become pure gauges as $R \to \infty$:

$$\underline{A}_\mu(\tau, \boldsymbol{r}) \stackrel{R\to\infty}{\to} -\frac{\mathrm{i}}{g}(\partial_\mu U)U^{-1}, \tag{27.4.47}$$

where here $U(\tau, \boldsymbol{r})$ depends on time as well.

Since the hypersurface $R = \infty$ in the four-dimensional Euclidean space $(\tau, \boldsymbol{r})$ is a three-sphere, i.e. S^3, the mappings generated by $U(\tau, \boldsymbol{r})$ are once again mappings $S^3 \to S^3$. Thus the finite action configurations can be classified by their winding numbers $\nu = 0, \pm 1, \pm 2, \cdots$

For these time-varying potentials, it can be shown that the winding number can be written as a Euclidean *space-time* integral

$$\nu = -\frac{g^2}{16\pi^2}\int \mathrm{d}\tau\, \mathrm{d}^3\boldsymbol{r}\, \mathrm{Tr}(\underline{G}^E_{\alpha\beta}\underline{\tilde{G}}^E_{\alpha\beta}). \tag{27.4.48}$$

This follows upon noting that one can define a current

$$K^\mu \equiv 2\epsilon^{\mu\nu\varrho\sigma}\mathrm{Tr}\{\underline{A}_\nu\partial_\varrho\underline{A}_\sigma - \tfrac{2}{3}\mathrm{i}g\underline{A}_\nu\underline{A}_\varrho\underline{A}_\sigma\} \tag{27.4.49}$$

for which, after some algebra, one can show that

$$\partial_\mu K^\mu = \mathrm{Tr}(\underline{G}_{\mu\nu}\underline{\tilde{G}}^{\mu\nu}). \tag{27.4.50}$$

Then by Gauss' theorem, using K^E_α as defined in Appendix 8, from 27.4.48 and 50)

$$\begin{aligned} \nu &= -\frac{g^2}{16\pi^2}\int \mathrm{d}\tau\, \mathrm{d}^3\boldsymbol{r}\, \partial^E_\alpha K^E_\alpha \quad (\alpha = x, y, z, \tau) \\ &= -\frac{g^2}{16\pi^2}\int_{r=\infty} \mathrm{d}S_\alpha K^E_\alpha, \end{aligned} \tag{27.4.51}$$

where the last integral is a volume integral over the three-dimensional hypersurface $R = \infty$. But at $R = \infty$ the finite action configurations $\underline{A}_{(\nu)}$ are pure gauges given by (27.4.47). When these are substituted for $\underline{A}_\mu$ in (27.4.49) both terms yield similar expressions and one gets (with Euclidean indices)

$$K^E_\alpha = \frac{2}{3g^2}\epsilon_{\alpha\beta\gamma\delta}\mathrm{Tr}\{(U\partial_\beta U^{-1})(U\partial_\gamma U^{-1})(U\partial_\delta U^{-1})\}. \tag{27.4.52}$$

By substituting this into (27.4.51) and by a suitable change of integration variables one recovers the expression (27.4.43) for the winding number.

One can also show for potentials of class ν that the Euclidean action is bounded below:

$$S_E[\underline{A}_{(\nu)}] \geq 8\pi^2\nu/g^2 \tag{27.4.53}$$

and that the minimum is achieved if the fields are 'self-dual' or 'anti-self-dual', i.e. $\underline{G}^E_{\mu\nu} = \pm\underline{\tilde{G}}^E_{\mu\nu}$. These correspond to having $E^a_j = \pm B^a_j$. The Euclidean field energy is

$$E_E = \tfrac{1}{2}\int \mathrm{d}^2\boldsymbol{r}\{(\boldsymbol{E}^a)^2 - (\boldsymbol{B}^a)^2\} \tag{27.4.54}$$

which therefore vanishes.

The simplest example of such a self-dual solution is the $\nu = 1$ instanton configuration given as follows. Let $x^{(0)} = (\tau_0, \boldsymbol{r}_0)$ be a fixed Euclidean four-vector, the *position* of the instanton, and let ϱ be a fixed length, the *size* of the instanton. Then

$$\underline{A}^E_\alpha(\tau, \boldsymbol{r}) = \frac{\mathrm{i}}{g}\left[\frac{(x - x^{(0)})^2}{(x - x^{(0)})^2 + \varrho^2}\right] U^{-1}\partial_\alpha U, \tag{27.4.55}$$

where $x = (\tau, \boldsymbol{r})$, x^2 is the squared Euclidean distance $x_\alpha x_\alpha$ and

$$U = \frac{(\tau - \tau_0) + \mathrm{i}\boldsymbol{\sigma}\cdot(\boldsymbol{r} - \boldsymbol{r}_0)}{\sqrt{(x - x^{(0)})^2}}. \tag{27.4.56}$$

This, of course, does not have $\underline{A}_0(\tau, \boldsymbol{r}) = 0$, but can be gauge transformed so as to do so. The position and size of the instanton are arbitrary so (27.4.55) is really an infinite family of instantons.

In summary, then, there is tunneling between the states $|\underline{\boldsymbol{A}}_n\rangle$ and we are forced to look for the approximate quantum ground state as a superposition of such states. To see why we require the Bloch form of the superposition (27.4.45), note firstly, from (27.4.40), that for a product of gauge transformations one has

$$U_m(\boldsymbol{r})U_n(\boldsymbol{r}) = U_{m+n}(\boldsymbol{r}). \tag{27.4.57}$$

Secondly, using (27.4.44 and 28), that under a transformation $U_m(\boldsymbol{r})$

$$\underline{\boldsymbol{A}}_n(\boldsymbol{r}) \to \underline{\boldsymbol{A}}_{n+m}(\boldsymbol{r}). \tag{27.4.58}$$

In particular then under the gauge transformation $U_1(\boldsymbol{r})$ the winding number increases by one unit. Now the theory is invariant under gauge transformations. Hence the operator $\hat{U}_1(\boldsymbol{r})$ which, in the quantized version of the theory, implements the transformation $U_1(\boldsymbol{r})$ will: (*a*) commute with the Hamiltonian, and (*b*) acting on an eigenstate $|\underline{\boldsymbol{A}}_n\rangle$ increase n to $n+1$. This is completely analogous to the translation operation x to $x+a$ in the ordinary quantum mechanical periodic potential example [see (27.3.5)]. By the same physical arguments, therefore, we require to take as physical vacuum state the Bloch type superposition (27.4.45). One then has that under a gauge transformation of class n

$$\hat{U}_n|\theta\rangle = \mathrm{e}^{-\mathrm{i}n\theta}|\theta\rangle. \tag{27.4.59}$$

27.5 Strong CP violation and the $U(1)$ problem

As mentioned earlier, an important element in quantum field theory is the transition amplitude to go from the ground state at time t_1 to the ground state at time t_2. It corresponds to the amplitude $\langle\mathrm{VAC}|\mathrm{e}^{-\mathrm{i}\hat{H}(t_2-t_1)}|\mathrm{VAC}\rangle$ and, as we have now learnt, in QCD we should utilize as ground state the θ-vacuum $|\theta\rangle$ given by (27.4.45). Going to imaginary time $t = -\mathrm{i}\tau$ we then have, with $T = \tau_2 - \tau_1$,

$$\langle\theta|\mathrm{e}^{-\hat{H}T}|\theta\rangle = \sum_{n,m} \mathrm{e}^{\mathrm{i}(n-m)\theta}\langle\underline{\boldsymbol{A}}_m|\mathrm{e}^{-\hat{H}T}|\underline{\boldsymbol{A}}_n\rangle. \tag{27.5.1}$$

The transition amplitudes on the RHS are given, in the path integral formalism, by a summation over *paths* going from $|\underline{\boldsymbol{A}}_n\rangle$ at $\tau = \tau_1$ to $|\underline{\boldsymbol{A}}_m\rangle$ at $\tau = \tau_2$, i.e. by the instanton configurations $\underline{A}_{(\nu)}(x)$ with $\nu = m - n$.

Since we need to sum over all n, m in (27.5.1) this is tantamount to saying that the θ-vacuum transition amplitude is given by a summation over *all* instanton configurations with $\exp\{-S_E[A]\}$ replaced by

$$e^{i\nu\theta}e^{-S_E[A]} = \exp\left\{-\int d\tau d^3r \left[\mathcal{L}_E - \frac{i\theta g^2}{16\pi^2}\mathrm{Tr}(\underline{G}^E_{\alpha\beta}\underline{\tilde{G}}^E_{\alpha\beta})\right]\right\}. \quad (27.5.2)$$

But this is equivalent to working with the new Lagrangian density (in Minkowski space)

$$\mathcal{L}_\theta = \mathcal{L} + \frac{\theta g^2}{32\pi^2}G^a_{\mu\nu}\tilde{G}^{\mu\nu}_a. \quad (27.5.3)$$

Alas, for $\theta \neq 0$, the new term violates CP invariance. To be consistent with the experimental bounds on CP violation one requires an incredibly small value, $\theta \lesssim 10^{-9}$.

Note, that as shown in (27.4.50), the new term is a total divergence. Normally such a term does not play any rôle since it does not affect the classical equations of motion (recall that in Hamilton's principle one varies paths with fixed end points) and in quantum mechanics it merely induces a unitary transformation which has no physical effect. But evidently this argument breaks down when the structure of the field configurations at space-time infinity has non-trivial topological properties.

An intriguing question, therefore, is what mechanism gives θ its tiny value. We shall return to this presently.

The existence of the θ-vacuum helps to resolve a long-standing puzzle known as the $U(1)$ *problem.* We shall discuss the simplest version of this problem. Consider the fundamental QCD Lagrangian constructed from the gauge fields and *massless* quark fields Ψ_j. Suppose that the Lagrangian is invariant under some infinitesimal flavour symmetry transformation

$$\delta\Psi_j = i\epsilon F_{jk}\Psi_k. \quad (27.5.4)$$

Then because of the absence of mass terms, the Lagrangian will also be invariant under the *chiral* transformation

$$\delta\Psi_j = i\epsilon\gamma_5 F_{jk}\Psi_k. \quad (27.5.5)$$

If the vacuum were invariant under the latter transformations one would find that the fermion spectrum is doubled. Every fermion would possesses a twin of opposite parity. Since no such parity-doubling is seen it is supposed that the chiral symmetry is spontaneously broken (see Chapter 3). For $SU(2)_F$ (isospin) flavour symmetry there will be 3 broken chiral generators resulting in 3 massless pseudoscalar Goldstone bosons. When allowance is made for small u, d quark masses, these Goldstone bosons acquire small masses and are recognized as the isotriplet of light pseudoscalar mesons $\pi^\pm, \pi^0$.

However, the theory appears also to be invariant under the $U(1)$ transformations:

$$\Psi_j \to \mathrm{e}^{\mathrm{i}\alpha}\Psi_j, \quad \Psi_j \to \mathrm{e}^{\mathrm{i}\alpha\gamma_5}\Psi_j, \tag{27.5.6}$$

where α is a parameter independent of j. Invariance under the first of these leads to baryon number conservation. If the chiral version is spontaneously broken it will give rise to a massless isosinglet Goldstone boson which for small u, d masses will turn into an isosinglet pseudoscalar meson with a mass m_0 comparable to m_π. In fact, Weinberg (1975) using current algebra techniques, showed that one would have

$$m_0 \leq \sqrt{3}\, m_\pi. \tag{27.5.7}$$

However, the lightest known $I = 0$ pseudoscalar is the η meson whose mass is 549 MeV/c^2, i.e. it is much too massive. This is the '$U(1)$ problem'. Hence the $U(1)$ chiral symmetry is an unwanted embarrassment.

We shall now see how the above topological considerations resolve this problem. The symmetry of the classical theory under the chiral $U(1)$ transformations gives rise to the Noether current

$$J^{(0)}_{\mu 5} = \sum_{j=u,d} \bar{\Psi}_j \gamma_\mu \gamma_5 \Psi_j \tag{27.5.8}$$

where we are considering just 2 flavours (the superscript indicates that it is an *isosinglet* current). Classically this current is, of course, conserved, but in the quantized theory it is affected by an *axial anomaly* (see Section 9.5) and one finds

$$\partial^\mu J^{(0)}_{\mu 5} = n_f \frac{g^2}{8\pi^2} \mathrm{Tr}(\underline{G}_{\mu\nu} \underline{\tilde{G}}^{\mu\nu}), \tag{27.5.9}$$

where, for us, the number of flavours is $n_f = 2$. It follows from (27.4.50) that the current

$$\tilde{J}_{\mu 5} \equiv J^{(0)}_{\mu 5} - n_f \frac{g^2}{8\pi^2} K_\mu \tag{27.5.10}$$

is conserved

$$\partial^\mu \tilde{J}_{\mu 5} = 0. \tag{27.5.11}$$

Associated with these currents we can, as usual, introduce charge operators, in this case the *chiral charges*

$$\hat{q}_5(t) = \int \mathrm{d}^3\boldsymbol{r}\, J^{(0)}_{05}(t, \boldsymbol{r}) \tag{27.5.12}$$

$$\hat{Q}_5 = \int \mathrm{d}^3\boldsymbol{r}\, \tilde{J}_{05}(t, \boldsymbol{r}). \tag{27.5.13}$$

Of these $\hat{q}_5$ is gauge invariant but it is not time independent (i.e. is not conserved) because $J^{(0)}_{\mu 5}$ is not a conserved current. On the other hand,

$\hat{Q}_5$ is conserved but is not invariant under *large* gauge transformations. Indeed, one finds after some algebra, that under a gauge transformation of the class U_n

$$\begin{aligned}\hat{Q}_5 &\rightarrow \hat{U}_n \hat{Q}_5 \hat{U}_n^{-1} \\ &= \hat{Q}_5 + 2nn_f \end{aligned} \qquad (27.5.14)$$

which is equivalent to

$$[\hat{U}_n, \hat{Q}_5] = 2nn_f \hat{U}_n. \qquad (27.5.15)$$

Now since the theory is gauge invariant and since $\hat{Q}_5$ is conserved, one must have

$$[\hat{H}, \hat{U}_n] = 0, \qquad [\hat{H}, \hat{Q}_5] = 0 \qquad (27.5.16)$$

but because $\hat{U}_n$ and $\hat{Q}_5$ do not commute we cannot diagonalize $\hat{H}, \hat{U}_n$ and $\hat{Q}_5$ simultaneously. The θ-vacuum $|\theta\rangle$ is an eigenstate of energy and of the gauge transformations $\hat{U}_n$. It is thus not an eigenstate of $\hat{Q}_5$, and from (27.5.15 and 16)

$$\mathrm{e}^{-\mathrm{i}\phi\hat{Q}_5}|\theta\rangle = |\theta + 2n_f\phi\rangle. \qquad (27.5.17)$$

Thus the vacuum is not chiral invariant and the $U(1)$ chiral symmetry is, in fact, not a symmetry of the theory. This resolves the $U(1)$ problem.

However, there remains the problem of the value of θ and the strong CP non-conservation. One might think one could simply postulate that $\theta = 0$ in the initial (bare) Lagrangian. But this will not work because of the mechanism for generating quark masses via spontaneous symmetry breaking. The key point is that in diagonalizing the mass-like terms arising from the coupling to the Higgs meson [see Section 9.7, especially eqn (9.7.12)] we have to apply *different* unitary transformations to the L and R parts of the quark fields Ψ. But this, as is easily checked, is tantamount to making a chiral transformation on Ψ itself. This chiral transformation seems to be permissible since, formally, all the other terms of the Lagrangian appear to be invariant under (27.5.6). But in the quantized theory, as discussed above, the chiral invariance is illusory and one finds that (27.5.6) generates a change in the Lagrangian equal to $(-2\alpha n_f \hbar g^2/32\pi^2) G^a_{\mu\nu} \tilde{G}^{\mu\nu}_a$. In other words, even if we start with $\theta = \theta_0 = 0$ in the bare Lagrangian the above mechanism will produce the unwanted CP violating term with some effective value of θ. Alternatively, in order to end up with a vanishing effective θ one would have to *fine-tune* the initial bare value of θ_0 so that it almost perfectly cancels the contribution coming from the diagonalization of the quark mass terms.

The mechanism for the almost vanishing value of $\theta_{\text{effective}}$ is not convincingly understood.

One intriguing suggestion due to Peccei and Quinn (1977) invokes an additional Higgs doublet in the electroweak Lagrangian. The two doublets are assumed to transform differently under the electroweak $U(1)$ transformations. In the SM the transformations (9.2.4 and 5) can be shown to involve a chiral transformation

$$u \to e^{i\alpha_u \gamma_5} u, \qquad d \to e^{i\alpha_d \gamma_5} d, \tag{27.5.18}$$

in which $\alpha_u = -\alpha_d$. With two Higgs doublets the Lagrangian is formally invariant for arbitrary α_u, α_d. This extra freedom allows one to perform a chiral rotation which, via the anomaly, can be adjusted to cancel the θ-term.

However, when the two Higgs doublets acquire non-zero vacuum expectation values and the symmetry is spontaneously broken one is left with a new pseudoscalar Goldstone boson — the *axion*, which, alas, has never been seen. Detailed analysis suggests that it could be very light — $m_a \gtrsim 180$ keV/c^2. But if light it would be produced in all sorts of decays and the calculated branching ratios are orders of magnitude larger than the experimental upper bounds. To escape this dilemma more exotic versions of the Peccei–Quinn symmetry have been suggested within the context of Grand Unified Theories.

For a detailed discussion of the strong CP problem see the review article of Cheng (1988). For the axion, see Mohapatra (1986) and Girardi (1982). A very careful and comprehensive treatment of the $U(1)$ problem (including its connection with so-called Kogut–Susskind ghost poles) is given in D'yakonov and Eides (1981).

27.6 Baryon and lepton non-conservations: sphalerons

Similarly to what happens in QCD the SM also possesses distinct classical vacua. But now the vacua will turn out to possess non-zero baryon and lepton numbers. Barrier penetration between vacua will therefore correspond to violation of baryon and lepton numbers ('tHooft, 1976). Early estimates of the cross-sections for B and L violating processes were ridiculously small, $\sigma \approx 10^{-129}$ pb! But the discovery of a field configuration called the *sphaleron* (Manton, 1983; Klinkhamer and Manton, 1984) and its use in reactions where many $W^{\pm}$ and Z^0 are produced results in a staggeringly large enhancement, $\sigma \approx 1$ pb (!) at energies of the order of 30 TeV (Ringwald, 1990). Ringwald's paper generated enormous excitement and there is much debate at present as to the reliability of the cross-section estimates.

27.6.1 Degenerate vacua in the SM

In the previous sections we discussed the non-trivial vacuum structure in QCD arising from the existence of topologically distinct classical vacuum configurations of the gluon field and the consequences of the existence of instanton solutions which tunnel between these vacua.

A somewhat similar situation arises in electroweak theory where the classical vacuum is again degenerate. In this case it is the classical vacuum configuration of the SM gauge boson potentials, i.e. $W^\pm$, Z^0, that is relevant, and the gauge transformations of interest are those of the electroweak $SU(2)_W$ group. Since, for simplicity, we discussed QCD in terms of an $SU(2)$ subgroup of colour $SU(3)$, all the statements and expressions about winding number etc. of the gluon field hold equally well for the $W^\pm$, Z^0 fields. We shall simplify slightly. Recall, from Section 4.2, that the weak $SU(2)$ really acts on $W^\pm_\mu$, W^3_μ *before* the electroweak mixing with B_μ. We ignore this because it has been shown that the mixing does not alter the qualitative aspects of the situation and, indeed, seems to have only a tiny effect quantitatively. Thus we deal with $W^\pm$, W^3 only.

However, there are two essential differences:

(i) Because of the Higgs mechanism that gives mass to the vector bosons, the instanton-type solutions (27.4.55), wherein $\underline{A}_\alpha$ should now be read as $\underline{W}_\alpha$, cannot be exact solutions of the Euclidean electroweak classical equations of motion. Heuristically it is easy to see why. When a field has mass M the solutions of the field equation typically behave like $\exp(-MR)$ as $R \to \infty$, where R is the Euclidean invariant distance $R^2 = \tau^2 + \boldsymbol{r}^2$. The instanton solutions, on the contrary, decrease like inverse powers of R as $R \to \infty$. Not surprisingly they *are* approximate solutions of the equations of motion for $R < M^{-1}$, i.e. they solve the equations up to terms of order R/M^{-1}. (More correctly, the error is of order R/v^{-1}. Recall that $M_W = gv/2$.)

Even though the instantons are not solutions of the SM equations of motion they do provide a set of field configurations which interpolate between one vacuum of winding number n $|\underline{\boldsymbol{W}}_n\rangle$ at $\tau = -\infty$ and the next vacuum $|\underline{\boldsymbol{W}}_{n+1}\rangle$ at $\tau = +\infty$. Moreover, it can be shown that the action along the instanton path is close to the value $8\pi^2/g^2$ which holds for the pure gauge theory, i.e. without Higgs field. (Here g is the weak interaction coupling constant so that $8\pi^2/g^2 = 2\pi/\alpha_W$ where $\alpha_W = \alpha/\sin^2\theta_W$ and α is the usual fine structure constant.) The instanton paths are thus important in the path integral approach to the SM.

(ii) Because the coupling of the leptons and quarks to the gauge fields involves left-handed doublets, i.e. a mixture of γ^μ and $\gamma^\mu\gamma_5$ couplings, even the *vector* currents are affected by anomalies and are not conserved at the quantum level ('tHooft, 1976).

Thus for each fermion doublet, leptonic (ν_e, e^-) etc. and hadronic (u, d) etc., one finds

$$\begin{aligned} \partial_\mu(\bar{\nu}\gamma^\mu\nu + \bar{e}\gamma^\mu e) &= \tfrac{1}{3}\textstyle\sum_{\text{colours}} \partial_\mu(\bar{u}\gamma^\mu u + \bar{d}\gamma^\mu d) \\ &= -\frac{g^2}{16\pi^2}\mathrm{Tr}(\underline{G}_{\mu\nu}\underline{\tilde{G}}^{\mu\nu}), \end{aligned} \tag{27.6.1}$$

where here $G_{\mu\nu}$ is the weak field tensor defined in (4.2.22).

It follows that both the *lepton number current*

$$J^L_\mu = \sum_{\text{generations } j} \{\bar{l}_j\gamma_\mu l_j + \bar{\nu}_j\gamma_\mu\nu_j\} \tag{27.6.2}$$

and the *baryon number current*

$$J^B_\mu = \sum_{\substack{\text{generations } j \\ \text{and colours}}} \left\{\tfrac{1}{3}\bar{u}_j\gamma_\mu u_j + \tfrac{1}{3}\bar{d}_j\gamma_\mu d_j\right\}, \tag{27.6.3}$$

where $u_1 = u$, $u_2 = c$, $u_3 = t$ etc. are not conserved:

$$\partial^\mu J^L_\mu = \partial^\mu J^B_\mu = -\frac{N_G g^2}{16\pi^2}\mathrm{Tr}(\underline{G}_{\mu\nu}\underline{\tilde{G}}^{\mu\nu}), \tag{27.6.4}$$

where N_G is the number of generations.

Note that because of the algebraic cancellation of charges in each generation, the electric current

$$J^{\text{em}}_\mu = \sum_{\text{generations } j} \left\{-\bar{l}_j\gamma_\mu l_j + \sum_{\text{colour}} \left(\tfrac{2}{3}\bar{u}_j\gamma_\mu u_j - \tfrac{1}{3}\bar{d}_j\gamma_\mu d_j\right)\right\} \tag{27.6.5}$$

is conserved:

$$\partial^\mu J^{\text{em}}_\mu = 0. \tag{27.6.6}$$

27.6.2 Baryon and lepton numbers of the vacua

We shall now argue that the field configurations corresponding to the non-perturbative classical vacua of the SM, i.e. those with non-zero winding, numbers, have non-zero lepton and baryon numbers even though they are configurations of a boson field. Indeed, if Q_B, Q_L are the baryon and lepton numbers then for the vacuum with winding number n

$$Q_B = Q_L = -nN_G. \tag{27.6.7}$$

To see this consider, for example, the baryon number

$$Q_B(t) = \int \mathrm{d}^3\boldsymbol{r}\, J^B_0(t, \boldsymbol{r}). \tag{27.6.8}$$

Because the current is not conserved $Q_B(t)$ will depend upon time.

Let us introduce the Euclidean baryon number $Q_B^{(E)}$ defined by

$$Q_B^{(E)} = \int d^3\boldsymbol{r} J_4^B(t, \boldsymbol{r}) \tag{27.6.9}$$

so that

$$Q_B = iQ_B^{(E)}. \tag{27.6.10}$$

Suppose we are in the usual perturbative vacuum with $n = 0$ and $Q_B = Q_B^{(E)} = 0$ at $\tau = -\infty$.

The rate of change of $Q_B^{(E)}$ is then

$$\frac{dQ_B^{(E)}}{d\tau} = \int d^3\boldsymbol{r} \frac{\partial J_4^B}{\partial \tau}. \tag{27.6.11}$$

The Euclidean version of (27.6.4) is

$$\partial_\alpha J_\alpha^B = -i\frac{N_G g^2}{16\pi^2}\mathrm{Tr}(\underline{G}_{\alpha\beta}^E \tilde{\underline{G}}_{\alpha\beta}^E) \tag{27.6.12}$$

so that (27.6.11) becomes

$$\frac{dQ_B^{(E)}}{d\tau} = \int d^3\boldsymbol{r}\left\{-\nabla\cdot\boldsymbol{J}^B - \frac{iN_G g^2}{16\pi^2}\mathrm{Tr}(\underline{G}_{\alpha\beta}^E \tilde{\underline{G}}_{\alpha\beta}^E)\right\}. \tag{27.6.13}$$

By Gauss' theorem the first term involves $\boldsymbol{J}_B$ at spatial infinity which, as usual, is assumed to be zero. (This is what is always done when, for a *conserved* current, one deduces that the 'charge' is time independent.)

We now consider the special case where we choose a *finite energy* configuration starting at the perturbative vacuum $|0\rangle$ where $W_\mu^a = 0$ at $\tau = -\infty$ and reaching the vacuum $|\underline{\boldsymbol{W}}_1\rangle$ at $\tau = \infty$, i.e. a configuration with winding number $\nu = 1$. Integrating (27.6.13) from $\tau = -\infty$ to ∞ yields

$$Q_B^{(E)}(\infty) - Q_B^{(E)}(-\infty) = -i\frac{N_G g^2}{16\pi^2}\int d\tau d^3\boldsymbol{r}\,\mathrm{Tr}(\underline{G}_{\alpha\beta}^E \tilde{\underline{G}}_{\alpha\beta}^E). \tag{27.6.14}$$

Now $Q_B^{(E)}(-\infty) = 0$ and the RHS of (27.6.14) is recognized from (27.4.48) as, up to a constant, the winding number ν (=1 in this case).

We conclude that $Q_B^{(E)}$ in the vacuum $|\underline{\boldsymbol{W}}_1\rangle$ has the value iN_G and hence, via (27.6.10) that the baryon number of $|\underline{\boldsymbol{W}}_1\rangle$ is $-N_G$. The generalization to the vacua $|\underline{\boldsymbol{W}}_n\rangle$ is trivial and the same results hold for lepton number. This justifies (27.6.7).

It is clear from (27.6.1) that actually the baryon and lepton numbers of *each generation* changes by one unit in going from one vacuum to the next. This will be important in deciding which processes will display B and L violations.

27.6.3 The sphaleron

It has been argued (Manton, 1983; Klinghamer and Manton, 1984) that the SM equations of motion possess a finite energy (and therefore localized in space) *static* solution, called a *sphaleron*. Because the only *stable* static configuration is the vacuum, the sphaleron must be an unstable solution corresponding to a maximum of the energy functional. Moreover, because it has no kinetic energy its total energy must correspond to a maximum in the potential energy. This maximum occurs as one varies the topological charge or winding number in going from one vacuum to the next. It thus corresponds to the height of the potential barrier between neighbouring vacua. (See Fig. 25.2.)

The barrier height, usually referred to as the sphaleron energy, is

$$E_{Sp} \approx 4M_W/\alpha_W \tag{27.6.15}$$

and depends weakly on the Higgs mass.

What is important is the scale involved, i.e.

$$E_{Sp} \approx 10 \text{ TeV}. \tag{27.6.16}$$

It is believed that in the directions in function space orthogonal to the winding number, the sphaleron corresponds to a minimum of the energy functional. It is thus a saddle-point.

A slight modifications of the arguments leading to (27.6.14), by choosing configurations which look like a sphaleron at $\tau = 0$, that is halfway between the neighbouring vacua, suggests that the sphaleron has $B = L = 1/2$. Finally, it turns out that the spatial size of the sphaleron is of order $2\pi\hbar c/M_W$.

We come now to the rôle of the sphaleron in baryon and lepton number violation. 'tHooft (1976) had shown, on the basis of instanton tunneling estimates, that the probability for B and L violating processes should be suppressed by a huge factor $\exp(-2S_{Inst})$ where $S_{Inst} = 2\pi/\alpha_W$ is the classical action of the instanton.

With $N_G = 3$ the simplest process that will manifest B and L non-conservation, is, at the quark level,

$$q_1 + q_2 \to \bar{q}_1 + 3\bar{q}_2 + 3\bar{q}_3 + \bar{l}_1 + \bar{l}_2 + \bar{l}_3 \tag{27.6.17}$$

where the subscripts refer to the generations and for each generation the colours of the three quarks or antiquarks must be different. Using instanton-like configurations in the path integral to estimate the amplitude for this process, Ringwald showed that it behaved like an effective point vertex for

$$qq \to 7\bar{q} + 3\bar{l} \tag{27.6.18}$$

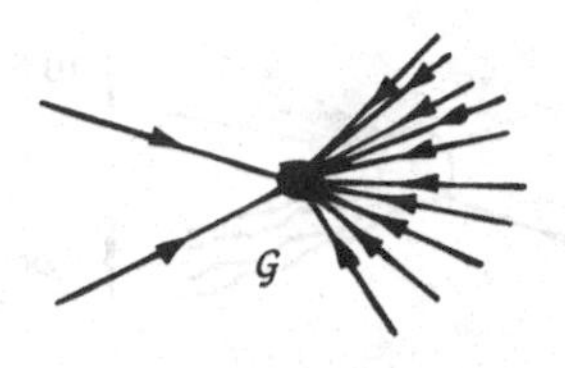

Fig. 27.8. Effective point vertex for $qq \to 7\bar{q} + 3\bar{l}$.

with a coupling constant

$$\mathcal{G} \approx 1.6 \times 10^{-101}\,(\text{GeV})^{-14} \tag{27.6.19}$$

as shown in Fig. 27.8.

Allowing for phase-space, but ignoring any factor coming from spin sums and averages, one finds for the partonic cross-section

$$\hat{\sigma}(qq \to 7\bar{q} + 3\bar{l}) \approx 2 \times 10^{-40}\mathcal{G}^2 \hat{s}^{13} \tag{27.6.20}$$

where $\sqrt{\hat{s}}$ in the qq CM energy. (The factor $\hat{s}^{13}$ can be deduced by dimensional arguments.) At $\sqrt{\hat{s}} = 10$ TeV this yields $\hat{\sigma} \approx 2 \times 10^{-129}$ pb!

Consider now the rôle of the sphaleron. Since the barrier height is given by (27.6.15 and 16) we might expect that for collision energies $\gtrsim 10$ TeV tunneling should not be necessary and the huge suppression factor $\exp(-2\pi/\alpha_W)$ should not be operative. Why then is the result (27.6.20) so tiny? It is suggested that to cross the barrier the fields need to form into a configuration close to that of the sphaleron — in a sense they produce a 'physical' sphaleron — which then decays. (This is analogous to resonance formation and decay.) Now the sphaleron has size $\approx 1/M_W$ so when it decays the momenta of its decay products should be $\approx M_W$. Since its energy is $\approx M_W/\alpha_W$ it could well decay into $\approx 1/\alpha_W$ weak gauge bosons ($W^\pm, Z^0$). Thus we might expect that the B and L violating processes which are not suppressed are of the type

$$q + q \to 7\bar{q} + 3\bar{l} + n \text{ gauge bosons.} \tag{27.6.21}$$

Indeed, Ringwald finds that the amplitude for this reaction is again point-like and contains a factor of order

$$n! \exp(-S_{Inst}) \tag{27.6.22}$$

for $n \gg 1$. The volume of phase-space now grows like $\hat{s}^{13+n}$ and one finds that

$$\hat{\sigma}(q + q \to 7\bar{q} + 3\bar{l} + n \text{ gauge bosons}) \approx 1 \text{ pb} \tag{27.6.23}$$

at $\sqrt{\hat{s}} \approx$ few tens of TeV for $n \approx 1/\alpha_W$!

However, there are two important caveats which suggest that this result should not be taken too literally:

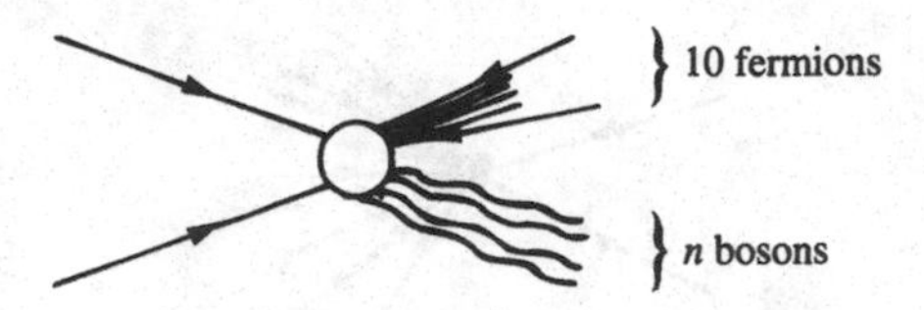

Fig. 27.9. Diagram for $qq \to 7\bar{q} + 3\bar{l} + n$ gauge bosons.

1. Because the amplitude turns out to be point-like the reaction takes place essentially through one partial wave and the cross-section must then not exceed the unitarity bound $\approx 16\pi/\hat{s}$. For $n = 50$, it is found that this bound is already exceeded at $\sqrt{\hat{s}} \approx 60$ TeV. The reason is that although we are in the non-perturbative instanton sector of the theory we are still working to lowest order in the quantum fields. Higher order corrections would presumably resolve this difficulty, but, as explained below, how to do this is non-trivial.

2. Consider Fig. 27.9 which represents the process (27.6.21) in lowest order perturbation theory in the instanton sector. If we include a radiative correction by exchanging a gauge boson between one pair of the final state bosons we will get a factor of α_W in the amplitude. But starting from a given boson the exchanged boson can be attached to $(n-1)$ other bosons. If $n \approx 1/\alpha_W$ this means that the total correction could be of order $\alpha_W \times 1/\alpha_W \approx 1$, i.e. not at all perturbatively small. Thus it may not make sense to be using perturbation theory for such large n.

The subject of B and L non-conservation is receiving much attention at present, because of its possible importance to physics at SSC energies. Improved estimates suggest that the *total* cross-section for B and L violation is of the form

$$\sigma_{tot}^{B,L\ viol} \propto \exp\left\{-\frac{4\pi}{\alpha_W} F\left(\frac{\sqrt{s}}{E_{Sp}}\right)\right\}, \tag{27.6.24}$$

where

$$F = 1 - 0.34\left(\frac{\sqrt{s}}{E_{Sp}}\right)^{4/3} + 0.09\left(\frac{\sqrt{s}}{E_{Sp}}\right)^2 + \cdots \tag{27.6.25}$$

Note that if F becomes negative as s increases unitarity will be violated. In the most optimistic scenario F would tend to zero as $\sqrt{s} \to \infty$.

For access to the most recent literature consult the rapporteurs, talk of Alvarez-Gaume at the 26th International Conference on High Energy Physics (Alvarez-Gaumé, 1992).

28
Beyond the standard model

28.1 Introduction

As we have constantly stressed, the great achievement of the SM model is the unification of the weak and electromagnetic interactions into a single gauge theory $SU(2)_L \times U(1)$. The strong interaction gauge theory, the $SU(3)_C$ of QCD, is totally separate, and the totality of weak, electromagnetic and strong interactions has been treated as the juxtaposition $SU(2)_L \times U(1) \times SU(3)_C$. There are many attempts to unify all these forces into larger schemes such as a grand unification theory (GUT), i.e. to look for a single semi-simple Lie group to describe all the interactions, and which would contain $SU(2)_L \times U(1) \times SU(3)_C$ as a subgroup.

There were early attempts to establish a 'baryon–lepton' (B–L) symmetry (Gamba, Marshak and Okubo, 1959). In more recent years, the outpouring of theoretical papers on grand unification theories, supersymmetric theories, supergravity theories and the boom of superstring theories is, perhaps, only exceeded by the frustration caused by the lack of any experimental indication that any one of these attempts is relevant to nature. The present limit of $> 10^{32}$ years on the proton lifetime has ruled out the so-called minimal $SU(5)$ (see Section 28.4) but this seems to be the only concrete assertion to emerge. It is very difficult at this time to imagine, for instance, that measurements of the proton lifetime will be able to discriminate among the various models. The rate of background due to cosmic rays will make it hard to detect proton decay if $\tau_p > 10^{33}$ years. Also the present rate of neutron–antineutron oscillations $\tau_{n\bar{n}} > 10^7 - 10^8$ s leaves little hope that this could be a viable approach to identifying the correct unification approach. Neutrino oscillations may provide the best clue to the problem if detectors separated by very large distances can ever be put into operation.

In this chapter we limit ourselves to a brief introduction to some of the motivations and ideas for going beyond the standard model. Since this is a totally open subject we shall confine ourselves to only the most popular possible extensions to the SM [see Okun (1991) for more exotic options]. The reader may find the following texts and review papers useful: Mohapatra (1986), Altarelli (1984), Barbieri (1988).

28.2 The 'missing links' of the SM

As we have already seen, there are at least two major missing links of the SM at the time of writing:
(i) the top quark has not yet been discovered;
(ii) the detailed Higgs mechanism of symmetry breaking has not yet been tested.

Assuming that the top quark will indeed be found where LEP results have indicated it should be ($m_t \simeq 130 \pm 40$ GeV/c^2), the major present question concerns the Higgs sector. As discussed in Chapter 6, this question is extremely elusive. It remains an entirely open problem not only where in mass to look for the Higgs (other than outside the mass region excluded by LEP) but also whether we should just have the minimal Higgs doublet demanded by the SM, or whether a richer Higgs spectrum should exist (required by larger models), or whether the Higgs should result from a dynamical effect or, finally, whether entirely different mechanisms should be at work (Okun, 1991).

All the above options have been advocated and the issue can, ultimately, only be answered on an experimental basis.

28.3 Criticisms of the SM

In spite of its successes, *the standard model is incomplete* (Okun, 1991) and many problems are left open; new physics can be expected at higher energies. The crucial question is whether these *higher energies* will ever become attainable in practice.

Let us briefly recall some of these problems (we ignore here standard questions such as: Is the colour symmetry exact? Is the theory confining? Can hadronization be understood? etc.).

28.3.1 The $U(1)$ and θ problems

If the masses of l light quarks are neglected, the QCD Lagrangian is invariant under a global $U(l)_{\mathrm{L}} \otimes U(l)_{\mathrm{R}}$ chiral group (where L=left, R=right). Such a symmetry would normally lead to pairs of particle multiplets with equal mass and opposite parity. However, no approximate parity doubling of light bound states is observed in nature which means that this symmetry must be broken (either spontaneously or dynamically). Normally, we take $l = 2$ and interpret the pion as the approximately massless Goldstone boson associated with the breaking of $U(2)_L \otimes U(2)_R$ down to $SU(2)_V \otimes U(1)_V \otimes U(1)_A$ where the labels V and A indicate vector and axial-vector. $SU(2)_V$ corresponds to the observed isospin symmetry and $U(1)_V$ to baryon conservation. The so-called $U(1)$ problem was discussed at length in Chapter 27 and is related to understanding the breaking mechanism of $U(1)_A$ for which instanton effects may be important.

The $U(1)$ problem is also linked to the so-called θ problem which arises because the baryon-number axial current

$$A^\mu = \sum_a \sum_f \bar{q}_f^a \gamma^\mu \gamma_5 q_f^a \tag{28.3.1}$$

(where f and a are flavour and colour indices respectively) is anomalously not conserved, i.e. one has

$$\partial_\mu A^\mu = \frac{g^2}{32\pi^2} \varepsilon_{\mu\nu\alpha\beta} G_a^{\mu\nu} G_a^{\alpha\beta}. \tag{28.3.2}$$

This leads to a degenerate vacuum structure for QCD. With the different vacuum (i.e. lowest energy) states labelled $|n\rangle$ ($n = 0, 1, \ldots$), the real vacuum must be a superposition of all these states with an arbitrary phase $e^{in\theta}$. n is known as the topological winding number associated with each vacuum solution of the classical equations of motion and is given by

$$\frac{g^2}{32\pi^2} \int \varepsilon_{\mu\nu\alpha\beta} G_a^{\mu\nu} G_a^{\alpha\beta} \equiv n \tag{28.3.3}$$

It can be shown that the above is equivalent to adding to the QCD Lagrangian a new term

$$\mathcal{L}_\theta = \theta \frac{g^2}{32\pi^2} \varepsilon_{\mu\nu\alpha\beta} G_a^{\mu\nu} G_a^{\alpha\beta} \tag{28.3.4}$$

The θ problem arises from the fact that the new term $\mathcal{L}_\theta$ in the QCD Lagrangian is P and CP odd giving rise to P and CP violation of the strong interaction. One observable consequence of the CP-violating Lagrangian is a non-zero electric dipole moment of the neutron. From the upper bound on the latter one derives $\theta \leq 10^{-9}$ and the question is why this parameter should be so small, i.e. why this strong CP violation should be so insignificant. Several different mechanisms have been invented to explain this, for example: degeneracies in the quark mass matrix, non-renormalization theorems from supersymmetry, and axions (a hypothetical very light pseudoscalar Goldstone boson). For details, see Mohapatra (1986).

28.3.2 Parameter counting

One unsatisfactory aspect of the SM is the large number of parameters, masses and couplings, left unspecified within the model. We count: three couplings (for the strong, the em and the weak interactions), six quark masses (one of these can actually be used to set the choice of units), one phase and three KM mixing angles, the Higgs mass, the value of the vacuum expectation value v, θ [introduced in (28.3.4)] and three to six lepton masses (according to how many neutrinos are assumed massless), i.e. at least 17–20 parameters. In addition, a lot of empirical facts are left unexplained in the sense that they do not follow from first principles, such as: Why are the charges quantized the way they are? Why are C, P and CP violated the way they are? Why do B, L_e, L_μ, L_τ appear to be conserved? Why are the neutrinos massless (if they are)? Why do the lepton and, even more so, the quark masses vary over so many orders of magnitude? etc.

Many more questions could be added to this list but the above is clearly sufficient to make the point: the SM cannot conceptually be the end of the story and some other larger unification is expected. Since the effects of any such larger symmetry are not manifested at present-day energies, one assumes that such unification holds at some higher energy scale. The further question is, of course, what the structure of this unification is and why the SM seems to work so well so far (at least in the electroweak sector).

28.4 Grand unification theories (GUT)

This is the attempt to unify electroweak and strong interactions (not gravity) under a single group G (this implies one coupling constant only) which *must* contain $SU(3)_{\mathrm{C}} \otimes SU(2)_{\mathrm{L}} \otimes U(1)$ as a subgroup in order to account for the *low-energy* (!) data (i.e. for the data in the energy region explored thus far). Aside from the known gauge bosons (W and Z^0) governing the SM interactions, other, heavier, gauge bosons are to be expected in larger symmetry schemes. The key question is the energy scale M_{GUT} at which this unification occurs, that is at which the strengths of the interactions known from *low-energy* physics all become equal, i.e. controlled by the single coupling constant of G.

$SU(5)$ (Georgi and Glashow, 1974) emerges as the minimal group that contains as a subgroup $SU(3)_{\mathrm{C}} \otimes SU(2)_{\mathrm{L}} \times U(1)$ and admits a spectrum of fermions with three (or more) generations as in the SM and no other fermions. Even though we now have good evidence that minimal $SU(5)$ cannot be the desired GUT (see below the discussion on the proton lifetime), it is nevertheless instructive to see how things work out in this

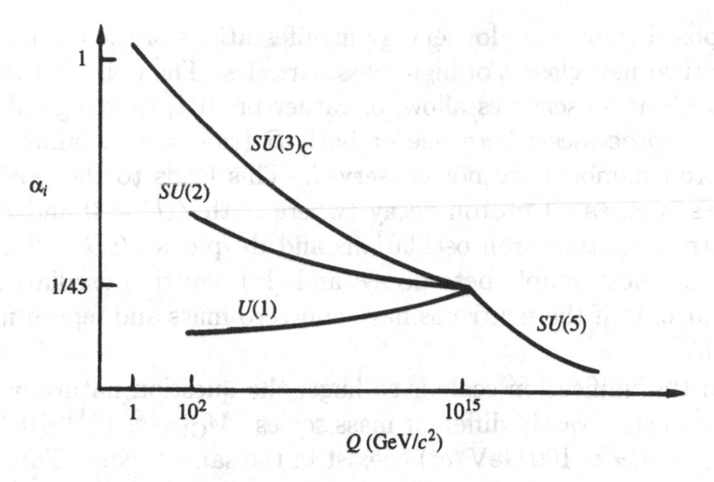

Fig. 28.1. Qualitative evolution with Q^2 of the couplings associated with $U(1), SU(2)_L$ and $SU(3)_C$ within the grand unification scheme $SU(5)$. (From Georgi and Glashow, 1974.)

simple case and the reader is urged to see, e.g., Altarelli (1984) for a very pedagogical discussion of how $SU(5)$ emerges as the simplest group possessing all natural requirements.

In the case of $SU(5)$ the unification of the various forces is indicated in Fig. 28.1. As shown, $SU(3)_C$ is more 'asymptotically free' than $SU(2)$ and the corresponding effective coupling constants fall off at a different rate to join the slowly increasing $U(1)$ coupling [$U(1)$ being 'asymptotically unfree', see Section 20.7] at a mass M.

If $\alpha_i(Q^2)$ ($i = 1, 2, 3$ for $U(1), SU(2)_L$ and $SU(3)_C$ respectively) are the coupling strengths [see (28.6.2) for a precise definition], the rate of approach of $\alpha_3(\equiv \alpha_s)$ to α_2 is, to first order, logarithmic and given by

$$\frac{1}{\alpha_3(Q^2)} - \frac{1}{\alpha_2(Q^2)} \simeq \frac{11}{12\pi} \ln(M^2/Q^2) \tag{28.4.1}$$

where M is the grand unification mass.

If we ignore the many subtleties involved in the problem, a rough evaluation of M gives

$$M \simeq 10^{15} \, \text{GeV}/c^2. \tag{28.4.2}$$

Taking it for granted that the energy domain involved in grand unification schemes will not be accessible to direct exploration in the laboratory for a long time to come, if ever, one restricts attention to indirect tests of the ideas underlying the grand unification schemes. All the tests so far proposed come from low energy manifestations of the exchange of the hypothetical new classes of high-mass particles. The point is that various grand unification schemes allow, or rather predict, exchange of particles leading to processes where one or both B (the baryon number) and L (the lepton number) are not conserved. This leads to the prediction of processes such as (i) proton decay (where both $\Delta B \neq 0$ and $\Delta L \neq 0$); (ii) neutron$\leftrightarrow$antineutron oscillations and d$\leftrightarrow$pions ($\Delta B = 2, \Delta L = 0$); (iii) neutrinoless double beta decay; and (iv) neutrino oscillations which can occur only if the neutrinos have non-zero mass and lepton number is violated.

When the unification scale is so huge, the question naturally arises as to why two such vastly different mass scales ($M_{\text{GUT}} \simeq 10^{14}$–$10^{15}$ GeV/c^2 and $M_W \simeq M_Z \simeq 100\,\text{GeV}/c^2$) coexist in the same theory. This is known as the *hierarchy problem*.

Within $SU(5)$, one can calculate both $\sin^2\theta_W$ and the proton lifetime. In the notation of Chapter 7, $\sin^2\bar{\theta}_W$ at the W mass is calculated (Marciano and Sirlin, 1981) to have the value

$$\sin^2\bar{\theta}_W(M_W) \simeq 0.214^{+0.004}_{-0.003} \tag{28.4.3}$$

where

$$\Lambda_{\text{QCD}} = 0.16^{+0.10}_{-0.08}\text{GeV}$$

in the $\overline{\text{MS}}$ scheme has been used. The above value compares very well with the experimental value of $\sin^2\bar{\theta}_W$ translated to the W mass in the same $\overline{\text{MS}}$ prescription, $\sin^2\bar{\theta}_W(M_W)_{\text{exp}} \simeq 0.215 \pm 0.04$.

The proton lifetime, on the other hand, is given by (Langacker, 1981)

$$\tau(\text{p}) \sim \frac{M^4_{\text{GUT}}}{\alpha^2_{\text{G}}(M_{\text{GUT}})}\frac{1}{m^5_{\text{p}}} \tag{28.4.4}$$

Using $M_{\text{GUT}} \simeq 10^{14}\,\text{GeV}/c^2$ [which was used to derive (28.4.3)], one finds

$$\tau(\text{p}) \simeq 10^{27} \div 10^{31}\,\text{years}. \tag{28.4.5}$$

which is ruled out by the present measurements since for the reaction $\text{p} \to \text{e}^+\pi^0$ (which should be the *dominant* proton decay channel in $SU(5)$), one has experimentally (Particle Data Group, 1992)

$$\tau(\text{p} \to \pi^0\text{e}^+) \geq 5.5 \times 10^{32}\,\text{years}. \tag{28.4.6}$$

When one goes beyond $SU(5)$ other possibilities are $SO(10), E_6$ etc. Longer lifetimes than (28.4.5) can then be achieved so that GUTs larger than minimal $SU(5)$ may not be ruled out by the present data.

A number of experiments are presently being prepared to analyse specifically the problem of the proton lifetime, with sensitivities up to $\approx 10^{33}$ years. The main difficulties in these experiments are the massiveness of the apparatus required and the need for very high background rejection. The background is mostly due to: (i) cosmic muons crossing the detector (this can be reduced by going deeper and deeper underground); (ii) neutrino interactions; and (iii) neutral particles produced by γ and μ outside the detector.

When $\Delta(B-L) \neq 0$ and $\Delta(B+L) \neq 0$ processes are allowed, in addition to proton decay one also expects neutrino$\leftrightarrow$antineutrino oscillations to occur as well as two nucleons$\leftrightarrow$pions. These processes are interesting because their presence or absence offers the possibility of discriminating among various grand unification schemes. For instance, the partial unification $SU(2)_\mathrm{L} \times SU(2)_\mathrm{R} \times SU(4)_\mathrm{L+R}$ leads naturally to $\Delta B = 2$ transitions; in the minimal $SU(5)$ scheme, a large $\Delta B = 2$ amplitude is in conflict with the present bounds on proton stability, whereas it is all right in the $SU(16)$ scheme.

The theoretical estimate of the $\mathrm{n} \leftrightarrow \bar{\mathrm{n}}$ oscillation period is based on the usual quantum mechanical technique of diagonalizing a mass matrix for a two-state system $n_{1,2} = (1/\sqrt{2})(n \pm \bar{n}), m_{1,2} = E_0 \pm \delta m$. Starting with a beam of pure ns and solving the time evolution equation of the wave function one gets an estimate of $\delta m < 10^{-21}\,\mathrm{eV}$ or $\tau_{\mathrm{n\bar{n}}} = \hbar/\delta m > 10^6\,\mathrm{s}$. If this estimate is taken at face value, one might worry what prevents matter from complete annihilation in a period of a few months. It turns out that the $\mathrm{n} \leftrightarrow \bar{\mathrm{n}}$ oscillations are quickly damped by the presence of an external field such as the Earth's magnetic field. Experimental searches thus require a substantial screening of the Earth's magnetic field (of $\sim 10^{-4}$). This is not difficult to realize and all methods utilize thermal and cold neutrons from working reactors, looking for energetic pions emitted by the annihilation with matter of the antineutrons originating from the expected $\mathrm{n} \leftrightarrow \bar{\mathrm{n}}$ oscillation. Some experiment should be able to detect the baryon non-conserving process $\mathrm{n} \leftrightarrow \bar{\mathrm{n}}$ even under the pessimistic assumption of $\tau_{\mathrm{n\bar{n}}}$ as high as 10^8–$10^{10}\,\mathrm{s}$. One expects the 'free' neutron oscillations in a shielded Earth's magnetic field to be a more promising candidate to signal $\Delta B = 2$ processes than the search for 'bound' $\Delta B = 2$ processes $N_1 + N_2 \to \pi\mathrm{s}$ in nuclei. The latter could, however, be investigated with the apparatus built to study proton decay by looking at multipion emission with total released energy around 2 GeV.

Turning now to the leptonic sector, the *neutrinoless* double β decay $(\beta\beta)_0$ can occur if the $(\mathrm{e}, \nu_\mathrm{e})$ current has a 'wrong helicity' component of amplitude η, i.e. is of the form $\bar{\mathrm{e}}\gamma_\mu[(1-\gamma_5) + \eta(1+\gamma_5)]\nu_\mathrm{e}$, and/or the neutrino has a finite mass (we shall not enter into technicalities as to whether the neutrino is a Dirac or a Majorana particle, and refer the reader to the specialized literature: Marshak, Mohapatra and Riazuddin, 1980). Results on the first direct measurement of double β decay with neutrinos by the Moe-Lowenthal-Irvine collaboration set the following limit on the ratio of $(\beta\beta)_0$ to $(\beta\beta)_{\rm all}$

$$R_{\rm exp} \equiv \frac{(\beta\beta)_0}{(\beta\beta)_{\rm all}} < 4.5 \times 10^{-3} \qquad (28.4.7)$$

leading to $\eta_{\rm exp} < 8 \times 10^{-5}$. The theoretical predictions with which these figures are to be compared depend on the details of the grand unification theory used, but, typically, are of the order $R_{\rm th} \simeq 3 \times 10^{-11}$, i.e. $\eta_{\rm th} \simeq 10^{-9}$. The present experimental limits are very far from the theoretical predictions and it will presumably be a long time before grand unification schemes will be tested by this method.

Neutrino oscillations can also occur if lepton number is violated and the neutrinos have a finite mass. The theoretical estimates of the probabilities for neutrino oscillations depend on the mass of the neutrinos. The experimental picture is confused, and conflicting limits exist from different sources, such as solar ν_e, deep mine ν_μ, reactor $\bar{\nu}_e$, LAMPF meson factory ν_e, CERN beam and accelerator data on $\nu_\mu \to \nu_e$ or ν_τ. At present, there is no compelling evidence for ν oscillations, but they are not excluded by the present data.

It should also be stated that within the same grand unification schemes, in the leptonic sector processes such as rare μ decay (i.e. $\mu \to e\gamma$) and various new types of CP violation are possible.

We end this brief comment with a review of the present situation as regards the mass of the neutrinos. We recall that the V–A theory is traditionally associated with left-handed massless neutrinos (γ_5 invariance of the Weyl equation) and that the standard WS model assumes $m_\nu = 0$ as a consequence of the absence of right-handed neutrinos. The motivations for attributing a mass to the neutrinos are not only of aesthetic origin (quark–lepton symmetry) but are supported by cosmological arguments and by indications from laboratory experiments.

In the cosmology arguments, one is led to attribute at least part of the missing nucleonic matter (as compared with what is predicted by the big bang theory) to massive neutrinos, and one obtains thereby constraints on the sum of the masses of all species of neutrinos (see Turner, 1981). From the fact that the universe is not manifestly closed, one obtains $\sum_i m_{\nu_i} \leq 200$ eV/c^2. In fact, if $\sum_i m_{\nu_i} \geq 3.5$ eV/c^2, the universe is neutrino dominated, and, if $\sum_i m_{\nu_i} \geq 100$ eV/c^2, it is closed. The present experimental limits give $m_{\nu_e} < 17$ eV/c^2 (95% confidence level) (from the shape of the spectrum in tritium β decay), $m_{\nu_\mu} < 0.27$ MeV/c^2 (from $\pi^+ \to \mu^+\nu_\mu$), $m_{\nu_\tau} < 35$ MeV/c^2 (from τ decay).

28.5 Compositeness

We have on several occasions mentioned that there is something rather *ad hoc* about the Higgs mechanism for generating the masses of the electroweak gauge bosons and fermions. In addition many people feel that there are too many particles to consider them all as fundamental or elementary. Attempts have therefore been made to generate the Higgs boson and the fermion families as bound states of a smaller number of genuinely elementary fermions. Several such schemes exist, and are based on an extention of the gauge theory concept in order to preserve the essential property of renormalizability. Perhaps the most popular is *technicolour.*

For details, references and a brief summary of the motivations for and models of composite quarks and/or leptons and of composite gauge bosons, see Altarelli (1984).

28.6 Supersymmetry and supergravity

Supersymmetry (SUSY) establishes a symmetry between bosons and fermions. It requires a Lagrangian which is invariant under transformations which mix the fermionic and bosonic degrees of freedom. In any supersymmetric scheme, all particles fall into supermultiplets with at least one boson and one fermion having the same gauge quantum numbers (Gel'fand and Likhtam, 1971; Volkov and Akulov, 1973; Wess and Zumino, 1974). Also, if the symmetry were unbroken, the pairs of bosons and fermions would have the same mass — in gross contradiction with experiment.

Hence, to each fermion and to each vector boson of a gauge theory Lagrangian there will correspond superpartners. For the SM Lagrangian $SU(3)_C \otimes SU(2)_L \otimes U(1)$ one must thus postulate spin 0 partners of quarks and leptons (called *squarks* for 'scalar quarks' and *sleptons*) and spin 1/2 partners of gauge bosons (called *gauginos*). Thus, one introduces *photinos* ($\tilde{\gamma}$), *gluinos* ($\tilde{g}$), *winos* ($\tilde{w}$), and *zinos* ($\tilde{z}$). One must assume them to have heavy masses because no supersymmetric particles has ever been detected and masses in the TeV range are expected. Also the Higgs sector has to be supersymmetrized and one needs a pair of Higgs doublet supermultiplets (Higgsinos $\tilde{h}$) with opposite hypercharge.

By definition of supersymmetry, in any supermultiplet there must be the same number of bosonic and fermionic degrees of freedom. In order to change fermions into bosons and vice versa, the generators of the SUSY transformations must be spin 1/2 'charges' Q_α, and one is led naturally to an extension of the Poincaré algebra (see, e.g., Sohnins, 1985). In

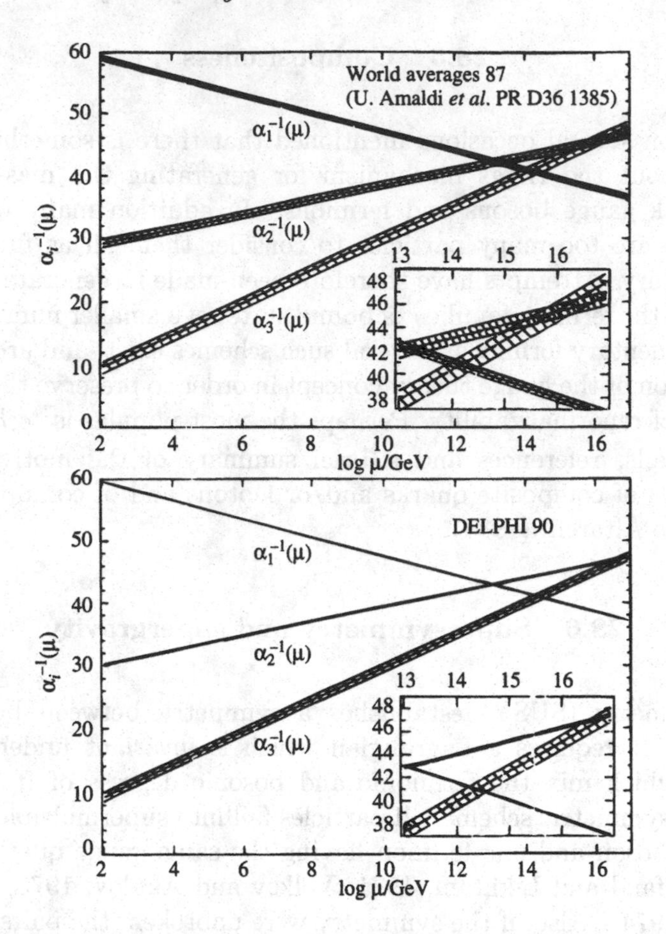

Fig. 28.2. Evolution of the couplings $\alpha_i(\mu)$ $(i = 1, 2, 3)$ in the SM as function of the logarithm of the scale μ (see text). (From *CERN Courier*, March 1991.)

addition to the normal algebraic relations between Q_α and the Poincaré generators, which specify the spinorial character of Q_α under Lorentz transformations and its invariance under translation, the essentially new relation is the anticommutator

$$\{Q_\alpha, \bar{Q}_\beta\} = -(\gamma_\mu)_{\alpha\beta} P^\mu \tag{28.6.1}$$

where P^μ is the energy momentum operator which generates space-time translations.

If in addition SUSY is a local gauge symmetry, like all fundamental symmetries, we are led to an exciting extension of the symmetry to gravity. The product of two local SUSY transformations, in fact, is a translation with space-time dependent parameters, as follows from (28.6.1), i.e. it is a general coordinate transformation! The extended version including gravity is called supergravity.

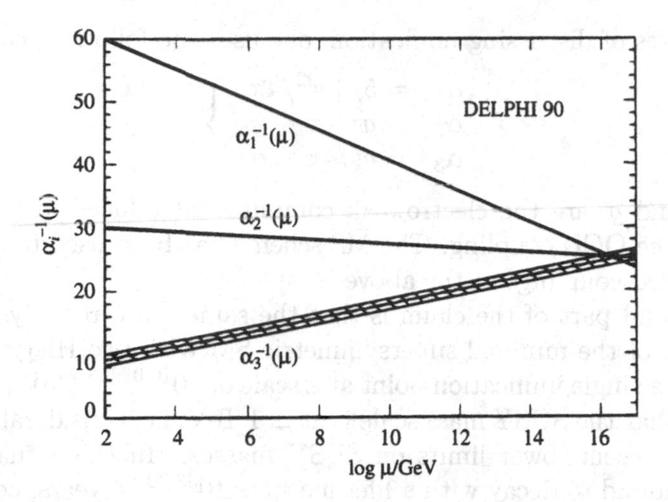

Fig. 28.3. Evolution of the couplings $\alpha_i(\mu)(i = 1, 2, 3)$ in the minimal SUSY SM with two Higgs doublets as function of the logarithm of the scale μ (see text). (From CERN Courier, April 1991.)

SUSY theories have many attractive features. First of all, SUSY is the maximum symmetry compatible with a non-trivial S matrix in a local relativistic field theory. Second, there are useful constraints between masses and couplings which reduce considerably the degree of singularity of the theory and powerful theorems exist about the non-renormalization of some physical parameters. Also the hierarchy and the θ problems mentioned earlier find a solution in SUSY theories. For instance, if one starts from a non-zero value of θ the theory is non-remnormalizable. A natural explanation is also offered for the observed vanishing (or nearly so) of the cosmological constant. In addition, though not free of problems, supergravity is far less singular than conventional quantum gravity. Extensions to space-time dimensions higher than $d = 4$ have also been suggested along lines similar to those pioneered long ago by Kaluza (1921) and Klein (1926).

Although no direct evidence of supersymmetric particles has yet been found (the best present limits place lower bounds on the masses of supersymmetric particles in roughly the LEP energy range), some excitement was caused in 1991 by the claim that DELPHI data give indirect evidence in favour of supersymmetric models.

First the more precise data on the couplings $\alpha_1, \alpha_2, \alpha_3$ introduced in Section 28.4 (see Fig. 28.1) now suggest (Amaldi *et al.*, 1992) that the running couplings in the SM do not, in fact, meet at a single point at some high energy; so the situation in Fig. 28.1 is now modified to what is shown in Fig. 28.2, which shows both 1987 and 1990 data. Note that for the purposes of discussing unification, one uses the following couplings

$$\left.\begin{aligned} \alpha_1 &= 5/3\, g'^2/4\pi \\ \alpha_2 &= g^2/4\pi \\ \alpha_3 &= g_s^2/4\pi \equiv \alpha_s \end{aligned}\right\} \tag{28.6.2}$$

where g and g' are the electroweak couplings introduced in Section 4.2 and g_s is the QCD coupling. The $\overline{\text{MS}}$ scheme has been used to define the renormalized couplings in the above.

The second part of the claim is that the same data, reanalysed in the framework of the minimal supersymmetric SM with two Higgs doublets, do lead to a single unification point at a scale of $10^{16.0\pm0.3}$ GeV (Fig. 28.3) provided that the SUSY mass scale is at $\simeq 1$ TeV, i.e. considerably higher than the present lower limits on SUSY masses. In this situation, the proton is found to decay with a lifetime $\tau_p \simeq 10^{33.2\pm1.2}$ years, compatible with the present limit and tantalizingly close to it [equation (28.4.6)]. Flaws can no doubt be found in the above conclusions due to the large extrapolation required but it is certainly nice to end on an optimistic note. We have just to wait and see: '*Se son rose fioriranno*'.

Appendix 1
Elements of field theory

The aim of this appendix is to illustrate a few of the techniques of perturbative field theory and to explain the derivation of some of the results that have been quoted in the text. (For the notational conventions see after the Preface.)

A1.1 Fields and creation operators

We consider first the expansion of a free field operator in terms of creation and annihilation operators. For a real scalar field describing quanta of mass μ we write

$$\phi(x,t) = \int \frac{\mathrm{d}^3\boldsymbol{k}}{(2\pi)^3 2\omega}[a(\boldsymbol{k})\mathrm{e}^{-\mathrm{i}k\cdot x} + a^\dagger(\boldsymbol{k})\mathrm{e}^{\mathrm{i}k\cdot x}], \tag{A1.1.1}$$

where $k_0 \equiv \omega = +\sqrt{\boldsymbol{k}^2+\mu^2}$.

The equal time commutation relations (20.2.5,6) then lead to

$$\begin{aligned}[a(\boldsymbol{k}), a(\boldsymbol{k}')] &= [a^\dagger(\boldsymbol{k}), a^\dagger(\boldsymbol{k}')] = 0 \\ [a(\boldsymbol{k}), a^\dagger(\boldsymbol{k}')] &= (2\pi)^3 2\omega\delta(\boldsymbol{k}-\boldsymbol{k}'),\end{aligned} \tag{A1.1.2}$$

showing that a and $a^\dagger$ are respectively destruction and creation operators.

The vacuum state $|0\rangle$ is normalized to one:

$$\langle 0|0\rangle = 1 \tag{A1.1.3}$$

and the one-particle states are defined by

$$|\boldsymbol{k}\rangle \equiv a^\dagger(\boldsymbol{k})|0\rangle. \tag{A1.1.4}$$

From (A1.1.2) follows

$$\langle\boldsymbol{k}|\boldsymbol{k}'\rangle = (2\pi)^3 2\omega\delta(\boldsymbol{k}-\boldsymbol{k}') \tag{A1.1.5}$$

and the very useful and simple result

$$\langle 0|\phi(x)|\boldsymbol{k}\rangle = \mathrm{e}^{-\mathrm{i}k\cdot x}. \tag{A1.1.6}$$

For *free fields* (A1.1.2) and (A1.1.1) allow the calculation of the commutator of $\phi(x), \phi^\dagger(y)$ for *arbitrary* times:

$$[\phi(x), \phi^\dagger(y)] = \mathrm{i}\Delta(x-y;\mu), \tag{A1.1.7}$$

where the singular function Δ is given by

$$\Delta(x;\mu) \equiv -\frac{\mathrm{i}}{(2\pi)^3}\int \mathrm{d}^4k\,\delta(k^2-\mu^2)\epsilon(k_0)\mathrm{e}^{-\mathrm{i}k\cdot x}, \tag{A1.1.8}$$

and the step function ϵ is defined by

$$\epsilon(k_0) = \pm 1 \text{ for } k_0 \gtrless 0. \tag{A1.1.9}$$

It should be noted that the fields satisfy *local commutativity* or *microscopic causality*, i.e.

$$[\phi(x), \phi^\dagger(y)] = 0 \qquad \text{if } (x-y)^2 < 0. \tag{A1.1.10}$$

For free fields the vacuum expectation value of the *time ordered product* of two bosonic fields is

$$\begin{aligned}\langle 0|T[\phi(x)\phi^\dagger(y)]|0\rangle &\equiv \theta(x_0-y_0)\langle 0|\phi(x)\phi^\dagger(y)|0\rangle \\ &\quad + \theta(y_0-x_0)\langle 0|\phi^\dagger(y)\phi(x)|0\rangle \\ &= \Delta_F(x-y),\end{aligned} \tag{A1.1.11}$$

where

$$\begin{aligned}\theta(x_0) &= 1 \text{ if } x_0 > 0 \\ &= 0 \text{ if } x_0 < 0\end{aligned} \tag{A1.1.12}$$

and Δ_F is the causal or Feynman propagator function:

$$\Delta_F(x;\mu) \equiv \frac{\mathrm{i}}{(2\pi)^4}\int \mathrm{d}^4k\frac{\mathrm{e}^{-\mathrm{i}k\cdot x}}{k^2-\mu^2+\mathrm{i}\epsilon}. \tag{A1.1.13}$$

Δ_F is a Green's function for the Klein–Gordon equation, i.e. it satisfies

$$(\Box_x + \mu^2)\Delta_F(x-y;\mu) = -\delta^4(x-y). \tag{A1.1.14}$$

For a simple product of fields it is easy to see the existence of a singularity as the space-time points approach each other. One has

$$\langle 0|\phi(x)\phi(y)|0\rangle = \int \frac{\mathrm{d}^3\boldsymbol{k}}{(2\pi)^3 2\omega}\mathrm{e}^{-\mathrm{i}k\cdot(x-y)} \tag{A1.1.15}$$

so that

$$\langle 0|\phi(x)\phi(x)|0\rangle = \int \frac{\mathrm{d}^3\boldsymbol{k}}{(2\pi)^3 2\omega} = \infty. \tag{A1.1.16}$$

Using (A1.1.1) and (A1.1.2) it is a straightforward matter to examine the behaviour of any kind of product, either as $x \to y$ or as $(x-y)^2 \to 0$, for free fields. This is the basis of the Wilson expansion.

For spinor fields of mass m we use

$$\psi(x) = \sum_{\substack{\text{spin projection} \\ r}} \int \frac{\mathrm{d}^3\boldsymbol{p}}{(2\pi)^3 2E} [a_r(\boldsymbol{p}) u_r(\boldsymbol{p}) \mathrm{e}^{-\mathrm{i}p\cdot x} + b_r^\dagger(\boldsymbol{p}) v_r(\boldsymbol{p}) \mathrm{e}^{\mathrm{i}p\cdot x}], \tag{A1.1.17}$$

the spinors being normalized so that

$$u_r^\dagger(\boldsymbol{p}) u_{r'}(\boldsymbol{p}) = v_r^\dagger(\boldsymbol{p}) v_{r'}(\boldsymbol{p}) = 2E\delta_{rr'}, \tag{A1.1.18}$$

where $E \equiv +\sqrt{\boldsymbol{p}^2 + m^2}$.

The canonical anticommutation relations lead to

$$\{b_r(\boldsymbol{p}), b_{r'}^\dagger(\boldsymbol{p})\} = \{a_r(\boldsymbol{p}), a_{r'}^\dagger(\boldsymbol{p})\} = (2\pi)^3 2E\delta_{rr'}\delta^3(\boldsymbol{p}-\boldsymbol{p}'), \tag{A1.1.19}$$

all other anticommutators vanishing. It can be shown that $a_r^\dagger(\boldsymbol{p})$ creates positive energy particles with momentum $\boldsymbol{p}$ and spin projection r, whereas $b_r(\boldsymbol{p})$ creates negative energy particles with momentum $-\boldsymbol{p}$ and spin projection $-r$, the latter being interpreted as the destruction of a positive energy particle of opposite charge, of momentum $\boldsymbol{p}$ and spin projection r.

The states for a single *particle* are defined by

$$|\boldsymbol{p}, r\rangle \equiv a_r^\dagger(\boldsymbol{p})|0\rangle \tag{A1.1.20}$$

and for *antiparticles* by

$$|\overline{\boldsymbol{p}, r}\rangle \equiv b_r^\dagger(\boldsymbol{p})|0\rangle. \tag{A1.1.21}$$

From (A1.1.19) the normalization is then

$$\langle \boldsymbol{p}', r'|\boldsymbol{p}, r\rangle = \langle \overline{\boldsymbol{p}', r'}|\overline{\boldsymbol{p}\,, r}\,\rangle = (2\pi)^3 2E\delta_{rr'}\delta^3(\boldsymbol{p}'-\boldsymbol{p}). \tag{A1.1.22}$$

From (A1.1.17) one then deduces the simple results

$$\begin{aligned} \langle 0|\psi(x)|\boldsymbol{p}, r\rangle &= u_r(\boldsymbol{p})\mathrm{e}^{-\mathrm{i}p\cdot x} \\ \langle \boldsymbol{p}, r|\overline{\psi}(x)|0\rangle &= \bar{u}_r(\boldsymbol{p})\mathrm{e}^{\mathrm{i}p\cdot x} \\ \langle 0|\overline{\psi}(x)|\overline{\boldsymbol{p}, r}\rangle &= \bar{v}_r(\boldsymbol{p})\mathrm{e}^{-\mathrm{i}p\cdot x} \\ \langle \overline{\boldsymbol{p}, r}|\psi(x)|0\rangle &= v_r(\boldsymbol{p})\mathrm{e}^{\mathrm{i}p\cdot x} \end{aligned} \tag{A1.1.23}$$

For a detailed exposition the reader is referred to Bjorken and Drell (1965). Note that, with our normalization, eqn (B.1) of their Appendix B holds for *all* particles irrespective of spin or mass. Care must be taken to use $\Lambda_\pm = m \pm \not{p}$ instead of their eqn (A.3).

For hermitian spin 1 vector fields we use generically

$$A_\mu(x) = \sum_\lambda \int \frac{d^3\boldsymbol{k}}{(2\pi)^3 2E} \{ \epsilon_\mu(\boldsymbol{k},\lambda) a_\lambda(\boldsymbol{k}) e^{-ik\cdot x} + \epsilon^*_\mu(\boldsymbol{k},\lambda) a^\dagger_\lambda(\boldsymbol{k}) e^{ik\cdot x} \} \quad (A1.1.24)$$

where λ = helicity and ϵ_μ the polarization vector.

One has for fixed $\boldsymbol{k}$ and λ

$$k_\mu \epsilon^\mu(\boldsymbol{k},\lambda) = 0, \qquad \epsilon^*_\mu(\boldsymbol{k},\lambda)\epsilon^\mu(\boldsymbol{k},\lambda) = -1 \quad (A1.1.25)$$

and for *massive* vector mesons of mass m

$$\sum_{\lambda=1,0,-1} \epsilon^*_\mu(\boldsymbol{k},\lambda)\epsilon_\nu(\boldsymbol{k},\lambda) = -g_{\mu\nu} + k_\mu k_\nu/m^2. \quad (A1.1.26)$$

The analogous result for physical photons and 'physical' gluons is

$$\sum_{\lambda=\pm1} \epsilon^*_\mu(\boldsymbol{k},\lambda)\epsilon_\nu(\boldsymbol{k},\lambda) = -g_{\mu\nu} + \frac{k_\mu n_\nu + k_\nu n_\mu}{k\cdot n} - \frac{k_\mu k_\nu}{(k\cdot n)^2} \quad (A1.1.27)$$

where n is the unit time-like vector $n^\mu = (1,0,0,0)$.

There are several cases of the use of (A1.1.24):

(*a*) for *photons* the sum over λ involves $\lambda = \pm1$ only.

(*b*) for *gluons* $\lambda = \pm1$ and in addition

$$A_\mu \to A^b_\mu, \qquad a_\lambda(\boldsymbol{k}) \to a^b_\lambda(\boldsymbol{k})$$

where b is the colour label.

(*c*) for *W bosons* $\lambda = 1, 0, -1$ and

$$A_\mu \to W^j_\mu, \qquad j = 1,2,3.$$

Recall that the physical W_μ are given by (Section 4.2)

$$W^\pm_\mu = \frac{1}{\sqrt{2}}\left(W^1_\mu \mp iW^2_\mu\right)$$

The commutation relations for photons and gluons are discussed in Section 21.2. For massive vector mesons see Gasiorowicz (1976), Chapter 3.

The analogue of (A1.1.23) is

$$\langle 0|A_\mu(x)|\boldsymbol{k},\lambda\rangle = \epsilon_\mu(\boldsymbol{k},\lambda)e^{-ik\cdot x}, \qquad \langle \boldsymbol{k},\lambda|A_\mu(x)|0\rangle = \epsilon^*_\mu(\boldsymbol{k},\lambda)e^{ik\cdot x} \quad (A1.1.28)$$

A1.2 Parity, charge conjugation and *G*-parity

A1.2.1 Parity

For the fields we have been considering the parity operator $\mathcal{P}$ has the following effect:

$$\mathcal{P}\phi(\boldsymbol{x},t)\mathcal{P}^{-1} = \pm\phi(-\boldsymbol{x},t) \quad \text{for scalar/pseudo-scalar fields,} \tag{A1.2.1}$$

$$\mathcal{P}\psi(\boldsymbol{x},t)\mathcal{P}^{-1} = \gamma_0\psi(-\boldsymbol{x},t)\,, \tag{A1.2.2}$$

$$\left.\begin{aligned}\mathcal{P}A_j(\boldsymbol{x},t)\mathcal{P}^{-1} &= -A_j(-\boldsymbol{x},t) \quad j=1,2,3\\ \mathcal{P}A_0(\boldsymbol{x},t)\mathcal{P}^{-1} &= A_0(-\boldsymbol{x},t)\end{aligned}\right\}\ \text{for photon or gluon fields} \tag{A1.2.3, A1.2.4}$$

It follows that $\overline{\psi}\gamma_\mu\psi$ is a vector, whereas $\overline{\psi}\gamma_\mu\gamma_5\psi$ is a pseudo-vector, i.e. if

$$\left.\begin{aligned}V_\mu(\boldsymbol{x},t) &\equiv \overline{\psi}(\boldsymbol{x},t)\gamma_\mu\psi(\boldsymbol{x},t)\\ A_\mu(\boldsymbol{x},t) &\equiv \overline{\psi}(\boldsymbol{x},t)\gamma_\mu\gamma_5\psi(\boldsymbol{x},t),\end{aligned}\right\} \tag{A1.2.5}$$

then

$$\left.\begin{aligned}\mathcal{P}V_j(\boldsymbol{x},t)\mathcal{P}^{-1} &= -V_j(-\boldsymbol{x},t) \quad j=1,2,3\\ \mathcal{P}V_0(\boldsymbol{x},t)\mathcal{P}^{-1} &= V_0(-\boldsymbol{x},t),\end{aligned}\right\} \tag{A1.2.6}$$

whereas

$$\left.\begin{aligned}\mathcal{P}A_j(\boldsymbol{x},t)\mathcal{P}^{-1} &= A_j(-\boldsymbol{x},t) \quad j=1,2,3\\ \mathcal{P}A_0(\boldsymbol{x},t)\mathcal{P}^{-1} &= -A_0(-\boldsymbol{x},t).\end{aligned}\right\} \tag{A1.2.7}$$

A1.2.2 Charge conjugation

The charge conjugation operator $\mathcal{C}$ has the following effect:

$$\mathcal{C}A_\mu(\boldsymbol{x},t)\mathcal{C}^{-1} = -A_\mu(\boldsymbol{x},t) \quad \text{for photons.} \tag{A1.2.8}$$

Thus an n-photon state is an eigenstate of $\mathcal{C}$ with eigenvalue $(-1)^n$, known as the charge parity.

For the neutral π^0 field one has

$$\mathcal{C}\phi^{(0)}(x)\mathcal{C}^{-1} = \phi^{(0)}(x), \tag{A1.2.9}$$

whereas for the charged fields

$$\mathcal{C}\phi(x)\mathcal{C}^{-1} = \phi^\dagger(x), \qquad \mathcal{C}\phi^\dagger(x)\mathcal{C}^{-1} = \phi(x). \tag{A1.2.10}$$

In terms of Hermitian fields $\phi_{1,2}(x)$

$$\left.\begin{aligned} \phi(x) &= \frac{1}{\sqrt{2}}(\phi_1 + \mathrm{i}\phi_2) \\ \phi^\dagger(x) &= \frac{1}{\sqrt{2}}(\phi_1 - \mathrm{i}\phi_2) \end{aligned}\right\} \tag{A1.2.11}$$

$$\mathcal{C}\phi_1(x)\mathcal{C}^{-1} = \phi_1(x), \qquad \mathcal{C}\phi_2(x)\mathcal{C}^{-1} = -\phi_2(x), \tag{A1.2.12}$$

so that $\mathcal{C}$ causes a reflection in the 1–3 plane of isospace. It can be seen that only an electrically neutral state, and in particular only a state with equal numbers of π^+ and π^-, can be an eigenstate of $\mathcal{C}$.

For spinors, if α, β label spin indices,

$$\left.\begin{aligned} \mathcal{C}\psi_\alpha(x)\mathcal{C}^{-1} &= (C\gamma^0)_{\alpha\beta}\psi^\dagger_\beta(x) \\ \mathcal{C}\overline{\psi}_\alpha(x)\mathcal{C}^{-1} &= -\psi_\beta(x)(C^{-1})_{\beta\alpha}, \end{aligned}\right\} \tag{A1.2.13}$$

where C is a 4×4 matrix

$$C = \mathrm{i}\gamma^2\gamma^0 = -C^{-1} = -C^T, \tag{A1.2.14}$$

which has the property of taking the transpose of the γ matrices

$$C\gamma_\mu C^{-1} = -\gamma_\mu^T. \tag{A1.2.15}$$

From (A1.2.13) and (A1.1.17) follows

$$\mathcal{C}b(\boldsymbol{p},r)\mathcal{C}^{-1} = a(\boldsymbol{p},r), \qquad \mathcal{C}a^\dagger(\boldsymbol{p},r)\mathcal{C}^{-1} = b^\dagger(\boldsymbol{p},r), \tag{A1.2.16}$$

so that $\mathcal{C}$ has the effect of interchanging particles and antiparticles.

It follows that the vector current built from coloured quarks (Section 21.3) behaves under charge conjugation as follows:

$$\begin{aligned} \mathcal{C}\overline{\psi}_i\gamma^\mu\left(\tfrac{1}{2}\boldsymbol{\lambda}^a\right)_{ij}\psi_j\mathcal{C}^{-1} &= (\mathcal{C}\overline{\psi}_i\mathcal{C}^{-1})\gamma^\mu\left(\tfrac{1}{2}\boldsymbol{\lambda}^a\right)_{ij}(\mathcal{C}\psi_j\mathcal{C}^{-1}) \\ &= -\psi_{i,\beta}(C^{-1})_{\beta\alpha}\gamma^\mu_{\alpha\delta}\left(\tfrac{1}{2}\boldsymbol{\lambda}^a\right)_{ij}(C\gamma^0)_{\delta\rho}\psi^\dagger_{j,\rho} \\ &= \psi_{i,\beta}(\gamma^{\mu^T})_{\beta\sigma}\left(\tfrac{1}{2}\boldsymbol{\lambda}^a\right)_{ij}\gamma^0_{\sigma\rho}\psi^\dagger_{j,\rho} \end{aligned}$$

by (A1.2.15)

$$= -\overline{\psi}_j\gamma^\mu\left(\tfrac{1}{2}\boldsymbol{\lambda}^a\right)_{ij}\psi_i, \tag{A1.2.17}$$

where the minus sign comes from the fact that ψ_i, $\overline{\psi}_j$ anticommute.

For leptons, with no colour, $\frac{1}{2}\boldsymbol{\lambda}^a$ is replaced by the unit matrix in the vector current, and one has

$$\mathcal{C}\overline{\psi}\gamma^\mu\psi\mathcal{C}^{-1} = -\overline{\psi}\gamma^\mu\psi, \tag{A1.2.18}$$

which, with (A1.2.8), shows that the electromagnetic coupling of photons and leptons is invariant under charge conjugation.

In order to make the coupling of gluons to coloured quarks invariant under charge conjugation, from (A1.2.17) we have to demand that

$$\mathcal{C}\left(\tfrac{1}{2}\boldsymbol{\lambda}^a\right)_{ij}\mathbf{A}^a_\mu\mathcal{C}^{-1} = -\left(\tfrac{1}{2}\boldsymbol{\lambda}^a\right)_{ji}\mathbf{A}^a_\mu. \tag{A1.2.19}$$

For a state of a fermion and its antiparticle with orbital angular momentum l and total spin S one has

$$\mathcal{C}|l,S\rangle = (-1)^{l+S}|l,S\rangle. \tag{A1.2.20}$$

It follows from this that the $^3S_1$ state of positronium decays into three photons whereas the $^1S_0$ state decays into two photons.

The situation is more complicated for $q\bar{q} \to$ gluons as a consequence of the more involved rule (A1.2.19), and was discussed in Section 11.6.

The consequences of CPT invariance on the structure of matrix elements is discussed in Appendix 6.

A1.2.3 G-parity

This is the combined operation of charge conjugation and a rotation of π about the '2' axis of isospace:

$$G \equiv \mathcal{C}e^{i\pi T_2}, \tag{A1.2.21}$$

where T_2 is the generator of rotations about the '2' axis.

The useful point is that, unlike $\mathcal{C}$, charged states can be eigenstates of G. For example, for pions

$$G\phi_i G^{-1} = -\phi_i, \tag{A1.2.22}$$

where i is the isospin label.

Thus the G-parity of an n-pion state is $(-1)^n$, and this leads to various selection rules for hadronic reactions.

For further details consult Gasiorowicz (1976), Chapters 16, 17 and 30.

A1.3 The S-matrix

The S-matrix in perturbation theory is given as follows.

The Hamiltonian in the so-called 'interaction picture' is first split into a free field part and an interaction (perturbative) part

$$H = H_0 + H_I. \tag{A1.3.1}$$

Then, short circuiting an involved and subtle argument, if H_I is considered as made up of an expression involving free field operators, the

S-operator is given by

$$\begin{aligned} S &= 1 - \mathrm{i}\int_{-\infty}^{\infty} \mathrm{d}t_1 H_I(t_1) + \frac{(-\mathrm{i})^2}{2!}\int_{-\infty}^{\infty} \mathrm{d}t_1 \mathrm{d}t_2 T[H_I(t_1)H_I(t_2)] + \cdots \\ &= \sum_{n=0}^{\infty} \frac{(-\mathrm{i})^n}{n!}\int_{-\infty}^{\infty} \mathrm{d}t_1 \cdots \mathrm{d}t_n T[H_I(t_1)\cdots H_I(t_n)]. \end{aligned} \tag{A1.3.2}$$

When one substitutes the actual form of H_I for a particular theory one has in (A1.3.2) a perturbative expansion for S. It is then not difficult to read off the rules for a diagrammatic representation of the perturbation series. The subtlety in a gauge theory is the problem of finding H. The Lagrangian contains redundant variables which have to be constrained by gauge fixing terms, and one is then dealing with the quantum version of a dynamical system subject to holonomic constraints—a non-trivial matter. It is partly for this reason that the Feynman integral approach is preferred for those theories.

Since the Hamiltonian is the space integral of the Hamiltonian density

$$H_I(t) = \int \mathrm{d}^3\boldsymbol{x}\, \mathcal{H}_I(t, \boldsymbol{x}) \tag{A1.3.3}$$

the expression for S is often written in the form

$$S = \sum_{n=0}^{\infty} \frac{(-\mathrm{i})^n}{n!}\int \mathrm{d}^4x_1 \mathrm{d}^4x_2 \cdots \mathrm{d}^4x_n T[\mathcal{H}_I(x_1)\mathcal{H}_I(x_2)\cdots\mathcal{H}_I(x_n)] \tag{A1.3.4}$$

known as the Dyson perturbative expansion.

Appendix 2
Feynman rules for QED, QCD and the SM

We give here, without derivation, the rules for calculating (up to a sign) what is known as the Feynman amplitude $\mathcal{M}$ in QED, QCD and the SM. We illustrate with a few topical examples. A detailed treatment can be found in Bjorken and Drell (1965) and in Cutler and Sivers (1978).

A2.1 Relation between S-matrix and Feynman amplitude

The rules for calculating the Feynman amplitude $\mathcal{M}$ are essentially universal and irrespective of the normalization convention for the states.

Clearly, however, the matrix elements $\langle f|S|i\rangle$ of the S-operator depend upon the normalization convention, so the relationship between $\langle f|S|i\rangle$ and $\mathcal{M}$ will be convention dependent.

With our relativistically invariant normalization

$$\langle \boldsymbol{p}'|\boldsymbol{p}\rangle = (2\pi)^3 2E\delta^3(\boldsymbol{p}'-\boldsymbol{p}) \tag{A2.1.1}$$

for all particles, one has the simple result

$$\langle f|S|i\rangle = \langle f|i\rangle + (2\pi)^4\delta^4(P_f - P_i)\mathcal{M}. \tag{A2.1.2}$$

Often it is useful to use other normalization conventions. To read off the analogue of (A2.1.2) it is simplest to avoid the complications of 'continuum normalization' and Dirac δ-function by quantizing the system in a finite volume V. The key connection is the equivalence

$$(2\pi)^3\delta^3(\boldsymbol{p}'-\boldsymbol{p}) \longleftrightarrow V\delta_{\boldsymbol{p}'\boldsymbol{p}} \tag{A2.1.3}$$

so that our normalization corresponds to

$$\langle \boldsymbol{p}'|\boldsymbol{p}\rangle = V2E\,\delta_{\boldsymbol{p}'\boldsymbol{p}} \quad \text{or} \quad \langle \boldsymbol{p}|\boldsymbol{p}\rangle = V2E. \tag{A2.1.4}$$

If we prefer to utilize states normalized to unity in the volume V, i.e.

$\langle \boldsymbol{p}|\boldsymbol{p}\rangle = 1$, then we will have

$$S_{fi} = \delta_{fi} + (2\pi)^4\delta^4(P_f - P_i) \prod_{\substack{\text{all} \\ \text{particles } j}} \left(\frac{1}{V2E_j}\right)^{\frac{1}{2}} \mathcal{M}. \tag{A2.1.5}$$

As a consequence of our normalization convention $u^\dagger u = v^\dagger v = 2E$ for spinors (see the notational conventions) the cross-section for any process $1 + 2 \to 3 + 4 + \cdots + n$ is expressed in terms of $\mathcal{M}$ by formula (B.1) of Appendix B of Bjorken and Drell (1964), for all particles, *mesons* or *baryons*

$$\begin{aligned} \mathrm{d}\sigma &= \frac{1}{|\boldsymbol{v}_1 - \boldsymbol{v}_2|}\frac{1}{2E_1}\frac{1}{2E_2}\overline{|\mathcal{M}|^2}\frac{\mathrm{d}^3\boldsymbol{p}_3}{(2\pi)^3 2E_3}\cdots\frac{\mathrm{d}^3\boldsymbol{p}_n}{(2\pi)^3 2E_n}\times \\ &\quad \times(2\pi)^4\delta^4(p_3 + p_4 + \cdots + p_n - p_1 - p_2) \end{aligned} \tag{A2.1.6}$$

where $\boldsymbol{v}_1$ and $\boldsymbol{v}_2$ are the velocities of the beam and target particles in a reference frame where their collision is collinear.

In (A2.1.6) $\overline{|\mathcal{M}|^2}$ is the modulus squared of the Feynman amplitude summed over final spins and colours and averaged over initial spins and colours.

For the differential decay rate of a particle of mass M in its rest frame, where its 4-momentum is $p = (M, 0, 0, 0)$,

$$\begin{aligned} \mathrm{d}\Gamma &= \frac{1}{2M}\overline{|\mathcal{M}|^2}\frac{\mathrm{d}^3\boldsymbol{p}_1}{(2\pi)^3 2E_1}\cdots\frac{\mathrm{d}^3\boldsymbol{p}_n}{(2\pi)^3 2E_n}\times \\ &\quad \times(2\pi)^4\delta^4(p - p_1 - p_2 - \cdots - p_n) \end{aligned} \tag{A2.1.7}$$

Note that in both (A2.1.6 and 7) care must be exercised when dealing with identical particles. When integrating over phase space, if there are n_j particles of type j in the final state then one must include a factor $1/(n_j!)$.

Many useful formulae needed in doing the spin sums implied in $\overline{|\mathcal{M}|^2}$, especially the traces of products of Dirac matrices, can be found in Bjorken and Drell (1964) Appendix A and Section 7.2.

A2.2 QCD and QED

In this section: Greek indices are Lorentz tensor indices; a, b, c, d are gluon colour indices; l, j are quark colour indices; k, p, q, r label 4-momenta. The f_{abc} and the matrices $\boldsymbol{t}^a$ are discussed in Section 21.3. No flavour indices are shown; all vertices are diagonal in flavour. The strong interaction coupling constant is g, and is related to the α_s used throughout the book by $\alpha_s = g^2/4\pi$. Q_f is the electric charge of fermion f in units of e.

The rules are given for two classes of gauges:

- the *covariant* gauges labelled by 'a', as discussed in Section 21.2 ($a = 1$ is the Feynman gauge; $a = 0$ the Landau gauge) in which the subsidiary condition, at least at the classical level, is $\partial^\mu A^c_\mu = 0$ for all values of the colour label c, and the gauge fixing term in the Lagrangian is $(-1/2a)\Sigma_c(\partial^\mu A^c_\mu)^2$.

- an *axial* gauge, one of a family again labelled by 'a', in which the subsidiary condition is $n^\mu A^c_\mu = 0$ for all c, where n^μ is a fixed space-like or null 4-vector, and where the gauge fixing term in the Lagrangian is $(-1/2a)\Sigma_c(n^\mu A^c_\mu)^2$.

We allow the quarks to have a mass parameter m which should be put to zero when working with massless quarks.

(a) The propagators

lepton (momentum p)
$$\frac{\mathrm{i}(\not p + m)}{p^2 - m^2 + \mathrm{i}\epsilon} \qquad \text{(A2.2.1)}$$

quark j  → l (momentum p)
$$\delta_{jl}\frac{\mathrm{i}(\not p + m)}{p^2 - m^2 + \mathrm{i}\epsilon} \qquad \text{(A2.2.2)}$$

In the above the arrow indicates the flow of fermion number and p is the 4-momentum in that direction. (Note: j, l are quark colour labels, b, c gluon and ghost colour labels.)

gluon b,β → c,γ (momentum k)
$$\delta_{bc}\frac{\mathrm{i}}{k^2 + \mathrm{i}\epsilon} \times \begin{cases} \text{Covariant gauges:} \\ \left[-g_{\beta\gamma} + (1-a)\frac{k_\beta k_\gamma}{k^2+\mathrm{i}\epsilon}\right] \quad \text{(A2.2.3)} \\ \text{Axial gauges with } a = 0\text{:} \\ \left[-g_{\beta\gamma} + \frac{n_\beta k_\gamma + n_\gamma k_\beta}{n\cdot k} - \frac{n^2 k_b k_\gamma}{(n\cdot k)^2}\right] \\ \qquad \text{(A2.2.4)} \end{cases}$$

Note that in the above axial gauges the propagator is orthogonal to n^β; and is orthogonal to k^β when $k^2 = 0$.

photon α → β (momentum k)
$$\mathrm{i}\frac{-g_{\alpha\beta}}{k^2 + \mathrm{i}\epsilon} \qquad \text{(A2.2.5)}$$

ghost b  - - - c (momentum p)
$$\delta_{bc}\frac{\mathrm{i}}{p^2 + \mathrm{i}\epsilon} \qquad \text{(Covariant gauges only)}$$
(A2.2.6)

(b) The vertices

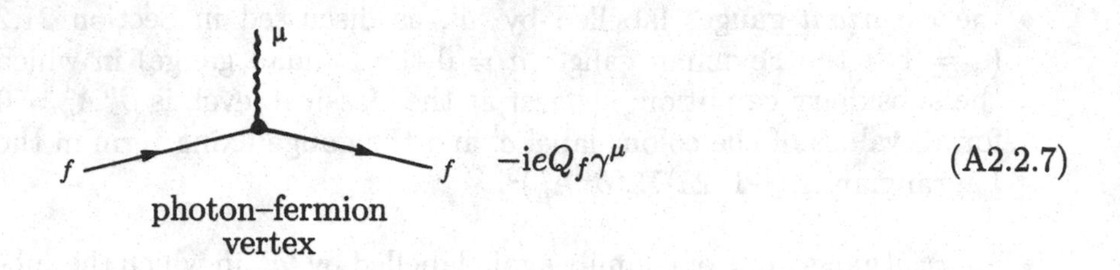

photon–fermion vertex

$$-ieQ_f\gamma^\mu \qquad \text{(A2.2.7)}$$

b, β
j
l

quark–gluon vertex

$$ig(t^b)_{lj}\gamma^\beta \qquad \text{(A2.2.8)}$$

a, α
p
r
q
c, γ
b, β

triple gluon vertex

$$gf_{abc}[g^{\alpha\beta}(p-q)^\gamma + g^{\beta\gamma}(q-r)^\alpha + g^{\gamma\alpha}(r-p)^\beta] \qquad \text{(A2.2.9)}$$

$$(p+q+r=0)$$

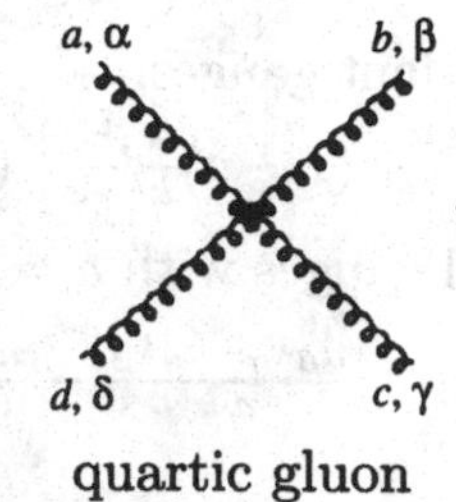

quartic gluon vertex

$$\begin{aligned}-ig^2\{&f_{eac}f_{ebd}(g^{\alpha\beta}g^{\gamma\delta}-g^{\alpha\delta}g^{\beta\gamma})\\ &+f_{ead}f_{ebc}(g^{\alpha\beta}g^{\gamma\delta}-g^{\alpha\gamma}g^{\beta\delta})\\ &+f_{eab}f_{ecd}(g^{\alpha\gamma}g^{\beta\delta}-g^{\alpha\delta}g^{\beta\gamma})\}\end{aligned} \qquad \text{(A2.2.10)}$$

a, α
p
c
r
q
b

gluon–ghost vertex

$$gf_{abc}q^\alpha$$

(Covariant gauges only) (A2.2.11)

$(p+r=q)$

Note that the ghosts are scalar fields, but a factor (-1) must be included for each closed loop, as is the case for fermions. Note also that the sign of $\mathcal{M}$, if it is important, has to be determined by comparing the order of the fermion operators in the diagram with their order in the expression for the S-operator.

A2.3 The SM

In this section: Greek indices are Lorentz indices; i, j label generations.

$u_i = (u, c, t)$, $d_j = (d, s, b)$; V_{ij} are the Kobayashi–Maskawa matrix elements [see eqns (9.2.6 and 7)].

For the $W^\pm$ and Z^0 propagator ($M = M_W$ or M_Z):

$$= \frac{\mathrm{i}[-g_{\alpha\beta} + k_\alpha k_\beta / M^2]}{k^2 - M^2 + \mathrm{i}\epsilon} \tag{A2.3.1}$$

For the vertices (all results are independent of colour):

1 Charged weak hadronic interactions.

Let us introduce generation labels $i, j = 1, 2, 3$ such that

$$(u_1, u_2, u_3) \equiv (u, c, t) \qquad \text{and} \qquad (d_1, d_2, d_3) \equiv (d, s, b)$$

then from (9.3.8), (9.3.10) and (9.2.1) one deduces

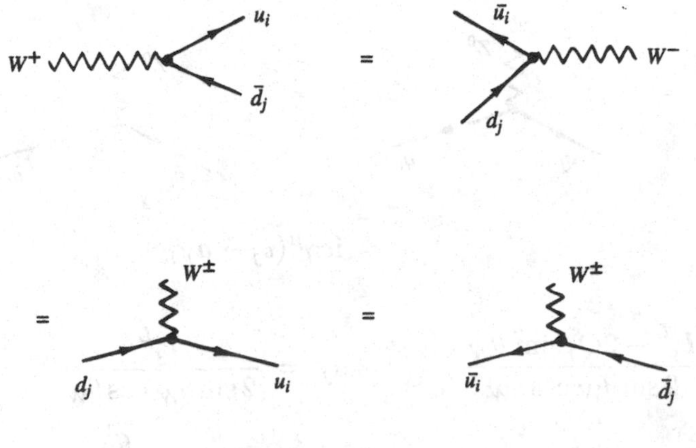

W^+ u_i $\bar{d}_j$ = $\bar{u}_i$ d_j W^-

= $W^\pm$ d_j u_i = $W^\pm$ $\bar{u}_i$ $\bar{d}_j$

$$= -\frac{\mathrm{ie}}{2\sqrt{2}\sin\theta_W} V_{ij}\gamma^\mu(1 - \gamma_5) \tag{A2.3.2}$$

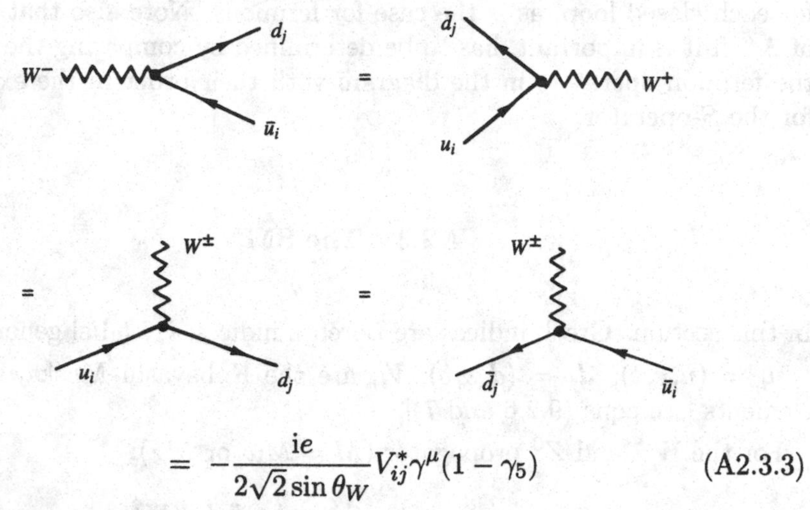

$$= -\frac{ie}{2\sqrt{2}\sin\theta_W} V_{ij}^* \gamma^\mu (1-\gamma_5) \qquad \text{(A2.3.3)}$$

By convention all the topologically similar vertices are given the same sign in the Feynman rules. The overall sign of a diagram has to be determined by comparing the order of the fermion operator in the diagram with their normal order in the S operator.

2 Neutral weak hadronic interactions.

From (9.3.1), (9.3.2) and (9.3.3) we get, with $q_j = u_j$ or d_j,

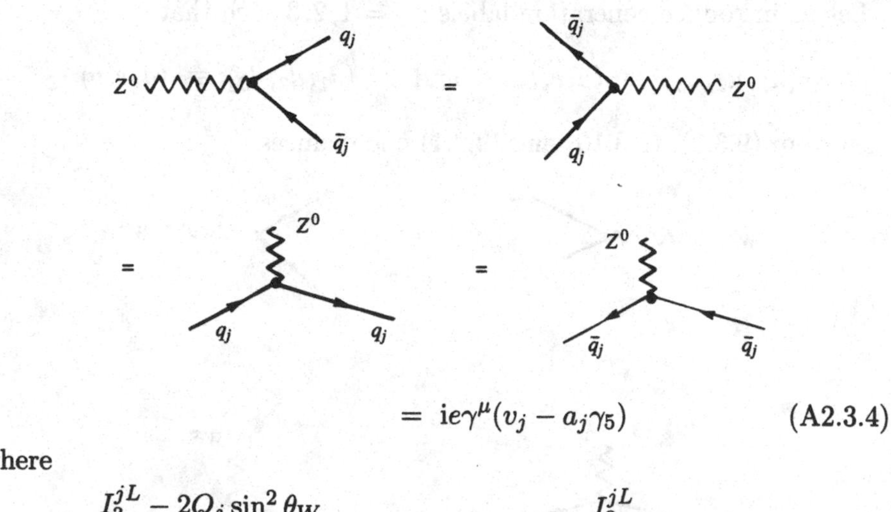

$$= ie\gamma^\mu(v_j - a_j\gamma_5) \qquad \text{(A2.3.4)}$$

where

$$v_j = \frac{I_3^{jL} - 2Q_j\sin^2\theta_W}{2\sin\theta_W\cos\theta_W} \equiv \frac{g_V^j}{2\sin\theta_W\cos\theta_W} \qquad a_j = \frac{I_3^{jL}}{2\sin\theta_W\cos\theta_W} \equiv \frac{g_A^j}{2\sin\theta_W\cos\theta_W} \qquad \text{(A2.3.5)}$$

where I_3^{jL} is the third component of weak isospin of the left-handed part of the quark j, i.e. $I_3^{jL} = \frac{1}{2}$ for u, c, t; $= -\frac{1}{2}$ for d, s, b. Concerning overall signs see comment in (1) above.

3 Higgs coupling to quarks.

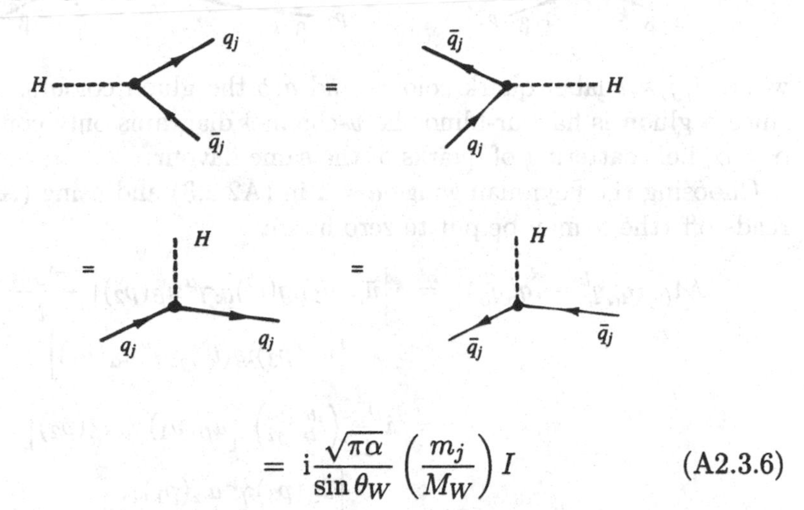

$$= \mathrm{i}\frac{\sqrt{\pi\alpha}}{\sin\theta_W}\left(\frac{m_j}{M_W}\right) I \tag{A2.3.6}$$

where m_j is the mass of quark j and I is the unit 4×4 matrix. Concerning overall signs see comment in (1) above.

A2.4 Some examples of Feynman amplitudes

A full and detailed calculation of a reaction rate, starting from a Feynman amplitude, is given in Section 8.6.

Here, for pedagogical reasons, we give a few examples of partonic Feynman diagrams and their corresponding amplitudes. As usual for any $2 \to 2$ partonic process

$$A(p_1) + B(p_2) \;\to\; C(p_3) + D(p_4) \tag{A2.4.1}$$

we define the partonic version of the Mandelstam variables:

$$\begin{aligned} \hat{s} &= (p_1 + p_2)^2 \\ \hat{t} &= (p_1 - p_3)^2 \\ \hat{u} &= (p_1 - p_4)^2. \end{aligned} \tag{A2.4.2}$$

1 Quark–quark scattering.

$$q_\alpha(p_1) + q_\beta(p_2) \;\to\; q_\alpha(p_3) + q_\beta(p_4) \tag{A2.4.3}$$

where α, β are *flavour* labels. The two lowest order diagrams are

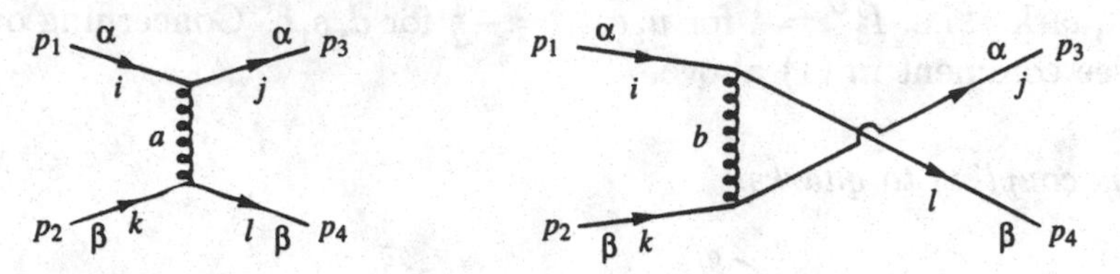

where i, j, k, l label quark colours and a, b the gluon colours. Note that since a gluon is flavour-blind the u-channel diagrams only contributes if $\alpha = \beta$, i.e. scattering of quarks of the same flavour.

Choosing the Feynman gauge $a = 1$ in (A2.2.3) and using (A2.2.7) one reads off (the $\mathrm{i}\epsilon$ may be put to zero here):

$$\begin{aligned} \mathcal{M}_{(t)}(q_\alpha^i q_\beta^k \to q_\alpha^j q_\beta^l) &= \left[\bar{u}_\beta(p_4)\mathrm{i}g(t^b)_{lk}\gamma^\mu u_\beta(p_2)\right]\frac{-\mathrm{i}g_{\mu\nu}}{\hat{t}} \times \\ &\quad\times \left[\bar{u}_\alpha(p_3)\mathrm{i}g(t^b)_{ji}\gamma^\nu u_\alpha(p_1)\right] \\ &= \mathrm{i}\frac{g^2}{\hat{t}}\left(t_{lk}^b t_{ji}^b\right)\left[\bar{u}_\beta(p_4)\gamma_\mu u_\beta(p_2)\right] \times \\ &\quad\times \left[\bar{u}_\alpha(p_3)\gamma^\mu u_\alpha(p_1)\right], \end{aligned} \tag{A2.4.4}$$

$$\begin{aligned} \mathcal{M}_{(u)}(q_\alpha^i q_\beta^k \to q_\alpha^j q_\beta^l) &= (-1)\delta_{\alpha\beta}\left[\bar{u}_\alpha(p_3)\mathrm{i}g(t^b)_{jk}\gamma^\mu u_\beta(p_2)\right] \times \\ &\quad\times \left(\frac{-\mathrm{i}g_{\mu\nu}}{\hat{u}}\right)\left[\bar{u}_\beta(p_4)\mathrm{i}g(t^b)_{li}\gamma^\nu u_\alpha(p_1)\right] \\ &= (-1)\delta_{\alpha\beta}\mathrm{i}\frac{g^2}{\hat{u}}\left(t_{jk}^b t_{li}^b\right)\left[\bar{u}_\alpha(p_3)\gamma_\mu u_\beta(p_2)\right] \times \\ &\quad\times \left[\bar{u}_\beta(p_4)\gamma^\mu u_\alpha(p_1)\right]. \end{aligned} \tag{A2.4.5}$$

The factor (-1) in (A2.4.5) does not come from the Feynman rules. It comes from comparing the order of the fermion operators underlying the two diagrams. Symbolically we have:

$$\begin{aligned} \mathcal{M}_{(t)}:&\quad (\overline{\psi}_3\psi_1)\,(\overline{\psi}_4\psi_2) \\ \mathcal{M}_{(u)}:&\quad (\overline{\psi}_4\psi_1)\,(\overline{\psi}_3\psi_2). \end{aligned} \tag{A2.4.6}$$

To rearrange the operators in $\mathcal{M}_{(u)}$ into the order occurring in $\mathcal{M}_{(t)}$ requires three interchanges (i.e. an odd number); hence the relative (-1).

In the above the colour b is, of course, summed over. Rules for doing colour sums are given in Section A2.5. To calculate the unpolarized cross-section involves forming

$$|\mathcal{M}|^2 \equiv |\mathcal{M}_{(t)} + \mathcal{M}_{(u)}|^2 \tag{A2.4.7}$$

summing over final spins and colours, averaging over initial colours and spins.

2 Quark-gluon scattering.

There are three diagrams for

$$q(p_1) + G_a(q_1) \rightarrow q(p_2) + G_b(q_2) \qquad \text{(A2.4.8)}$$

The amplitude does not depend on the quark flavour so it is not indicated. (Of course initial and final quarks have the same flavour.)

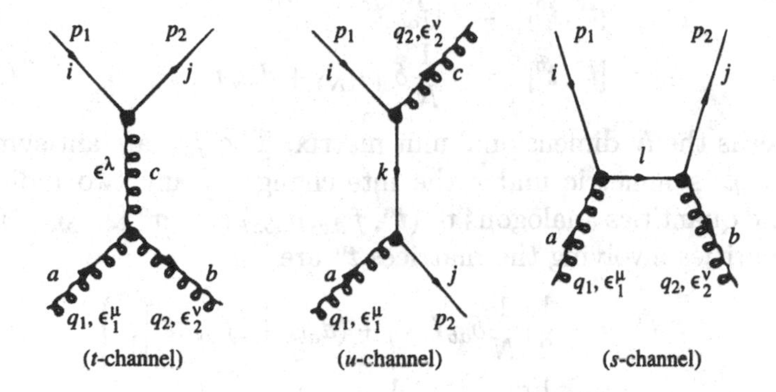

(*t*-channel) (*u*-channel) (*s*-channel)

Here

$$\hat{s} = (p_1 + q_1)^2, \qquad \hat{t} = (p_1 - p_2)^2, \qquad \hat{u} = (p_1 - q_2)^2. \qquad \text{(A2.4.9)}$$

We shall only deal with the t-channel amplitude. From (A2.2.3) in Feynman gauge ($a = 1$) and using (A2.2.8 and 9) we get

$$\begin{aligned} \mathcal{M}_{(t)}(q_i G_a \rightarrow q_j G_b) &= \left[\epsilon_\sigma^*(q_2) g f_{cba} C^{\mu\rho\sigma} \epsilon_\rho(q_1)\right] \times \\ &\times \left(\frac{-ig_{\mu\nu}}{\hat{t}}\right) \left[\bar{u}(p_2) ig (t^c)_{ji} \gamma^\nu u(p_1)\right] \\ &= \frac{g^2}{\hat{t}} \left(f_{cba} t^c_{ji}\right) \left[\epsilon_\sigma^*(q_2) C^{\mu\rho\sigma} \epsilon_\rho(q_1)\right] \times \\ &\times \left[\bar{u}(p_2) \gamma_\mu u(p_1)\right] \end{aligned} \qquad \text{(A2.4.10)}$$

where

$$\begin{aligned} C^{\mu\rho\sigma} &= g^{\mu\rho}(p_1 - p_2 + q_2)^\sigma + g^{\rho\sigma}(-q_2 - q_1)^\mu + \\ &+ g^{\sigma\mu}(q_1 - p_1 + p_2)^\rho. \end{aligned} \qquad \text{(A2.4.11)}$$

A2.5 Colour sums

We list here some identities, generalized to $SU(N)$ where appropriate, which are useful in performing the sums over initial and final colour states. The summation convention is assumed throughout this discussion.

The qqG vertex involves a factor of $\boldsymbol{t}^a$:

$$\boldsymbol{t}^a \equiv \tfrac{1}{2}\boldsymbol{\lambda}^a \tag{A2.5.1}$$

where the $SU(3)$ matrices $\boldsymbol{\lambda}^a$ are those introduced by Gell-Mann. The commutation relations for the $\boldsymbol{t}^a$ are given by the structure constants of the group,

$$[\boldsymbol{t}^a, \boldsymbol{t}^b] = \mathrm{i} f_{abc} \boldsymbol{t}^c \tag{A2.5.2}$$

$$[\boldsymbol{t}^a, \boldsymbol{t}^b] = \frac{1}{N}\delta_{ab} I_{(N)} + d_{abc}\boldsymbol{t}^c, \tag{A2.5.3}$$

where $I_{(N)}$ is the N-dimensional unit matrix. The f_{abc} are antisymmetric and the d_{abc} symmetric under the interchange of any two indices. In $SU(2)$, the quantities analogous to $(\boldsymbol{t}^a, f_{abc}, d_{abc})$ are $(\boldsymbol{\sigma}^a/2, \epsilon_{abc}, 0)$. Some useful identities involving the matrices $\boldsymbol{t}^a$ are

$$\left.\begin{aligned}
\boldsymbol{t}^a \boldsymbol{t}^b &= \frac{1}{2}\left[\frac{1}{N}\delta_{ab}\boldsymbol{I}_{(N)} + (d_{abc} + \mathrm{i} f_{abc})\boldsymbol{t}^c\right], \\
\boldsymbol{t}^a_{ij}\boldsymbol{t}^a_{kl} &= \frac{1}{2}\left[\delta_{il}\delta_{jk} - \frac{1}{N}\delta_{ij}\delta_{kl}\right], \\
\mathrm{Tr}\,\boldsymbol{t}^a &= 0, \\
\mathrm{Tr}(\boldsymbol{t}^a\boldsymbol{t}^b) &= \frac{1}{2}\delta_{ab}, \\
\mathrm{Tr}(\boldsymbol{t}^a\boldsymbol{t}^b\boldsymbol{t}^c) &= \frac{1}{4}(d_{abc} + \mathrm{i} f_{abc}), \\
\mathrm{Tr}(\boldsymbol{t}^a\boldsymbol{t}^b\boldsymbol{t}^a\boldsymbol{t}^c) &= -\frac{1}{4N}\delta_{bc}.
\end{aligned}\right\} \tag{A2.5.4}$$

It is sometimes profitable to define the (N^2-1)-dimensional matrices $\boldsymbol{F}_a$ and $\boldsymbol{D}_a$;

$$\left.\begin{aligned}
(\boldsymbol{F}_a)_{bc} &= -\mathrm{i} f_{abc}, \\
(\boldsymbol{D}_a)_{bc} &= d_{abc}.
\end{aligned}\right\} \tag{A2.5.5}$$

The Jacobi identities are

$$\left.\begin{aligned}
f_{abe}f_{ecd} + f_{cbe}f_{aed} + f_{dbe}f_{ace} &= 0, \\
f_{abe}d_{ecd} + f_{cbe}d_{aed} + f_{dbe}d_{ace} &= 0,
\end{aligned}\right\} \tag{A2.5.6}$$

or, equivalently,

$$\left.\begin{aligned}
[\boldsymbol{F}_a, \boldsymbol{F}_b] &= \mathrm{i} f_{abc}\boldsymbol{F}_c \\
[\boldsymbol{F}_a, \boldsymbol{D}_b] &= \mathrm{i} f_{abc}\boldsymbol{D}_c.
\end{aligned}\right\} \tag{A2.5.7}$$

A generalization of the $SU(2)$ relation

$$\epsilon_{ijm}\epsilon_{klm} = \delta_{ik}\delta_{jl} - \delta_{il}\delta_{jk} \tag{A2.5.8}$$

is

$$f_{abe}f_{cde} = \frac{2}{N}(\delta_{ac}\delta_{bd} - \delta_{ad}\delta_{bc}) + (d_{ace}d_{bde} - d_{bce}d_{ade}). \tag{A2.5.9}$$

Some further identities, written in both notations, are

$$\left.\begin{array}{lll} f_{abb} = 0, & \mathrm{Tr}\,\boldsymbol{F}_a = 0, & \\ d_{abb} = 0, & \mathrm{Tr}\,\boldsymbol{D}_a = 0, & \\ f_{acd}f_{bcd} = N\delta_{ab}, & \mathrm{Tr}(\boldsymbol{F}_a\boldsymbol{F}_b) = N\delta_{ab} & \\ & \boldsymbol{F}_a\boldsymbol{F}_a = N\boldsymbol{I}_{(2N-1)} & \\ f_{acd}d_{bcd} = 0, & \mathrm{Tr}(\boldsymbol{F}_a\boldsymbol{D}_b) = 0 & \\ & \boldsymbol{F}_a\boldsymbol{D}_a = 0 & \\ d_{acd}d_{bcd} = \dfrac{N^2-4}{N}\delta_{ab}, & \mathrm{Tr}(\boldsymbol{D}_a\boldsymbol{D}_b) = \dfrac{N^2-4}{N}\delta_{ab}, & \\ & \boldsymbol{D}_a\boldsymbol{D}_a = \dfrac{N^2-4}{N}\boldsymbol{I}_{(2N-1)}. & \end{array}\right\} \tag{A2.5.10}$$

Specializing to the matrix notation, one has

$$\left.\begin{array}{rcl} \mathrm{Tr}(\boldsymbol{F}_a\boldsymbol{F}_b\boldsymbol{F}_c) &=& \mathrm{i}\dfrac{N}{2}f_{abc}, \\ \mathrm{Tr}(\boldsymbol{D}_a\boldsymbol{F}_b\boldsymbol{F}_c) &=& \dfrac{N}{2}d_{abc}, \\ \mathrm{Tr}(\boldsymbol{D}_a\boldsymbol{D}_b\boldsymbol{F}_c) &=& \mathrm{i}\dfrac{N^2-4}{2N}f_{abc}, \\ \mathrm{Tr}(\boldsymbol{D}_a\boldsymbol{D}_b\boldsymbol{D}_c) &=& \dfrac{N^2-12}{2N}d_{abc}. \end{array}\right\} \tag{A2.5.11}$$

The above relations can be used to show

$$\mathrm{Tr}(\boldsymbol{F}_a\boldsymbol{F}_b\boldsymbol{F}_a\boldsymbol{F}_c) = \frac{N^2}{2}\delta_{bc}. \tag{A2.5.12}$$

We now illustrate the use of these relations by calculating some colour sums representative of those required in Section A2.4.

Consider the $|\mathcal{M}_{(t)}|^2$ term for $qq \to qq$ in (A2.4.4). $\overline{|\mathcal{M}_{(t)}|^2}$ will involve a colour factor

$$\frac{1}{3}\sum_i \frac{1}{3}\sum_k \sum_{j,l} (t^b_{lk}t^b_{ji})\,(t^a_{lk}t^a_{ji})^*$$

which, since the $\boldsymbol{t}^a$ are Hermitian,

$$\begin{aligned}
&= \tfrac{1}{9}\sum_{i,j,k,l}\left(t^b_{lk}t^a_{kl}\right)\left(t^b_{ji}t^a_{ij}\right)\\
&= \tfrac{1}{9}\,\mathrm{Tr}\,(\boldsymbol{t}^b\boldsymbol{t}^a)\mathrm{Tr}\,(\boldsymbol{t}^b\boldsymbol{t}^a)\\
&= \tfrac{1}{9}\left(\tfrac{1}{2}\delta_{ab}\right)\left(\tfrac{1}{2}\delta_{ab}\right) = \tfrac{2}{9} \qquad \text{(A2.5.13)}
\end{aligned}$$

where we have used (A2.5.4) and the fact that $a, b = 1, 2, \cdots, 8$.

As a final example consider $|\mathcal{M}_{(t)}|^2$ in $qG_a \to qG_b$ in (A2.4.10). $\overline{|\mathcal{M}_{(t)}|^2}$ will involve a colour factor

$$\begin{aligned}
&\frac{1}{8}\sum_a \frac{1}{3}\sum_i \sum_{b,j}\left(f_{cba}t^c_{ji}\right)\left(f_{dba}t^d_{ji}\right)^*\\
&= \frac{1}{24}\sum_{abij} f_{cba}f_{dba}t^c_{ji}t^d_{ij}\\
&= \frac{1}{24}\sum_{a,b} f_{cba}f_{dba}\,\mathrm{Tr}\,(\boldsymbol{t}^c\boldsymbol{t}^d)\\
&= \frac{1}{24}\left(N\delta_{cd}\right)\left(\frac{1}{2}\delta_{cd}\right) = \frac{1}{2} \qquad \text{(A2.5.14)}
\end{aligned}$$

where we have used (A2.5.10) with $N = 3$, and (A2.5.4).

A2.6 The Gell-Mann $SU(3)$ matrices

The Hermitian generators of $SU(3)$ are represented by the matrices $\boldsymbol{\lambda}^a/2$ ($a = 1, \cdots, 8$) where the $\boldsymbol{\lambda}^a$ are the Gell-Mann matrices which are the analogue for $SU(3)$ of the Pauli matrices for $SU(2)$. We list some of their properties.

They satisfy the commutation relations

$$\left[\frac{\boldsymbol{\lambda}^a}{2}, \frac{\boldsymbol{\lambda}^b}{2}\right] = \mathrm{i}\sum_c f_{abc}\frac{\boldsymbol{\lambda}^c}{2} \qquad \text{(A2.6.1)}$$

$a, b, c = 1, 2, \cdots, 8$.

The same commutation rules are obeyed by the $n^2 - 1$ generators of any $SU(n)$ group where, however, $a, b, c = 1, 2, \cdots, n^2 - 1$.

The 'structure constants' of the group f_{abc} are antisymmetric with respect to interchange of any two indices. The non-zero constants are given in Table A2.1.

The anticommutation rules for the $SU(3)$ matrices are also useful:

$$\{\boldsymbol{\lambda}^a, \boldsymbol{\lambda}^b\} \equiv \boldsymbol{\lambda}^a\boldsymbol{\lambda}^b + \boldsymbol{\lambda}^b\boldsymbol{\lambda}^a = 2\sum_c d_{abc}\boldsymbol{\lambda}^c + \tfrac{4}{3}\delta_{ab} \qquad \text{(A2.6.2)}$$

where the coefficients d_{abc}, symmetric under permutation of any two indices, are given in Table A2.2.

abc	f_{abc}	abc	f_{abc}
123	1	345	1/2
147	1/2	367	−1/2
156	−1/2	458	$\sqrt{3}/2$
246	1/2	678	$\sqrt{3}/2$
257	1/2		

Table A2.1. Non-zero f_{abc} for $SU(3)$.

abc	d_{abc}	abc	d_{abc}
118	$1/\sqrt{3}$	355	1/2
146	1/2	366	−1/2
157	1/2	377	−1/2
228	$1/\sqrt{3}$	448	$-1/2\sqrt{3}$
247	−1/2	558	$-1/2\sqrt{3}$
256	1/2	668	$-1/2\sqrt{3}$
338	$1/\sqrt{3}$	778	$-1/2\sqrt{3}$
344	1/2	888	$-1/\sqrt{3}$

Table A2.2. Non-zero d_{abc}.

From the fact that the $\boldsymbol{\lambda}$s are traceless matrices, using (A2.6.1 and 2) one gets

$$\begin{aligned} \mathrm{Tr}(\boldsymbol{\lambda}^a\boldsymbol{\lambda}^b) &= 2\delta_{ab} & \text{(A2.6.3)}\\ \mathrm{Tr}\Big(\boldsymbol{\lambda}^a[\boldsymbol{\lambda}^b,\boldsymbol{\lambda}^c]\Big) &= 4\mathrm{i}f_{abc} & \text{(A2.6.4)}\\ \mathrm{Tr}\Big(\boldsymbol{\lambda}^a\{\boldsymbol{\lambda}^b,\boldsymbol{\lambda}^c\}\Big) &= 4d_{abc} & \text{(A2.6.5)} \end{aligned}$$

A2.7 The Fierz reshuffle theorem

It sometimes happens, when dealing with the matrix element corresponding to a Feynman diagram involving spin $\frac{1}{2}$ particles, that it is convenient to rearrange the order of the spinors compared with the order they acquire directly from the Feynman diagram. An example of this occurred in Section 5.1 where it was helpful to go from the form (5.1.17) to (5.1.18).

In general, let $\Gamma^i (i = 1, \cdots, 16)$ stand for any one of the independent combinations of unit matrix and γ matrices: $\boldsymbol{I}, \gamma^\mu, \sigma^{\mu\nu} = \frac{1}{2}\mathrm{i}[\gamma^\mu, \gamma^\nu]$ with $\mu > \nu, \mathrm{i}\gamma^\mu\gamma_5, \gamma_5$.

Let Γ_i stand for the above set of matrices with their Lorentz indices lowered where relevant, i.e. Γ_i contains for example γ_μ whereas Γ^i contains γ^μ etc.

As a result of the algebraic properties of the set Γ^i it can be shown that

$$\tfrac{1}{4}\textstyle\sum_i(\Gamma_i)_{\alpha\beta}(\Gamma^i)_{\gamma\delta} = \delta_{\alpha\delta}\delta_{\beta\gamma}. \tag{A2.7.1}$$

If now **A** and **B** are any 4×4 matrices, then on multiplying (A2.7.1) by $A_{\rho\alpha}B_{\nu\gamma}$ we obtain

$$\begin{aligned}\tfrac{1}{4}\textstyle\sum_i A_{\rho\alpha}(\Gamma_i)_{\alpha\beta}B_{\nu\gamma}(\Gamma^i)_{\gamma\delta} &= A_{\rho\delta}B_{\nu\beta}\\ A_{\rho\delta}B_{\nu\beta} &= \frac{1}{4}\sum_i(\mathbf{A}\Gamma_i)_{\rho\beta}(\mathbf{B}\Gamma^i)_{\nu\delta}.\end{aligned} \tag{A2.7.2}$$

Since the 16 Γ^i are a complete set of 4×4 matrices, each product $\mathbf{A}\Gamma_i$ etc. will reduce to a sum of Γ_i.

After some labour one can obtain the following relation

$$\begin{aligned}&[\gamma^\mu(1-\gamma_5)]_{\rho\delta}[\gamma_\mu(1-\gamma_5)]_{\nu\beta}\\ &\qquad = -[\gamma^\mu(1-\gamma_5)]_{\rho\beta}[\gamma_\mu(1-\gamma_5)]_{\nu\delta},\end{aligned} \tag{A2.7.3}$$

which when sandwiched between spinors leads from (5.1.17) to (5.1.18).

Clearly, analogous relations can be worked out for any product of the Γ matrices. Results may be found in Section 2.2B of Marshak, Riazuddin and Ryan (1969).

A2.8 Dimension of matrix elements

A knowledge of the physical dimensions of a matrix element is often very useful in assessing its possible kinematical behaviour.

For cross-sections and differential cross-sections the dimension counting is trivial: (we are using $\hbar = c = 1$ natural units)

$$\begin{aligned}[\sigma] &= [\mathrm{AREA}] = [\mathrm{MASS}]^{-2}\\ \left[\frac{d\sigma}{d\Omega}\right] &= [\mathrm{MASS}]^{-2}\\ \left[\frac{d\sigma}{dt}\right] &= [\mathrm{MASS}]^{-4}.\end{aligned}$$

For a Feynman amplitude, as computed directly from a Feynman diagram *but without any spinors for the external lines*, one has the following dimensional factors:

Internal boson, photon, gluon line: $[\mathrm{MASS}]^{-2}$
Internal fermion line: $[\mathrm{MASS}]^{-1}$
Integration over four-momentum of *each* loop: $[\mathrm{MASS}]^{4}$

The dimensions of Green's functions were discussed in Section 20.6.

Appendix 3
Conserved vector currents and their charges

In Section 2.3.1 we showed how the invariance of a Lagrangian under an infinitesimal global gauge transformation

$$\delta\phi_j(x) = -\mathrm{i}\theta q_j \phi_j(x) \qquad \text{(not summed)} \tag{A3.1}$$

led to the existence of a vector current $J^\mu(x)$, the Noether current, which is conserved

$$\partial_\mu J^\mu = 0 \tag{A3.2}$$

and a 'charge operator'

$$\hat{Q} = \int \mathrm{d}^3\boldsymbol{x}\ J_0(\boldsymbol{x}, t) \tag{A3.3}$$

which is independent of time. We also mentioned that the q_j are the 'charges', i.e. the eigenvalues of $\hat{Q}$. Moreover it can be shown that $\hat{Q}$ is the generator of the transformations (A3.1), i.e.

$$\phi'_j = \mathrm{e}^{\mathrm{i}\theta\hat{Q}}\phi_j \mathrm{e}^{-\mathrm{i}\theta\hat{Q}} \tag{A3.4}$$

and

$$[\hat{Q}, \phi_j] = -q_j\phi_j \qquad \text{(not summed)}. \tag{A3.5}$$

If the q_j are the *electric* charges of the particles then J^μ is the electromagnetic current.

In the more general, non-Abelian case (2.3.22), where ϕ is a column vector composed of n fields, and where

$$\delta\phi = -\mathrm{i}\boldsymbol{L}\cdot\boldsymbol{\theta}\phi = -\mathrm{i}L^j\cdot\theta_j\phi \tag{A3.6}$$

with the L^j $(j = 1, \cdots, N)$ being $n \times n$ matrices representing the N group generators T_j and satisfying the group commutation relations (2.3.21), one finds a set of N conserved Noether currents

$$J^\mu_j = -\mathrm{i}\frac{\delta\mathcal{L}}{\delta(\partial_\mu\phi_r)}L^j_{rs}\phi_s \tag{A3.7}$$

and one has a set of N 'charge operators'

$$\hat{Q}_j = \int \mathrm{d}^3\boldsymbol{x}\; J_0^j(\boldsymbol{x}, t) \tag{A3.8}$$

which are independent of time.

Moreover one can show that the $\hat{Q}_j$ are the generators of the transformations (A3.6).

Thus

$$\hat{Q}_j = T_j \tag{A3.9}$$

and the eigenvalues of the $\hat{Q}_j$ are the eigenvalues of the group generators. Also one has, analogous to (A3.5),

$$[T_j, \phi_r] = -L^j_{rs}\phi_s \tag{A3.10}$$

As an example consider the group of isospin rotations, $SU(2)$. In this case the L^j $(j = 1, 2, 3)$ are the isospin matrices; for example, for an isospin-half field they are $\boldsymbol{\tau}/2$ where the $\boldsymbol{\tau}$ are the Pauli matrices, and the diagonal charge $\hat{Q}_3$ is T_3, the operator whose eigenvalues are the third components of isospin.

In electroweak theory the matrix elements of conserved vector currents taken between hadronic states are of great importance.

Consider now the matrix elements of such a set of $J_j^\mu(x)$ and their associated 'charge operators' $\hat{Q}_j$. We shall study matrix elements between sets of states of definite momentum $\boldsymbol{P}$ which transform under the action of the group generators as

$$T_j|\boldsymbol{P}, \rho\rangle = M^j_{\rho'\rho}|\boldsymbol{P}, \rho'\rangle \qquad (\rho = 1, \cdots, n) \tag{A3.11}$$

where the $\boldsymbol{M}^j$ are an n-dimensional representation of the T_j. For example for the group of isospin rotations and for a set of isodoublet states (A3.11) would read

$$\boldsymbol{T}|\boldsymbol{P}; I = \tfrac{1}{2}, I_3\rangle = \left(\tfrac{\boldsymbol{\tau}}{2}\right)_{I_3' I_3}|\boldsymbol{P}; I = \tfrac{1}{2}, I_3'\rangle. \tag{A3.12}$$

We shall now prove the remarkable result that

$$\langle\boldsymbol{P}; \rho'|J_j^\mu(0)|\boldsymbol{P}; \rho\rangle = 2M^j_{\rho'\rho}P^\mu \tag{A3.13}$$

i.e. the matrix elements are known *exactly* without any dynamical calculation.

On grounds of Lorentz invariance we can write

$$\langle\boldsymbol{P}; \rho'|J_j^\mu(0)|\boldsymbol{P}; \rho\rangle = 2a^j_{\rho'\rho}P^\mu \tag{A3.14}$$

where the $a^j_{\rho'\rho}$ are scalars (numbers). So we must show that

$$a^j_{\rho'\rho} = M^j_{\rho'\rho} \tag{A3.15}$$

Consider, using (A3.9) and (A3.11),

$$\begin{aligned}\langle \boldsymbol{P}';\rho'|\hat{Q}_j|\boldsymbol{P};\rho\rangle &= M^j_{\rho''\rho}\langle \boldsymbol{P}';\rho'|\boldsymbol{P};\rho''\rangle \\ &= M^j_{\rho'\rho}(2\pi)^3 2P_0\delta^3(\boldsymbol{P}'-\boldsymbol{P}). \qquad \text{(A3.16)}\end{aligned}$$

From the definition of $\hat{Q}_j$ the left-hand side equals

$$\begin{aligned}&\int \mathrm{d}^3\boldsymbol{x}\ \langle \boldsymbol{P}';\rho'|J^0_j(\boldsymbol{x},0)|\boldsymbol{P},\rho\rangle \\ &= \int \mathrm{d}^3\boldsymbol{x}\ \mathrm{e}^{\mathrm{i}\boldsymbol{x}\cdot(\boldsymbol{P}-\boldsymbol{P}')}\langle \boldsymbol{P}';\rho'|J^0_j(0)|\boldsymbol{P},\rho\rangle \qquad \text{(A3.17)}\end{aligned}$$

where we have used the translation property

$$\mathrm{e}^{\mathrm{i}\hat{P}\cdot x}\hat{O}(0)\mathrm{e}^{-\mathrm{i}\hat{P}\cdot x} = \hat{O}(x). \qquad \text{(A3.18)}$$

Thus, by (A3.14), the right-hand side of eqn (A3.17) becomes

$$\begin{aligned}&(2\pi)^3\delta^3(\boldsymbol{P}'-\boldsymbol{P})\ \langle \boldsymbol{P};\rho'|J^0_j(0)|\boldsymbol{P};\rho\rangle \\ &= (2\pi)^3\delta^3(\boldsymbol{P}'-\boldsymbol{P})\ 2a^j_{\rho'\rho}P^0. \qquad \text{(A3.19)}\end{aligned}$$

Comparing (A3.19) with (A3.16) yields the desired result (A3.15).

As an example, if J^μ_j is the electromagnetic current and q the charge of the hadron, then from (A3.13),

$$\langle \boldsymbol{P};q|J^\mu_{\mathrm{em}}(0)|\boldsymbol{P};q\rangle = 2qP^\mu. \qquad \text{(A3.20)}$$

For $\mu = 0$ this simply states that the expectation value of the charge density (in units of e) at the origin is $2P^0q$ for a plane wave. The factor $2P^0$ reflects the relativistic normalization of the states.

Similarly, if J^μ_j are the isotopic spin currents

$$\langle \boldsymbol{P};I,I_3'|J^\mu_j(0)|\boldsymbol{P};I,I_3\rangle = 2(I_j)_{I_3'I_3}P^\mu \qquad \text{(A3.21)}$$

where the $(2I+1)\times(2I+1)$ matrices I_j represent the isospin generators for isospin I.

If we take the isospin raising current $J^\mu_+ = J^\mu_1 + iJ^\mu_2$, the associated 'charge operator' is now the isospin raising operator T_+ (see Section 1.3) whose effect on a state of definite isospin is given in (1.3.15). From the above one deduces

$$\langle \boldsymbol{P};I,I_3+1|J^\mu_+(0)|\boldsymbol{P};I,I_3\rangle = \sqrt{(I-I_3)(I+I_3+1)}\ 2P^\mu \qquad \text{(A3.22)}$$

It is important to appreciate the miraculous power of (A3.13). The currents could be made of quark and gluon fields so that all the complications of the strong interactions intervene in relating them dynamically to the hadron states. Nonetheless if the group of transformations is a symmetry group of the physical hadrons we are able to get an exact result for the hadronic matrix elements of the currents! Such considerations were extensively utilized in the appendix to Chapter 16 (Section 16.9).

It is also important to separate the dynamical properties from the group theoretical properties. An example from $SU(3)$ will illustrate this very well. Let V_j^μ $(j = 1, \cdots, 8)$ be any set of currents or operators which transform as an octet under $SU(3)$, i.e.

$$[T_i, V_j^\mu] = i f_{ijk} V_k^\mu \tag{A3.23}$$

where the f_{ijk} are the $SU(3)$ structure constants defined in Appendix 2. From this alone follows a 'Wigner–Eckart' theorem which for the baryon octet $|\boldsymbol{P}; i\rangle$ $(i = 1, \cdots, 8)$ reads

$$\langle \boldsymbol{P}; i|V_j^\mu|\boldsymbol{P}; k\rangle = 2P^\mu \{i f_{ijk} F_V + d_{ijk} D_V\} \tag{A3.24}$$

where the group constants d_{ijk} are defined in Appendix 2.6 and F_V and D_V are *unknown* constants. Eqn (A3.24) is itself miraculous — 8^3 matrix elements are expressed in terms of two numbers, F_V and D_V. But the values of F_V and D_V depend upon the detailed dynamics. Only if the V_j^μ form a set of *conserved* currents can we deduce that $F_V = 1$, $D_V = 0$.

If we consider non-diagonal matrix elements of vector currents between a spin-half baryon octet the most general form allowed by Lorentz invariance and parity conservation is

$$\begin{aligned}\langle \boldsymbol{P}'; i|V_j^\mu|\boldsymbol{P}; k\rangle = \bar{u}(\boldsymbol{P}')\{G^1_{ijk}(q^2)\gamma^\mu &+ G^2_{ijk}(q^2)\sigma^{\mu\nu}q_\nu \\ &+ G^3_{ijk}(q^2)q^\mu\}u(\boldsymbol{P})\end{aligned} \tag{A3.25}$$

where the Gs are form factors depending on the momentum transfer.

Now because for $\boldsymbol{P}' = \boldsymbol{P}$,

$$\bar{u}(\boldsymbol{P})\gamma^\mu u(\boldsymbol{P}) = 2P^\mu \tag{A3.26}$$

when we let $q^2 \to 0$ we find, via (A3.24),

$$G^1_{ijk}(0) = i f_{ijk} F_V + d_{ijk} D_V. \tag{A3.27}$$

If the vector currents are *conserved* we find in addition, via (1.1.4), for all q^2

$$G^3_{ijk}(q^2) = 0 \tag{A3.28}$$

and by (A3.13)

$$F_V = 1, \qquad D_V = 0. \tag{A3.29}$$

For any octet of *axial-vector* currents A_j^μ with matrix elements taken between a spin-half baryon octet whose polarization is specified by the covariant spin vector S^μ, one has an expression analogous to (A3.24):

$$\langle \boldsymbol{P}, \boldsymbol{S}; i|A_j^\mu|\boldsymbol{P}, \boldsymbol{S}; k\rangle = 2m_B S^\mu\{i f_{ijk} F_A + d_{ijk} D_A\} \tag{A3.30}$$

with F_A, D_A unknown constants, and m_B the baryon mass. The Lorentz structure arises from the fact that

$$\bar{u}(\boldsymbol{P}, s)\gamma^\mu\gamma_5 u(\boldsymbol{P}, s) = 2m_B S^\mu. \tag{A3.31}$$

The above type of matrix element occurs in the analysis of hyperon β-decay, where, if $SU(3)_F$ invariance holds, each transition amplitude is expressed in the form

$$(KM \text{ matrix element}) \times (V^\mu + A^\mu) \tag{A3.32}$$

the vector and axial-vector currents being certain linear combinations of the octet members V_j^μ and A_j^μ. In this way *all* hyperon β-decays are described in terms of a KM matrix element and two constants F_A and D_A. Details may be found in Chapter 4 of Bailin (1982).

Appendix 4
Operator form of Feynman amplitudes and effective Hamiltonians

The Feynman amplitude, aside from proportionality constants, corresponds to a matrix element between initial and final states of the S-operator which is given in terms of the field operators by the Dyson expansion (A1.3.4). In the quark–parton model the Feynman amplitudes describe transitions between quark and gluon states whereas physically we are concerned with transitions between states of hadrons or leptons. In this context it is sometimes more useful to express a Feynman amplitude $\mathcal{M}_{fi}$ as a matrix element of a function of the field operators $\hat{\mathcal{M}}$, i.e. write

$$\mathcal{M}_{fi} = \langle f|\hat{\mathcal{M}}|i\rangle \tag{A4.1}$$

The form of this operator expression, for a given Feynman diagram, can of course be derived from the Dyson expansion, but this is tantamount to re-deriving the Feynman rules. It can be found more easily if we bear in mind that the final step in obtaining the Feynman rules involves the use of operations like (A1.1.23)

$$\psi(x)|\boldsymbol{p}, r\rangle = u_r(\boldsymbol{p})\mathrm{e}^{-\mathrm{i}p\cdot x}|0\rangle \tag{A4.2}$$

to replace the initial and final states by the vacuum state $|0\rangle$ which then disappears via $\langle 0|0\rangle = 1$. Hence we can find the relevant operator by starting with the Feynman amplitude and judiciously reversing this procedure. For example a factor $u_r(\boldsymbol{p})$ belonging to an incoming fermion line can be replaced by $\psi(0)|\boldsymbol{p}, r\rangle$ etc. via (A4.2).

In this fashion the following rules emerge for associating field operators with fermion lines in a given Feynman diagram:

$$
\begin{array}{lcl}
\mathbf{p}, r & u_r(\boldsymbol{p}) & \rightarrow \quad \psi(0) \\
\mathbf{p}, r & \bar{u}_r(\boldsymbol{p}) & \rightarrow \quad \overline{\psi}(0) \\
\mathbf{p}, r & \bar{v}_r(\boldsymbol{p}) & \rightarrow \quad \overline{\psi}(0) \\
\mathbf{p}, r & v_r(\boldsymbol{p}) & \rightarrow \quad \psi(0)
\end{array}
\qquad \text{(A4.3)}
$$

Similar rules follow for scalar and vector particles, based upon (A1.1.6 and 28).

It sometimes happens that the operator identified, $\hat{O}$, contributes more than once to the Feynman diagram under consideration. If it contributes N_0 times then the correct operator corresponding to the Feynman diagram is clearly

$$\hat{\mathcal{M}} = \hat{O}/N_0$$

An example of this type occurs in Section 19.3.

When colour is involved some care is necessary. For example in the SM even if the above diagrams refer to a *particular* colour i the operators involved must nonetheless have come from colour singlet terms of the form $\sum_{\text{colours } j} \overline{\psi}_j(x)\psi_j(x)$ in the electroweak Hamiltonian and $\hat{\mathcal{M}}$ should reflect this.

As an example, let us evaluate the operator expression $\hat{\mathcal{M}}$ associated with the Feynman amplitude $\mathcal{M}$ for the diagram of Fig. A4.1 contributing to $s\bar{d} \rightarrow c\bar{c}$ (colour labels i, j).

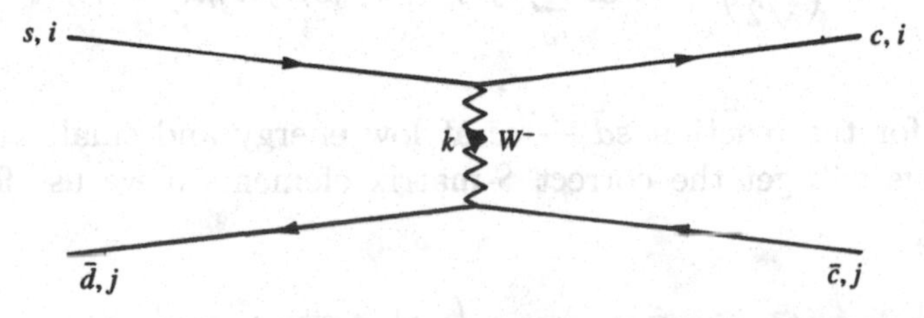

Fig. A4.1. A Feynman diagram contributing to $s\bar{d} \rightarrow c\bar{c}$.

We have from Section A2.3, independent of i, j

$$\begin{aligned}\mathcal{M} &= \left(\frac{-\mathrm{i}e}{2\sqrt{2}\sin\theta_W}\right)^2 V_{cs}V_{cd}\Big[\bar{v}(\bar{d})\gamma^\mu(1-\gamma_5)v(\bar{c})\Big]\,\frac{\mathrm{i}(-g_{\mu\nu}+k_\mu k_\nu/M_W^2)}{k^2-M_W^2+\mathrm{i}\epsilon}\\ &\qquad \Big[\bar{u}(c)\gamma^\nu(1-\gamma_5)u(s)\Big]\\ &= -\mathrm{i}\left(\frac{e}{2\sqrt{2}\sin\theta_W}\right)^2 V_{cs}V_{cd}\frac{(-g_{\mu\nu}+k_\mu k_\nu/M_W^2)}{k^2-M_W^2+\mathrm{i}\epsilon}\\ &\qquad \langle c\bar{c}|\overline{\psi}_d\gamma^\mu(1-\gamma_5)\psi_c\overline{\psi}_c\gamma^\nu(1-\gamma_5)\psi_s|s\bar{d}\rangle \end{aligned} \tag{A4.4}$$

so that the associated operator, disregarding colour, is

$$\begin{aligned}\hat{\mathcal{M}}(k) &= -\mathrm{i}\Big(\frac{e}{2\sqrt{2}\sin\theta_W}\Big)^2 V_{cs}V_{cd}\frac{(-g_{\mu\nu}+k_\mu k_\nu/M_W^2)}{k^2-M_W^2+\mathrm{i}\epsilon}\\ &\qquad \bar{d}\gamma^\mu(1-\gamma_5)c\bar{c}\gamma^\nu(1-\gamma_5)s\\ &\equiv f_{\mu\nu}(k^2)\bar{d}\gamma^\mu(1-\gamma_5)c\bar{c}\gamma^\nu(1-\gamma_5)s\end{aligned} \tag{A4.5}$$

where we have used the abbreviations for the fields $\psi(x)$ like $s=\psi_s(x=0)$ etc.

The topology of the diagram clearly indicates what the correct colour structure should be:

$$\hat{\mathcal{M}}(k) = f_{\mu\nu}(k^2)\sum_{i,j}\bar{d}_j\gamma^\mu(1-\gamma_5)c_j\bar{c}_i\gamma^\nu(1-\gamma_5)s_i \tag{A4.6}$$

The operator $\hat{\mathcal{M}}(k)$ which is a function of the momentum transfer k is not helpful since it depends on the momenta of the particles in the reactions. The above procedure becomes useful when one is considering reactions where the momenta are small on the scale of the propagator masses; M_W in the above example. For this case, ignoring the k^2 dependence in the propagator, we get, using (4.2.32),

$$\hat{\mathcal{M}} = -\mathrm{i}\left(\frac{G}{\sqrt{2}}\right)^2 V_{cs}V_{cd}\sum_{i,j}\bar{d}_j\ \gamma^\mu(1-\gamma_5)c_j\bar{c}_i\gamma_\mu(1-\gamma_5)s_i \tag{A4.7}$$

Hence for the reaction $s\bar{d}\to c\bar{c}$ at low energy and small momentum transfer we will get the correct S-matrix elements if we use for the S-operator

$$\hat{S} = 1-i\int \mathrm{d}^4x\mathcal{H}^{\text{eff}}(x) \tag{A4.8}$$

where the 'effective' *local* Hamiltonian density is

$$\begin{aligned}\mathcal{H}^{\text{eff}}(x) &= \left(\frac{G}{\sqrt{2}}\right)^2 V_{cs}V_{cd}\left[\sum_j \bar{d}_j(x)\gamma^\mu(1-\gamma_5)c_j(x)\right] \times \\ &\qquad \left[\sum_i \bar{c}_i(x)\gamma_\mu(1-\gamma_5)s_i(x)\right]\end{aligned} \tag{A4.9}$$

which, not surprisingly, is of the form of a local current–current interaction as in the pre-gauge theory of weak interactions. (See Section 1.2).

Similar considerations hold for other reactions at low energy and momentum transfer.

Appendix 5
S-matrix, T-matrix and Feynman amplitude

It is sometimes helpful to introduce a transition operator T defined by

$$\langle f|S|i\rangle = \langle f|\mathrm{i}\rangle - i2\pi\delta(E_f - E_i)T_{fi} \quad \text{(A5.1)}$$

The motivation for the factors and signs in (A5.1) is so that in lowest order perturbation theory the T_{fi} are just the matrix elements of the Hamiltonian. But the reader is warned that in treatises on general scattering theory the minus sign is sometimes not used.

If in (A1.3.4) we use the time-translation property

$$H_I(t) = \mathrm{e}^{\mathrm{i}H_0(t)}H_I(0)\mathrm{e}^{-\mathrm{i}H_0(t)} \quad \text{(A5.2)}$$

and utilize the step function $\theta(t)$ defined by

$$\begin{aligned} \theta(t) &= 1 \qquad \text{for } t \geq 0 \\ &= 0 \qquad \text{for } t < 0 \end{aligned} \quad \text{(A5.3)}$$

to effect the time ordering, the time integrations can be carried out using the representation

$$\theta(t) = \frac{1}{2\pi\mathrm{i}}\int_{-\infty}^{\infty}\mathrm{d}z\frac{\mathrm{e}^{\mathrm{i}zt}}{z - \mathrm{i}\epsilon} \quad \text{(A5.4)}$$

where $\epsilon > 0$ is infinitesimal. One finds eventually

$$T_{fi} = \langle f|\hat{T}(E = E_i)|i\rangle \quad \text{(A5.5)}$$

where the operator $\hat{T}(E)$ has the perturbative expansion

$$\begin{aligned} \hat{T}(E) &= H_I + H_I\frac{1}{E - H_0 + \mathrm{i}\epsilon}H_I + \\ &+ H_I\frac{1}{E - H_0 + \mathrm{i}\epsilon}H_I\frac{1}{E - H_0 + \mathrm{i}\epsilon}H_I + \cdots \\ &\equiv \hat{T}^{(1)} + \hat{T}^{(2)} + \hat{T}^{(3)} + \cdots \end{aligned} \quad \text{(A5.6)}$$

where H_I is short for $H_I(0)$. This equation is the basis for the so-called 'old-fashioned perturbation theory'.

Inserting a complete set of *physical* states $|n\rangle$ we get, for example,

$$T_{fi}^{(2)} = \sum_n \frac{\langle f|H_I|n\rangle \langle n|H_I|i\rangle}{E_i - E_n + \mathrm{i}\epsilon} \tag{A5.7}$$

where, mathematically, one has

$$\frac{1}{E_i - E_n + \mathrm{i}\epsilon} = \mathcal{P}\frac{1}{E_i - E_n} - \mathrm{i}\pi\delta(E_i - E_n) \tag{A5.8}$$

and $\mathcal{P}$ stands for the principal value.

If we deal with states normalized to unity in volume V then comparing (A5.1) and (A2.1.2) yields, on replacing $(2\pi)^3\delta^3(\boldsymbol{P}_f - \boldsymbol{P}_i)$ by $V\delta_{\boldsymbol{P}_f\boldsymbol{P}_i}$,

$$T_{fi} = \mathrm{i}V\delta_{\boldsymbol{P}_f\boldsymbol{P}_i} \prod_{\substack{\text{all}\\ \text{particles } j}} \left(\frac{1}{V2E_j}\right)^{\frac{1}{2}} \mathcal{M} \tag{A5.9}$$

In comparing 'old fashioned' with covariant perturbation theory, note the Feynman amplitude conserves both energy and momentum whereas the operator $\hat{T}(E)$ given by (A5.6) conserves 3-momentum but has non-zero matrix elements between states of different energy.

In fact, by translational invariance, the Hamiltonian H_I conserves 3-momentum so the intermediate states $|n\rangle$ in (A5.7) have

$$\boldsymbol{P}_n \equiv \boldsymbol{P}_i$$

but can have $E_n \neq E_i$. By contrast the intermediate 'states' in a Feynman diagram have the same 3-momentum and energy as the initial state, but they need not contain particles which are on the mass-shell, i.e. E^2 need not equal $\boldsymbol{p}^2 + m^2$ for each particle.

The separation (A5.8) into principal value and δ-function parts corresponds to splitting T_{fi} into its *dispersive* and *absorptive* parts respectively,

$$T_{fi} \equiv D_{fi} - \frac{\mathrm{i}}{2}A_{fi} \tag{A5.10}$$

and is of important physical significance.

In many physical situations the absorptive part of T_{fi} is just equal to $-2\mathrm{Im}\ T_{fi}$, with the above definition of T_{fi}.

Appendix 6
Consequences of CPT invariance for matrix elements

The operators $\mathcal{P}$ and $\mathcal{C}$ which generate parity inversion and charge conjugation (Section A1.2) are linear and unitary. Bearing in mind the forms of the relationship between the $\hat{S}$ operator or the transition operator $\hat{T}$ in terms of the Hamiltonian [see (A1.3.4) and (A5.1)] we have the following consequences of invariance:

$$\begin{array}{lcl} \text{P invariance} \Longrightarrow & \mathcal{P}^{-1}H\mathcal{P} = H & \Longrightarrow \mathcal{P}^{-1}\hat{T}\mathcal{P} = \hat{T}, \\ \text{C invariance} \Longrightarrow & \mathcal{C}^{-1}H\mathcal{C} = H & \Longrightarrow \mathcal{C}^{-1}\hat{T}\mathcal{C} = \hat{T}. \end{array} \qquad \text{(A6.1)}$$

However the generator $\Im$ of time reversal is a unitary but anti-linear operator, i.e. $\Im^\dagger\Im = 1$ but $\Im(\alpha|\phi\rangle + \beta|\psi\rangle) = \alpha^*\Im|\phi\rangle + \beta^*\Im|\psi\rangle$. Such operators are rather confusingly sometimes called 'anti-unitary'. Hence because $\Im i = -i\Im$ we will have

$$\text{T invariance} \Longrightarrow \Im^{-1}H\Im = H \Longrightarrow \Im^{-1}\hat{T}\Im = \hat{T}^\dagger. \qquad \text{(A6.2)}$$

The combined symmetry operation

$$\theta = \Im\mathcal{C}\mathcal{P} \qquad \text{(A6.3)}$$

is much more fundamental than each individual symmetry. It is trivial to write down Hamiltonians that violate $\mathcal{P}$ or $\mathcal{C}$ or $\mathcal{CP}$ or $\Im$. It is *impossible* to construct a local causal quantum field theory that violates $\Im\mathcal{CP}$! If nature violates $\Im\mathcal{CP}$ we shall face a truly major crisis in the theory of elementary particles.

Thus we always assume that

$$\theta^{-1}H\theta = H \quad \text{and} \quad \theta^{-1}\hat{T}\theta = \hat{T}^\dagger. \qquad \text{(A6.4)}$$

One consequence of $\Im\mathcal{CP}$ invariance is that the masses of a particle and its antiparticle should be equal. It is therefore fascinating to note the present status of measurements to compare the proton and antiproton masses (Gabrielse *et al.*, 1990):

$$\frac{m_{\bar{\mathrm{p}}}}{m_{\mathrm{p}}} = 0.999\,999\,977 \pm 0.000\,000\,042. \tag{A6.5}$$

Consider now the consequence of symmetry conservation upon transition matrix elements. We have

$$\begin{aligned} \text{P invariance}: \quad \langle f|\hat{T}|i\rangle &= \langle f|\mathcal{P}^{-1}\hat{T}\mathcal{P}|i\rangle = \langle f|\mathcal{P}^{\dagger}\hat{T}\mathcal{P}|i\rangle \\ &= \langle f^P|\hat{T}|i^P\rangle \end{aligned} \tag{A6.6}$$

where $|i^P\rangle$ means the state obtained from $|i\rangle$ by the action of $\mathcal{P}$.

Similarly

$$\text{C invariance}: \quad \langle f|\hat{T}|i\rangle = \langle f|\mathcal{C}^{\dagger}\hat{T}\mathcal{C}|i\rangle = \langle f^C|\hat{T}|i^C\rangle \tag{A6.7}$$

where $|i^C\rangle$ is the state $|i\rangle$ with all particles replaced by antiparticles, i.e. $|\bar{i}\rangle$.

Because of its antilinearity

$$\begin{aligned} \text{T invariance}: \quad \langle f|\hat{T}|i\rangle &= \langle f|\Im^{\dagger}\hat{T}^{\dagger}\Im|i\rangle = \langle f^T|\hat{T}^{\dagger}|i^T\rangle^* \\ &= \langle i^T|\hat{T}|f^T\rangle \end{aligned} \tag{A6.8}$$

where $|i^T\rangle$ is the time-reversed version of the state $|i\rangle$.

It follows that

$$\text{CPT invariance}: \quad \langle f|\hat{T}|i\rangle = \langle i^{\mathrm{CPT}}|\hat{T}|f^{\mathrm{CPT}}\rangle \tag{A6.9}$$

Let us now examine the consequences of these results for the matrix elements which appear in the discussion of CP violation in Section 19.1, in particular for the structure of H in (19.1.2) in the weak interactions.

Use of CPT invariance in (19.1.3) and the fact that for a pseudoscalar particle at rest $|P^{0\ \mathrm{CPT}}\rangle = -|\bar{P}^0\rangle$ immediately yields $H_{11} = H_{22}$ which is then written as $M - \mathrm{i}\,\Gamma/2$ with M, Γ real. Use of CPT in H_{12} yields no information. It is useful to write

$$\hat{T}_w = \tfrac{1}{2}\left\{(\hat{T}_w + \hat{T}_w^{\dagger}) - (\hat{T}_w^{\dagger} - \hat{T}_w)\right\} \tag{A6.10}$$

where the subscript w reminds us that we are dealing with the weak interaction Hamiltonian, and to define

$$M_{12} \equiv \frac{1}{2}\langle P^0|\hat{T}_w + \hat{T}_w^{\dagger}|\bar{P}^0\rangle \tag{A6.11}$$

$$\Gamma_{12} \equiv \frac{1}{\mathrm{i}}\langle P^0|\hat{T}_w^{\dagger} - \hat{T}_w|\bar{P}^0\rangle \tag{A6.12}$$

so that

$$H_{12} = M_{12} - \mathrm{i}\Gamma_{12}/2 \tag{A6.13}$$

where, in general, M_{12} and Γ_{12} can be complex.

Then by CPT invariance

$$\begin{aligned} H_{21} &= \langle \bar{P}^0 | \hat{T}_w | P^0 \rangle = \langle \bar{P}^0 | \theta^{-1} \hat{T}_w^\dagger \theta | P^0 \rangle = \\ &= \frac{1}{2} \langle \bar{P}^0 | \theta^{-1} (\hat{T}_w^\dagger + \hat{T}_w) \theta | P^0 \rangle - \frac{1}{2} \langle \bar{P}^0 | \theta^{-1} (\hat{T}_w - \hat{T}_w^\dagger) \theta | P^0 \rangle \\ &= \frac{1}{2} \langle P^0 | \hat{T}_w + \hat{T}_w^\dagger | \bar{P}^0 \rangle^* - \frac{1}{2} \langle P^0 | \hat{T}_w - \hat{T}_w^\dagger | \bar{P}^0 \rangle^* \\ &= M_{12}^* - \mathrm{i}\Gamma_{12}^*/2 \end{aligned} \tag{A6.14}$$

This explains the form

$$\boldsymbol{H} = \begin{pmatrix} M - \mathrm{i}\Gamma/2 & M_{12} - \mathrm{i}\Gamma_{12}/2 \\ M_{12}^* - \mathrm{i}\Gamma_{12}^*/2 & M - \mathrm{i}\Gamma/2 \end{pmatrix} \tag{A6.15}$$

used in (19.1.5).

Appendix 7
Formulae for the basic partonic 2 ⟶ 2 processes

Because of their fundamental importance we list here the formulae for the CM unpolarized differential cross-sections for all 2 → 2 partonic and Compton-like reactions. The results are taken from Krawczyk and Żarnecki (1991).

Interesting plots of the angular distributions are given in Figs. 24.17 and 24.18.

The cross-sections are summed and averaged over colour and spins, and in this section i and j label the quark *flavour*. The partonic Mandelstam variables $\hat{s}, \hat{t}$ and $\hat{u}$ are defined in (A1.4.24). We subdivide into processes with 0,1 or 2 photons.

A7.1 Reactions with only quarks and gluons

The CM cross-section is

$$\frac{\mathrm{d}\sigma}{\mathrm{d}\hat{t}} = \frac{\pi\alpha_s^2}{\hat{s}^2}\overline{|\mathcal{M}|^2}. \tag{A7.1.1}$$

The Feynman diagrams and the results for $|\mathcal{M}|^2$ are as follows:

1 $q_iq_j \to q_iq_j \quad (i \neq j)$; $\quad \bar{q}_i\bar{q}_j \to \bar{q}_i\bar{q}_j \quad (i \neq j)$; $\quad q_i\bar{q}_j \to q_i\bar{q}_j$

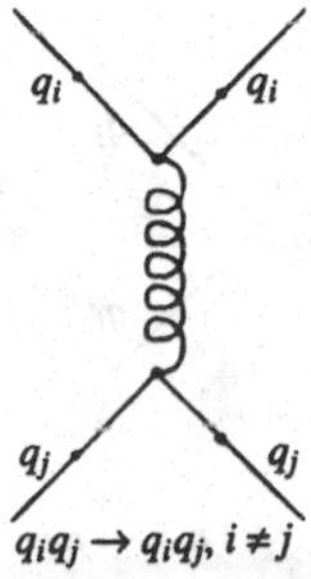

$q_iq_j \to q_iq_j,\ i \neq j$

$$\overline{|\mathcal{M}|^2} = \frac{4}{9}\,\frac{\hat{s}^2 + \hat{u}^2}{\hat{t}^2} \tag{A7.1.2}$$

2 $q_i q_i \rightarrow q_i q_i$ *and* $\bar{q}_i \bar{q}_i \rightarrow \bar{q}_i \bar{q}_i$

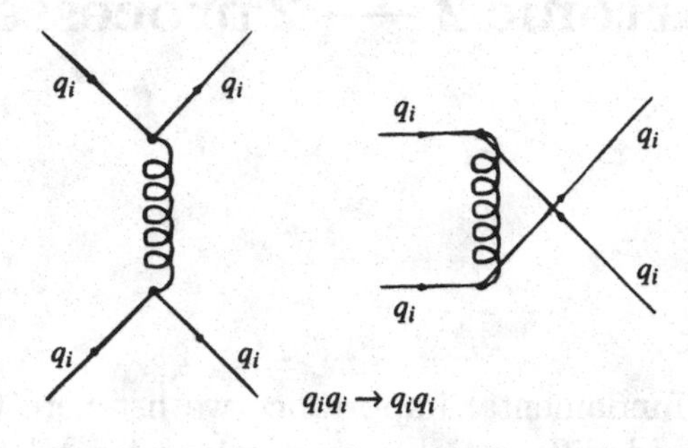

$q_i q_i \rightarrow q_i q_i$

$$\overline{|\mathcal{M}|^2} = \frac{4}{9}\left(\frac{\hat{s}^2 + \hat{u}^2}{\hat{t}^2} + \frac{\hat{s}^2 + \hat{t}^2}{\hat{u}^2}\right) - \frac{8}{27}\,\frac{\hat{s}^2}{\hat{t}\hat{u}} \tag{A7.1.3}$$

3 $q_i \bar{q}_i \rightarrow q_j \bar{q}_j \quad (i \neq j)$

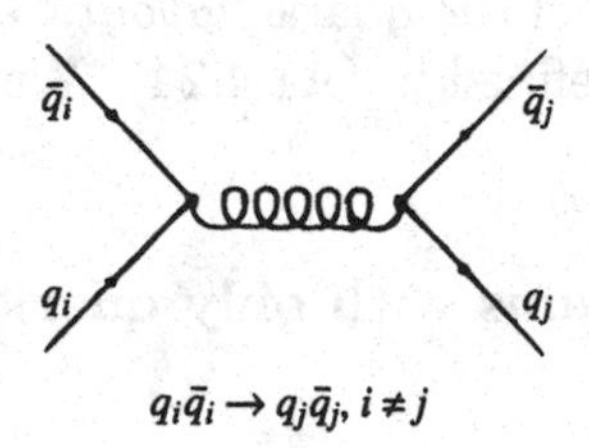

$q_i \bar{q}_i \rightarrow q_j \bar{q}_j,\ i \neq j$

$$|\mathcal{M}|^2 = \frac{4}{9}\,\frac{\hat{t}^2 + \hat{u}^2}{\hat{s}^2} \tag{A7.1.4}$$

4 $q_i \bar{q}_i \rightarrow q_i \bar{q}_i$

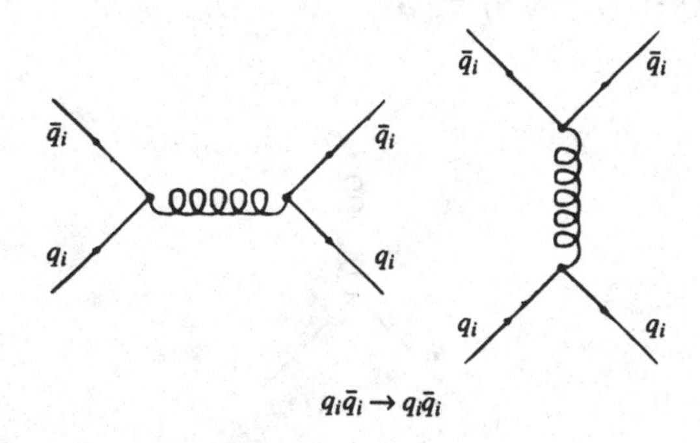

$q_i \bar{q}_i \rightarrow q_i \bar{q}_i$

$$\overline{|\mathcal{M}|^2} = \frac{4}{9}\left(\frac{\hat{s}^2+\hat{u}^2}{\hat{t}^2} + \frac{\hat{u}^2+\hat{t}^2}{\hat{s}^2}\right) - \frac{8}{27}\frac{\hat{u}^2}{\hat{s}\hat{t}} \qquad \text{(A7.1.5)}$$

5 $q_i\bar{q}_i \to GG$

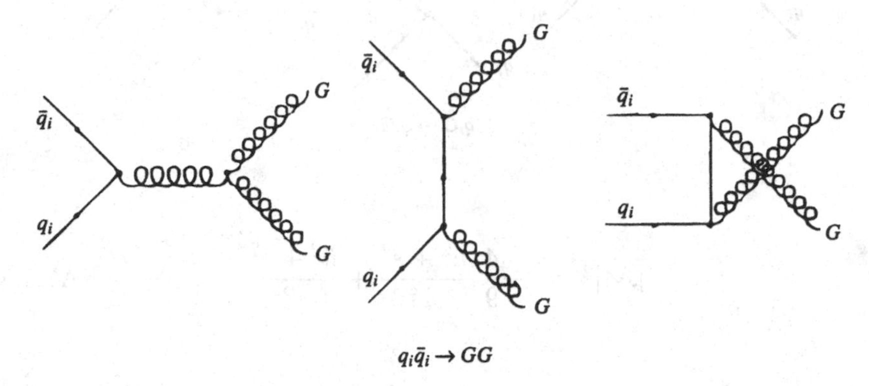

$q_i\bar{q}_i \to GG$

$$|\mathcal{M}|^2 = \frac{32}{27}\frac{\hat{u}^2+\hat{t}^2}{\hat{u}\hat{t}} - \frac{8}{3}\frac{\hat{u}^2+\hat{t}^2}{\hat{s}^2} \qquad \text{(A7.1.6)}$$

6 $GG \to q_i\bar{q}_i$

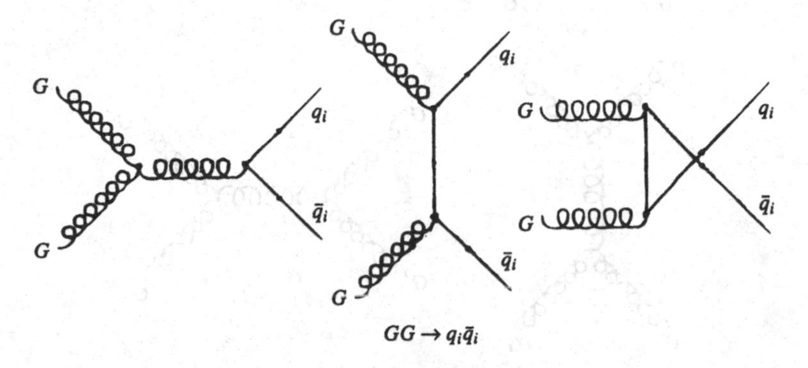

$GG \to q_i\bar{q}_i$

$$\overline{|\mathcal{M}|^2} = \frac{9}{64}\left[\frac{32}{27}\frac{\hat{u}^2+\hat{t}^2}{\hat{u}\hat{t}} - \frac{8}{3}\frac{\hat{u}^2+\hat{t}^2}{\hat{s}^2}\right] \qquad \text{(A7.1.7)}$$

Note that this equals 9/64 times (A7.1.6).

7 $q_iG \to q_iG$ *and* $\bar{q}_iG \to \bar{q}_iG$

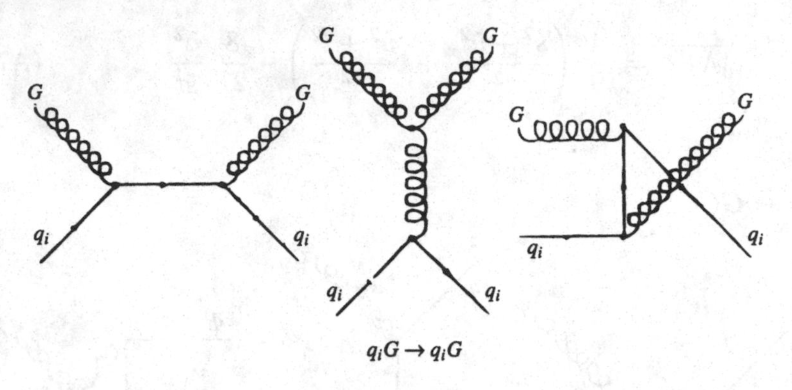

$$\overline{|\mathcal{M}|^2} = -\frac{4}{9}\,\frac{\hat{u}^2+\hat{s}^2}{\hat{u}\hat{s}} + \frac{\hat{u}^2+\hat{s}^2}{\hat{t}^2} \qquad \text{(A7.1.8)}$$

8 $GG \to GG$

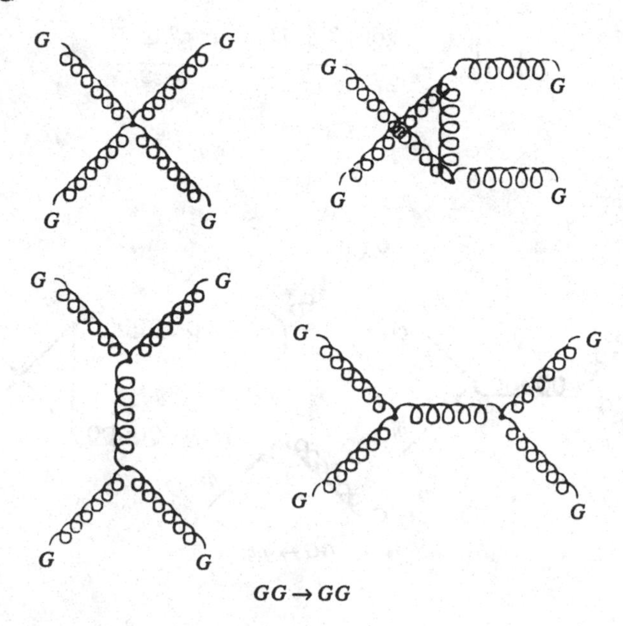

$$\overline{|\mathcal{M}|^2} = \frac{9}{2}\left(3 - \frac{\hat{u}\hat{t}}{\hat{s}^2} - \frac{\hat{u}\hat{s}}{\hat{t}^2} - \frac{\hat{s}\hat{t}}{\hat{u}^2}\right) \qquad \text{(A7.1.9)}$$

A7.1.1 Comparison of parton cross-section at 90°

It is useful to compare the sizes of various partonic cross-section at 90° in the parton–parton CM where $\hat{t} = \hat{u} = -\hat{s}/2$.

One finds:

$$\frac{d\sigma}{d\hat{t}}(q_i q_j \to q_i q_j)\Big|_{90^\circ} = \frac{\pi\alpha_s^2}{\hat{s}^2}(2.22 + 1.04\ \delta_{ij}) \qquad \text{(A7.1.10)}$$

$$\frac{d\sigma}{d\hat{t}}(q_i \bar{q}_j \to q_k \bar{q}_l)\Big|_{90^\circ} = \frac{\pi\alpha_s^2}{\hat{s}^2}\Big[2.22\ \delta_{ik}\delta_{jl} + + (0.22 + 0.15\ \delta_{ik})\delta_{ij}\delta_{kl} \qquad \text{(A7.1.11)}$$

$$\frac{d\sigma}{d\hat{t}}(qG \to qG)\Big|_{90^\circ} = \frac{d\sigma}{d\hat{t}}(\bar{q}G \to \bar{q}G)\Big|_{90^\circ} = \frac{\pi\alpha_s^2}{\hat{s}^2} \times 6.11 \qquad \text{(A7.1.12)}$$

In (A7.1.12) the dominant part of the cross-section comes from the t-channel gluon exchange diagram. Naive generalization of QED, where only s-channel and u-channel quark exchanges exist, can thus be misleading.

$$\frac{d\sigma}{d\hat{t}}(GG \to GG)\Big|_{90^\circ} = \frac{\pi\alpha_s^2}{\hat{s}^2} \times 30.4 \qquad \text{(A7.1.13)}$$

$$\frac{d\sigma}{d\hat{t}}(GG \to q\bar{q})\Big|_{90^\circ} = \frac{\pi\alpha_s^2}{\hat{s}^2} \times 0.15 \qquad \text{(A7.1.14)}$$

$$\frac{d\sigma}{d\hat{t}}(q\bar{q} \to GG)\Big|_{90^\circ} = \frac{64}{9}\frac{d\sigma}{d\hat{t}}(GG \to q\bar{q}) = \frac{\pi\alpha_s^2}{\hat{s}^2} \times 1.04 \qquad \text{(A7.1.15)}$$

It is seen that $GG \to GG$ followed by $qG \to qG$ (and $\bar{q}G \to \bar{q}G$) have the biggest cross-section at large momentum transfer.

A7.2 Reactions with one photon

The cross-section is

$$\frac{d\sigma}{d\hat{t}} = \frac{\pi\alpha_s\alpha Q_i^2}{\hat{s}^2}\overline{|\mathcal{M}|^2} \qquad \text{(A7.2.1)}$$

The Feynman diagrams and results for $\overline{|\mathcal{M}|^2}$ follow:

1 $q_i\bar{q}_i \to G\gamma$

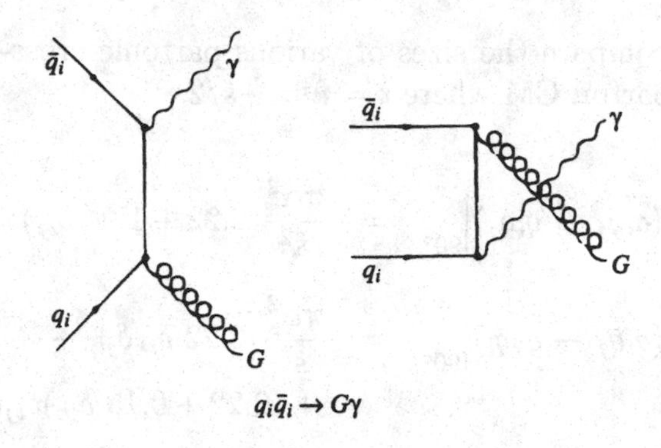

$q_i\bar{q}_i \to G\gamma$

$$\overline{|\mathcal{M}|^2} = \frac{8}{9}\,\frac{\hat{u}^2 + \hat{t}^2}{\hat{u}\hat{t}} \qquad \text{(A7.2.2)}$$

2 $G\gamma \to q_i\bar{q}_i$

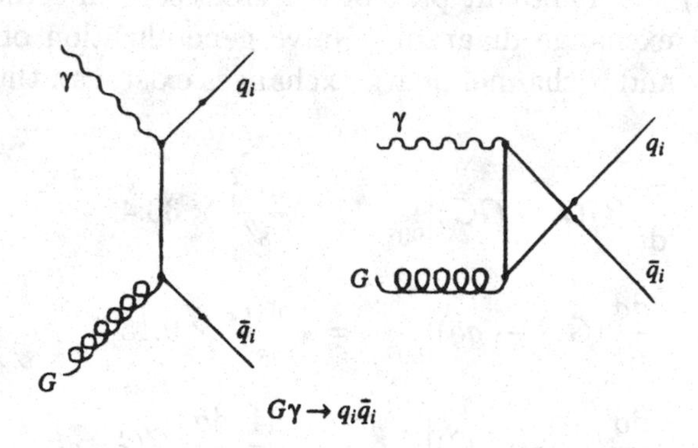

$G\gamma \to q_i\bar{q}_i$

$$\overline{|\mathcal{M}|^2} = \frac{\hat{u}^2 + \hat{t}^2}{\hat{u}\hat{t}} \qquad \text{(A7.2.3)}$$

3 $q_iG \to q_i\gamma$ *and* $\bar{q}_iG \to \bar{q}_i\gamma$

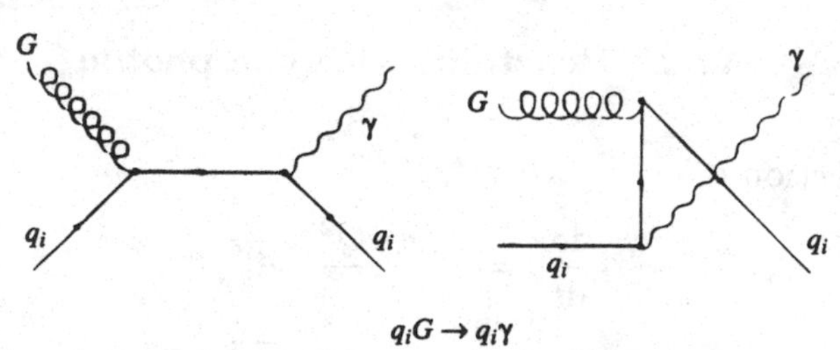

$q_iG \to q_i\gamma$

$$\overline{|\mathcal{M}|^2} = -\frac{1}{3}\frac{\hat{u}^2+\hat{s}^2}{\hat{u}\hat{s}} \qquad \text{(A7.2.4)}$$

(the reader must not forget that $\hat{u}$ is negative in this case).

4 $q_i\gamma \to q_iG$ *and* $\bar{q}_i\gamma \to \bar{q}_iG$

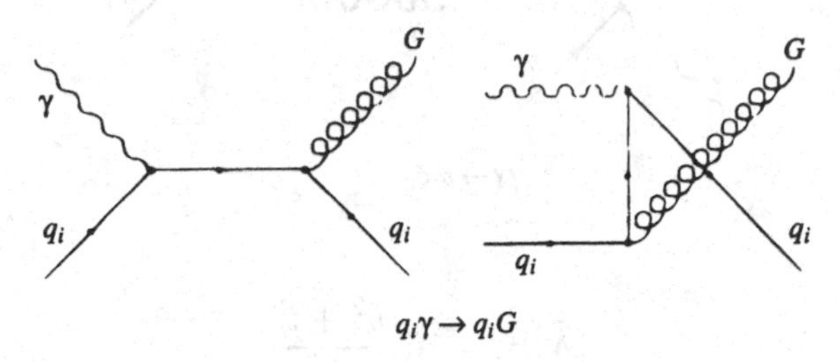

$q_i\gamma \to q_iG$

$$\overline{|\mathcal{M}|^2} = -\frac{8}{3}\frac{\hat{u}^2+\hat{s}^2}{\hat{u}\hat{s}} \qquad \text{(A7.2.5)}$$

A7.3 Reactions with two photons

The cross-section is

$$\frac{d\sigma}{d\hat{t}} = \frac{\pi\alpha^2 Q_i^4}{\hat{s}^2}\overline{|\mathcal{M}|^2} \qquad \text{(A7.3.1)}$$

Feynman diagrams and results for $\overline{|\mathcal{M}|^2}$ follow:

1 $q_i\bar{q}_i \to \gamma\gamma$

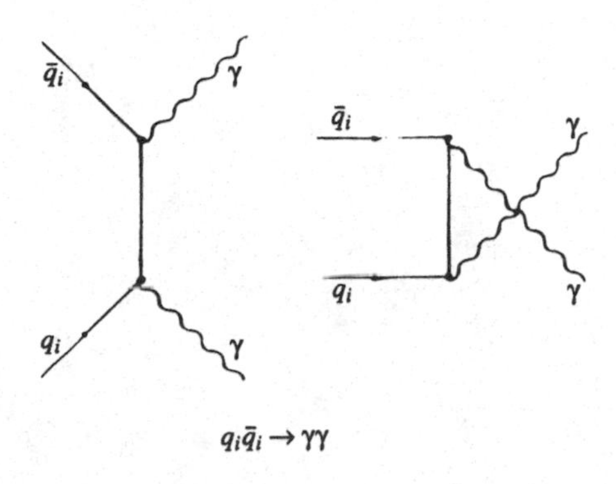

$q_i\bar{q}_i \to \gamma\gamma$

$$\overline{|\mathcal{M}|^2} = \frac{2}{3}\frac{\hat{u}^2+\hat{t}^2}{\hat{u}\hat{t}} \qquad \text{(A7.3.2)}$$

2 $\gamma\gamma \rightarrow q_i\bar{q}_i$

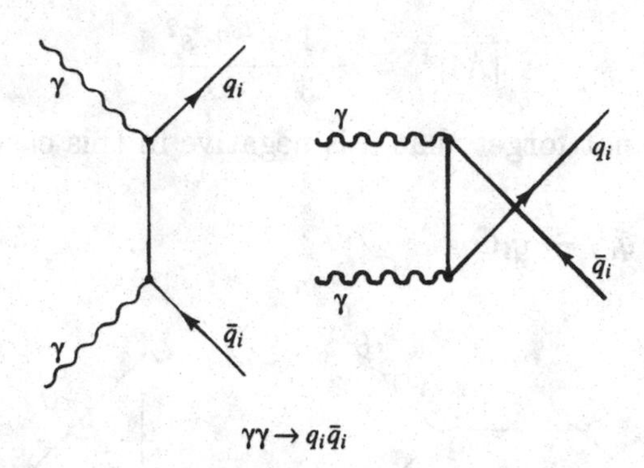

$\gamma\gamma \rightarrow q_i\bar{q}_i$

$$\overline{|\mathcal{M}|^2} \;=\; 6\,\frac{\hat{u}^2+\hat{t}^2}{\hat{u}\hat{t}} \tag{A7.3.3}$$

3 $q_i\gamma \rightarrow q_i\gamma$ and $\bar{q}_i\gamma \rightarrow \bar{q}_i\gamma$

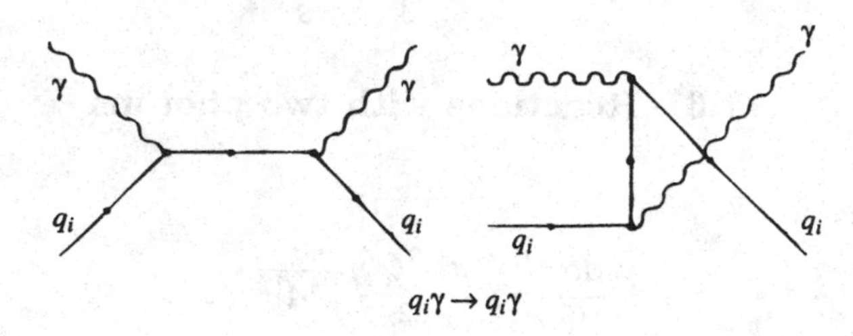

$q_i\gamma \rightarrow q_i\gamma$

$$\overline{|\mathcal{M}|^2} \;=\; -2\,\frac{\hat{u}^2+\hat{s}^2}{\hat{u}\hat{s}} \tag{A7.3.4}$$

Appendix 8
Euclidean space conventions

In Sections 27.3–27.6 we frequently made use of the Euclidean version of various field theory expressions. Here we specify the conventions in some detail.

One defines x^E_α, $(\alpha = 1, 2, 3, 4)$ such that

$$x^E_4 \equiv \tau = \mathrm{i}t, \quad x^E_1 = x, \quad x^E_2 = y, \quad x^E_3 = z \tag{A8.1}$$

and A^E_α such that

$$A^E_4 = \mathrm{i}A^0 \quad , \quad A^E_j = A^j. \tag{A8.2}$$

Also, for the field tensor, one takes

$$F^E_{\alpha\beta} = \frac{\partial}{\partial x^E_\alpha} A^E_\beta - \frac{\partial}{\partial x^E_\beta} A^E_\alpha \tag{A8.3}$$

so that

$$F^E_{ij} = -F^{ij}, \qquad F^E_{4j} = -\mathrm{i}F^{0j} \tag{A8.4}$$

It then follows that

$$F^E_{\alpha\beta}\, F^E_{\alpha\beta} = F_{\mu\nu}F^{\mu\nu}. \tag{A8.5}$$

The dual field tensor is defined by

$$\tilde{F}^E_{\alpha\beta} = \tfrac{1}{2}\epsilon_{\alpha\beta\gamma\delta}F^E_{\gamma\delta} \qquad (\epsilon_{1234} = 1) \tag{A8.6}$$

If follows that

$$\tilde{F}^E_{ij} = -\mathrm{i}\tilde{F}^{ij}, \qquad \tilde{F}^E_{4j} = \tilde{F}^{0j} \tag{A8.7}$$

and hence

$$F^E_{\alpha\beta}\tilde{F}^E_{\alpha\beta} = \mathrm{i}F_{\mu\nu}\tilde{F}^{\mu\nu} \tag{A8.8}$$

The relevant term in the Feynman path integral is then

$$\begin{aligned} iS &= i\int d^4x\mathcal{L} = -\frac{i}{4}\int dtd^3\mathbf{r}\ F_{\mu\nu}F^{\mu\nu} \\ &= -\frac{i}{4}\int(-id\tau)\ d^3\mathbf{r}\ F^E_{\alpha\beta}F^E_{\alpha\beta} \qquad \text{by (A8.5)} \\ &= -S^E \end{aligned} \tag{A8.9}$$

where the Euclidean action is

$$S^E = \frac{1}{4}\int d\tau\ d^3\mathbf{r}\ F^E_{\alpha\beta}F^E_{\alpha\beta}. \tag{A8.10}$$

Also useful is the result

$$\partial^E_\alpha A^E_\alpha = \partial_\mu A^\mu. \tag{A8.11}$$

Finally, for K^E_α one defines

$$K^E_\alpha = \epsilon_{\alpha\beta\gamma\delta}\ \mathrm{Tr}\ \{\underline{A}^E_\beta\partial^E_\gamma\underline{A}^E_\delta - \frac{2}{3}ig\underline{A}^E_\beta\underline{A}^E_\gamma\underline{A}^E_\delta\} \tag{A8.12}$$

and then finds that

$$K^E_4 = -K^0, \qquad K^E_j = iK^j \tag{A8.13}$$

$$\partial^E_\alpha K^E_\alpha = i\partial_\mu K^\mu. \tag{A8.14}$$

References

Ademollo, M. and Gatto, R. (1964). *Phys. Lett.* **13**, 264.

AFS (1982a): Åkesson, T. *et al.*, *Phys. Lett.* **B118**, 185 and 193.

AFS (1982b): Åkesson, T. *et al.*, *Phys. Lett.* **B118**, 178.

AFS (1983a): Åkesson, T. *et al.*, *Phys. Lett.* **B123**, 133.

AFS (1983b): Åkesson, T. *et al.*, *Phys. Lett.* **B123**, 367.

Albrecht, H. *et al.* (1990). *Phys. Lett.* **B234**, 409.

ALEPH (1991): Decamp, D. *et al.*, *Phys. Lett.* **B258**, 236.

Altarelli, G. (1979). In *Proceedings of the EPS Conference on High Energy Physics*, Geneva, 1979, p. 726. Geneva: CERN Scientific Information Service.

Altarelli, G. (1982). *Phys. Rep.* **81**, 2.

Altarelli, G. (1984). *Riv. N. Cim.* **7**, no. 3.

Altarelli, G. (1989). *Ann. Rev. Nucl. Part. Sci.* **39**, 357.

Altarelli, G. (1990). In *Proceedings of Present and Future of Collider Physics*, Rome, 1990. CERN-Th 5955/90.

Altarelli, G. and Martinelli, G. (1978). *Phys. Lett.* **76B**, 89.

Altarelli, G. and Parisi, G. (1977). *Nucl. Phys.* **B126**, 298.

Altarelli, G., Ellis, R. K. and Martinelli, G. (1979). *Nucl. Phys.* **B157**, 461.

Altarelli, G., Kleiss, R. and Verzegnassi, C., editors (1989). *Z Physics at LEP1*. CERN 89-08.

Altarelli, G. *et al.* (1982). *Nucl. Phys.* **B208**, 365.

Alvarez-Gaumé, L. (1992). In *Proceedings of the 26th International Conference on High Energy Physics*, Dallas, 1992.

Amaldi, U., de Boer, W. and Fürsterman, H. (1992). *Phys. Lett.* **B260**, 447.

Angelis, A. L. S. *et al.* (1979). *Physica Scripta* **19**, 116.

Anselmino, M. and Predazzi, E. (1992). *Phys. Lett.* **B254**, 203.

Anselmino, M. Eriksson, F., Lichtenberg, D. B., Predazzi, E. and Ekelin, S. (1993). *Rev. Mod. Phys.*, **65**, 1199.

Applequist, T. and Carazzone, J. (1975). *Phys. Rev.* **D11**, 2856.

ARGUS (1987a): Albrecht, H. *et al.*, *Phys. Lett.* **B185**, 218; **B197**, 452; **B199**, 451.

ARGUS (1987b): Albrecht, H. *et al.*, *Phys. Lett.* **B192**, 245.

Arneodo, M. (1994). *Phys. Rep.* **240**, 301.

Arnold P. and Reno, M. H. (1989). *Nucl. Phys.* **B319**, 37. [Erratum: *Nucl. Phys.* **B330** (1990), 284.]
Artru, X. (1983). *Phys. Rev.* **97**, 47.
Augier, C. *et al.* (1993). *Phys. Lett.* **B316**, 448.
Aurenche, P. *et al.* (1984). *Phys. Lett.* **B140**, 87.
Aurenche, P. *et al.* (1989). *Phys. Rev.* **D39**, 3275.
Aversa, F., Chiappetta, P., Greco, M. and Guillet, J. Ph. (1989). *Nucl. Phys.* **B327**, 105; (1990), *Phys. Rev. Lett.* **65**, 401.
Bailin, D. (1982). *Weak Interactions*, 2nd edn. Sussex University Press.
Barbieri, R. (1988). *Riv. N. Cimento* **11**, no. 4.
Bardeen, W. A., Buras, A. J., Duke, D. W. and Muta, T. (1978). *Phys. Rev.* **D18**, 3998.
Barnett, M. J. (1976). *Phys. Rev.* **D14**, 70.
Barone, V. and Predazzi, E. (1987). *Ann. de Physique Fr.* **12**, 525.
Barone, V. *et al.* (1991). *Phys. Lett.* **B268**, 279.
Bartels, J. (1991). *Particle World* **2**, 31.
Bartels, J. (1992). Unitary corrections to the Lipatov Pomeron. DESY Preprint 92-114.
Basham, C. *et al.* (1979). *Phys. Rev.* **D19**, 2018.
Bazizi, K., Wimpenny, S. J. and Sloan, T. (1990). Contributed paper to the *25th International Conference on High Energy Physics*, Singapore, 1990. CERN-PRE 90-175.
BCDMS (1985): Bari, G. *et al.*, *Phys. Lett.* **B163**, 282; see also Voss, R., talk at the *11th International Conference on Neutrino Physics and Astrophysics*, Dortmund, 1984.
BCDMS (1989): Benvenuti, A. C. *et al.*, *Phys. Lett.* **B223**, 485.
Bengtson, M. and Sjöstrand, T. (1987). *Nucl. Phys.* **B289**, 810.
Berman, S. M., Bjorken, J. D. and Kogut, J. (1971). *Phys. Rev.* **D4**, 3388.
Bitar, K., Johnson, P. W. and Wu-Ki Tung (1979). *Phys. Lett.* **83B**, 114.
Bjorken, J. D. and Drell, S. C. (1964). *Relativistic Quantum Mechanics.* New York: McGraw-Hill.
Bjorken, J. D. and Drell, S. C. (1965). *Relativistic Quantum Fields.* New York: McGraw-Hill.
Blankenbecler, R., Brodsky, S. J. and Sivers, D. (1976). *Phys. Rep.* **23C**, 1.
Boulder (1990): *Proceedings Theoretical Advanced Study Institute in Elementary Particle Physics*, Boulder, 1989. Singapore: World Scientific.
Bourrely, C., Leader, E. and Soffer, J. (1980). *Phys. Rep.* **59**, no. 2.
Bozzo, M. *et al.* (1984a,b). *Phys. Lett.* **B147**, 385, 392.
Brandelik, R. *et al.* (1980). *Phys. Lett.* **97B**, 448.
Brandt, S. and Dahmen, H. D. (1979). *Z. Phys.* **C1**, 61.
Brodsky, S. J. and Farrar, G. (1975). *Phys. Rev.* **D11**, 1304.
Brodsky, S. J. and Lepage, G. P. (1979). *Phys. Lett.* **B87**, 359.
Brown, C. N. *et al.* (1989). *Phys. Rev. Lett.* **63**, 2637.
Buchalla, G., Buras, A. J. and Harlander, M. K. (1990). *Nucl. Phys.* **B337**, 313.
Buchalla, G., Buras, A. J. and Harlander, M. K. (1991). *Nucl. Phys.* **B349**, 1.
Budny, R. (1973). *Phys. Lett.* **45B**, 340.

Buras, A. J. (1980). *Rev. Mod. Phys.* **52**, 200.
Buras, A. J. and Gaemers, K. J. F. (1978). *Nucl. Phys.* **B132**, 249.
Burchell, M. (1992). In *Proc. 12th Int. Conf. on Physics in Collision*, Boulder, Colorado, 1992.
Carosi, R. *et al.* (1990). *Phys. Lett.* **B237**, 303.
Caswell, W. (1974). *Phys. Lett.* **33**, 244.
Catani, S. (1992). In *QCD at 200 TeV*, eds. L. Cifarelli and Yu. L. Dokshitzer, p. 21. New York: Plenum.
Cavasinni, V. (1990). *Riv. N. Cim.* **13**, no. 7.
CCFR (1990b): Foudas, C. *et al.*, *Phys. Rev. Lett.* **64**, 1207.
CCOR (1981): Angelis, A. L. S. *et al.*, *Phys. Lett.* **B98**, 115.
CDF (1988): Abe, F. *et al.*, *Phys. Rev. Lett.* **61**, 1819.
CDF (1989a): Abe, F. *et al.*, *Phys. Rev. Lett.* **63**, 1447.
CDF(1989b): Huth, J. In *Proceedings of the Workshop on Calorimetry for the SSC*, Tuscalosa, Alabama, 1989. Fermilab–Conf. 89/117-E.
Chanowitz, M., Furman, M. and Hinchcliffe, I. (1979). *Nucl. Phys.* **B159**, 225.
Cheng, H-Y. (1988). *Phys. Rep.* **158**, 1.
CLEO (1989): Artuso, M. *et al.*, *Phys. Rev. Lett.* **62**, 2233.
Collins, J. (1984). *Renormalization.* Cambridge: Cambridge University Press.
Collins, J. C. and Soper, D. E. (1987). *Ann. Rev. Nucl. Part. Sci.* **37**, 383.
Collins, P. D. B. (1977). *Regge Theory and High Energy Physics.* Cambridge: Cambridge University Press.
Collins, P. D. B. and Martin, A. (1984). *Hadron Interactions.* Bristol: Adam Hilger.
Combridge, B., Kripfganz, J. and Ranft, J. (1977). *Phys. Lett.* **70B**, 234.
Covolan, R. and Predazzi, E. (1991). In *Hadronic Physics with Multi-GeV Electrons*, eds. B. Desplanques and D. Goutte, p. 199. New York: Nova Scientia.
Creutz, M. (1983). *Quarks, Gluons and Lattices.* Cambridge: Cambridge University Press.
Creutz, M., Jacobs, L. and Rebbi, C. (1983). *Phys. Rep.* **95**, 201.
Curci, G., Furmanski, W. and Petronzio, R. (1980). *Nucl. Phys.* **B175**, 27.
Cutler, R. and Sivers, D. (1978). *Phys. Rev.* **D17**, 196.
d'Agostini, G., de Boer, W. and Grindhammer, G. (1989). *Phys. Lett.* **B229**, 160.
de Alfaro, V., Fubini, S., Furlan, G. and Rossetti, C. (1973). *Currents in Hadron Physics*, ch. 11. Amsterdam: North Holland.
de Boer, W. (1991). In *Proceedings of the 18th SLAC Summer Institute on Particle Physics*, CERN-PPE/90-161.
Desgrolard, P., Giffon, M. and Predazzi, E. (1994). *Zeit. F. Phys.* **C63**, 241.
Diemoz, M. *et al.* (1986). *Phys. Rep.* **130**, 293.
Diemoz, M. *et al.* (1988). *Z. Phys.* **C39**, 21.
Dokshitzer, Ya. L. *et al.* (1985). *Phys. Lett.* **B165**, 147.
Donnachie, A. and Landshoff, P. V. (1983). *Phys. Lett.* **B123**, 345; (1984). *Nucl. Phys.* **B244**, 342; (1985a). *Phys. Lett.* **B152**, 256; (1985b). *Nucl. Phys.* **B267**, 690.

Donoghue, J. F., Holstein, B. R. and Klimt, S. W. (1987). *Phys. Rev.* **D35**, 934.
Duke, D. and Owens, J. (1984). *Phys. Rev.* **D30**, 49.
D'yakonov, D. I. and Eides, M. I. (1981). *Sov. Phys. JETP* **54(2)**, 232.
Dydak, F. (1991). In *Proceedings of the 25th International High Energy Conference*, Singapore, 1990, Vol. 1, p. 3. Singapore: World Scientific.
E710 (1992). Shukla, S. *et al.*, In *Proceedings of the IV Blois Workshop on Elastic and Diffractive Scattering*, La Biodola, Elba, 1991. *Nucl. Phys. B Proc. Suppl.* **25B**, 11.
E731 (1990). Winstein, B. *et al.*, in *Proceedings of the XIV International Symposium on Lepton and Photon Interactions*, Stanford, 1989, ed. M. Riordan, p. 155. Teaneck, New Jersey: World Scientific.
Eden, R. J. (1967). *High Energy Collisions of Elementary Particles.* Cambridge: Cambridge University Press.
Efremov, A. V. and Radyushkin, A. V. (1980). *Riv. N. Cim.* **3**, no. 2, 1.
Eichten, E. *et al.* (1984). *Rev. Mod. Phys.* **56**, 579.
Eichten, E., Hinchcliffe, I., Lane, K. and Quigg, C. (1984). *Rev. Mod. Phys.* **56**, 579. Erratum **58** (1986), 1065.
Ellis, R. K. and Stirling, W. J. (1990). QCD and collider physics, FERMILAB-Conf-90/164-T.
Ellis, S., Kunszt, Z. and Soper, D. (1989). *Phys. Rev.* **D40**, 2188; (1990). *Phys. Rev. Lett.* **64**, 2121.
Feinberg, G. and Weinberg, S. (1959). *Nuovo Cimento* **16**, 571.
Floratos, E. G., Lacaze, R., and Kounnas, C. (1981). *Phys. Lett.* **98B**, 89.
Fulton, R. *et al.* (1990). *Phys. Rev. Lett.* **64**, 16.
Furman, M. (1982). *Nucl. Phys.* **B197**, 413.
Furmanski, W. and Petronzio, R. (1982). *Z. Phys.* **C11**, 293.
Gabrielse, G. *et al.* (1990). *Phys. Rev. Lett.* **65**, 1317.
Gaillard, M. K. and Lee, B. W. (1974). *Phys. Rev.* **D10**, 897.
Gaisser, T. K. (1976). In *Proceedings of the 7th International Colloquium on Multiparticle Reactions*, Tutzing, 21–25 June, 1976, ed. J. Benecke *et al.*, p. 521. Munich: Max Planck Institute for Physics.
Gamba, A., Marshak, R. E. and Okubo, S. (1959). *Proc. Nat. Acad. Sci.* **45**, 881.
Gasiorowicz, S. (1976). *Elementary Particle Physics.* New York: Wiley.
Gasser, J. and Leutwyler, H. (1982). *Phys. Rep.* **87**, 77.
Gauron, P., Leader, E. and Nicolescu, B. (1985). *Phys. Rev. Lett.* **54**, 2656; **55**, 639; (1990). *Phys. Lett.* **B238**, 406.
Gel'fand, Y. and Likhtam, E. (1971). *JETP Lett.* **13**, 323.
Gell-Mann, M. and Low, F. (1954). *Phys. Rev.* **95**, 1300.
Georgi, H. and Glashow, S. L. (1974). *Phys. Rev. Lett.* **32**, 438.
Georgi, H. and Politzer, H. D. (1976). *Phys. Rev.* **D14**, 1829.
Giffon, M. *et al.* (1984). *Riv. N. Cim.* **7**, no. 5, 1.
Giovannini, A. (1973). *Nuovo Cimento* **15A**, 573.
Giovannini, A. and Van Hove, L. (1986). *Z. Phys.* **C30**, 39.
Girardi, G. (1982). In *Proceedings of the 17th Rencontre de Moriond*, Les Arcs, Savoie, France, Vol. I, p. 361. Gif-sur-Yvette: Editions Frontieres.

Gluck, M. (1987). In *Proceedings of the HERA Workshop*, Hamburg, 1987, ed. R. D. Peccei, Vol. I, p. 119. Hamburg: DESY.
Gluck, M. and Reya, E. (1978). *Nucl. Phys.* **B145**, 24.
Gluck, M., Hoffman, E. and Reya, E. (1982). *Z. Phys.* **C13**, 119.
Gorishny, S. G., Kataev, A. L. and Larin, S. A. (1991). *Phys. Lett.* **B259**, 144.
Gribov, L. V., Levin, E. M. and Ryskin, M. G. (1982). *Phys. Rep.* **100**, 1.
Gribov, V. N. and Lipatov, L. N. (1972). *Sov. J. Nucl. Phys.* **15**, 438 and 675.
Gross, D. J. (1976). Applications of the renormalization group. In *Methods in Field Theory*, Les Houches, 1975, Session XXVIII, eds. R. Bailin and J. Zinn-Justin. Amsterdam: North Holland.
Halzen, F. and Scott, D. M. (1978). *Phys. Rev. Lett.* **40**, 1117.
Hanson, G. (1976). In *Proceedings of the 7th International Colloquium on Multiparticle Reactions*, Tutzing, 21–25 June, 1976, ed. J. Benecke *et al.*, p. 313. Munich: Max Planck Institute for Physics.
Hanson, G. *et al.* (1975). *Phys. Rev. Lett.* **35**, 1609.
Harriman, P. N., Martin, A. D., Roberts, R. G. and Stirling, W. J. (1991). *Nucl. Phys. B. Proc. Suppl.* **18C**, 17.
Heusch, C. (1989). In *Proceedings of the Tau-Charm Factory Workshop*, Stanford 1989, ed. Lydia Beers, p. 528. SLAC–343.
Horgan, R. and Jacob, M. (1981). *Nucl. Phys.* **B179**, 441.
Huth, J. (1991). In *Proceedings Particles and Fields*, Fermilab, 1991. Fermilab Conf. 91/223-E.
Inami, T. and Lim, C. S. (1981). *Prog. Th. Phys.* **65**, 297.
Ingelman, G. and Schlein, P. (1985). *Phys. Lett.* **B152**, 256.
Isgur. N. and Llewellyn-Smith, C. H. (1984). *Phys. Rev. Lett.* 52, 1080.
Itzykson, C. and Zuber, J-B. (1980). *Quantum Field Theory.* New York: McGraw-Hill.
Jackiw, R. (1980). *Rev. Mod. Phys.* **52**, 661.
Jacob, M. (1979). In *Proceedings of the EPS Conference on High Energy Physics*, Geneva, 1979, p. 473. Geneva: CERN Scientific Information Service.
Jacob, M. (1990). In *Proceedings of the 25th International Conference on High Energy Physics*, Singapore, 1990, Vol. I, p. 174. Singapore: World Scientific.
Jacob, M. and Landshoff, P. (1976). *Nucl. Phys.* **B113**, 395.
Jones, D. R. T. (1974). *Nucl. Phys.* **B75**, 531.
Kaluza, Th. (1921). *Sitzungsber. Preuss. Akad. Wiss. Phys. Mat.* **1**, 966.
Karsten, L. H. and Smit, J. (1981). *Nucl. Phys.* **B183**, 103.
Kawai, H., Nakayama, R. and Seo, K. (1981). *Nucl. Phys.* **B189**, 40.
Kim, C. S. and Martin, A. D. (1989). *Phys. Lett.* **B225**, 186.
Klein, O. (1926). *Z. Phys.* **37**, 895.
Kleiss, R. (1989). In *Z Physics at LEP1*, eds. G. Altarelli, R. Kleiss and C. Verzegnassi, Vol. 3, p. 1. CERN 89-08.
Klinkhamer, F. and Manton, N. (1984). *Phys. Rev.* **D30**, 2212.
Koba, Z., Nielsen, H. B. and Olesen, P. (1972). *Nucl. Phys.* **B40**, 317.
Kogut, J. (1983). *Rev. Mod. Phys.* **55**, 775.
Krawczyk, M. and Zarnecki, A. F. (1991). Formulae for basic two body QCD processes, Warsaw University Preprint IFT08/91.

Kryzywicki, A. (1985). *Nucl. Phys.* **A446**, 135.

Kunszt, Z. and Nason, P. (1989). In *Z Physics at LEP1*, eds. G. Altarelli, R. Kleiss and C. Verzegnassi, Vol. 1, p. 373. CERN 89-08.

L3 (1990a): Adeva, B. *et al.*, *Phys. Lett.* **B247**, 473.

L3 (1990b): Adeva, B. *et al.*, *Phys. Lett.* **B248**, 225.

L3 (1990c): Adeva, B. *et al.*, *Phys. Lett.* **B250**, 199.

L3 (1991): Adeva, B. *et al.*, *Phys. Lett.* **B252**, 703.

Langacker, P. (1981). *Phys. Rep.* **72**, 183.

Lattice (1989): *Nucl. Phys. Proc. Suppl.* **B17** (1990).

Lattice (1990): *Nucl. Phys. Proc. Suppl.* **B20** (1991).

Leader, E. (1992). In *Proceedings of the IV Blois Workshop on Elastic and Diffractive Scattering*, La Biodola, Elba, 1991. *Nucl. Phys. B Proc. Suppl.* **25B**, 299.

Levin, E. and Ryskin, M. (1990). *Phys. Rep.* **189**, 267.

Leutwyler, H. and Roos, M. (1984). *Z. Phys.* **C25**, 91.

Lipatov, L. N. (1974). *Yad. Fiz.* **20**, 181.

Lipatov, L. N. (1992). In *Proceedings of the IV Blois Workshop on Elastic and Diffractive Scattering*, La Biodola, Elba, 1991. *Nucl. Phys. B Proc. Suppl.* **25B**, 139.

Lubrano, P. (1990). In *Proceedings of the Workshop on Collider Physics*, La Thuile, 1990. Gif-sur-Yvette: Editions Frontières.

Lusignoli, M., Maiani, L., Martinelli, G. and Reina, L. (1992). *Nucl. Phys.* **B369**, 139.

Maiani, L. (1991). *Helv. Phys. Acta* **64**, 853.

Manton, N. (1983). *Phys. Rev.* **D28**, 2019.

Marchesini, G. and Webber, B. R. (1984). *Nucl. Phys.* **B238**, 1.

Marchesini, G. and Webber, B. R. (1988). *Nucl. Phys.* **B310**, 461.

Marciano, W. J. (1989). *Nucl. Phys.* **B** (Proc. Suppl.) **11**, 5.

Marciano, W. J. and Sirlin, A. (1981). *Phys. Rev. Lett.* **46**, 163.

Marciano, W. J. and Sirlin, A. (1986). *Phys. Rev. Lett.* **56**, 22.

Marshak, R. E., Mohapatra, R. N. and Riazuddin. (1980). Majorana Neutrinos and Neutron Oscillations: Low Energy Test of Unified Models. Virginia Polytechnic Institute Preprint VPI-HEP 80/7.

Marshak, R. E., Riazuddin and Ryan, C. P. (1969). *Theory of Weak Interactions in Particle Physics.* New York: Wiley-Interscience.

Martin, A. D., Roberts, R. G. and Stirling, W. J. (1989). *Mod. Phys. Lett.* **A4**, 1135; *Phys. Lett.* **B228**, 149.

Martinelli, G. (1991). Università di Roma 'La Sapienza' Preprint n. 799.

Matano, T. *et al.* (1968). *Can. J. Phys.* **46**, S56.

Matano, T. *et al.* (1975). In *Proceedings of the 14th International Cosmic Ray Conference*, Munich, vol. 12, p. 4364. Munich: Max Planck Institute for Extraterrestial Physics.

Mohapatra, R. N. (1986). *Unification and Supersymmetry.* New York: Springer Verlag.

Morfin, J. G. and Tung, W. K. (1990). In *Proceedings of the Workshop on Hadron Structure Functions and Parton Distributions*, Fermilab, 1990, p. 18. Singapore: World Scientific.
Morfin, J. G. and Tung, W. K. (1991). *Z. Phys.* **C52**, 13.
NA31 (1988): Burkhardt, H. *et al.*, *Phys. Lett.* **B206**, 169.
Nachtmann, O. (1973). *Nucl. Phys.* **B63**, 237.
Nir, Y. and Quinn, H. R. (1992). 'CP violation in B physics', *Ann. Rev. Nucl. and Part. Sci.* **42**, 211.
NMC (1991): Amaudruz, P. *et al.*, *Phys. Rev. Lett.* **66**, 2712.
Okun, L. (1991). In *Proceedings of the 25th International Conference on High Energy Physics*, Singapore, 1990, eds. K. K. Phua and Y. Yamaguchi, Vol. 1, p. 319. Singapore: World Scientific.
OPAL (1990): Akrawy, M. Z., *Phys. Lett.* **B241**, 133.
Owens, J. F. and Tung, W. K. (1992). *Ann. Rev. Nucl. and Part. Sci.*, 291.
Paige, F. (1989). In *Lectures at the Theoretical Advanced Summer Institute*, Boulder, Co. 1989, BNL-43525.
Palladino, V. (1985). In *Proceedings of the Workshop on Elastic and Diffractive Scattering*, Blois, France, 1985, eds. B. Nicolescu and J. Tran Thanh Van, p. 79. Gif sur Yvette: Edition Frontières.
Particle Data Group (1992). *Phys. Rev.* **D45**, N. 11, Part II.
Paschos, E. A. and Türke, U. (1988). *Phys. Rep.* **178**, 145.
Pauli, W and Villars, F. (1949). *Rev. Mod. Phys.* **21**, 434.
Peccei, R. D. and Quinn, H. R. (1977). *Phys. Rev.* **D16**, 1791.
Politzer, H. D. (1976). *Nucl. Phys.* **B117**, 397.
Preparata, G., Ratcliffe, P. G. and Soffer, J. (1990). *Phys. Rev.* **D42**, 930.
Rabi, S., West, G. B. and Hoffman, C. M. (1989). *Phys. Rev.* **D39**, 828.
Rebbi, C. (1983). *Lattice Gauge Theories and Monte Carlo Simulations.* Singapore: World Scientific.
Reinders, L. J., Rubinstein, H. and Azachi. S. Y. (1985). *Phys. Rep.* 127.
Ringwald, A. (1990). *Nucl. Phys.* **B330**, 1.
Ryder, L. H. (1985). *Quantum Field Theory.* Cambridge: Cambridge University Press.
Schwitters, R. F. *et al.* (1975). *Phys. Rev. Lett.* **35**, 1320.
Shifman, M. A., Vainshtein, A. I. and Zakharov, V. I. (1978). *Nucl. Phys.* **B147**, 385, 448, 519.
Sirlin, A. and Zucchini, R. (1986). *Phys. Rev. Lett.* **57**, 1994.
Sivers and Cutler (1978) *Phys. Rev.* **D17**, 196.
Sjöstrand, T. (1989). In *Z Physics at LEP1*, eds. G. Altarelli, R. Kleiss and C. Verzegnassi, Vol. 3, p. 143. CERN 89-08.
Sohnins, M. F. (1985). *Phys. Rep.* **128**, 39.
Stack, J. (1984). *Phys. Rev.* **D29**, 1213.
Stodolski, L. (1967). *Phys. Rev. Lett.* **18**, 135.
Strocchi, F. and Wightman, A. S. (1974). *J. Math. Phys.* **15**, 2198.
Stueckelberg, E. C. G. and Peterman, A. (1953). *Helv. Phys. Acta.* **26**, 499.
Stuller, R. L. (1987). *Phys. Rev.* **D35**, 4034.
Surguladze, L. R. and Samuel, M. A. (1991). *Phys. Rev. Lett.* **66**, 560.

TASSO (1980): Brandelik, R. *et al. Phys. Lett.* **B97**, 453.
Taylor, J. C. (1976). *Gauge Theories of Weak Interactions.* Cambridge: Cambridge University Press.
'tHooft, G. (1973). *Nucl. Phys.* **B61**, 455.
'tHooft, G. (1976). *Phys. Rev. Lett.* **37**, 8.
'tHooft, G. and Veltman, M. (1972). *Nucl. Phys.* **B44**, 189.
Tung, W. K. *et al.* (1989). In *Proceedings of the 1988 Summer Study on High Energy Physics in the 1990s*, Snowmass, Colorado, 1989.
Turner, M. S. (1981). In *AIP Conference Proceedings No. 72: Weak Interactions as Probes of Unification*, Viginia Polytechnic Institute, 1980, eds. G. B. Collins, L. N. Chang and J. R. Ficenec. New York: American Institute of Physics.
UA1 (1985): Arnison, G. *et al., Phys. Lett.* **B158**, 494.
UA1 (1986): Arnison, G. *et al., Phys. Lett.* **B172**, 224.
UA2 (1982): Banner, M. *et al., Phys. Lett.* **B118**, 203.
UA2 (1984): Bagnaia, P. *et al., Phys. Lett.* **B138**, 430.
UA2 (1985): Appel, T. A. *et al., Z. Phys.* **C30**, 341.
UA4 (1987): Bernard, D. *et al., Phys. Lett.* **B198**, 583.
UA5 (1983): Alpguard, K. *et al., Phys. Lett.* **B121**, 209.
UA5 (1984): Alner, G. J. *et al., Phys. Lett.* **B138**, 304.
UA5 (1985): Alner, G. J. *et al., Phys. Lett.* **B160**, 193.
UA5 (1986): Alner, G. J. *et al., Phys. Lett.* **B167**, 476.
Ukawa, A (1991). In *Proceedings of the 25th International Conference on High Energy Physics*, Singapore, 1990, p. 79. Singapore: World Scientific.
Vilain, P. (1990). *14th International Conference on Neutrino Physics and Astrophysics*, Geneva. CERN Scientific Information Service.
Volkov, D. and Akulov, V. (1973). *Phys. Lett.* **B46**, 109.
Voss, R. (1986). Contributed paper to the *23rd International Conference on High Energy Physics*, Berkeley, 1986.
Wegner, F. J. (1971). *J. Math. Phys.* **12**, 2259.
Watson, K. M. (1954). *Phys. Rev.* **95**, 228.
Weinberg, S. (1975). *Phys. Rev.* **D11**, 3583.
Weingarten, D. (1989). *Nucl. Phys. B Proc. Suppl.* **9**, 447.
Wess, J. and Zumino, B. (1974). *Nucl. Phys.* **B70**, 39.
Wilson, K. G. (1974). *Phys. Rev.* **D10**, 2445.
Wilson, K. G. (1975). *Phys. Rep.* **23**, 331.
Withlow, L. C. (1990). Ph.D. thesis and SLAC rep. 357 and references therein.
Wolf, G. (1979). In *Proceedings of the EPS International Conference on High Energy Physics*, Geneva, 1979, p. 220. Geneva, CERN Scientific Information Service.
Wolf, G. (1980). In *Proceedings of the JINR-CERN School of Phyics*, Dobagokö, Hungary, 1979, Vol. 1, p. 192. Budapest: Hungarian Academy of Science, Central Research Institute of Physics.
Wu, T. T. and Yang, C. N. (1964). *Phys. Rev. Lett.* **13**, 380.

Analytic subject index for vols. 1 and 2

Bold numbers refer to major references, for example definitions

Printed in the United States
By Bookmasters